EVOLUTIONARY PALEOECOLOGY OF THE MARINE BIOSPHERE

EVOLUTIONARY PALEOECOLOGY OF THE MARINE BIOSPHERE

James W. Valentine

University of California, Davis

Prentice-Hall, Inc., *Englewood Cliffs, New Jersey*

Library of Congress Cataloging in Publication Data

Valentine, James W.
Evolutionary paleoecology of the marine biosphere.

Bibliography: p. 472
1. Paleoecology. I. Title.
QE720.V34 560 72-14116
ISBN 0-13-293720-4

Printed in the United States of America

10 9 8 7 6 5 4 3 2 1

Prentice-Hall International, Inc., *London*
Prentice-Hall of Australia, Pty. Ltd., *Sydney*
Prentice-Hall of Canada, Ltd., *Toronto*
Prentice-Hall of India Private Limited, *New Delhi*
Prentice-Hall of Japan, Inc., *Tokyo*

To my family, colleagues, and students. Each has made his own peculiar demands upon my resources, thereby seriously interfering with the prosecution of this book. Still, the diversity of their demands and the multiplicity of their interactions have had many salutory effects as well, perhaps contributing to my stability. And after all, each is merely functioning in his own niche.

CONTENTS

PREFACE

Evolutionary Paleoecology of the Marine Biosphere is intended as a reference and advanced text. In the sense that theory is a body of ideas that attempt to explain a set of observations, this book represents a theoretical review of biological aspects of the marine fossil record. The emphasis has been on reviewing or introducing biological concepts and applying them to fossil data.

In many ways this is a propitious time to attempt such a review. Advances in population biology and in our understanding of the structure, function, and natural regulation of ecosystems now permit paleontologists to study the qualities of fossil populations and their patterns of association and of distribution from a new perspective. With these changes paleoecology, once a loose collection of techniques and environmental interpretations, has become a focused discipline with its own problems, values, and goals. And the development of the revolutionary concepts of global tectonics has provided paleontologists with a set of tools for environmental interpretation more powerful than any they possessed previously. For the first time the way is open to erect a model of the historical sequences of change in the environmental framework of the planet.

Certainly, paleoenvironmental interpretations have often been well founded in the past. But the reasons why any major environmental configurations were developed, or why they changed in any particular direction, could not usually be formulated. Now it appears likely that we shall soon be able to reconstruct the sequence of past environmental regimes from first principles, independently of the fossil evidence. And then we may trace the interactions of biota and environment and interpret the processes of evolution and of ecology that have been at work during the history of life.

Of course, the fossil record is spotty and we shall never have anything approaching a complete ecological history, but this does not prevent paleoecological research from being conducted within an integrated theoretical framework. This book presents preliminary models that may contribute to such a framework, and it attempts to review especially the paleoecological methods and concepts that appear at this time to hold promise for eventual achievement of a paleoecological theory.

Despite the generally favorable climate for paleoecological theorizing, there are drawbacks to attempting a review at this time. These are due chiefly to the present rapid advances in the main fields that are concerned. New concepts associated with global tectonics and major advances in our knowledge of continental history steadily accrue. Understanding of the ecology of populations and communities is also progressing. And finally, the field of paleoecology itself is increasingly vigorous, even though the applications of global tectonics have not yet reached their full impact. It is likely that only a few years will witness a quantum advance in our knowledge of ecological history. Thus it has seemed prudent to concentrate here on processes and on restricted models that may serve as exemplars, although the final chapter attempts a generalized application of the models to the fossil record.

The ferment in earth sciences has required revisions of early chapters as the book has progressed, and I am much indebted to colleagues and graduate students for their criticisms, insights, arguments, and general running commentaries and dialogues, which have added up to a large contribution to this work. Most of these individuals were at Oxford University, Yale University, and the University of California, Davis. To all these stimulating associates I am most grateful. I must particularly thank Drs. Richard Cowen, W. H. Hamner, and E. M. Moores, University of California, Davis, and J. R. Beerbower, State University of New York, Binghamton, for their aid.

Davis, California *James W. Valentine*

CHAPTER ONE

THE DISCIPLINE OF PALEOECOLOGY

Questions are the engines of intellect, the cerebral machines which convert energy to motion, and curiosity to controlled inquiry.
—D. H. Fischer, *Historians' Fallacies,* Harper & Row, Publishers, New York, 1970.

Life has existed on the earth for about the last 3.5 billion years, developing from little more than large organic molecules into vast arrays of complex beings. There are a few million separate species living, ranging from relatively simple to exceedingly complex life forms. Life today is richly diverse in form and function. This great diversity is accommodated within a wide variety of biotic associations that range from those which include millions of individuals belonging to thousands of different species' populations, to those which include only hundreds of individuals representing only a few different species. Biotic organization today is richly diverse also.

It is possible to account very well for the complexity of life forms themselves and of their biotic associations by the processes of evolution through natural selection. The evolution of biotic associations appears to be part of the same process which has produced the evolution of lineages of specific organisms. Although the process is of almost inconceivable complexity in detail, it is based on the properties of life that require or permit reproduction of offspring containing heritable modifications of parental characters. As some organisms prove more successful at reproduction than others, any modifications that these successful ones contain are propagated in relatively higher numbers. And as some modifications contribute directly to reproductive success, they help ensure their own propagation. Such reproductive success usually occurs because the modification results in an improved adaptation of organisms to their environment; here environment is used in a broad sense to include both the physical and the biological environments, including the effects of other organisms.

Clearly, the relationships between an organism and its environment are critical to determining the success or failure of a modification to establish itself and to endure. These relationships are the province of ecology. Since ecology is of fundamental significance in determining the pathways of evolution, it is logical to join the studies of ecology and of evolutionary biology into a single discipline in order to investigate the history of life. The interrelationships of these fields are intimate. The biological environment is itself a product of previous evolution, and yet is a major determinant of subsequent evolution. The physical environment plays a major role in determining the structure of the biological environment at any time and also in determining the subsequent evolutionary events directly. Furthermore, the physical environment is greatly affected by the activities of organisms, and environmental changes caused by the organisms affect, in turn, the state and future of the biological systems. The entire system of evolving lineages and changing environments feeds back upon itself in complex ways. This provides the potential for self-organization of the biological systems. It is the study of the evolution of biological organizations that forms the chief subject of evolutionary paleoecology.

The evidence as to the processes and the course of the evolution of ecological systems is drawn from two chief sources. First and most obviously there are the living ecological systems themselves, which stand as a sort of progress report on the status of organic evolution today. From observation and experiment, biologists have learned a great deal about the way in which ecological systems work and about the processes that regulate them. The second important source of evidence is the fossil record of ecological systems, from which much evidence of biological processes can be extracted. A synthesis of biological and paleontological evidence should provide a sound basis for advancing our knowledge of the history of ecological systems.

Paleoecology Concerns the Life Processes of Ancient Organisms

Fossils are commonly defined as traces or remains of once-living beings. Some authors make further demands—that fossils be naturally preserved in the rock record, or that they be of a certain age. Rather young organic debris, such as is represented by the food refuse of ancient man, is sometimes called "subfossil." However, the processes operating upon organic remains that are accumulating now, both "in nature" and in the garbage dump, are extremely pertinent to our inquiry into their occurrences in older deposits. To draw the line between nonfossils, subfossils, and fossils at any given stage in the history of their preservation is entirely arbitrary.

The bodies of organisms are composed of highly organized materials in forms that range from large proteins with their amino acid components to simpler molecules that sometimes contain unusual elements. Energy may be extracted from the processes of degradation of many of these materials. The existence of these materials is the result not only of the expenditure of the energy immediately associated with their synthesis but in a very real sense of a long preliminary period of inorganic and organic evolution. Apart from solar radiation, they represent the most valuable of biological resources. Indeed, the securing of energy sources is surely the most fundamental theme of organic evolution.

The sources of energy and materials represented by the bodies of organisms have been tapped again and again by other organisms in a bewildering variety of ingenious methods, as testified by the complex systems of energy transfer within communities. Rapid utilization of organic remains, chiefly by scavengers and decomposers, is the rule. The chief parts of organisms that ordinarily survive degradation for any length of time are mineralized skeletons that have been created by evolutionary processes for their strength and durability. In addition to skeletons, many of the burrows, trails, and other structures made by organisms during their lives are easily preserved, and these form an important fraction of the fossil record.

Figure 1-1 categorizes various stages in the history of organisms and of their traces and remains, and shows that there is a separate subdiscipline of interpretive paleontology for each stage. Paleoecology, the subject at hand, deals with the environmental relations of organisms during their lifespans. The relations of the traces and remains of dead organisms to the environment are studied by the discipline of *biostratinomy.* Once fossils are buried they may continue to interact by being dissolved or recrystallized or replaced or deformed, as a product of diagenesis or metamorphism, until they are finally discovered or completely destroyed. The study of all the processes that affect organisms or their traces from death to discovery is sometimes called *taphonomy* (Fig. 1-1). Taphonomy is clearly an important study, but it is not our primary concern, for we are focused upon the processes that occur during the lives of organisms.

The reconstruction of these living processes can involve every discipline in the earth sciences and in most of biology. The interpretation of ancient environmental conditions, for example, employs evidence from sedimentology and stratigraphy, from mineralogy and petrology, and from structural geology and tectonics, to establish the depositional conditions of a fossiliferous unit, the local paleogeographic framework, and the regional and planetary setting.

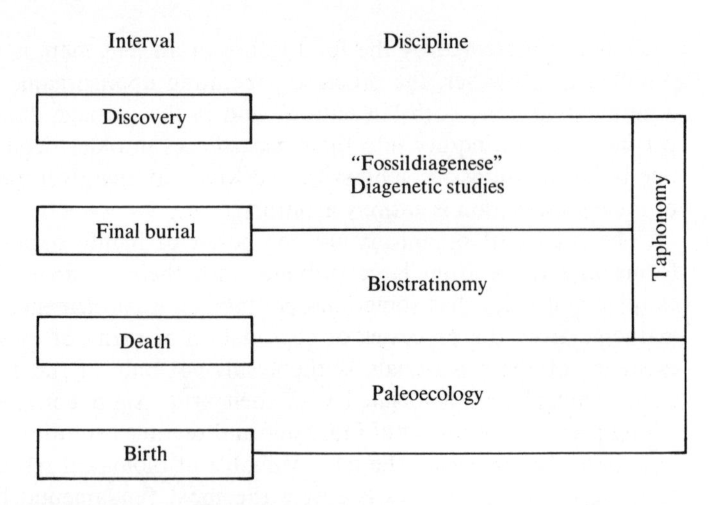

Fig. 1-1 Four key events in the history of a fossil, and the paleontological disciplines that deal with the intervals between these events. (After Lawrence, 1968.)

Some lines of evidence from these and other fields will be considered in many of the discussions and examples that follow, but they are not our primary concern, and the methodologies of environmental interpretation in these disciplines are not addressed here. Nevertheless, they are extremely important and must be employed regularly in order to prosecute successful paleoecological research. Introductions to varied methods of environmental reconstruction are given by Ager (1963), Imbrie and Newell (1964), and Laporte (1968).

The Fossil Record Is Poor but Deserving

The fossil record, then, is made up of skeletal parts together with impressions and petrifactions, tracks and burrows, pellets and other scraps and marks left by organisms. It is on these that our direct evidence of ancient life is based. Not many more than 6 million marine species have probably existed and become extinct (Valentine, 1970). There may be about 100,000 marine fossil invertebrate species described (see Easton, 1960, and Durham, 1967). Thus perhaps 1 in 60 of these species is known to science. Certainly there is a great number of undescribed species preserved in the fossil record, perhaps as many more than have already been described. Ultimately, then, we may look forward to knowing 1 in 30, or perhaps more, of all the extinct marine invertebrate species.

This does not sound very encouraging, even if these estimates are rather pessimistic—and they may well be rather optimistic. The present biosphere is very heterogeneous; animals differ in form and function from habitat to habitat and region to region. If we had only a regular sample of 1 in 30 of the 400,000 or so marine species extant, our knowledge of present marine life would be inadequate to support significant conclusions about the structure and dynamics of the present marine biosphere. There are, for example, about 4000 species of shelled mollusks living along the northwest coast of America between the equator and the pole. Mollusks are one of the dominant animal groups in the bottom communities, and the composition and functions of the molluscan fauna vary from place to place. If we knew only about 1 in 30, or 133 species scattered up and down the coast, we should be virtually ignorant of the pattern of variation of the molluscan fauna.

The imperfection of the geological record has been a favorite subject in evolutionary literature since Darwin's famous discussion (1859, Chapter X). The fossil record is certainly a poor record of the past, and it is badly biased, with very uneven representation taxonomically and with large gaps in space and time that are simply not represented at all.

Figure 1-2 gives some idea of the geographic patchiness of the record. It shows the outcrop pattern of Ordovician sedimentary rocks in parts of North America (black areas). Other Ordovician rocks are present in this region but are concealed beneath younger strata (stippled areas); for practical purposes they are unavailable for sampling, although some small amount of data may be recovered from drill holes. The current theories of ocean-floor spreading imply that little or no Paleozoic sediment is preserved beneath the ocean basins. The Ordovician fossils that lie close to the surface on the continents and around continental margins may be about all that we shall ever see. The Ordovician Period lasted approximately 60 million years, so that the record of a great deal of time is lumped together in Fig. 1-2.

It is likely that in any given sequence of Ordovician rocks only a fraction of Ordovician time is represented. This is true for most rock sequences of any duration, although there are some that may be fairly continuous over moderate intervals of time, such as 10 million years, or perhaps in unusual cases even more. Some of the factors contributing to the intermittent character of the rock record have been graphically portrayed by Barrell (1917). Figure 1-3 depicts the accumulation of a sedimentary sequence during a period of general transgression at a given locality. The trend of this rise in sea level relative to the original land surface is shown by curve *A-A'*. However, there is imposed upon this transgression, from another cause, a sea-level change that has a shorter frequency and amplitude, and that results in sea level following the curve *B-B'*. A third factor causes even smaller, higher frequency, sea-level oscillations, resulting in the curve *C-C'*, which is the actual course of sea level in this example. At any locality, sea level does, in fact, vary in response to a wide variety of factors, from local tectonic movements to changes in ocean basin volume, isostasy, entrapment and release of ocean water in continental ice, ocean currents, water temperature, atmospheric pressure, and many others.

In Barrell's example, sediments are supplied to the sea floor as sea level rises, and they accumulate in increasing thickness until sea level begins to decline. As the water shoals and turbulence increases, the upper sediments are progres-

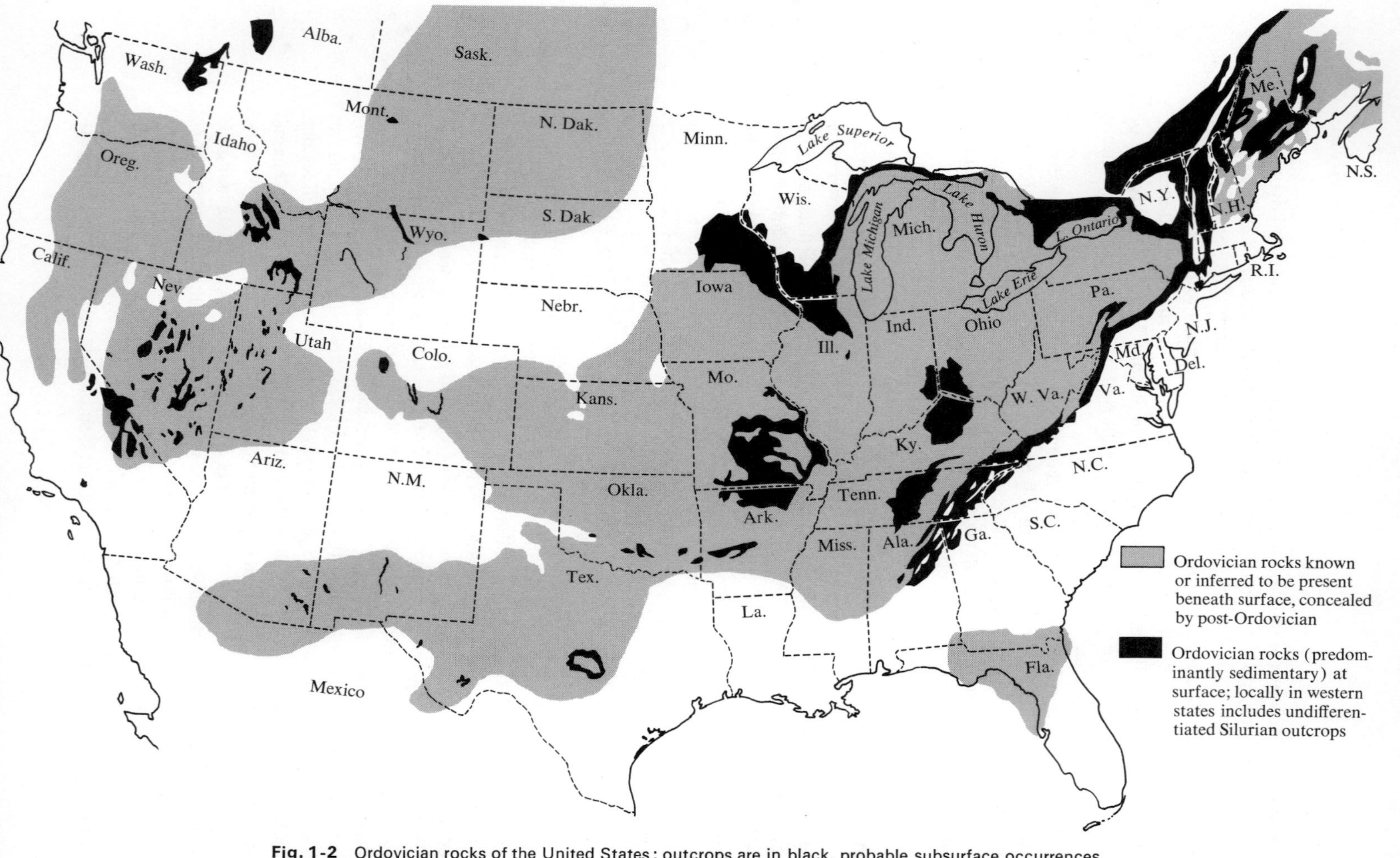

Fig. 1-2 Ordovician rocks of the United States; outcrops are in black, probable subsurface occurrences in stipple. (After Moore, 1958.)

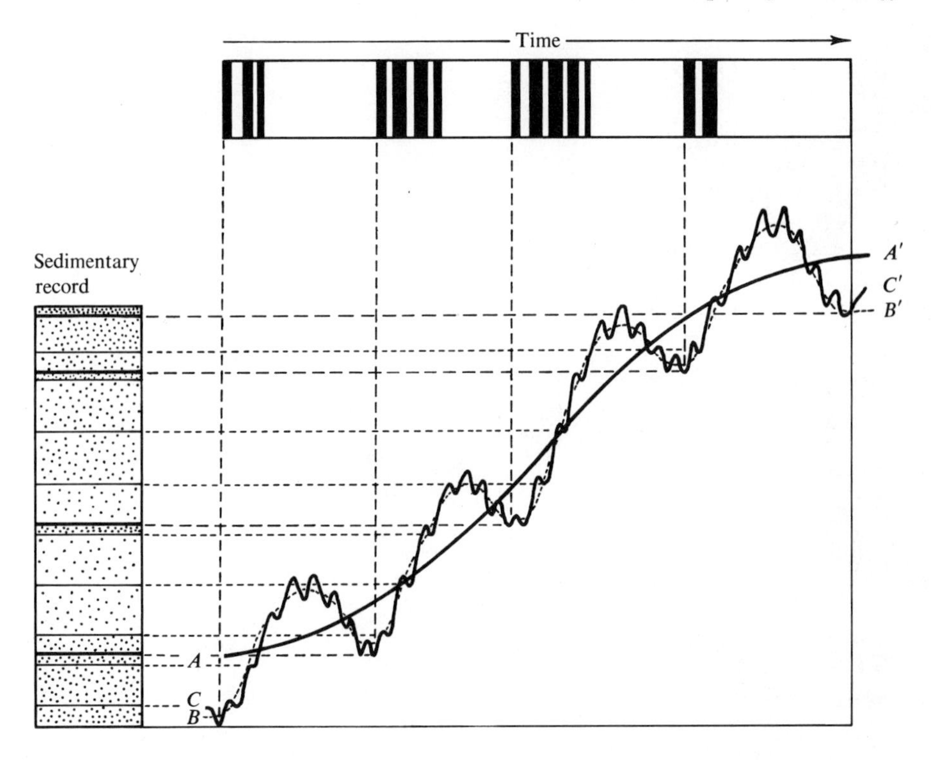

Fig. 1-3 The sedimentary record resulting from three sets of base-level oscillations that each permit alternate deposition and erosion. The black bars at the top indicate the time intervals that would be represented by sediments under this regime. (After Barrell, 1917.)

sively eroded. When sea level again rises, additional sediments are deposited; the continual interplay of deposition and erosion results in the sequence shown on the left of Fig. 1-3. The time that is actually represented by preserved sediments is shown by the vertical black bars in the panel at the top of the figure. In nature, the sea-level fluctuations would usually be unsymmetrical and would result in a time record of much greater irregularity. Also in nature, additional factors, especially those causing a fluctuation of sediment supply, would likewise affect the sedimentary sequence.

If a map were presented for a small interval of Ordovician time, say 5 million years, most of the black areas in Fig. 1-2 would be reduced or disappear; the total area of sedimentary outcrop representing an average 5-million-year period during the Paleozoic is relatively small. During the Ordovician, the oceans were more widespread over continents than at any other time during the Paleozoic Era, so that the Ordovician is better represented by sediments than the average period. We are concerned with fossils, and not sediments per se, so that our interest lies primarily in the geographic and stratigraphic distribution of fossiliferous sediments. Figure 1-4 is from a paper on the strati-

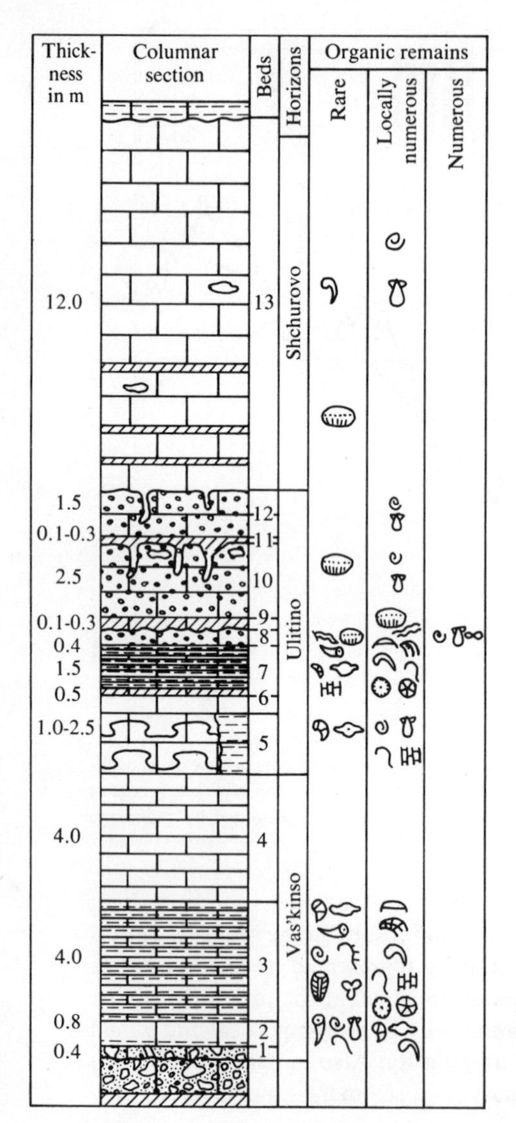

Fig. 1-4 The stratigraphic occurrence of fossils in a section of Carboniferous rocks in the Moscow Basin. [After Hecker, 1957 (English translation, 1965), from Ivanova and Khvorova, 1955.]

graphy of Carboniferous rocks in the Moscow Basin (Ivanova and Khvorova, 1955). It was employed by Hecker (1957; English translation, 1965) in a Soviet textbook on paleoecology to demonstrate a nice technique of illustrating fossil occurrences, as indeed it is. It is also interesting in that, although these rocks are in general quite fossiliferous, there are numerous intervals and horizons within the 29.7 meters (m) depicted that are barren of fossils, especially the limestones of unit 4 and the dolomites of unit 13. (The column is, of course, rather diagrammatic.) The sequence contains unconformities.

Thus, representation of the geographic distribution of all fossils during a

5-million-year interval of Ordovician time would require the subtraction of unfossiliferous areas from the map (which would be numerous); indication of the time represented by fossils on Barrell's figure would require removing those bars or parts of bars that are unfossiliferous (which, on the average, would be well over half). Furthermore, when this is done the resulting maps and bar figures show nothing of the incompleteness with which the once-living assemblages that have contributed the fossils have been represented. It is really not necessary to continue these examples further, for the gaps and bias in the record are evident. To be an unbiased sample of past life in many of its most important aspects, the fossil record would have to include representatives of all past forms of organisms, in proportion to their former abundance, and distributed in patterns that fairly reflect their living patterns. Clearly the record is very far from this ideal.

It is true enough that if, starting with the beginning of life, we had each three-hundredth skeletonized species preserved, counting from the first species, we could tell only a little about past patterns of species associations and distributions. But this is not how the record is. What we have are large gaps where we know of no fossils but also certain intervals of time represented at certain localities by a record of certain animal groups that is truly superb. Of course, there are also many places and times represented by records that run the gamut from slightly less than superb to slightly better than nothing. Nevertheless, the record is locally although sporadically good for certain animal groups, which makes paleoecology possible. For the places and times that the record is good, we can gather critical data on ecological units—on populations, communities, and provinces. We may then extrapolate between the units represented at these localities, and attempt to discover the principles accounting for the spatial patterns and the processes that can lead from one to another. The fossil record deserves to be taken very seriously.

To Reconstruct Past Events We Must Reconstruct Past Configurations as Well as Past Processes

To make reconstructions of past events, we must make some assumptions, some ground rules relating to the bases of the methodology of historical science and their limitations. Geologists have long recognized the need for working principles to guide their researches into past conditions. The most famous such principle is the doctrine of *uniformitarianism.*

The modern tradition of uniformitarianism originated with James Hutton, an eighteenth-century Scot who looked at geological phenomena as the products of ordinary processes such as are going on in the world today. In Hutton's time, many people sought catastrophic explanations for geological features. The great age of the earth was not appreciated, and the historical outlook of most educated people tended to be, by today's standards, shortsighted. Supernatural agencies were often invoked to explain geological and paleontological phenomena. Although Hutton's views conformed with the more advanced thinking of his time, the catastrophist position was still strongly held in the first half of the nineteenth century, when Sir Charles Lyell marshaled much evidence to support the idea of gradual change by natural processes. In fact, the term "uniformitarian" was first applied to this position in a review of one of Lyell's books. The doctrine of uniformitarianism has been epitomized by the memorable maxim of Sir Archibald Geikie, "the present is the key to the past." Frequently the doctrine is eulogized as a fundamental, inescapable basis of geological and paleontological interpretations, whereas, on the other hand, it is sometimes dismissed as having simply outlived its usefulness (see Simpson, 1963, and Gould, 1965). As paleoecology has arisen as a branch of geology and is almost entirely concerned with the state of conditions and the operation of processes in the past, it is appropriate here to examine the assumptions of uniformitarianism with care.

Uniformitarianism has been applied to two rather distinct concepts, which has led to some confusion (Gould, 1965). In the first place, many workers have interpreted this doctrine to mean that natural laws operating today have functioned in similar fashion in the past. The laws work because of the intrinsic properties of energy and matter, which Simpson (1963) has called *immanent* qualities. These qualities cannot change in time, by their definition. Examples include the hardness and cleavage of minerals of particular compositions, based on their internal structures; the structure of water based on its molecular composition; and enzymatic functions of nucleic acids and proteins, based on their compositions and structures. These properties are all based on the immanent properties of components of the materials, and so are not fundamental in the sense that they are elementary qualities, but are derivative; they are nonetheless immanent. Under specific conditions, the immanent properties of matter require that certain processes occur, and these processes are therefore immanent also. Thus, these "timeless" immanent events will occur whenever the necessary and sufficient conditions occur, and they will recur whenever these conditions recur. The application of the doctrine of uniformitarianism to these timeless qualities is certainly logical but it is also quite redundant. This concept has been called *methodological* uniformitarianism (Gould, 1965).

The other concept of uniformitarianism has been called *substantive* uniformitarianism (Gould, 1965) and holds that the *action* of the processes has been uniform, that they have gone on in the past much as they go on today, with similar rates and intensities. But this is clearly not so; we are now well aware that the past was very different from the present. Glaciers are a common example. They have been much more widespread at times in the past than today, and perhaps they have been totally absent at times. The presence and intensity of glacial processes depend upon the state of the environment. Another example is afforded by the theory of plate tectonics, which indicates that the geographic patterns and sizes of continents and oceans have varied greatly through time, implying vast changes in the environment.

Thus, in contrast to the timelessness of immanent processes stands the transient nature of the state of the earth. The biosphere is especially ephemeral. Myriads of great systems of life processes are constantly in operation across the face of the earth, so that the state of the biosphere is constantly changing. If we imagine that these changing systems could be frozen and examined as if they were static, the state which we would see may be called the *configuration* of the biosphere (Simpson, 1963). The history of life is the product of the configurational flux, but each succeeding configuration is unique. Since the precise outcome of the operation of a process depends upon the configuration in which it occurs, and since the configuration of the biosphere is constantly changing, it is unsafe to assume that the effects of a given process or system of processes will be the same at different times. Processes may have very different outcomes when operating in different environments and involving different biological entities.

Thus, of these two concepts of uniformitarianism, one states that immanent processes are timeless, which is not very helpful today, and the other is invalid. Nevertheless it seems true that a wide knowledge of the present is a great help in interpreting the past. Why is this?

Partly it is because there are a great number of derivative immanent properties, and we can investigate them directly only in the present biosphere. Partly also it is because the present configuration is extremely rich in variety, so that ideas as to the effects of processes in many different possible configurations of the past may be gained from studying the present. And partly it appears that some aspects of the state of the biosphere change only very slowly, and for these parts the present is a key to much of the past. We must learn so much about the present that we can extrapolate processes to configurations different from the present one and evaluate their effects. The important trend in modern paleontology, for workers to invade the present marine realm and attempt to discover the principles and processes at work there, can thus provide significant gains to the science, so long as the present systems are not automatically and uncritically referred to the past when it comes to paleoecological interpretation. To interpret a past configuration we must commonly recreate a whole "new" world, or rather an old one, and not try to force the present on the past.

Models Are Powerful Tools in the Interpretation of Ecological Units

The method of employing models to explain paleontological data has come into common use in recent years. This method is old, to be sure, but the rise and spread of the discipline of cybernetics and of computerized systems and operations research have popularized this approach and widely demonstrated its power. The models may be mathematical, or qualitative, or even actual physical working models. They are all really working hypotheses, or systems of hypotheses, and are constrained in the ways that ordinary working hypotheses and theories are constrained. One of the chief reasons for modeling in paleoecology is to overcome the differences between past and present configurations. Using the principles of ecology and of evolution and the information that can be gleaned from study of the stratigraphic context of fossils, we can construct conceptual models of the dynamics of ecological units and of their evolution. The state of the units that is determined at each sampling station must represent a state in the model at its place and time—a sample of the configuration of an ancient biosphere.

The detailed techniques of model building are variable, depending upon circumstances, but a few methodological approaches are common to most modeling (Fig. 1-5). It is best to begin the modeling process with questions upon

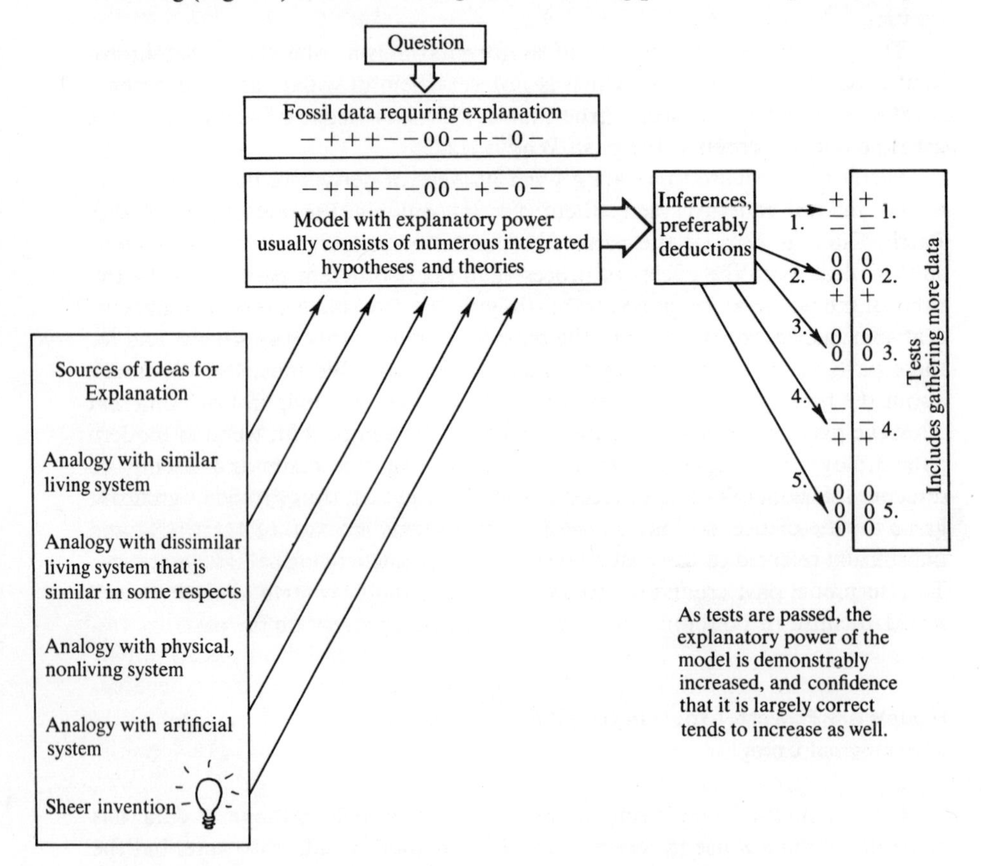

Fig. 1-5 Common sources of ideas and the course of modeling procedure in paleontology.

which the model is to bear. Ideally questions should precede data gathering and guide the techniques of the investigation. Usually the questions are modified or expanded as data are collected. Ideas for answers to questions about ecological and evolutionary conditions and processes originate most frequently from a knowledge of the present biosphere (see Rudwick, 1961, 1964). One obvious source of ideas to explain fossil systems is in living systems that closely resemble them, and which are therefore presumed to function in a similar manner (Levins, 1968). It is certainly worth searching for a close modern analogue for an ancient system. Another common source of ideas lies in biological systems that generally are not much like the fossil systems under study, but that nevertheless resemble them in some particular features, for which they may serve as analogues. Still another source is in physical systems that lack biological components entirely but suggest ways in which biological systems might function. An example is an electrical circuit, which has suggested ways in which energy might be routed within ecosystems (Odum, 1960). Completely artificial systems, such as engineering or economic systems, may also serve (Simon, 1970). Finally, ideas may spring from "pure invention," whatever that may be. It is sometimes impossible to identify accurately the source of ideas; we are all familiar with the indefinable intuitive component that so often accompanies the achievement of new personal insights.

As most ecological systems are complex, even a simplified model will often require many hypotheses concerning individual processes to be woven into a larger system in which they interact. There are no special rules of logic to guide the construction of these models, although they commonly involve inductive processes. A helpful rule of thumb in constructing ecosystem models is to build them so as to be as simple as possible considering the nature of the fossil record, and as efficient as possible considering the nature of the paleoenvironmental regime.

Models are only curiosities unless they can be tested, which is actually the critical part of the modeling process (Popper, 1961). The testing is simple in principle. If the model is useful it will inevitably entail consequences in nature. These are the predictions of the model, and they can be tested by making observations that confirm or deny them. If the observations are contrary to the prediction, the model is wrong. Of course, a complicated model need not be completely wrong because of one unfulfilled prediction, and some judicious tinkering can sometimes make faulty models work. If the prediction is fulfilled by observation, the model is not proven, but neither is it falsified, and so it remains tenable. If a large number of independent predictions are verified, the model obviously has much explanatory power, and may eventually come to be considered as correct in all probability.

No matter how perfectly a paleoecological model may fit the data, it can never be more than an oversimplified explanation of events, a shadow of the substance of history. Furthermore, models inevitably contain features that did not, in fact, occur in history. No matter how well they "work," historical models are only simplified approximations that extrapolate in reasonable but inevitably somewhat erroneous ways between datum points. In addition to these problems in internal verisimilitude, models are affected by the external scientific perspective of the modeler, by his understanding of the natural world. Usually, his is a view that is accepted by the science of the time. But this conventional perspective of nature is actually a model itself, built up of numerous interpretations into a more or less consistent supermodel that "explains" the empirical observations. Kuhn (1962) has suggested that most scientific activity is devoted to elaborating these supermodels. This "normal" science is not aimed at shaking the major assumptions of supermodels, but at extending or modifying the models to remove internal inconsistencies or explain new data. Darwinian evolution is a good example of a supermodel (see Ghiselin, 1969). However, occasionally some data cannot be accommodated into the model, which suggests a basic flaw; then an entirely new supermodel may be constructed, one which rivals its predecessor for acceptance as a common framework for normal science. The supermodel that better explains the empirical observations will eventually become accepted. The replacement of one supermodel by another is not normal science, but constitutes a scientific revolution (Kuhn, 1962).

Geology has recently been confronted by two such supermodels of global processes: an earlier one, assuming more or less permanent continental geographies and accommodating geosynclines as primarily continental phenomena; and a newer one, proposing lithospheric plates that move by sea-floor spreading and explaining geosynclines as plate-interaction features. Plate tectonics has clearly the greater explanatory power and is rapidly eclipsing the earlier supermodel in an uncanny demonstration of a scientific revolution, à la Kuhn.

The Past Is a Key to the Present

Clearly, despite their flaws, models are powerful devices for explaining past conditions and events; indeed, a discussion of modeling is really only a way of explaining a scientific method. This book contains numerous examples of models of ecological units, some simple and some intricate, some good and some bad. Some of them are untested, but they appear to be *testable*. None of them can be precisely correct. They are elaborated within certain current supermodels of evolution, ecology, and global history, and they all suffer from the internal deficiencies inherent in historical models. They may be divided into two rather arbitrary parts. One concerns dynamic models of the ecological functions of ancient organisms; this is paleoecology in the strict sense. The other concerns the evolution of these functions. The fossil record is most complete for marine invertebrates of shallow seas, and we shall be concerned primarily with these animals.

To construct models of the evolutionary paleoecology of marine organisms, we require certain basic concepts and data. The following three chapters are concerned primarily with this basic information. The first topic covered is the present understanding of evolution (Chapter 2), both to model the evolution of ancient lineages and to serve as a source of analogies when considering changes in ecological units. A model of the ecological architecture of the present marine biosphere is necessary in order to provide a description of the sorts of ecological units found in the sea and of their modes of change through time (Chapter 3). The environmental parameters of the sea that are of greatest ecological consequence are reviewed, together with some information on the physiological effects making them important (Chapter 4). The treatment of these fields here can only be sketchy in the extreme.

With these basic concepts and data, we may proceed to examine paleoecological configurations and their development. It is reasonable to begin at the level of the individual organism and to see how ecological functions can be inferred for them, and how this may lead to evolutionary interpretations that have ecological foundations (Chapter 5). However, individuals do not evolve, and so we must proceed to the level of the population to examine the full evolutionary implications of the individual adaptations and to investigate the development of population parameters (Chapter 6). Communities draw their properties from the properties of the populations that compose them, and their structures and functions can be investigated, and their evolution understood, in this light (Chapter 7). As communities of organisms are restricted, not only to certain habitats but also to certain geographic regions, it is necessary to investigate the factors controlling the geographic distributions of ecological units and regulating the ecological differences between biotic provinces (Chapter 8). The boundaries, numbers, and ecological configurations of provinces change in time, and the effects of temporal environmental changes on these major divisions of the biosphere are considered in Chapter 9.

Finally, it is incumbent upon paleoecologists to attempt to extend their investigations beyond the ecological units themselves and to consider the significance of paleoecological processes for the larger aspects of the history of life. Paleoecology certainly does have its descriptive side, devoted to the delimitation and description of ecological units, but its main purpose must surely be to employ the configurational descriptions as a basis for investigating the processes that brought them about, and for examining their consequences. Chapter 10 is a provisional attempt to construct a model of the consequences of the configurational flux on invertebrate evolution. It is necessarily rather speculative, but it is hoped that it will at least stimulate the construction of better models.

It has sometimes been asserted that paleontology is not a discipline from which important scientific discoveries may be expected—that it can serve only to indicate the history of processes that must be discovered by other disciplines, notably those of modern biology. But this contention is clearly false in principle. For the processes that have significantly affected the configuration of the fossil record, and that operate on the levels documented therein, explanations may be formulated from fossil as well as from recent evidence. Ideas for process models may be derived from the past as well as from the present; moreover, the tests of these models may be conducted in the fossil record as well as among the living biota. The fossil record contains ecological configurations that are very unlike those at present, and so they provide configurational tests which ecological and evolutionary process models must satisfy, regardless of their sources.

For whatever it can be taken to mean, uniformitarianism must work both ways, from the past to the present as well as from the present to the past. This is certainly true of methodological uniformitarianism, which asserts that natural laws are timeless. For substantive uniformitarianism, it is possible to substitute the proposition that the present is a good place to look for ideas in building process models of the past; this proposition must work in the opposite temporal direction as well. There is, it is true, more evidence available for the present biosphere than for any past configuration, so the temporal symmetry is by no means perfect. Nevertheless, the outlook for paleontological research is quite promising. Even if the record were absolutely complete, we should be unable to describe it except by sampling and constructing models, with their built-in defects, to explain the sample data. This is, in fact, what is done to explain the present biosphere. Paleoecological problems are made somewhat difficult by the bias that does exist in the fossil record, but still the record is remarkable. By modeling, it is possible to perform sorts of experiments on historical questions, to recreate the ways in which factors might have interacted in the past, and to test the relevance of these experiments to life history.

CHAPTER TWO

FUNDAMENTAL EVOLUTIONARY PRINCIPLES AND PROCESSES

The problem is the tendency in organic beings descended from the same stock to diverge in character as they become modified. . . . The solution, as I believe, is that the modified offspring of all dominant and increasing forms tend to become adapted to many and highly diversified places in the economy of nature.

—Charles Darwin, *The Life and Letters of Charles Darwin, Including an Autobiographical Chapter,* John Murray, London, 1887.

The discovery of the principles of evolution forms by far the most dramatic chapter in the history of science, and perhaps in the entire history of ideas. The famous episodes connected with the establishment and verification of the theory of evolution, especially those involving Darwin and Wallace, epitomize the rise of a scientific understanding of the world.

The processes that formed the cornerstone of the evolutionary theory of Darwin and Wallace, the processes of *natural selection*, have been amply confirmed and form a central theme of modern evolutionary theory. Briefly, Darwin and Wallace observed that a certain amount of variation among individuals of the same kind of organism—even among brothers or sisters—is found nearly everywhere in nature, and some of the variation seemed to be heritable. Organisms as a rule can produce many more offspring than can be supported by natural resources. As populations do not grow in size indefinitely, they must be limited so that only a fraction of the offspring survive. It is likely that the individuals that do survive are on the average those that vary so as to be better fitted to their environments. The survivors breed and furnish offspring from which survivors of the next generation are selected. Thus, there is a possibility of maintenance or improvement of the fitness of a stock for an environment by "natural selection," when there are appropriate heritable variations from which to select.

The intimate relation between heredity and ecology is at once obvious. The environment is a major factor in determining changes in the hereditary makeup of descendant populations by natural selection of the most advantageous inheritances, as represented in vigorous, well-adapted individuals. Organisms are fitted to their environments by selection, and at the same time the environments help to determine the course of future evolution. Most organisms are what they are chiefly because of the nature of their environments and of the environments through which their ancestors have passed. If we know all about an organism, ancient or modern, we know a very great deal about its environment as well. To understand why and how an environment-organism system has developed, and thus to gain any fundamental insight into the principles of paleoecology, we must understand the principles of evolution.

The processes of heredity provide evolution with its basic machinery. The subject of genetics is far too vast to be treated here except in barest outline. Nevertheless, a brief account of some principles must be given here in order to introduce the more basic terms and concepts required in subsequent discussions.

The Darwinian hypothesis of natural selection implies that hereditary materials have three important properties (Muller, 1966). First, genetic material must be capable of influencing *other* materials—in other words, of influencing the development and functioning of an organism. Second, genetic materials must be able to reproduce themselves. Third, they must be capable of changing in some way, and then of reproducing the change. Darwin himself proposed a genetic model, but we know today that it was quite wrong. The beginnings of the genetic supermodel that is accepted today were published by Mendel in 1865, but did not come to the attention of the scientific community until 1900, 42 years after the publication of the theory of natural selection. A model of heredity was gradually elaborated by generations of scientists from observation and experiment; then, in 1953 Watson and Crick defined the chemical structure of deoxyribonucleic acid (DNA) and showed that it possesses many of the properties required of genetic material. Today we have achieved a preliminary understanding of how organisms can reproduce themselves and their genetic material and of how changes that are themselves heritable may occur in genetic material. We now have some inkling of the ways in which genetic material influences the development of organisms.

The Genotype Coded by DNA Constructs a Phenotype

DNA is believed to be responsible for most heredity, although some viruses employ ribonucleic acid, or RNA, rather than DNA as their primary genetic material, and the other cell contents of the zygote, which are also inherited, may affect ontogeny. The evidence that has led to an understanding of the properties of DNA is not reviewed here; good accounts are available elsewhere (especially Watson, 1965).

DNA is organized as macromolecules, which are often up to several microns long. Each molecule ordinarily consists of two coiled strands or helices composed of subunits called *nucleotides*. A nucleotide consists of a phosphate group and a sugar group (deoxyribose), with each sugar group bearing a nitrogen base (Fig. 2-1). Each of the bases of each helix is chemically bonded to bases of the other helix, connecting the two helices like rungs on a spiral ladder (Fig. 2-2). Four bases are common in DNA: the purines *adenine* and *guanine* and the pyradmidines *thymine* and *cytosine*. The structure of these bases is such that bonds are ordinarily formed only between adenine and thymine, or between guanine and cytosine. Thus, for any given succession of bases in one of the helices, the succession of bases in the other helix is nearly completely specified.

The order of the bases in the DNA helix represents a sort of code that directs the synthesis, organization, and regulation of the compounds forming the material structure of an organism. This is owing to the ability of the DNA to control the order in which long series of amino acids are bonded together into strands. A given sequence of bases along the DNA helices controls the sequence of amino acids; information coded in the base sequence is copied by the nucleic acid RNA, which transmits the code to a part of the cell, the ribosome, wherein proteins are manufactured.

The bonds that unite amino acids are known as *peptide bonds*, and amino acid strands are called *polypeptides*. Polypeptide strands are proteins, or may unite to form multistrand proteins. Therefore, DNA possesses the ability to direct the synthesis of proteins. The enzymes that catalyze biochemical reactions are proteins, as are many of the materials of cell walls and other structures. Different proteins are composed of different combinations of amino acids, and, indeed, must owe their properties to the kinds and ordering of the amino acids, although some substitutions and irregularities are possible without impairing some protein functions. Three consecutive bases code for an amino acid; therefore, a long sequence of such triplets determines the sequence of amino acids. There is some redundancy in the code, for different triplets may code for the same amino acid.

Inasmuch as proteins that are pure enzymes, and some of the "structural" proteins as well, can catalyze certain biochemical reactions, a number of enzymes may react in a given sequence to catalyze a number of related steps along a biochemical pathway leading to the synthesis of a particular substance which has a place in the development of an organism. In such a case, the substance

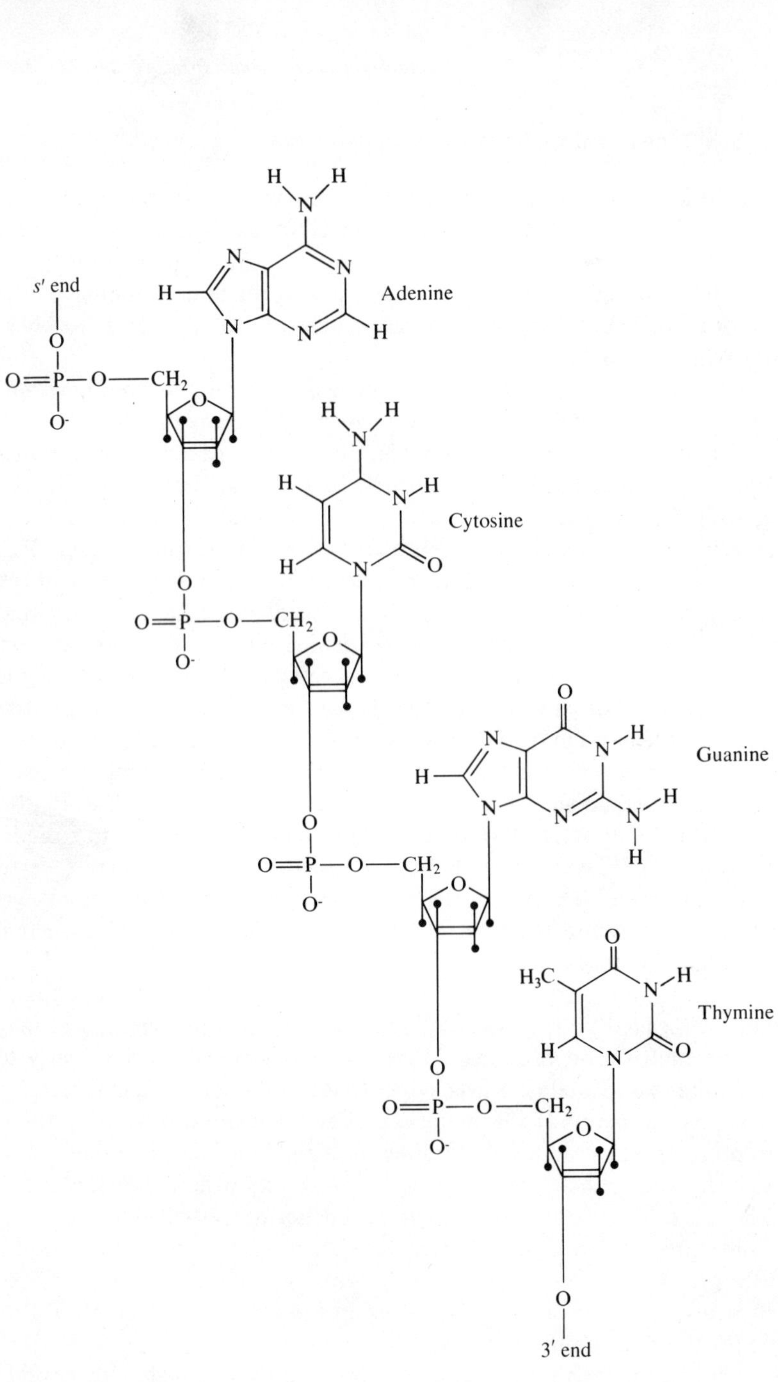

Fig. 2-1 The structure of a segment of a deoxyribonucleic acid (DNA) molecule, showing four nucleotides bonded to the backbone in a chain. (After James D. Watson, *Molecular Biology of the Gene,* copyright © 1965 by J. D. Watson; W. A. Benjamin, Inc., Menlo Park, Calif.)

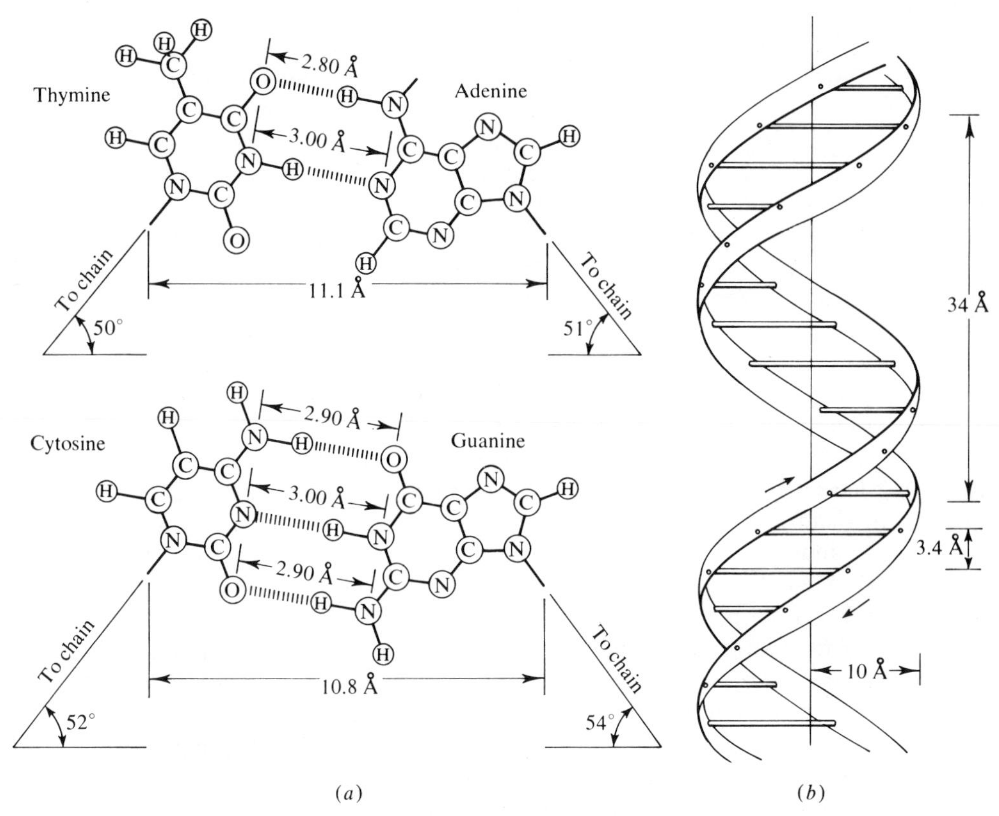

Fig. 2-2 (*a*) Hydrogen bonding between base pairs in DNA; thymine and adenine form bonds, as do cytosine and guanine, owing to the complementary configurations of these pairs of molecules. (*b*) Two DNA strands are folded spirally and connected by hydrogen bonds between base pairs; the backbones are represented by the spiral sides, with the base pairs bonded between them like rungs. (After James D. Watson, *Molecular Biology of the Gene*, copyright © 1965 by J. D. Watson; W. A. Benjamin, Inc., Menlo Park, Calif.)

has been generated along a *biosynthetic pathway*. In other cases, developmental activity is carefully sequenced by the switching on and off of different genes in a precise order, called an *epigenetic sequence* (Stebbins, 1968). The switching action may be mediated by gene products or by substances derived from them.

However, DNA does not regulate the entire developmental process through only these sorts of biosynthetic pathways and epigenetic sequences. Still another important series of processes occurs, consisting of interactions entirely between gene products or their derivatives, which also leads to the synthesis of substances that function in the organization or metabolism of an organism. Chains of such processes, which do not directly involve gene action beyond the preliminary stage, have been called *informational relays* (Stebbins, 1968). The products of informational relays range from proteins that contain more than one peptide chain to tissues or organs. The information in these relays may often be transmitted by the molecular structure of gene products and their derivatives, so that the relays are integrated by remote control (see Stebbins, 1968, for examples).

Development of the zygote into an adult obviously requires the operation of powerful and complex chemical processes, elegantly integrated so as to function as a harmonious system that is viable throughout the extensive transformation from egg to adult and beyond, called the *ontogeny*. Organisms do not inherit from their parents just the characters of their vigorous adult lives, but a range of potential characters from those of the zygote through senescence. The parental gametes must carry all the information necessary to elaborate a whole organism from a fertilized cell, a formidable amount, and also they must possess biochemical machinery adequate to this task or at least the potential to manufacture such machinery from materials in their environment.

The sort of organism constructed by the action of genetic material is thus dependent chiefly upon the information coded in the DNA contributed by the parents and upon a supply of appropriate building block substances and energy. In addition, reactions along some biochemical pathways are activated by environmental stimuli. For example, environmental temperatures of a given level for a given period of time may be required before a sequence of biochemical events leading to, say, metamorphosis, will be initiated or will proceed to completion. The advantage of these environmental cues is obvious; the processes they set in motion tend to take place under the most favorable environmental conditions. The cue is therefore a harbinger of these optimum conditions, although cues and target conditions are not always causally related; nevertheless, the utilization of these cues is a consequence of evolution. Another sort of environmental influence on ontogeny occurs when alternative pathways of development are available, leading to somewhat different individuals, and the pathway that is followed is selected by an environmental cue or stimulus. Again, the advantage is obvious; the individual most appropriate to a given environment is selected from among all the prospective types of individuals coded by the DNA.

Thus the organism that is eventually produced from a zygote is the product of an interaction between the environment and genetic material and its products. The environment provides building materials, energy, and information of the sort called "intelligence" by military establishments, which may act as a developmental cue. The genetic material oversees the organization and transformation of these environmental contributions, although often this occurs along informational relays. This individual organism is called a *phenotype*—the visible product of reproduction to which both environment and heredity have contributed. The potential that resides in the genetic material for creation of this individual is called the *genotype*. Several phenotypes, even though constructed from identical genotypes, will nevertheless vary according to the contributions of their environments during their development.

The relation between these genetic units (that is, functional DNA sequences or genes) and the physiological functions and morphologic characters observed in whole organisms must be exceedingly varied. Most phenotype characters are built by numerous biochemical reactions and thus require the operation of numerous gene products. These characters are termed *polygenic*. On the other hand, many gene products are employed again and again, now in one biochemical sequence, and now in another sequence leading to a different end-product. Thus, the gene product affects more than one character, and the gene is said to be *pleiotropic*. Finally, it is possible that different genes may have similar products which are functionally identical or at least selectively neutral in some gene systems (King and Jukes, 1969). If this proves to be true, then genes are not unique and the amount of flexibility between the base-pair sequences, the biochemical structure of organisms, and the phenotypes is greater than previously suspected. At any rate, various hereditary units and various phenotypic characters are connected by a complex web of interactions, so that their relationships are often indirect.

Thus, the causes of differences or changes in morphology observed within a given fossil group cannot with certainty be assigned to any particular genetic mechanism, even when the changes can be assumed to be genotypically and not phenotypically based. The "functional units" we shall be mostly concerned with in fossils are either measurable morphological characters or inferred physiological tolerances; either of these sorts of units is usually genetically complex.

Cell Division May Duplicate or Halve the DNA in Daughter Cells

An enormous number of cell divisions is required to produce a mature individual of a moderately large species from a zygote. An adult human, for example, contains on the order of 10^{12} cells, so there must be many more cell divisions in a human ontogeny, for replacements are constantly required for worn-out or damaged cells. Except for reproductive cells (and for certain minor exceptions), each of these cells contains the same genetic information as the zygote, insofar as DNA is concerned. All these cells arise by division of a parent into daughter cells; division that produces cells with identical copies of the DNA of the parent nucleus is called *mitosis*. The duplication of DNA involves the separation of the paired helical strands of nucleotides into single strands by a loosening of the bonds between the nitrogen bases. As each of the original strands becomes independent of the other, it helps to direct the synthesis of a new strand with which it forms a paired helix. Since the order of the bases in any one strand is specified by the order in the strand with which it is paired, it is evident that a newly synthesized strand must be an exact duplicate of the former partner of the old strand. Actually, mistakes and substitutions do occur, perhaps about once in 10^5 times. Usually the two new pairs, each containing one of the old pair of strands, will be exactly alike, and will be exact copies of the original pair. When mitotic (nonreproductive) cell division occurs, the nucleus divides, and one of the daughter strands goes with each nucleus. Obviously if all the DNA strands in a nucleus duplicate themselves before a nuclear division, the DNA of each daughter nucleus will duplicate the original parental DNA. However, the formation of reproductive products, or gametes, involves a different type of cell division. Instead of daughter cells containing the full complement of DNA of their mother cell, each gamete (sperm or egg) ordinarily has only half the complement. Thus when gametes unite, each carries half the DNA complement that is normal in the cells (excepting other reproductive cells) of their parents, and so the zygote (fertilized egg) has a full complement composed of about equal contributions from each parent. The cell division that leads to the production of daughter cells with only half the parental complement of DNA is called *meiosis*. It is accomplished by a sequence of two cell divisions during which the parental DNA is duplicated only once (Fig. 2-3).

It has been known for many decades that units of heredity occur on chromosomes, which are rod- or dot-shaped bodies that appear in cell nuclei during the division process. We now know that chromosomes are composed of DNA, RNA, and proteins, and that the hereditary units are segments of the DNA (or RNA in some viruses). Just preceding mitosis, the amount of DNA in cell nuclei doubles; and as mitosis begins, the DNA condenses into chromosomes that have divided into two halves. As the cell divides, each daughter cell bears one-half of each chromosome; the mother cell has been duplicated. Thus, the ability of DNA to reproduce itself is illustrated in the manufacture of new chromosomes during mitosis. Actually, some DNA also occurs in certain organelles, such as mitochondria, that lie outside the cell nucleus. This extrachromosomal DNA is responsible for some protein synthesis and interacts with nuclear DNA. Its origin and evolutionary significance is not yet understood.

Segregation, Assortment, and Crossing Over Can Produce Great Genetic Variation Among Recombinants

The behavior of chromosomes during meiotic cell division accounts for most of the heritable changes that occur from one generation to the next. A full complement of chromosomes consists of pairs of similarly appearing chromosomes except that one pair, the sex chromosomes, are normally distinct in morphology. When during meiosis the cells divide without a corresponding doubling of DNA, the daughter cells each end up with half of the chromosomes of the mother cell. Because the number of chromosomes in a given species is constant, the number of chromosomes in the daughter cells following meiotic

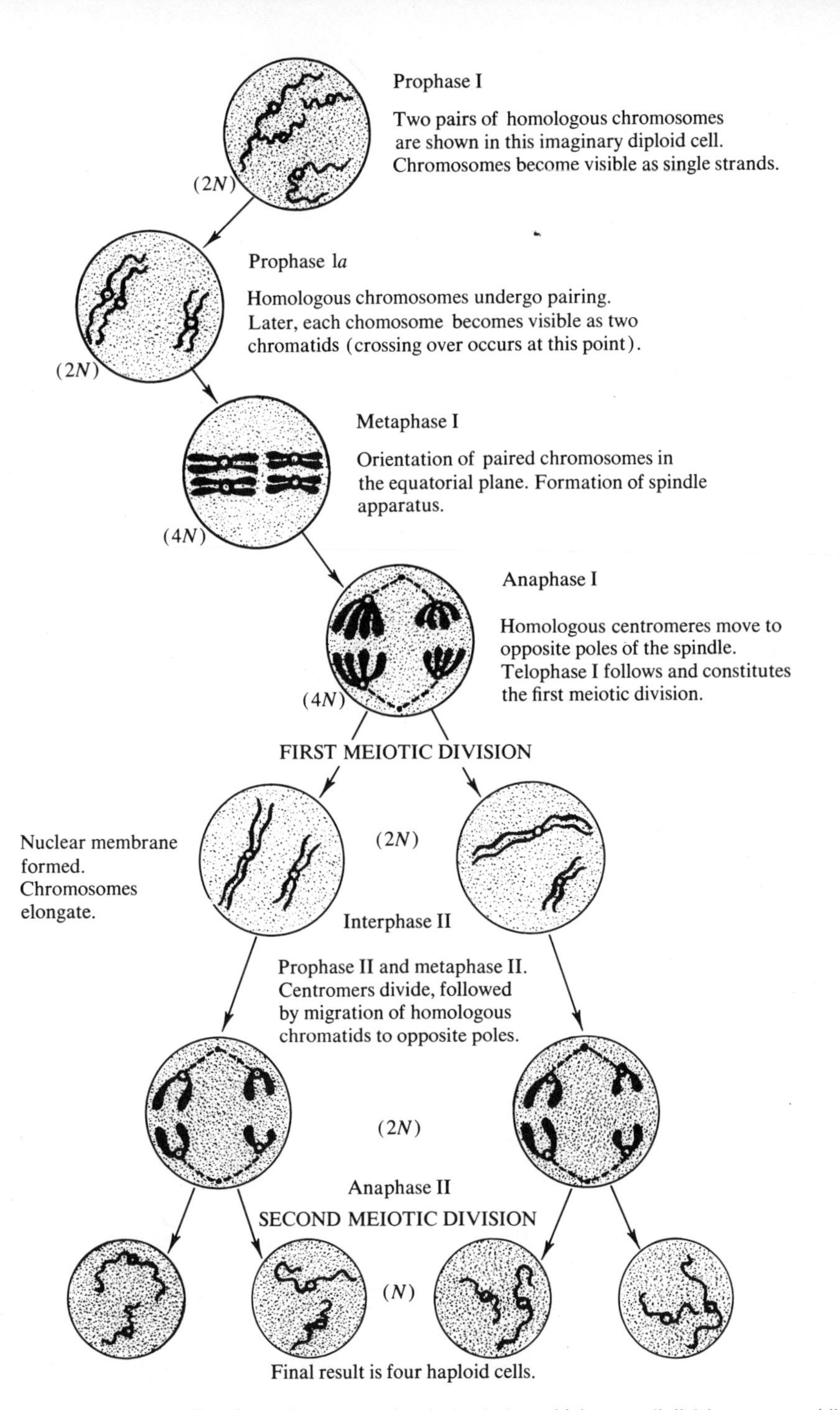

Fig. 2-3 The phases of meiosis, during which two cell divisions occur while the chromosomes are duplicated only once; following the second division, the daughter cells are haploid. (After James D. Watson, *Molecular Biology of the Gene,* copyright © 1965 by J. D. Watson; W. A. Benjamin, Inc., Menlo Park, Calif.)

division is constant also, and is termed the *haploid number*, or *N*. The full complement as represented in zygotes is the *diploid number*, or *2N*, and the zygote contains about twice as much DNA as either gamete.

Ordinarily, each one of the paired chromosomes contains genetic information for the same set of ontogenetic processes as the other; in other words, there are usually two DNA sequences for each biochemical step, one on each of the paired chromosomes. Both sequences may be the same, a condition termed homozygous, or they may differ, a condition termed heterozygous. Different sequences for the same step are called alleles. Many alleles control the formation of phenotypic characters that are observably different from those controlled by their sister alleles—in color, dimensions, activity, or other qualities. If a pair of chromosomes is heterozygous for some sequence, the product that is actually manufactured may be controlled by only one of the DNA sequences. In peas, for example, the allele for yellow seed color will produce yellow seeds even if an allele for green seed color is also present. The allele for yellow seeds is said to be *dominant*, and that for green seeds, *recessive*. Recessive alleles are expressed in the homozygous condition, when no dominant is present; therefore, for a phenotype to express a recessive allele, both parents must donate it to the offspring. However, a character need not have been expressed in the phenotype of either parent, for both may have been heterozygous for the character. Thus, genes in offspring may be found in combinations not present in either parent; offspring possessing gene combinations that are not parental are called *recombinants*.

Recall that during meiosis one of each pair of chromosomes is incorporated in the nucleus of each daughter cell. So long as heterozygosity exists for any gene on those chromosomes, each chromosome of a parental pair will carry alleles that the other does not have. Thus, daughter cells will not have identical genotypes. This phenomenon is called the Law of Segregation, which we owe to Mendel (1866). Furthermore, the chances of a given chromosome of one pair being included in a daughter cell with either chromosome of another pair are equal, so that the genes on these chromosomes have only a 50 percent chance of ending up in the same daughter cell, a situation Mendel called the Law of Independent Assortment.

When genes on the same chromosome are transmitted *en bloc* and do not segregate, they are said to be *linked*. So although variation between daughter cells is introduced by segregation and independent assortment during meiosis, the combination of alleles in gametes will still be relatively limited when linkage is complete within each chromosome. For an organism with $2N = 8$ (that is, 8 chromosomes in the diploid state) 16 combinations of chromosomes are possible in the haploid daughter cells, and 120 combinations are possible among F_1 zygotes, not a very great number. Among humans, with a chromosome number of $2N = 46$, there are 8,388,608 possible chromosome combinations in descendant zygotes. This is a large number, but obviously does not begin to account for the great inherited variability within the human population. There must be another process operating to increase genetic variability among progeny.

In fact, there are several; one of the most important, quantitatively, is *crossing over*.

During meiosis, homologous chromosomes (that is, both members of a similar pair) come to lie against each other at one stage, and they often exchange segments. Breakage commonly occurs at the same positions on each chromosome and the broken ends join the opposite chromosomes. Now, genes within the exchanged segments, instead of being linked to the other genes in the chromosome from which they have become separated, segregate. Linkage is thus not complete. Crossing over is common and nearly random along chromosomes, so that any two genes on the same chromosome may be separated by crosses at one of a great number of loci. In general, the closer two genes are positioned in a chromosome, the less often they will segregate, since the chances of a break occurring between them for crossing over diminishes with the distance between them. Nucleotides on opposite ends of an average chromosome have only about a 50 percent chance of remaining on the same chromosome after meiosis. Obviously, the number of possible gene combinations due to crossing over is vastly in excess of the number that can be expected from simple chromosome assortment without crossing over. For example, it has been shown that breaks are probably possible between nearly every base pair along a DNA helix (see Benzer, 1962). It is easy to understand why two individuals that originate from separate zygotes have only an insignificantly small chance of being genetically identical.

Mutations Create Genetic Novelties

Although the continuous recombination of genes that results from assortment and crossing over provides a vast potential of genetic variability within populations and between generations, it cannot provide for the continual genetic changes that must occur during the evolution of new sorts of organisms along extensive adaptive pathways. During extensive evolutionary transformations, a supply of wholly new genes is required; genotypes must appear which are neither parental nor recombinant but are subsequently heritable. The process responsible for this supply of novel genes is *mutation*; it involves the substitution, rearrangement, duplication, or loss of parts of the DNA sequence. Mutations commonly cause a change in the developmental steps and therefore in the phenotype of the organism.

Many mutations involve structural changes in the chromosomes. Sometimes segments of a chromosome are lost, giving rise to a *deficiency*, or segments are repeated, giving rise to a *duplication*. Rearrangements within chromosomes also occur and seem to owe their origin to the sort of chromosome breakage that normally results in crossing over. Chromosomal segments sometimes rotate end-for-end and then reform with their original chromosome, so that the order of bases becomes reversed and the genes are in reverse order; this is called an *inversion*. Finally, segments may be interchanged between nonhomologous chromosomes, forming chromosomes with new gene combinations; this is *translocation.* These intrachromosomal changes can occur with chromosomal segments large enough to be observed under the light microscope, or they can involve only one or a few nucleotides. Deficiencies and duplications often have obvious phenotypic effects that may be advantageous or, more usually, may not. Inversions and translocations affect external morphology less often, and in plants may have little effect at all; in animals, they may be responsible for physiologically different strains having different relative advantages under different conditions (Dobzhansky, 1951).

Mutations may also involve the alteration of one or more of the bases along a DNA strand; this changes the genetic code, and the product of the gene of which the new base is a part may be altered. The code may be changed by base replacement or by deletion or addition of a nucleotide which not only alters the code triplet in which it occurs but which throws off the order of counting, affecting all subsequent triplets in the gene. The causes of mutations are not completely established, although it is well known that certain types of radiation, heat, and some chemical stimuli will raise mutation rates. Some mutations may arise from mispairing of the bases in DNA, which is known to occur under certain conditions (see Stebbins, 1971), and some may be due to small chromosomal additions, subtractions, and rearrangements that arise during crossing over and have the effect of altering the number or sequence of the bases so that different information is present.

The information carried in a gene altered by a mutation may not "make sense," so that the mutation does not produce a polypeptide product; or the premutation product of the original gene may be formed but only in reduced quantities; or there may be no functional change ("neutral mutation"); or the mutant gene may produce a new polypeptide, which may or may not harmonize with other gene products with which it is associated in ontogeny. More often than not, it will not harmonize. Most changes are deleterious; even if they will allow the formation of a viable organism, they will usually confer a disadvantage on their bearers with respect to the gene they replaced. However, occasionally a change may prove advantageous, and the new gene may eventually become common among descendants of the mutant, which multiply extensively.

Under Certain Conditions Gene Frequencies Tend to Remain Constant Although Gene Combinations Change

Consider a population of 2000 individuals that is brought together by some unusual agency, such as a scientist, and in which all individuals are heterozygous for two alleles, *A* and *a*. The same number of gametes will be formed with *A* as with *a*. Assuming that there is no special advantage in possessing either allele, and that mating is random, zygotes will be formed with the combinations *AA*, *aA*, *Aa*, and *aa* in equal numbers, giving rise to the genotype proportion, 1*AA*:

$2Aa$: $1aa$. If the population size remains stable so that the first descendant generation (F_1) is composed of 2000 individuals, there will be about 500 AA, 1000 Aa, and 500 aa individuals. If this generation then proceeds to reproduce, the same number of gametes will again be formed with allele A as with allele a. That is, all gametes of AA individuals and half the gametes of Aa individuals will be A, representing half the gamete production of the population; the other half of the gamete production will, of course, contain a. The second descendant (F_2) generation will have the same genotype ratio as the F_1—that is, $1AA$: $2Aa$: $1aa$. In fact, this ratio is an equilibrium ratio, which may be maintained indefinitely.

The principle of an equilibrium genotype ratio for two-allele genes was proposed independently in 1908 by Hardy and by Weinberg, and is known as the Hardy-Weinberg law. It may be stated in the following form:

$$\text{Genotype equilibrium} = p^2 + 2pq + q^2$$

where p is the proportion of one allele, q the proportion of the other, and $p + q = 1$. It is evident that the ratio of the alleles need not be 1 : 1 as in the previous example. Weinberg extended the law to cover cases of more than two alleles; for example, when there are three alleles, A, a, and a', in the proportions of p, q, and r, respectively, and $p + q + r = 1$, then the equilibrium expression is:

$$\text{Genotype equilibrium} = p^2 + q^2 + r^2 + 2pq + 2pr + 2qr$$

Among the population of 2000 individuals we considered previously, the genotype ratio in the parental generation, which was entirely heterozygous, was $0AA$: $1Aa$: $0aa$. However, it took only a single generation for the ratio to become $1AA$: $2Aa$: $1aa$, the equilibrium ratio. This is the general case; equilibrium can be established for most alleles in a single generation. An important exception involves genes on the sex chromosomes. Because males and females have different combinations of sex chromosomes, equilibrium may not be established for several generations; the course of the approach to equilibrium will depend upon the nature of the chromosomal mechanisms governing sex. In animals, such as man, in which the male has dissimilar sex chromosomes (XY) and the female similar ones (XX), equilibrium for an allele on the X chromosome is approached rapidly but with fluctuating proportions from generation to generation.

Each member of a moderately large, freely interbreeding population of sexually reproducing, multicellular organisms will have a considerable genetic resemblance to every other member. On the other hand, no two members will be precisely alike genotypically (unless they are "identical" siblings that do not happen to carry new mutations) or phenotypically (in any event). Always excepting some identical twins, each individual will possess a gene combination not present in any other, and the total number of alleles in the population is much greater than the number that can be present in any single individual. The total genetic variability present in the population can be thought of as the pooled genetic variability of all its members—the *gene pool.* The sort of population that we are dealing with is termed a *Mendelian population,* defined as "a reproductive community of sexual and cross-fertilized individuals which share a common gene pool" (Dobzhansky, 1951).

Frequencies of all the genes in a large pool will tend to remain relatively constant because of the Hardy-Weinberg effect, when one gene has no special advantage over another, and when mating is random.

Evolution Is a Change in Gene Frequencies

At the genic level, evolution has been defined as a change in the genetic composition of populations. A number of processes can cause such changes; these include mutation, drift, selection, and migration. Selection is the classic Darwinian process, which can now be stated in genetic terms. Numerous examples of such processes in animals are given by Dobzhansky (1970) and Mayr (1963).

In large populations, the Hardy-Weinberg effect will tend to maintain equilibrium so long as each allele has an equal opportunity of being incorporated into descendant organisms that themselves reproduce at average rates. However, if a given allele is usually included in a genotype that produces a superior phenotype—a phenotype which usually gives rise to more than an average number of offspring—the frequency of that allele will increase. The alternative allele(s) necessarily found in phenotypes with fewer than average offspring will decrease in frequency. The process that operates to change the allele frequencies in this event is *selection.* Under the influence of selection, populations may evolve.

Selection is commonly represented as a coefficient, s, which is a kind of measure of the intensity of selection. Basically it measures the differential perpetuation of genotypes (Mayr, 1963). A useful coefficient of selection is the complement of the fraction of the decrease in frequency of the less favored allele relative to the more favored, or, symbolically,

$$s = 1 - \frac{L}{M}$$

where L is the descendant/ancestor proportion of abundance of the less favored allele and M the similar proportion of abundance of the more favored allele.

The proportion expressed by the coefficient of selection can be represented as

$$A = 1, \qquad a = 1 - s$$

where A and a represent proportionate allele frequencies between ancestral and descendant populations. The ratio $1 : 1 - s$ is sometimes called the fitness or survival ratio. When the ratio is 1: 1, there is no selection and s must be 0, but if one allele should be lethal, the ratio would be indeterminant (1: 0) and s would be 1.

As nearly all (or perhaps all) genes affect several characters, and each character is probably affected by at least several genes, the interrelations and interactions among genes in the same genotype—*epistatic interactions*—are intricate. It is a combination of polygenes whose products are jointly expressed as the advantageous character that is being selected. Genes that are successful in phenotype after phenotype will persist or increase in frequency. Therefore, those genes are most likely to persist which contribute to robust phenotypes in the greatest variety of genotypes. Such genes are "good mixers," as Mayr (1954) has put it. Gene pools will thus have a tendency to harbor many of these genes, which produce fit phenotypes in most combinations, and immigrant or mutant genes will be winnowed by selection for adaptation either to the more frequent alleles with which they must combine in heterozygotes or to polygene systems of which they form a part.

Many alleles act so as to interfere with the coordination of metabolic activities, destroying the integration or coadaptation of cellular functions. Selection against such alleles, which must arise as mutants, is of a special sort called *internal selection* (Stebbins, 1968) or *hard selection* (Wallace, 1968). It is probably not difficult for populations to cope with the lethal load of such selection through reproduction. It is *external* or *soft selection* by the environment, which permits the more "fit" organisms to produce a disproportionate number of offspring, that is of chief interest in evolutionary ecology.

In small populations, gene frequencies may change even in the absence of differential selective advantages among different genotypes. As a special case we may take two castaways. Their first child will bear only half the chromosomes of either of them, so that some of their genetic variability will be lost in the F_1 generation, assuming some original heterozygosity in the parents. Even with several children, the chances of every parental allele being transmitted to one or another are small. The mechanisms of independent assortment and crossing over that make for genetic variability will also prevent complete sampling of the parental gene pool. The gene pool of the children will contain fewer alleles, and the frequency of those alleles present will have changed from the parental frequencies. This is a special case of evolution. The changes in gene frequency resulting from the chance failure of a descendant population to sample the parental alleles in their original proportions is called *genetic drift*, or the *Sewall Wright effect*, after the geneticist who developed expressions to demonstrate this process.

The effects of drift on gene frequencies are quite different in large and small populations, because as population size becomes smaller and smaller, the relative importance of each single allele increases. If the frequency of allele A is 0.1 in a population of 100, then there are 10 of these alleles present in the population, and the chance loss of 5 of them between the parental and F_1 generations would change the frequency of A from 0.1 to 0.05. However, in a population of 1000, 100 of these alleles are present when the allele frequency is 0.1, and loss of 5 of them would lower the frequency only to 0.095.

In a large population, where changes in allele frequencies resulting from drift tend to be small, it is expected that the direction of drift of an allele will change randomly from increase to decrease, and so tend to cancel drift effects over a number of generations. A few genes, however, may by chance drift in one predominant direction. In small populations, a gene drifting in one predominant direction for very many generations will either be lost or attain a frequency of 1. Thus, small populations will tend to have their genetic variability reduced in time, just as with the descendants of our castaways.

The castaways illustrate another evolutionary effect that commonly occurs when a species invades a new region. When the colonists are only a few individuals, they will carry only a fraction of the genetic variation of the population from which they came; therefore, the new population will have a restricted gene pool that may be quite different from the original one. This is the *founder principle* (Mayr, 1942). Many local populations may owe their peculiarities to their founders rather than to selection for a distinctive phenotype. These peculiarities may even persist up to the species level.

Local populations are not free from outside genetic influence, but may receive a steady supply of immigrants bred elsewhere. These immigrants may introduce new alleles; in any event, they will tend to change allele frequencies. For fairly mobile animals, the numbers of immigrants may approach 50 percent of the population regularly. Furthermore, when two local populations that can interbreed border on each other, a few individuals from each may breed with individuals of the other. Back-crossing of the progeny with members of either population will then introduce genes from one gene pool into the other.

Emigration may also alter gene frequencies and may play an important role in the pattern of the gene pool. If the differential migration of certain genotypes is environmentally controlled, it can be regarded as a sort of selection, whereas if individuals come and go randomly, their genetic effects are owing to a sort of drift.

The Evolutionary Factors of Selection, Mutation, Recombination, Migration, and Drift Interact

The effects of selection, mutation, and recombination combine with drift to produce a given F_1 gene pool from a parental gene pool. The changes in gene frequencies occurring between generations are compounded of these factors, and together represent a sort of measure of evolution. The statistics of gene frequency change in any one generation will not always tell us much about the nature of the evolutionary forces actually at work, but if we have averages for many generations, so that relatively minor and ephemeral effects imposed by more or less random environmental fluctuations tend to cancel out, an evolutionary trend may become apparent.

In addition to the changes from selection, mutation, recombination and drift, the breeding stock in many populations is constantly altered by migration, so that the offspring from one parental stock are not the precise parental stock of the next generation. Immigration probably often accounts for the introduction of well over 90 and perhaps as much as 99 percent of the new genes. Of course, these genes have ultimately arisen by mutation elsewhere. When the effects of migration are balanced with those of other factors, we have an idea of the total changes in gene frequencies arising from all sources, and thus of the evolutionary progression within a population.

As previously noted, population size has a considerable bearing on the effects of drift; gene frequencies are more easily altered and random sampling effects are greater in small than in large populations. Wright (1931) has calculated that in small populations the effect of drift can override the effects of selection and mutation. When population sizes (N) are over $N = 1/2s$, selection predominates in changing allele frequencies. However, in populations as small as $N = 1/4s$, or smaller, drift predominates. Between these sizes, both factors are effective to a degree. The effective population size, whether "large" or "small," depends upon the coefficient of selection. No population is so small that drift will alter the effects of a lethal gene. On the other hand, if selection is low, moderately large populations may drift; for example, if $s = 0.0001$, drift may be an important cause of gene frequency change in populations of up to 5000. But this is such a large population that drift will usually affect gene frequencies only slightly, although if all the factors affecting gene change remained constant for a long enough time, gene loss or fixation would eventually occur through drift. An s of 0.0001 is therefore to be considered quite small.

The interactions of drift with mutation, recombination, and migration are of the same sort. When populations are larger than $N = 1/2\mu$ (where $\mu =$ the mutation rate, the fraction of times that a mutation occurs out of the total number of chances it has to occur), or $N = 1/2m$ (where $m =$ the migration rate, the fraction of occurrences of a gene that are due to immigration), then the genes introduced by mutation or migration will on the average not have their frequencies significantly altered by drift; in this case gene loss or fixation may not occur. In populations smaller than $N = 1/4\mu$, or $N = 1/4m$, drift rather than mutation or migration will determine gene frequencies. When μ or m is on the order of 0.0001, they are relatively ineffective by themselves in altering gene frequencies in even moderately large populations. However, if selection strongly favors a given gene, even a small supply of that gene, brought into the gene pool by mutation or migration, may be ample for selection to enlarge into significantly high frequencies. Mutation rates themselves are subject to control by natural selection (see Dobzhansky, 1970).

Population-Environment Systems Are Stabilized by Selection

Imagine an isolated population living in such a benign environment that there are no selection pressures, yet wherein population size is regulated at a steady state. The regulatory process would have to be completely nonselective—perhaps a steady lethal rain of randomly disposed meteorites, from which no protection was available, would do the trick. Gene frequencies would change even in the absence of selection and gene migration, owing solely to effects of drift and mutation.

In a real but fairly constant environment that is devoid of important lethal extraterrestrial interference but that imposes ordinary physical and biological limitations on its inhabitants, selection will produce differential reproduction between more fit and less fit genotypes. In real populations of any size, the direction of evolution is therefore ordinarily under the chief control of the environment. When a gene pool has been subjected to the selective pressures arising from a given environmental range for a long time, many of the deleterious genes will be culled and a sort of standard phenotype mix will develop. After a very long time, most of the possible mutations which produce phenotypes that vary from the standard mix will have occurred, and most mutations that are sufficiently favorable in the given genetic environment and in the external environment will have been incorporated into the gene pool. Subsequently, nearly all mutations will be either neutral or deleterious. Selection is now helpless to provoke evolution in this population, but it is active in the elimination of variant phenotypes arising by mutation and recombination. Instead of altering gene frequencies, selection is operating in a way that tends to preserve the status quo. It is helping to prevent evolution. This aspect of selectional operations has been called *stabilizing selection* (Schmalhausen, 1949).

Even while stabilizing selection is at work to eliminate genes that produce poorly adapted, deviant phenotypes, new genes and gene combinations may appear that act to produce phenotypes in the normal, well-adapted range, although based upon novel genotypes. Evolution may still be going on under the surface of selection, so to speak. New genes that provide new paths to old phenotypes, or that form alternate steps in paths which do not effect the standard phenotype, remain in the gene pool without selective disadvantage because selection works on phenotypes. Assuming that the new and old developmental pathways themselves do not have differential advantages, selection here is permissive, allowing the employment of some alternative developmental pathways to the achievement of a standard product. This sort of selection is possible partly because development is often channeled by a buffering system to lead to a certain narrow adult phenotype, even though gene substitutions (or environmental changes) result in minor differences in ontogeny. Waddington especially has explored this aspect of stabilizing selection, which he calls *canalizing*.

Thus, the gene pools of many populations are partly coadapted and affected by a reduction in variability owing to stabilizing selection, and their phenotypes are often stabilized by canalization of development. Populations may therefore be relatively invariant and form a stable system throughout their geographic ranges under conditions of relative environmental stability. The gene pools form the genetic basis of the phenotypic population-environment system, which in ecologic terms is called a *niche*.

Gene Pool Diversity Is Maintained by Modifications of Gene Activity and by Environmental Heterogeneity

A number of mechanisms discussed previously, such as crossing over, have the effect of increasing the basic genetic variability of a population. Yet if the processes of stabilizing selection were able to eliminate all genes that produce any slight lowering of fitness under the average ambient conditions, the variability of gene pools might eventually be very slight. It seems evident that for a population to persist for a great many generations in the face of inevitable environmental fluctuations, it must be able to change its character to retain adaptation. A large pool of genetic variability is a useful and probably necessary attribute of nearly all populations. The causes and uses of genetic diversity are particularly well-discussed by Dobzhansky (1970).

Means of accumulating and preserving genetic variability include such genetic mechanisms as dominance and the selection of modifiers. Dominance has the effect of protecting fully recessive alleles from selection in heterozygotes, since they are not expressed phenotypically. Furthermore, if recessive alleles are not neutral but are actually valuable in heterozygotes, they may be actively retained in a population by selection, even if they are lethal as homozygotes. The heterozygote may, in fact, be the best combination of all, conveying more fitness than the homozygous conditions. In this situation (termed *overdominance*), a balance is struck in the frequency of the alleles which depends upon their relative fitnesses as homozygotes. This creates *balanced genetic polymorphism*, which appears to be very common.

Protection of alleles against selection can also be achieved through modification of their effects by other genes (that is, genes at other loci) with which they interact. Genes having effects that are or become disadvantageous may be difficult to eliminate from the gene pool; any other gene acting so as to limit or offset their deleterious effect would be favored by selection insofar as this aspect of its activities was concerned. Inasmuch as most genes have multiple effects, any deleterious effects may be modified by other genes, leaving the advantageous effects alone to be expressed. Also, some genetic variability is maintained simply because canalized development permits a variety of genotypes to be present if their differences are not strongly expressed in the ambient environment.

Environmental variety is expressed partly in space, where differences in local habitats will bring different selective pressures to bear on different local populations, and partly in time, wherein fluctuating environments will favor now one, now another, allele, and thus may allow several competing alleles or genotypes to coexist. Even within local populations it has been shown (Ludwig, 1950; Levene, 1953; Levins and MacArthur, 1966) that it is possible for a gene pool to include alleles which form variant sorts of genotypes having special utility in limited parts of the population's habitat or at limited times. This results in *genetic polymorphism*, the presence of alternate genotypes related to the heterogeneity of the environment and to the associated balance of selective pressures. Selection that produces genetic polymorphs within the same population as adaptations to different aspects of a heterogeneous environment is called *disruptive selection.*

Genetic diversity may also be maintained by the sorting out of genotypes into the most appropriate environments among those available owing to active habitat selection by the organisms themselves. Many invertebrate planktonic larvae exercise some selectivity in settling; they test the bottom, and if it seems unsuitable they postpone settling and metamorphosis and remain floating in the water, to be transported to new sites. In many plants, genotypic variability may be considerable and may be highly correlated with differential genotype fitness in a wide variety of microhabitats. This is due to the combing of an immense original number of zygotes by the environment, which allows only appropriately endowed plants to develop in each habitat.

The various devices that maintain genic diversity operate widely and effectively. Most populations seem to be at least moderately polymorphic, and thus most gene pools are rather large. Therefore, much evolution is simply based on genetic materials that are already present, causing more or less rare alleles to increase to high frequencies as conditions change so as to favor new gene combinations in which they are involved. A chief function of mutation, then, is to create a supply of novelties to stock the pool with varied genes.

Many Different Patterns of Phenotypic Variation Are Found Among Potentially Interbreeding Populations

Populations that are continuous over long distances may have some of the properties of local populations and some of the properties of two or more separate populations. They are composed of individuals that interbreed freely with their neighbors, but they stretch continuously over an area which significantly exceeds the area that any two individuals ordinarily visit or throughout which they ordinarily disseminate gametes in an overlapping pattern. Thus, direct gene exchange between distant members of the population is not possible, although it is possible for them to transmit genes, through intermediaries, to each other's descendants. Gene flow is continuous from margin to margin of the population, but since there is a lag in gene exchange between opposite ends of the population, the gene pool may contain different genes and markedly different gene frequencies in different regions. These differences may be maintained by selective differences that in turn reflect environmental differences. When the differences are connected by a gradient, phenotypic characters may display a geographic gradation from one state near one population margin to another state near the opposite margin. Such gradients in characters are termed *clines* (Huxley, 1939). Abrupt changes in the environment within the region that do not disrupt the continuity of the breeding populations may or may not be reflected in localized phenotypic or genotypic change. Certainly, this will depend upon the degree to which the phenotype is responsive to the environment, for intricately coadapted gene pools and unresponsive phenotypes presumably tend not to reflect local departures from regional environmental trends.

Between different populations, gene flow is somewhat (or completely) restricted, by definition. Nevertheless, some population systems, even when composed of many local populations, have rather invariant phenotypes. Presumably this results from selection for a valuable standard phenotype, or possibly from the presence of a cohesive gene pool highly coadapted to average conditions throughout the range of all the local populations.

Another common pattern is for each separate population in an interbreeding population system to have a distinctive phenotype frequency based in part on distinctive gene pools. Variation is thus discontinuous, and each population will have a distinctive phenotypic norm. Distinctness may depend upon the magnitude of the environmental differences between their habitats, upon the amount of gene flow between their gene pools, and/or upon the extent of a founder effect. In very small populations, drift will affect gene frequencies also, thus contributing to their distinctiveness.

A Species Is:

Populations between which gene exchange occurs commonly in nature may be grouped into a larger biological entity, the *species*. Species are thus composed of all the individuals that derive their genotypes from, and ordinarily contribute gametes to, a given gene pool, within which gene flow occurs. The gene pool of the species is the pooled gene pools of all the interbreeding populations composing it. Gene flow does not ordinarily occur between species; if it does, it is sporadic and limited. When populations exchange genes regularly they are considered to belong to the same species. Species may consist of a single, small, geographically restricted population, such as the California condor, or, at the other extreme, of large numbers of small and large populations and clines scattered over much of the earth's surface in appropriate habitats, such as man.

The great variability in population patterns makes it difficult to speak of species in general terms. Most species are composed of a number of local populations. Populations may coalesce in time when barriers to gene exchange are reduced, or they may split, resulting in the partitioning of a single gene pool. Clines may fragment into smaller clines or into local populations. The entire system of interbreeding local populations, each of which is based on a gene pool representing a collection of diverse genotypes, and among which marked genetic discontinuities frequently exist, forms a sort of evolutionary unit. To be sure, each local population is also a sort of evolutionary unit, with local selection pressures interacting with gene sources to produce harmonious local gene pools. However, so long as genetic continuity is maintained within the entire population system, each local population is part of a greater whole. Each local population may evolve through alteration of gene frequencies in its gene pool, but the results of that evolution also affect the gene frequency in the entire species. Furthermore, novel genes or gene combinations arising in a local population are available for incorporation in the gene pools of other populations of the same species under appropriate circumstances. Species with two or more fairly distinctive gene pools are termed *polytypic*, those with only one, *monotypic*. Probably most species are polytypic.

Variation in modal genotypes between local populations is sometimes so great that distant populations are not interfertile and can exchange genes only via an intermediate series of populations, each of which interbreeds with its neighbors. However distinctive the end members of such series may be, they do both draw genes from the same pool, within which gene flow occurs, so that such a population system is best regarded as comprising a single polytypic species.

Two or more populations that share parts of their geographic ranges are said to be *sympatric*. Two or more populations that do *not* share parts of their geographic ranges are termed *allopatric*, and gene flow does not ordinarily occur between them. However, the gene pools of many allopatric populations are quite similar and the populations may be interfertile, and thus be potentially members of the same species. If allopatric populations are incapable of interbreeding because of cytogenetic or other differences, they obviously belong to different species. If they remain separated for long enough, the natural differences between their environments may lead to growing differences between their gene pools (owing to differential selection) and presumably to the eventual establishment of infertility between them. Allopatric populations are therefore under suspicion of having gene pools between which gene flow will never again occur and which henceforward will evolve independently. Whether this will occur cannot usually be predicted for any given allopatric population at any time.

Usually, as a practical matter, an investigator will consider such populations as separate species if they are as distinctive phenotypically as are other closely related populations known to represent distinct species. Conversely, they are lumped into the same species if they are closely similar phenotypically. In either case, events may prove the decision to be in error. Furthermore, many populations differ from each other by intermediate amounts and cannot be separated or combined by the criterion just mentioned. An alternative way of dealing with such perplexing allopatric populations is to assign them to a category different from species, the *semispecies* (Mayr, 1940; Vern Grant, 1963). About all we can do with them is to watch them patiently for thousands or perhaps millions of years. If they maintain the separateness of their gene pools, we may conclude they have been species; if they begin to exchange genes, we may conclude they have only been temporarily isolated local populations of the same species. If one of them should become extinct, we shall never know what it might have become.

Biospecies Are Populations or Population Systems Surrounded by Reproductive Barriers and Are Often but Not Always Morphologically Distinctive

It thus appears that the key event in the formation of a new species is the stoppage of gene flow with other populations. The stoppage may occur between very distinctive populations, which have arisen by selection acting over a mosaic of habitats, or even between essentially identical populations, which have become separated by the rise of an environmental barrier. The degree of difference between populations that become isolated provides a sort of founder effect, and subsequent evolution must act upon organisms with different genetic backgrounds in the separate populations. When separated populations are initially identical, any eventual genetic distinctiveness simply arises after rather than before the separation, owing to the inevitably different environments and histories to which they are subjected.

This is not to imply that original genetic distinctiveness in founder populations is not an important contributor to the evolution of new species. Even if populations are distinctive and are adapted to rather different habitats, they may nevertheless interbreed frequently if interfertile. However, their hybrids will require a hybrid habitat in order to survive, and if none is available, a hybrid population will not develop and gene exchange may not occur. Furthermore, fertility is commonly lower between genetically distinctive populations for a variety of other reasons. In short, barriers to gene flow are naturally more easily erected and maintained between more genetically distinctive populations, and thus the sorts of differential specializations represented by selectively evolved adaptive modes within species can lead to speciation.

In most cases, stoppage of gene flow is partly due to geographic separation. When conspecific populations that are isolated by geographic barriers give rise to new species, it is termed *allopatric speciation* [Fig. 2-4(*a*)]. This is probably most common among populations that live at the outer margins of a species range. Some species, on the other hand, live in environments which are hetero-

BIOSPECIES

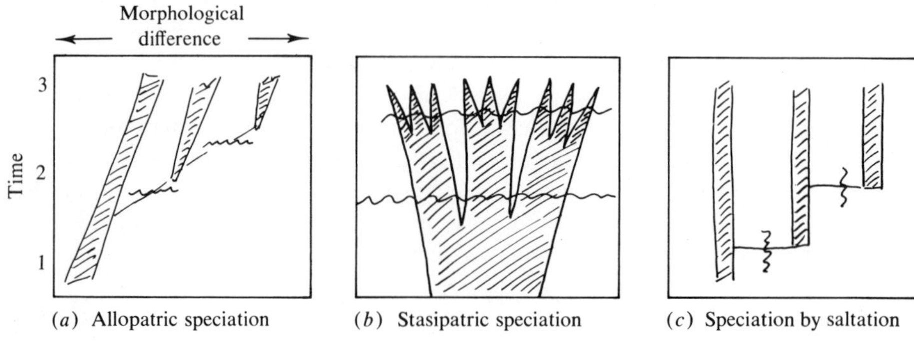

CHRONOSPECIES

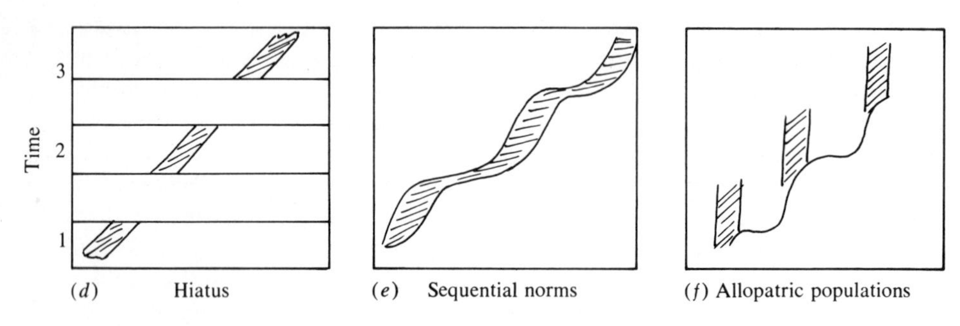

Fig. 2-4 Some modes of formation of biospecies and chronospecies.

geneous on such a scale that local populations are subjected to significantly different habitat conditions even well within the species range. If other conditions are favorable (such as a stable food supply), the populations may each become rather specialized, with hybridization between separate populations becoming a relatively restricted event. Eventually some of these populations may become a species [Fig. 2-4(*b*)]; this is termed *stasipatric speciation* (White, 1968). It is also possible for speciation to occur within a more homogeneous area inhabited by the parental population; this is called *sympatric speciation.* One sympatric mode is the appearance of a chromosomal aberration (such as a multiplication of chromosomes) that renders its bearer infertile with the parent population [Fig. 2-4(*c*)]. This phenomenon is reported to occur among plants, which may fertilize themselves and therefore need only one founding individual, in theory, to develop a species; it is not known among marine invertebrates.

So long as reproductive isolation between populations is not based on cytogenetic infertility, it is possible that events will occur which permit separated populations to rejoin. The phenotypes of two conspecific populations may become quite dissimilar, especially if genetic differences are emphasized by environmentally directed production of extreme phenotypes. Yet even great phenotypic differences may be lost as a new combined gene pool is formed, following the broaching of partial reproductive barriers. On the other hand, phenotypes produced from two rather different gene pools may resemble each other closely. Furthermore, the differences between gene pools that are responsible for infertility may concievably arise either early or late in the differential evolution of two isolated populations. Thus, perfectly distinct species may resemble each other closely. In fact, their resemblance may be so close that specialists may not recognize that more than one species is present. Distinct species with such closely similar phenotypes are termed *sibling species*. A celebrated case concerns certain populations of Australian frogs that do not interbreed but differ so slightly phenotypically that identifications based on morphology are largely uncertain; however, their croaks are quite distinctive and serve to separate them easily. It is quite understandable that croaks could be involved in reproductive isolation because they are mating signals, but this is not very helpful to the paleontologist since croaks do not fossilize at all well. The degree of phenotypic difference between allied populations is obviously not an exact indication of their status as species. Indeed, the genetic differences between sibling species have been investigated in fruit flies (*Drosophila*), and it turns out that they are rather significant, probably amounting to thousands of genes (see Dobzhansky, 1970).

Chronospecies Are Arbitrary Divisions of Evolving Lineages

All species are parts of ancestor-descendant population successions, or lineages, and the definition of species based on contemporaneous, reproductively isolated populations is not applicable to separate, successive populations within their own evolving lineages, no matter how far removed they may be in time.

For example, consider a fairly large population within which gene frequencies are more or less stabilized by selection under a persistent habitat regime. Now suppose that the environment changes in a gradual, even manner in one direction for a very long time. The selective value of certain genes will gradually change, and mutations and gene combinations that had always been deleterious may begin to contribute to fitness. They will replace genes and gene combinations that become less and less fit as the environment shifts. If all the new genes are harmonious with the general genetic background or can be easily coadapted to the gene pool with a minimum of reconstruction, the population may evolve gradually, keeping step with the environmental shift and maintaining its general size and biological position within its environment. If such an environment-species linked or yoked evolutionary pattern were persistent, it would generate a lineage within which change was gradual and regular. Identification of "species" within such a lineage is not really possible, even though morphological change may have been great. These populations represent a sort of cline through time, and are called *chronoclines* [Fig. 2-4(*d*)].

It is difficult to discover actual examples of this sort of even progression of successive populations. Usually the fossil record of closely allied species in time discloses a temporal replacement of one modal type by another, some of the types being closely similar whereas others are rather distinctive. This situation has often been regarded as an artifact of the incompleteness of the fossil record. That is, it has been assumed that if the record were really intact, our collections would show complete intergradations along all lineages, and we would be unable to distinguish successive morphological types. Only the points of branching of lineages would provide "objective" positions at which to draw natural taxonomic boundaries. It is often asserted that the fossil species which we see are members of an intergrading pattern of evolution, and they are either successive forms owing their distinctiveness to a gap or hiatus in the record which conceals the intermediates, or branches from a trunk lineage which had been evolving elsewhere and is not preserved.

Other evolutionary models are possible, however. Even a moderately large population faced with a steady, slow, environmental shift may respond in quite a different manner. For a speculative example, suppose that as the environment shifted away from the modal adaptation of the population, the genes which were then selectively favored did *not* immediately become coadapted with the gene pool; instead, the relatively inferior genes already in the pool were retained for a while because of their superiority in coadaptation. Retention of the original gene pool in a changed environment would probably result in partial loss of fitness of some of the population, with perhaps a concomitant reduction in population size. As the environmental shift continued, the selective value of genes that were adapted to the new conditions would increase to the point at which it would exceed the disadvantages resulting from their more or less disharmonious incorporation into the gene pool. As these genes increased in frequency, the population size might continue to shrink or at least to be held at a relatively low level, since fitness would not be promoted by a somewhat incoherent gene pool. Selection would favor genes and gene combinations possessing activities that were coordinated with the newly frequent genes, and eventually restoring coadaptation. The descendant populations with reconstructed gene pools might increase their sizes and regain a share of environmental resources commensurate with that of the ancestral populations. Continued environmental shifting would lead to other partial breakdowns in coadaptation and reconstruction of the gene pools. The evolutionary pattern would not be that of a chronocline but rather of a time series of larger, slowly evolving populations connected by smaller, rapidly evolving populations [Fig. 2-4(*e*)]. These populations would be united through time as a continuous lineage, but are partially distinctive in that they represent different levels—or, more accurately, different states—of coadaptation.

Still other biological explanations may be found for the discontinuous nature of lineages. Small, episodically isolated populations that develop in habitats on the margins of a species range, and are therefore adapted to an atypical environment, may commonly undergo modifications that make them superior to the modal type of population (Eldredge, 1971). In this event, these new genotypes may replace those of the modal population rather rapidly during an appropriate environmental fluctuation, and the shift in population modes could appear to be instantaneous [Fig. 2-4(*f*)]. As small isolated populations can evolve rapidly, their fossil records would be correspondingly rare. Nevertheless, fossil examples of this mode of population replacement are actually known (Eldredge, 1971).

In any case, any temporal discontinuities that may be generated by biological processes merely provide a natural boundary on which an evolving lineage may, if desired, be separated into nominal "species." They do not actually separate entities that can be considered as equivalent to the reproductively isolated, contemporaneous lineages that we have been discussing (see Beerbower, 1968). In practice, it is possible to adopt some arbitrary morphological difference, as based on experience with valid contemporary species, to indicate a degree of distinctiveness sufficient to assign different species names to different segments of an evolving lineage. The assumption is that, on the average, the populations will have had sufficiently distinctive gene pools that they would have been reproductively isolated, even if contemporary. However, these distinctions are strictly arbitrary, and we should never lose sight of the fact that the existence of separate lineages is one thing and the existence of separate stages in an evolving lineage is quite another. The former include biological species and the latter have been called *chronospecies.*

There Are Multiple Modes of Species Formation

Thus, among organisms that Mendelize, species are thought to originate in the following manner. First, the genetic mechanisms regulate ontogeny in concert with the environment, and are capable of undergoing heritable changes. The species gene pool is continually supplied with new genetic material by mutation, and genetic variety is maintained by a number of genetic and ecologic mechanisms. The gene pool provides the genetic variability, which responds to natural selection chiefly by changes in gene proportions that alter the functional mode of the species populations.

Some species are relatively homogeneous (monotypic) genetically, some others contain genetic gradients (clines), others have polytypic populations containing two or more genotypic and phenotypic modes, and still others contain numbers of separate populations that are genetically distinctive. If gene flow between populations is eliminated, these isolated populations are thereafter on independent evolutionary pathways. Reproductive isolation probably occurs most commonly between populations under certain particular conditions. For example, populations that live on the margins of a species geographic range are more likely to become isolated than those near the center. Populations that live in subhabitats and have distinctive genotypic modes are more likely to become isolated than those that are nearly homogeneous, genetically and ecologically, with their neighboring populations.

At any rate, once isolation is established, evolutionary divergence will often lead to the establishment of cytogenetic infertility between the isolates and other populations from the same lineage. The isolates are then *biospecies*. In a few cases, infertility occurs in a single step, as when chromosomal aberrations arise and reproductive isolation is established from the start. Chronospecies must be formed by the division of continuous lineages into arbitrary segments. The boundary may be chosen at an artificial horizon that merely represents an hiatus in the record of the lineage, or at a horizon where some natural, evolutionary event has changed the character of the lineage. Some of the processes of biospecies formation, and some of the events that lead to the recognition of chronospecies, are depicted in Fig. 2-4.

Genetic Systems Vary According to the Complexity and Strategy of Adaptation

Heretofore, our discussion of genetic mechanisms in evolution has been based on a kind of diploid life cycle that is evidently nearly universal among living larger invertebrates and vertebrates and is rather common among many plants. This is the genetic system of paired chromosomes that undergo frequent crossing over at meiosis and segregate to produce recombinants. However, many organisms possess other sorts of life cycles and other genetic systems, which confer upon them more or less different adaptive advantages and evolutionary potentials.

The different genetic systems were presumably evolved in response to different requirements of inheritance and reproduction that are associated with different *adaptive strategies*. Although the term "adaptive strategy" sounds as if we were endowing populations with the ability to plan ahead, we use it merely to denote the results of natural selection for adaptation to an environmental pattern, such as a pattern of habitat dispersion or of temporal fluctuations in food or other factors (see Levins, 1968). In the same way, we may speak of a *genetic strategy*, which merely implies that a certain genetic system and structure has evolved as appropriate to the mode of life and environmental regime of the lineage.

The most basic division of organisms with cells (which excludes the viruses, which have no cells and have RNA-based genetic systems) is into those without well-defined nuclei, and therefore without chromosomes, called *prokaryotic* organisms (bacteria and blue-green algae), and those with nuclei and chromosomes, the *eukaryotic* organisms (the remaining plants and all animals). The prokaryotes obviously lack meiosis, but do exchange genetic material between individuals, although in a less organized manner than the eukaryotes.

Among the eukaryotes there is a variety of chromosomal cycles, nine of which are illustrated in Fig. 2-5. Type II is the diploid cycle that has been assumed in previous discussions; the haploid stage is represented only by gametes and forms a small although important portion of the life cycle. At the other end of the spectrum of chromosomal cycles is that represented by type VI. This is an entirely asexual haploid cycle wherein reproduction is accomplished by fission, and recombination is unknown. For organisms with such a cycle, all progeny have genotypes identical with the parent except when mutation occurs. Mutations introduce the only changes in gene frequencies that are possible in these lineages. Populations of such organisms are certainly not Mendelian.

Consider a haploid asexual population that develops from a founding individual, assuming that no viable mutations occur; all the F_1 progeny will possess identical genotypes, all the F_2 progeny will have the same genotypes also, and so on, until within successful "families" great alliances of genetically identical individuals will be found. Such lineages are called *clones*. Even should some mutants arise, *their* descendants will be genetically identical, and two or more mutant strains will then be present within the family, each traceable to a mutant founder (or possibly to more than one founder if the same mutation arises more than once). Any selective pressure that favors one genotype over another will then, in fact, favor one whole clone (or group of genetically identical clones) over another. Evolution proceeds by a selection of viable and fit clones, but genetic variability is minimized. Because each clone is reproductively isolated, the criterion of fertility breaks down when one attempts to define species among these organisms. Each clone cannot be considered a separate species, for this would be rather like considering each genotype to be a species among sexual organisms. A practical solution is simply to group clones by physiological, morphological, or biochemical similarities. Obviously such groups, made up of a number of lineages (clones) that are entirely independent as far as their evolutionary futures are concerned, are not the same things as sexually reproducing species.

Various combinations of sexual and asexual phases in populations are found in nature. Chromosomal cycles of types III, IV, and V, Fig. 2-5, involve alternation of (1) a haploid generation produced by asexual fission and meiotic divisions of diploid individuals with (2) a diploid generation produced by sexual fusion of haploid gametes that arise by mitotic divisions of haploid individuals. For some groups of organisms the haploid generation gives rise to the characteristic individuals (cycle IV, Fig. 2-5), and for others the diploid generation is dominant (cycle V, Fig. 2-5). In many foraminifera and algae, the haploid and

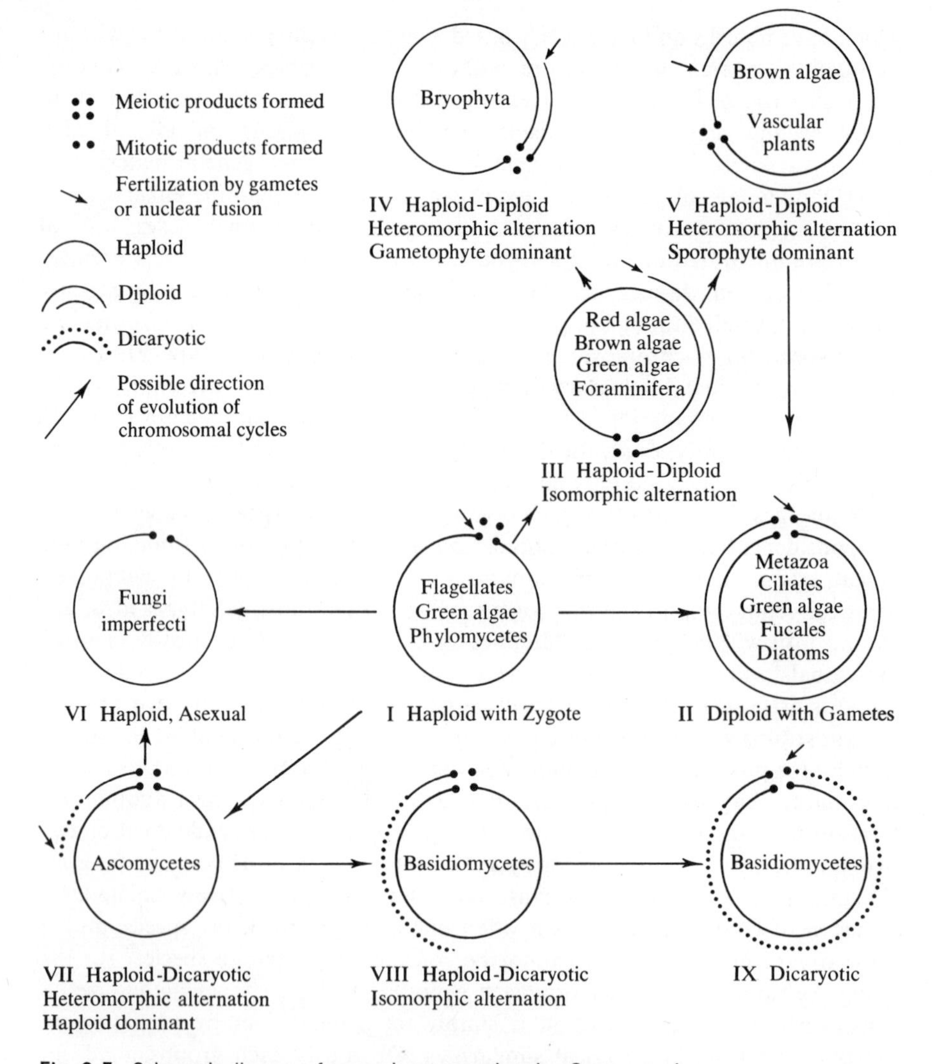

Fig. 2-5 Schematic diagram of some chromosomal cycles. Some organisms can prolong certain phases of complex cycles. For example, some foraminifera may continue indefinitely to reproduce asexually in the haploid phase of what appears on the chart as an isomorphic alternation of haploid-diploid phases, whereas the Basidiomycetes with "isomorphic alternation" of haploid-dicaryotic phases may vegetate in either phase indefinitely. (Chiefly after Stebbins, 1960, copyright © 1960 by the University of Chicago; with data from Raper, 1954.)

diploid generations are nearly equal. Some foraminifera have rather complicated cycles (see the examples in Loeblich and Tappan, 1964).

In addition to variations in haploidy and diploidy, diversity in chromosomal cycle is attained in fungi by the development of a genetic system wherein two nuclei (of opposite mating types) are present in each cell. Such a condition is termed *dicaryotic*. There is interaction among alleles, just as there is in diploid organisms, even though the chromosomes are in separate nuclei. Even during the "haploid" phase of the chromosomal cycle of dicaryotic fungi (Fig. 2-5, cycles VII, VIII, and IX), when cells may be multinucleate but all nuclei are of the same mating type, genetic differences among nuclei may be expressed in the phenotype (Raper, 1954).

Another chromosomal condition that is especially common among higher plants is *polyploidy*, the possession of a chromosomal complement larger than some ancestral diploid number. This can arise through suppression of meiosis during gametogenesis to give rise to diploid gametes, which then form tetraploid ($4N$) zygotes. It can also arise through suppression of cell division after mitotic chromosomal doubling, so that such cells contain four sets of chromosomes ($4N$). This is followed by the formation of gametes through meiotic division of the $4N$ cells into $2N$ gametes, which may then unite to form $4N$ zygotes. Polyploids also exist as triploids ($3N$), hexaploids ($6N$), octoploids ($8N$), and so on. Odd-numbered chromosome sets (such as triploids) usually arise through crossing of a diploid (with haploid gametes) with a polyploid (for example, a tetraploid, with diploid gametes, thus producing a triploid zygote). Many other methods of chromosome duplication and reduction have been described (see Stebbins, 1950).

It is evident that between organisms with fully Mendelian populations and those with non-Mendelian asexual populations there lie a number of somewhat intermediate population types. Take, for example, a population of a foraminifer that has alternating sexual and asexual generations. It would in general consist of a mixture of haploid and diploid individuals, so that *two* reproductive associations may be present, one of which reproduces sexually and one asexually. In the simplest case, one of these associations is responsible for segregation and the other for recombination, and the entire chromosomal cycle must occur to approximate Mendelian reproduction.

An additional complication in genetic systems arises when self-fertilization occurs. Self-fertilization is found in representatives of all the chromosomal cycles illustrated in Fig. 2-5 (excluding, of course, the totally asexual cycle, VI). If a form is entirely self-fertilizing, then clonelike families develop that consist of a founder and its offspring. In nature, occasional cross fertilization seems to occur in most or perhaps all groups that are essentially self-fertilizing (Stebbins, 1950). Nevertheless, heredity within these populations is far from the Mendelian ideal but may be called quasi-Mendelian.

For quasi-Mendelian populations, the gene pool concept may be retained as a basis of definition of species, with slight modification for self-fertilizing species. We may define species among predominately self-fertilizing forms as all populations between which gene flow is possible when cross fertilization does occur. All the individuals who draw their gametes from these populations are conspecific.

From this brief review, which is by no means comprehensive, we see that a fairly broad variety of genetic systems is known. What are their various purposes, and how and why have they evolved? These problems have been investigated by a number of workers; the present account is taken chiefly from works by Darlington (1958) and Stebbins (1960), which include good bibliographies.

In short, it seems likely that there are two end-points to the range of strategies of adaptation that have been developed in genetic systems. At one end-point, exemplified by haploid asexual organisms, a premium is placed upon immediate, short-term adaptation of large numbers of individuals to the ambient environment. Whenever a genotype is especially favored, it will be duplicated in a whole population—clonally—instead of being broken up by recombination. When conditions change, however, the entire clone can become poorly adapted and face extinction. At the other extreme, a premium is placed upon genetic flexibility and upon the ability to maintain adaptation over the long term. This strategy is exemplified by diploid genotypes. There is a range of fitness present in any population, so that less than optimally fit individuals are common. But when conditions change, genotypes are likely to be available that are well adapted to the new conditions, and by reaching new genic equilibria and perhaps by achieving new coadaptation within the gene pool the population may adapt. The maintenance of a reservoir of genetic variability in recessive alleles and by other means is obviously a great advantage to a population following this strategy. Inasmuch as heterozygosity frequently results in superior fitness, there is an added advantage in the diploid (or the dicaryotic) chromosome system. Probably one of the advantages of polyploidy is that recessive alleles will rarely be homozygous yet can be retained easily in the gene pool for their advantage in the heterozygous state. For example, in tetraploids, four recessives would have to be present in a genotype for homozygosity to occur, and the chances of this are only one-quarter as great as in diploids, other things being equal.

The organisms possessing genetic systems that follow either of two extreme strategies have general biological attributes which seem quite appropriate to their genetic strategies and surely have evolved as coadaptations. Haploid asexual forms are small and tend to be simple morphologically and physiologically and to reproduce rapidly. Diploid sexual forms are generally large, morphologically and/or physiologically complex, and commonly reproduce more slowly. Organisms that have chromosomal cycles which are mixtures of haploid and diploid states follow an intermediate genetic strategy. This strategy is part of their more or less harmonious coadaptation of intermediate biological attributes and adaptive requirements.

Since many diploid, sexually reproducing populations that live in widely different environments clearly have different adaptive strategies, it has been suggested that they may have different genetic strategies as well, as one component of their adaptive difference. It seems possible that under some conditions diploid genetic systems might evolve mechanisms that optimize fitness, whereas under other conditions, mechanisms would be favored that optimize flexibility (see the theoretical explorations of Carson, 1960, and Levins, 1968). For example, favorable gene combinations might be held together to favor fitness by suppression of crossing over, or by a high degree of homozygosity. However, there is as yet only a little evidence that these features occur in a manner which is consistently associated with environmental patterns and which could be considered to be a strategy. Some types of chromosomal inversions suppress crossing over and thus may lock up favorable gene combinations in some populations (Dobzhansky and Epling, 1944), but these inversions are not known to occur more frequently in any given type of environment. There is some slight evidence that heterozygosity may increase with environmental instability within some species (for example, Staiger, 1956; Lewontin, 1958), and homozygosity has been shown to be higher in experimental populations maintained in more constant environments (Powell, 1971).

However, heterozygosity appears to be rather high in most of the populations tested. These include man and several species of fruit flies (*Drosophila*) among terrestrial organisms (see O'Brien and MacIntyre, 1969) and the horseshoe crab (Selander *et al.*, 1970) and two species of ectoprocts (Gooch and Schopf, 1970) among marine organisms. These data show that the genetic structure and variability of Mendelian populations with widely differing modes of life—as different as motile, noncolonial terrestrial organisms with separate sexes, and sessile, colonial marine organisms that are mostly hermaphroditic—are much alike (Gooch and Schopf, 1970). The processes of evolution in marine organisms are probably similar to those of the terrestrial organisms such as *Drosophila*, which have provided the bulk of the evidence on genetics in animal evolution. We cannot yet reach any definite conclusions on whether different diploid genetic strategies occur and whether they are important in evolution. This question is of particular interest to paleoecologists because of the possibility that species may become inadaptive when their environmental patterns change, not only because the phenotypic components of their adaptive strategies become less fit, but because their genetic systems are poorly structured to the new environmental pattern and thus their genetic strategies fail. Bretsky and Lorenz (1969, 1970) have suggested that such failure may underlie some of the mass extinctions in geological history.

Ideally, Taxa Are Monophyletic Branches of the Phylogenetic Tree

Organisms are classified in a hierarchical system. Each level is called a *taxonomic category* (Table 2-1); a group of organisms classified as a unit within a category forms a *taxon* (plural, *taxa*). For example, "species" is a category, but any particular species is a taxon; "phylum" is a category, whereas "Arthropoda" is a taxon.

Ideally, the members of a taxon have all descended from some common

Table 2-1 The Taxonomic Hierarchy

Category	*Example of taxon*	*Rank*
Phylum	Arthropoda	High
Class	Trilobita	↑
Order	Redlichiida	│
Family	Olenellidae	│
Genus	*Olenellus*	│
Species	*Olenellus thompsoni*	Low

ancestor that is the earliest member of the taxon. Thus, a genus is a group of species that are all descended from an ancestral species which is the first member of the genus. Similarly, a family is composed of genera which have all descended from a common ancestral genus, and because that genus includes a common ancestral species, that species has, in fact, given rise to the entire family. When on this basis families are grouped into orders, orders into classes, and classes into phyla, and so forth, a classification is created which is phylogenetic, that is, which mirrors the actual evolutionary relations of the lineages, and the phylogenetic tree becomes a sort of family tree of life.

Unfortunately, the relationships between many taxa cannot be determined well enough to permit us to draw perfectly accurate phylogenetic trees. The trees that we do draw represent hypotheses as to what the true phylogenetic relations may be. Commonly, more than one phylogenetic hypothesis is tenable from the evidence at hand, so that for some taxa several different trees, each supported by the opinions of a different investigator, may exist at the same time. As new data come to light, the trees must be revised; sometimes some of them are eliminated as possibilities.

It is not uncommon that a recognized taxon turns out to include organisms which did not descend from a common ancestor within the taxon. Frequently, investigators have been misled because lineages derived from very different ancestors come to resemble each other owing to their evolutionary convergence toward similar adaptations. If the intermediate members of these lineages are not discovered, it may be mistakenly assumed that they are closely allied, since they closely resemble each other. Difficulties of this sort have so discouraged some workers that they prefer not to attempt a classification based upon phylogenetic relationships. Instead, they prefer to classify simply on the basis of morphological (or *phenetic*) resemblance. Thus, phenetics comes to replace genetics as the criterion of relationship. However, for our purposes, which involve the interpretation of pathways of evolution, the phylogenetic schemes are clearly the best, but we must not be deluded into believing that the current phylogenetic hypotheses are necessarily correct. A taxon that has descended from a common ancestral species within the taxon is called *monophyletic*, whereas a taxon that includes lineages with no common ancestors within the taxon is called *polyphyletic*.

Although taxa are real and can be arranged into a hierarchy, their limits are arbitrary except, theoretically, for biospecies. How many species to include within a genus, for example, is somewhat a matter of tradition, experience, and taste. If investigators working with one phylum tend to split their taxa into small groups, whereas investigators working with another phylum employ large inclusive groups, then taxa in identical categories that belong to these different phyla do not have the same significance. Much splitting and lumping of this sort exists, even between the work of different specialists within a single group. For that matter, any concept of the natural equivalence of any two taxa is ambiguous, since taxa all differ qualitatively.

Higher Taxa Arise as Adaptations in Response to Novel Environmental Opportunities

Individuals belonging to different conspecific populations usually resemble each other closely, and individuals belonging to different species of the same genus also resemble each other closely, but usually differ consistently in some morphological details. Similarly, individuals belonging to the same family but to different genera normally have a general resemblance owing to their common inheritance, but may differ in many characters that have arisen since their lineages diverged. On the average, the morphological gaps between genera are larger than those between species. The same generality can be applied at higher levels of the taxonomic hierarchy, and when we reach the level of the phylum, morphological differences are associated with basic architectural characteristics. The distinctive assemblage of structural characters that underlies each phylum is said to be its *ground plan*, and sets its members apart from representatives of other phyla. Typical ground-plan characters include type of symmetry, presence and number of tissue layers, presence of a body cavity, and presence and character of segmentation. Classes and sometimes orders also differ in rather fundamental structural ways. To account for the origin of taxa in the higher categories, then, one must account for the origin and association of the characters that comprise their ground plans.

There is no reason to believe that any unusual evolutionary mechanisms are involved in the origin of higher taxa. Some workers have suggested that special mutations or unusual processes are required to account for the appearance of new types of organisms. The existence of these special processes has not been demonstrated, however, and has come to seem less likely as understanding of the molecular basis of heredity has grown. Considering the efficacy of well-known evolutionary mechanisms such as operate in speciation, the amount of geological time that is available, and the sequence of events known to have occurred in organic evolution, it appears that the well-known mechanisms are sufficient to account for the evolution of life as we know it. Of course, we do not yet understand all the processes associated with evolution.

The evolution of wholly new ground plans appears to occur in response to the invasion of a new environmental realm. This usually comes about when an important environmental opportunity becomes available to organisms that have what is at the time a rather unusual way of life which just happens to preadapt them for life in the new environment. Simpson (1944, 1953) developed a creative approach to this problem. He defined the pathways along which taxa evolve as *adaptive zones*, regions of the environment that are actually or potentially inhabited by organisms representing an *adaptive type*. Adaptive types have modal morphologies and functions that together form a distinctive way of life. Cats, dogs, and bats each are distinctive adaptive types, and each occupies a distinctive adaptive zone. Occasionally relatively unrelated forms, such as ichthyosaurs and dolphins, marsupial "wolves" (thylacines) and placental wolves, may belong to similar adaptive types; in these cases, both morphological and functional convergence has occurred. Usually, however, a similarity of adaptive types indicates a close relationship.

Adaptive zones may be broad or narrow and may include high taxonomic categories or individual species lineages, or they may even be unoccupied by organisms, being represented by only the environment, which provides potentialities for adaptation. The adaptive zone of a taxon includes not only all the realized adaptations of the taxon, and all the prospective adaptations, but also all the possible future adaptations that could be evolved—the "phylogenetically prospective" adaptations. Discontinuities between zones correspond to discontinuities in nature, as between dogs and cats. Most adaptive zones are visualized as being subdivided into generally discrete bands occupied by adaptive subtypes. If a zone represents an order, each family may occupy a distinctive subzone, and genera or other lower taxa occupy distinctive bands within the subzones.

Figure 2-6 depicts some of the adaptive zones involved in penguin evolution (Simpson, 1953). A broad adaptive zone of aerial flight is subdivided into aquatic and terrestrial parts. Another adaptive zone represents a way of life involving both aerial flight and "submarine flight"—swimming underwater by a sort of flying motion with wings. This zone is now occupied by pelicans and some other diving birds. A third adaptive zone involves only submarine flight, and is now inhabited by penguins. Presumably the penguins, descended from birds with aerial flight, first occupied the zone of both aerial and submarine flight and finally shifted into their present zone from there. Once the protopenguin lineage had reached the point where selection for adaptations to exclusively submarine flight was higher than selection for adaptations to aerial flight, it had crossed a threshold, bridging the discontinuity between these two adaptive zones. Numerous changes in adaptation that coordinated form and function to improve efficiency in the new zone would "postadapt" the lineage to its new zone and

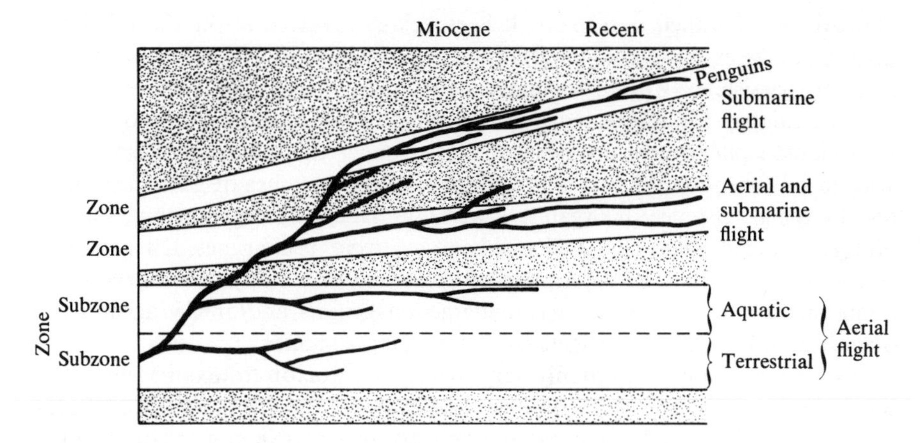

Fig. 2-6 Grid diagram of adaptive zones associated with penguin evolution. (After Simpson, 1953.)

increase its adaptive distinctness from the ancestral condition, re-establishing the interzonal threshold. Another example of the invasion of a new adaptive zone involves the rise of the tetrapods, which evolved as amphibians from early lobe-finned lungfish that lived in lakes and ponds. These fish could breathe atmospheric oxygen and had some terrestrial locomotory ability. They were bizarre and highly specialized as fish, perhaps, but nicely preadapted for invasion of the terrestrial adaptive zone.

In both of these examples, the invasion of the new adaptive zone has involved a change in locomotion. This is very commonly the case with the appearance of new higher taxa, for the movement of an organism presents a major problem, the evolutionary solution of which tends to involve the body architecture. The modifications required to achieve an efficient locomotory mode in a novel environment have often given rise to new body plans. Of course, many other functional changes eventually become associated with these major evolutionary events. We shall return to these points again in Chapters 3 and 10.

Lineages May Split, Improve, or Merely Persist

Huxley (1957) has recognized three distinctive evolutionary "processes." One is diversification resulting from the splitting of lineages into separate populations, which may eventually become species and in some cases may represent the foundations of separate higher taxa, up to and including phyla or kingdoms. The splitting is called *cladogenesis;* its effects can be seen in the splitting of populations, and in the radiation of the major branches of the phylogenetic tree. A second process is improvement, which is the tendency for a lineage to achieve greater functional efficiency owing to natural selection. Evolutionary improvement is called *anagenesis*. It is especially involved in the rise of higher taxa and in the evolution of new grades of organization. Finally, the third process is simply persistence with little change, called *stasigenesis*. This is the expression of the tendency for integrated and harmoniously coadapted gene pools with well-adapted phenotypic products to persist under the influence of stabilizing selection. It is shown by taxa that endure for long stretches of geological time, and by the persistence of organisms with low grades of organization, such as blue-green algae, that originated well back in Precambrian times. Each of these three "processes" can be descriptive of evolution within any level of the taxonomic hierarchy. A phylum may undergo anagenesis and/or cladogenesis or may display a period of stasigenesis, and so may a species.

Evolutionary rates commonly vary greatly from taxon to taxon even within the same category, and from time to time even within the same taxon. Rates of evolution are judged in more than one way, and involve more than one phenomenon. A lineage may be said to evolve more rapidly when it displays a higher rate of morphological change, implying extensive changes within the gene pools involved. This sort of morphological rate of evolution is an anagenetic process and is reflected taxonomically, for the changing lineages are regarded as sequences of ancestral and descendant taxa, and high morphological rates of change will lead to high taxonomic rates of evolution—that is, the rates will be high when judged by the rate of turnover in nominal taxa. These are really "chronotaxa," arbitrary segments of branches of the tree of life.

Rates of evolution that are judged on the basis of the rate of appearance of lower taxa within a lineage—for example, the rate at which new species appear within an order—may be due to processes other than taxonomic turnover. The appearance of new species may result not only from the appearance of sequential chronospecies within lineages but often from diversification, the multiplication of the numbers of biospecies present at any time, a cladogenetic process. A burst of diversification on the species level may be owing to the appearance of numerous opportunities for the isolation of populations, as might occur when an extensive seaway breaks up into segments after the rise of land barriers, creating numerous allopatric populations which eventually become separate species. In this case, the modal morphology of the order need not have been significantly altered, nor the general ecology of the order necessarily changed in any significant way. Nevertheless, evolution could be said to be "rapid" because of the appearance of numerous species. If in our example the land barriers are repeatedly overlapped by transgressions or are in tectonic flux, species could appear and disappear at a considerable rate as opportunities for increased diversity were alternately created and destroyed. Barriers that give rise to this effect are of numerous kinds, and will be discussed in later chapters. The principle is simply that when the number of opportunities for genetic isolation fluctuates over appropriate lengths of times, the number of species fluctuates.

Some lineages appear to have slow evolutionary rates at all their taxonomic levels for long periods of time; they exhibit stasigenesis. The inarticulate brachiopods are the classic invertebrate example. After a relatively brief period of diversification in the Cambrian, which may represent a period of rapid evolution or may be an artifact of the record, the inarticulate lineages have changed little either in morphology or in diversity, and therefore display a slow rate of taxonomic evolution. We might say that their chief evolutionary characteristic is simply persistence.

The relative evolutionary pathways of different taxa are commonly described in terms of increases or decreases in their similarity of adaptation. By the word "adaptation" is meant the acquisition of heritable characteristics that are advantageous to an individual and a population (see Simpson, 1953, p. 160). A given adaptation, then, is one such characteristic. Usually the direction of evolution of adaptations is judged by changes in morphology. Convergent evolution or *convergent adaptation* is the evolution of two or more lineages toward similar adaptations, usually measured by a morphological convergence. A classic example of convergence is the similarity in form of such phylogenetically diverse large swimming animals as sharks (fish), ichthyosaurs (reptiles), and dolphins (mammals). Numerous cases of this sort occur among invertebrates; the "limpet" form in mollusks has been evolved independently in three orders of the class Gastropoda and in the class Monoplacophora. Within the Gastropoda, low trochoid shells that resemble each other have been evolved separately by the archaeogastropod trochids such as *Tegula* and the mesogastropod littorinids such as *Bembicium*, both of which occupy wave-swept rocky shores. Convergence may affect only one or two characters or it may result in phylogenetically distinct skeletons that are extremely similar morphologically; such skeletons are called *homeomorphs*. Convergent adaptation is a widespread and common pattern of evolution.

Iterative adaptation is the repetitive production of an adaptive type; it is a repetitive convergence in time. It would be represented in the fossil record by a skeletal form that is generated again and again from the same evolving stock, presumably in response to fluctuating, repetitive, adaptive opportunities. Clear examples of this pattern are rare. Among planktonic foraminifera, iteration of trochoid planktonic forms with globose chambers seems to have occurred in the Late Cretaceous and Cenozoic.

Parallel adaptation is a similar change in separate lineages, presumably owing to similar adaptive responses to some environmental change or opportunity. An example would be the evolution of larger body size in different lineages. A classic example of parallelism is found in graptolites, which display an independent reduction of thecal rows from two to one per stipe in different evolving lineages (George, 1948).

Divergent adaptation is the pattern produced every time a lineage undergoes cladogenesis. When a number of stocks diverge from some ancestral condition to give rise to a pattern of lineages radiating from a common origin, the pattern is termed *adaptive radiation*. This is sometimes said to be the most basic of all patterns of evolution. It evidently occurs in all major taxa, for whenever a major taxon is subdivided into a number of lesser but still important taxa, it follows that their patterns are of diverging branches from ancestral stems, which is the pattern of adaptive radiation. Similarly, the phylogenetic trace of a cluster of contemporaneous species such as are classed into a genus is usually that of a number of divergent forms and can be called an adaptive radiation.

CHAPTER THREE

THE ECOLOGICAL ARCHITECTURE OF LIFE

Some of the great controversies about evolution have had little more basis than this: that different theories were based on one factor or another when the essential point was really the relationship between the two.

—From George Gaylord Simpson, *The Major Features of Evolution,* Columbia University Press, New York, 1953.

Evolution Has Interacting Biological and Environmental Components and Is Therefore an Ecological Process

Evolution can be viewed as an ecological process. It involves on the one hand the genotype and all the array of phenotypes that a given genotype has the potential to produce, and on the other hand the environment, which through its contribution of materials, mediation of physiological rates, and provision of developmental cues, leads the genotypic potential to a particular phenotypic expression throughout a life cycle, which usually determines reproductive success. The contribution of a genotype to succeeding generations is controlled chiefly through the relationship of the organism to its environment, which is by definition an ecological relationship.

Since the changes that constitute genic evolution affect gene frequencies within populations, population-environment relations are exceedingly important. Within a population, individuals are often in competition with one another for food, shelter, and other necessities of existence, and successful competitors will tend to perpetuate their genes. Populations may also compete with other populations for a share of the food, living space, and other environmental resources. Thus, among the many selective pressures that operate on gene pools are those arising from the necessity to live among populations of other species. Gene pools of numbers of species will tend to become coadapted to each other's presence, in the sense that they will be adapted to compete successfully with or prey upon one another. Another way of becoming coadapted is for a species to evolve so that interactions are minimized or eliminated; that is, species may evolve special habits so that they compete or otherwise interact with only a very few other species. Competition and other interspecific interactions are thus reduced to a few, for which a species may become highly specialized. Such a specialized species may then live in perfect harmony with great numbers of other species, with which it has little or nothing to do.

In Chapter 2, we were concerned with basic factors of evolution, especially the genetic mechanisms and genetic compositions of individuals and populations. In this chapter, we shall examine the ecological organization of life. It is the intertwined and interacting mechanisms of evolution and ecology, each of which is at the same time a product and a process, that are responsible for life as we see it, and as it has been.

Form and Function Provide Homeostasis

All organisms must maintain a myriad of coordinated physiological processes throughout their lifetimes. The reactions associated with these processes can usually occur only under relatively narrow ranges of conditions. For example, temperature is an important physical parameter that has much effect upon metabolism. In fact, many organic compounds are stable only under relatively narrow temperature ranges. Furthermore, the rates at which biochemical reactions occur are usually temperature dependent. These reaction rates must be closely correlated over numerous biochemical pathways to achieve a balanced metabolism. For most organisms there is probably only a relatively narrow range of temperature at which all the reaction rates are harmoniously correlated.

Chemical conditions are of utmost importance in physiological processes, most of which are, after all, biochemical, and nearly all chemical factors of the internal environment (and for marine organisms of the external environment as well) affect the life processes in significant ways. Such factors as pH, eH, specific ion concentrations, and the concentrations of all the substances directly involved in any process, must be within certain critical limits if a process is to occur and to have the required effects (Chapter 4). Yet environments, internal as well as external, tend to fluctuate. Each organism, then, is faced with the problem of regulating its internal conditions so as to provide a stable physical-chemical environment suitable for proper, coordinated functioning of its physiological processes; some of the processes themselves must be concerned with this regulation. Organisms are self- or auto-regulating. This balancing of conditions by the coordination of complex physiological processes in order to approximate steady-state conditions in the face of fluctuating environments is called homeostasis.

Another aspect of the organism is its physical structure, developed through and coordinated with physiology so that form and function are indivisible biological entities. There is certainly a harmony among structures as there is among physiological processes. It was partly to this sort of structural harmony that the term *coadaptation* was first applied by Charles Darwin (1859). One of Darwin's examples is the woodpecker, which has its feet, tail, beak, and tongue all harmoniously adapted to catch insects under the bark of trees. Obviously the morphology of the woodpecker, as of other organisms, is in general adaptive to a particular way of life, and provides the means of achieving a regimen of behavior within its regime of habitat.

The study of the adaptive utility of the structures of organisms is called functional morphology. It is directed more toward understanding the relations between an organism and its external rather than its internal environment. For obvious reasons, the physiological study of fossils is largely precluded, but the functional morphology of preserved hard parts can sometimes provide suggestions as to former physiological processes in organisms whose remains are fossilized.

Each Unique Phenotype Has a Unique Functional Range

Well over 100 years ago, J. Leibig began to formulate the relations between the functions of an organism and its viable functional range in a given environment. "Leibig's law of the minimum," proposed by others but based on Leibig's work, states that *those environmental factors which are in minimum supply are limiting*. From this simple but basic beginning, modern insight into the complexities of organism-environment interactions has developed. An expansion of Leibig's law was proposed by Shelford (1913); he included consideration of the entire range of factors that are limiting at upper as well as at lower levels, and that furthermore may interact so as to produce limits different from when the factors operate separately. Thus, in Shelford's approach, limiting factors are expanded to levels of tolerance within a particular environment. Odum and Odum (1959) have expanded these ideas still further to form a combined concept of limiting factors, which considers both the quantity and variability of environmental factors and the tolerance limits of the various organisms.

A useful way of envisioning these relations is suggested by the work of Hutchinson (1957a, 1967). He has described a conceptual model in which the environment is imagined to be a multidimensional region, with each dimension representing a single environmental variable. For example, one dimension may represent some environmental aspect of temperature. This dimension can be drawn as a line [Fig. 3-1(*a*)], with absolute zero at one end and some very high

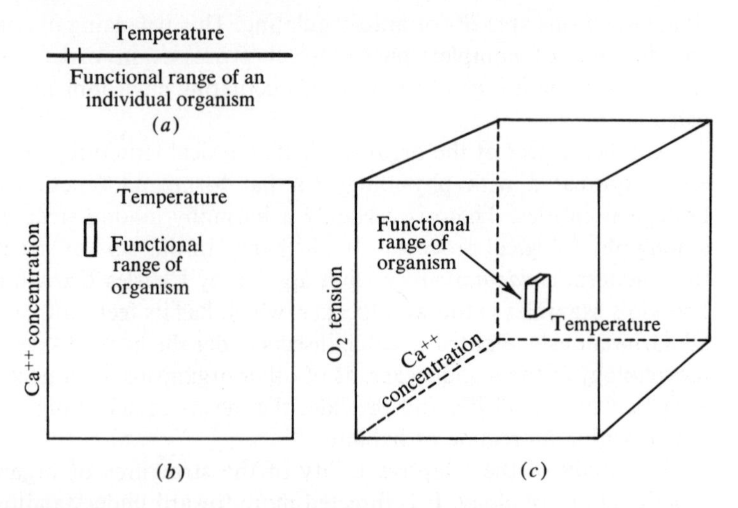

Fig. 3-1 A representation of biospace. (*a*) The range of tolerance of a marine organism for temperature, plotted on a temperature scale. (*b*) The range of tolerance of the organism for temperature and Ca^{++} concentration, plotted on a two-dimensional diagram. (*c*) The range of tolerance of the organism for temperature, Ca^{++} concentration, and O_2 tension, represented as in three-dimensional space. Further dimensions cannot be diagrammed, but may be easily conceptualized and are amenable to mathematical representation.

temperature at the other. Any point along the line represents some environmental temperature, and every segment of the line represents some environmental temperature range.

Another dimension of the model may represent, say, calcium concentration [Fig. 3-1(*b*)]. This dimension is plotted at right angles to the temperature dimension and ranges from zero to the highest possible calcium concentration. We now have an area representing two environmental parameters, and any point on the area is some unique combination of these two. A third parameter, perhaps oxygen tension, may now be plotted at right angles to the previous dimensions, with an appropriate range [Fig. 3-1(*c*)]. The next parameter takes us into a fourth dimension that cannot be represented geometrically, but can be imagined; as all the environmental parameters are eventually added, the model becomes a hypervolume with as many dimensions as there are parameters. The parameters include all the biological factors in the environment, such as the presence of various food resources, predators, competitors, and the like, as well as the physical factors. Any point within the final hypervolume represents a unique combination of environmental parameters, and all possible combinations of parameters are represented. It is assumed throughout this book that the multidimensional environmental model is standardized by having each parameter allotted an arbitrary but permanent dimension. Thus, the model may be regarded as a fixed reference lattice.

We have said that each individual has a unique range of response to environmental parameters. To describe this range in terms of the concept of environmental hyperspace, we can say that there is a region within that hyperspace which represents the functional range of any given individual. This region is composed of the collection of all points representing all combinations of parameters that form all the possible environments wherein the individual can live. This collection of points will form a hypervolume. Actually, this hypervolume changes throughout the life of the individual, and in some cases the change is great, for quite different environments are often required by young than by adults; for example, oyster larvae are pelagic, whereas adults are sessile benthic forms.

From the discussion of development in Chapter 2, it is evident that any given genotype bears the potential for the construction of quite a number of different phenotypes; the precise ontogenetic history of any individual depends partly upon the environment and will usually be different for different individuals. The phenotypes resulting from alternate histories would each have slightly different environmental responses. Thus, the total functional range that can be controlled by a genotype, which we shall call the *prospective* functional range of the genotype, is greater than the prospective functional range of the phenotype that actually develops. If the prospective functional range of the genotype is visualized as a hypervolume within the environmental hyperspace, then the realized functional range of the phenotype is only a part of that hypervolume. Figure 3-2 is an attempt to depict some of these relations in two dimensions.

Of course, the environment that is actually realized at any time does not include all the possible combinations of environmental factors. The total range of possible combinations we shall call the *prospective environment*, and the combination that actually exists at a given time or place is the *realized environment*. The environmental hyperspace representing the realized environment is only a small part of the prospective environmental hyperspace. A special term for "realized environmental hyperspace" is useful in order to avoid a constant repetition of this clumsy phrase. Accordingly, it will be referred to hereafter as *biospace* (employed in a more limited sense by Doty, 1957). Biospace can be conceived as a multidimensional model of the environmental parameters that actually exist. Some of the dimensions of biospace have special properties, especially: (1) dimensions of *real* space, along which the shape of the biospace varies and where ecological discontinuities (such as barriers) occur; and (2) time, in which the changing shapes and sizes of biospace are perceived.

It usually happens that many of the possible environmental interactions of an individual do not actually occur, simply because the environment does not evoke them. For example, if a coral has a temperature tolerance that ranges from 20 to 30°C, and the environmental temperature during its lifetime ranges only from 20 to 25°C, some of the possible environmental responses of that coral never occur. In other words, the realized environment overlaps with only a part of the prospective functional range of the phenotype. We shall call this

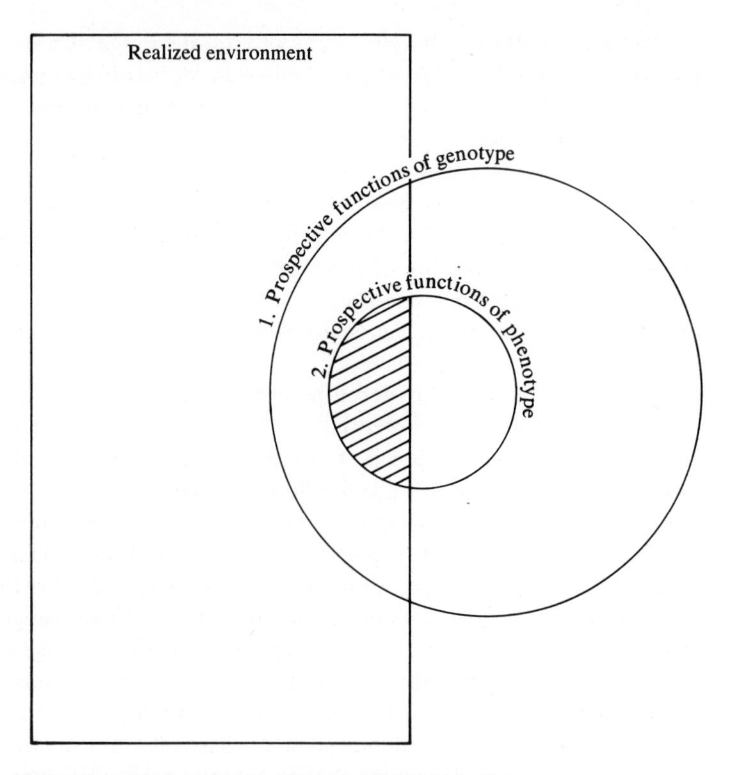

Fig. 3-2 The prospective functional range of a given genotype (circle 1) is limited during development, since the possible functions are reduced by developmental "choices" of various kinds, to a smaller range of functions in the phenotype (circle 2). Of these, only a portion are actually employed (shaded region), for some are not required by the environment. (See Simpson, 1953.)

overlap region the *realized* functional range of the phenotype. In our multidimensional conceptual model, the realized functional range is the portion of the prospective functional range of the phenotype that overlaps with local biospace. The realized functional range will fluctuate as the biospace fluctuates, so as to overlap different parts of the prospective functional range. The organism must adjust to these fluctuations to maintain internal homeostasis and to maintain external adjustment and safety. If biospace changes so that it no longer overlaps with an organism's prospective functional range, the organism will die.

Each Population Has a Unique Functional Range, Which Forms a Niche

A population is a group of individuals of the same species, and as the term is commonly used in ecology, and as it will be used here, it refers to a group that inhabits some specific region. That is, the population has a more or less well-defined geographic range at any given time, and there is more interaction between individuals within a population (especially as regards breeding) than between individuals in separate populations. The boundary zones of populations represent discontinuities in the distribution of the species.

Population ecology involves two main classes of properties and processes. One class is formed by the collection of the functions of all the individuals that compose the population. Since no two individuals are alike, the range of functions in the population is far wider than that of any individual. The other class arises from relations among the individuals, and therefore does not exist on the individual level at all, but is composed strictly of group properties. These properties include population size, birth and death rates, population growth rates, and so on.

As indicated in Chapter 2, a population is the phenotypic expression of a *gene pool*, which is simply the collection of genes in the genotypes of all the members of the population. For any given gene pool there are a vast number of potential phenotypes, consisting of the phenotypic potential of all the possible genotypes. Many of the phenotypes have more or less different ranges of prospective functions. A real population of realized phenotypes in various stages of ontogeny is equipped to contend with a certain range of physical and biological conditions, and it interacts with its environment, for example, by consuming resources and furnishing itself as a resource to its predators; it has a place in the economy of its surroundings. This place has been termed a *niche* (Elton, 1927); as employed here, the niche is the total functional aspect of a population-environment system.

The prospective functions of a population can be regarded in several ways (Fig. 3-3). First, there is the pool of prospective functions of all potential genotypes. This is the widest possible functional range of any given population, without considering the possibilities inherent in potential mutations and the like. Second, there is a potential functional range composed of the prospective functions of all the realized genotypes, and, third, potential functional range composed of the prospective functions of all the realized phenotypes. Of course, there are the realized functions that actually occur. For the most part, we shall be interested in realized population functions, which comprise what we shall call the *realized niche*. The widest possible potential functional range of a population—the prospective functions of all potential genotypes—comprises what we shall call the *prospective niche*.

It is useful to again visualize the environment as a hyperspace, each dimension of which is an environmental variable. Every point within the hyperspace corresponds to a particular environmental state. For any population at a given time, there is a hypervolume within this hyperspace that corresponds to all environmental states which can overlap the prospective functions of all possible genotypes; this is the prospective niche. There is a smaller region within this one containing the collection of all points that actually do overlap with environmental states occupied by the realized population; this is the realized niche (Fig. 3-3). Whenever the unqualified term "niche" is employed in this book, it implies the realized niche.

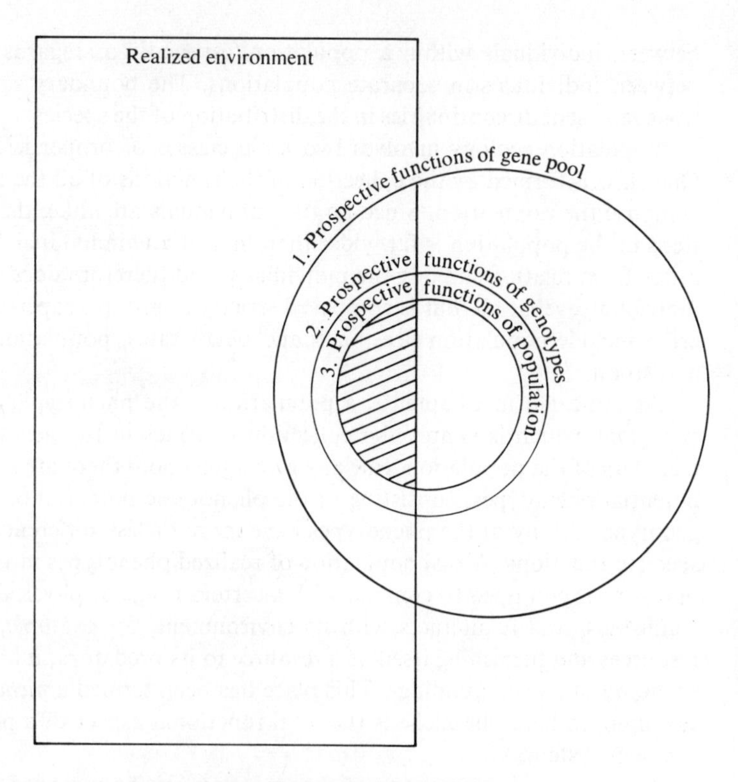

Fig. 3-3 The prospective functions that are inherent in a gene pool (circle 1) are reduced to the functions that are inherent in the gene combinations that actually exist at any time (circle 2), and these are limited further since some are not employed in the environment inhabited by the population (circle 3). The realized functions are depicted by the shaded region, which represents the realized niche of the population.

Species Niches Consist of Population Niches: Both Have Many Structural Variations

Since species are collections of populations that can exchange genes, the niche of a species is represented by the combined niches of its populations. Any prospective functions that occur within one population may be transferred to another via gene flow. Thus, when we view the living world in an instant of time, it is the species niches that form the fundamental units of adaptation, for they are all isolated from each other by barriers to gene exchange. However, changes in prospective niche dimensions, based on changes in gene frequencies, can occur independently within the individual populations forming the species. Furthermore, since different populations tend to live under more or less different conditions, the changes occurring in their population gene pools are indeed commonly different, so that each population may have a somewhat different niche. Therefore, even within an instant of time, there are local differences within the species niche from place to place, both within and between populations. The pattern of these differences may be called the *niche structure*.

Several modes of niche structure exist (Fig. 3-4). Within a given population,

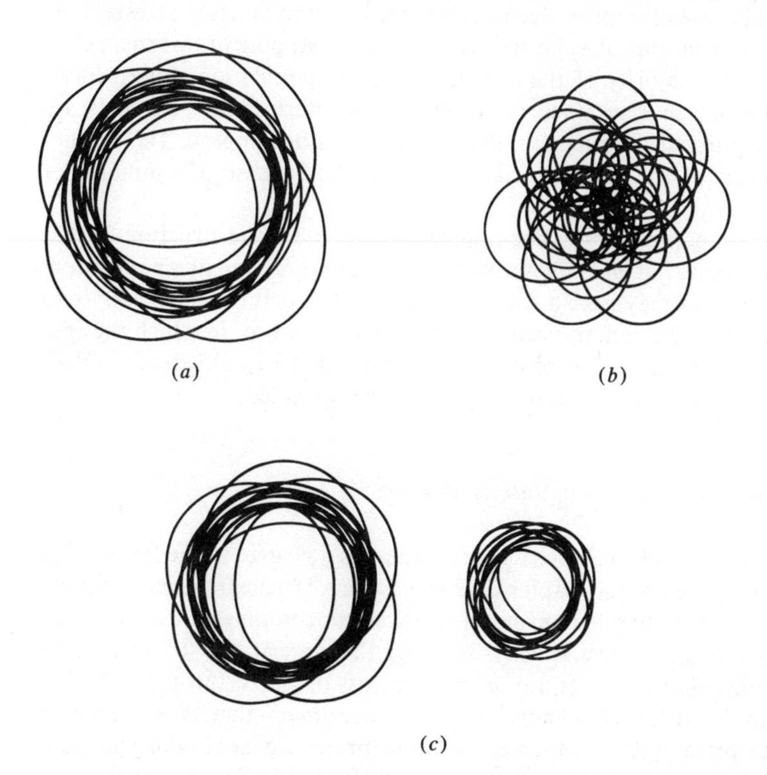

Fig. 3-4 Some possible patterns of niche occupation. (*a*) Most individuals have similar functional ranges. (*b*) Most individuals have somewhat different functional ranges, so the niche is much larger than the functional range of any individual. (*c*) Two modes of function are present that are not contiguous along many biospace dimensions.

individuals may each possess a similar range of environmental interactions, all nearly as broad as the realized niche, so that they broadly overlap functionally [Fig. 3-4(*a*)]. Another pattern is for each individual to occupy only a small segment of one or more dimensions of the realized niche, so that there is less functional overlap [Fig. 3-4(*b*)]. A population may also be made up chiefly of individuals with broadly overlapping functions but may contain a minority of individuals that share a special set of functions, and thus have a special subniche of their own [Fig. 3-4(*c*)]. If we are considering prospective niches, then differences such as those depicted by the patterns of Fig. 3-4, parts (*a*) through (*c*), would usually depend upon the genetic structure of the population; broadly overlapping similar genotypes, as in Fig. 3-4(*a*), suggest an emphasis on a genetic strategy leading to immediate fitness, whereas the pattern of Fig. 3-4(*b*) suggests much genetic flexibility, which might mean a genetic system with large numbers of genes and lots of crossing over and recombination. The pattern of variation in Fig. 3-4(*c*) implies the maintenance of a minor class of especially useful gene combinations, despite the tendency for swamping of this small class by genes from the majority of the individuals in the population. Intermediate gene combinations that produce individuals with intermediate functions are obviously at a great disadvantage in the figure, since none appear. This implies that maintenance of the small special class is at the expense of a number of genetic deaths.

A species composed of only one population has the niche structure of that population. However, most species are composed of more than one population, and each population may have a somewhat different structural pattern. Species niche structure may be summarized and defined as "the way in which the species' niche hypervolume is occupied by populations and by individuals." Obviously, species niche structures can be extremely complicated.

Niches Change in a Bewildering Variety of Ways

Populations are said to evolve whenever there is a change in the frequencies of genes in their gene pools (Chapter 2). Three aspects of gene frequency changes are particularly interesting for our purposes, even though they all usually occur together: (1) change in gene *quality*—that is, the appearance of new genes (through immigration or mutation) or the total loss of genes (through selection or emigration or drift); (2) change in gene *proportions*—that is, the relative increase or decrease in the representation of one or more genes in the gene pool (through selection, migration, mutation, or drift); and (3) change in gene *diversity*—that is, change in the total number of genes in the gene pool, through the addition of new genes without corresponding loss, or losses of old genes without corresponding additions. The prospective function, and therefore the prospective niche, of a population will ordinarily change as these changes occur in the gene pool.

Niches may change even though gene frequencies are constant, although these sorts of changes are not usually considered to be evolutionary. To bring some order into our discussion, we find it convenient to divide all niche changes into the same three aspects we used to summarize the changes in gene frequencies: (1) those that alter the *quality* of the individuals in the population; (2) those that alter the *proportions* of different types of individuals (phenotypes) present; and (3) those that alter the absolute number of individuals. We shall be interested in the functional features and consequences of these changes.

Changes in the quality of individuals result from genetic changes, to be sure, but they may also stem from environmental change. For example, the environment may fluctuate so that different parts of a prospective niche are overlapped by the realized environment at different times. Furthermore, environmental fluctuations may evoke different phenotypes among the growing organisms at different times owing to environmental involvement in the developmental pathways, and the realized niches of succeeding generations will therefore change. Such new phenotypes, which owe their peculiarities to environmental and not to genetic changes, are called *ecophenotypes*. For that matter, each new birth and death affects the prospective niche of a Mendelian population. As genes are constantly recombined, and as each ontogenetic pathway is unique and varies in time, the sum total of the functional ranges of all the individuals in a population is in constant flux. Thus, realized niches change incessantly, not necessarily because evolution is occurring on the genetic level, but because changes are occurring in the relation among genes—their combinations are changing and so are their effects. All these changes affect the *quality* of the individuals that make up the population, even though they do not directly affect the prospective niche. Nevertheless, they are important ecologically and in evolution.

Changes in the proportions of phenotypes present, although they commonly involve changes in gene proportions, may involve only a shift in the relative frequency with which one or several ecophenotypes are represented, or a shift in genotype proportions that involves recombination without significant changes in individual gene frequencies. Such changes are represented in biospace chiefly by shifts in the frequency of occupation of certain regions, and not by much alteration in the shape or size of the population ecospace.

Changes in the number of individuals present alter the density with which the biospace is occupied, and, as we have seen, population size affects importantly the evolutionary potential of selection, mutation, and drift.

Thus, changes in either the quality, proportion, or number of individuals will usually affect both the prospective and realized niches if they involve genetic changes, but only the realized niche if they do not. Clearly, populations with very large prospective niches are very flexibly adapted. But their realized niches may be narrow at any time, so that if the environmental regime changes, the realized niche may change considerably yet still fall well within the prospective niche. Few or no genes will appear or disappear. Such populations might persist in highly fluctuating environments without novel gene input.

Communities Are Associations of Populations, Some of Which Interact

A great number of environmental divisions may be recognized within the oceanic realm. Environments within the water column are inhabited by swimming and floating organisms, and numbers of distinctive environments exist on sea floors, such as in the deep sea and on rocky shores and floors of bays. Each environment contains different ecological conditions, and therefore each occupies a somewhat different portion of environmental hyperspace. Although there is overlap along some dimensions, parts of other dimensions are unique to each environment. Opportunities for adaptation vary widely between environments, and indeed different associations of species are found in different environments. If one of these environments, such as a rocky shore, is examined in detail, it turns out to be composed of numerous habitats, each with a characteristic assemblage of organisms. For example, there are differences between the exposed and protected faces of rocks, between the tops and the dark undersides of rocks, between smooth and pitted faces; there are special associations among algal holdfasts and among tufts and clumps of smaller algae and forms that live chiefly in crevices. Since some organisms live on others, organisms are usually habitats themselves.

Suppose that a coastline several tens of miles long composed chiefly of sandy beaches is interrupted at a number of places by rocky points. The rocky shores at these points may be rather similar, each having about the same range of habitats as the others. All other things being equal, the points would normally be inhabited by rather similar associations of organisms. Any two points would contain great numbers of species in common, but it is likely that each point would support populations of some species not found at all of the other points, and that some species would be found at only a few of the points. These recurring but not identical population associations belong to a rocky shore *community*.

When populations of different species live in close association, they commonly interact. An interaction may be critical to the continued existence of a population, such as when one species uses another for food, or when two species compete for the same limited food supply. Volterra (1926) has postulated and Gause (1935) has demonstrated experimentally for some cases that two species utilizing the same limited resources cannot coexist indefinitely. If within a community two different species populations are ecologically identical or even very similar, one will ordinarily prove superior, however slightly, in reproduction, and the other will eventually be eliminated. Thus, each population in a community should have its separate, distinctive role—its own niche. Whether this hypothesis is strictly true is not certain. In some cases, spatial heterogeneity of the environment seems to permit some different species to coexist even though they share limiting resources, whereas perhaps in environments that are extremely stable temporally, coexistence can also occur. However, it is a fact of experience that species within communities tend to have distinctive niches. Many interspecific interactions are relatively insignificant, such as two species avoiding each other in "traffic." Species may even be adapted to completely ignore each other.

Since all individuals must have a source of energy in order to live, such as food or sunlight, the population interactions related to food are of special ecological importance within communities. These interactions are called *trophic* interactions, and most communities have basically similar trophic patterns, as first stressed by Elton (1927). The plants, most of which obtain their energy from the sun and nutrients from sea water or soil, form the lowest trophic level. They synthesize complex organic componds from simple nutrients with the aid of an energy source, usually the sun. Much of their energy is expended in carrying out their life processes, such as respiration or reproduction, but some of it is locked up in compounds within their bodies. Field and experimental work on this stored energy gives variable results; it may amount to 10 or 15 percent of the energy that is taken in, and it is available to plant-eating organisms. These herbivores use much of the energy they receive from the plants for *their* life processes, but there is some (evidently often around 10 to 15 percent) available to predators that feed on the hervibores. Other predators may feed on these predators. If any of these organisms die before being consumed, detritus feeders, scavengers, or decomposers may utilize their stored energy.

Thus, energy flows through a community, from trophic level to trophic level. At the base are forms that can synthesize protoplasm from inorganic materials, called *autotrophs*. Next come a series of organisms that require other organisms as food, called *heterotrophs*. A great deal of energy is expended in metabolism at each level, so the amount of living matter being produced decreases at each higher level. On a yearly basis, a trophic structure will have the outward form of a pyramid. Much of the energy is finally liberated from compounds wherein it is stored and is dissipated, although some may be stored in sediments. The sorts of internal complexities found within trophic pyramids will be discussed in Chapter 7. Obviously, many animals employ a variety of foods, sometimes including both plants and animals, so that their biomass has to be assigned partly to the herbivore level and partly to one or more of the carnivore levels. Bookkeeping can be very difficult in tracing the energy distribution and flow within a community. Figure 3-5 presents a simplified version of feeding relations within a rocky-shore community of only moderate diversity.

The nutrient chemicals also flow along pathways within communities. That is, plants take up inorganic nutrients and trace elements, and some of these materials are passed up the trophic pyramid from prey to predator. Animals take up some simple inorganic materials themselves, and some nutrients and trace elements are returned to the inorganic environment at each trophic level in the form of metabolic waste or in the bodies of dead organisms. Complex organic compounds are broken down, often by decomposer organisms such as bacteria, and the simpler nutrient chemicals again became available for uptake into living organisms. While energy flows through the pyramid and is dissipated, the nutrient chemicals can be recycled. Communities and their environments thus form rather open natural systems, called *ecosystems*.

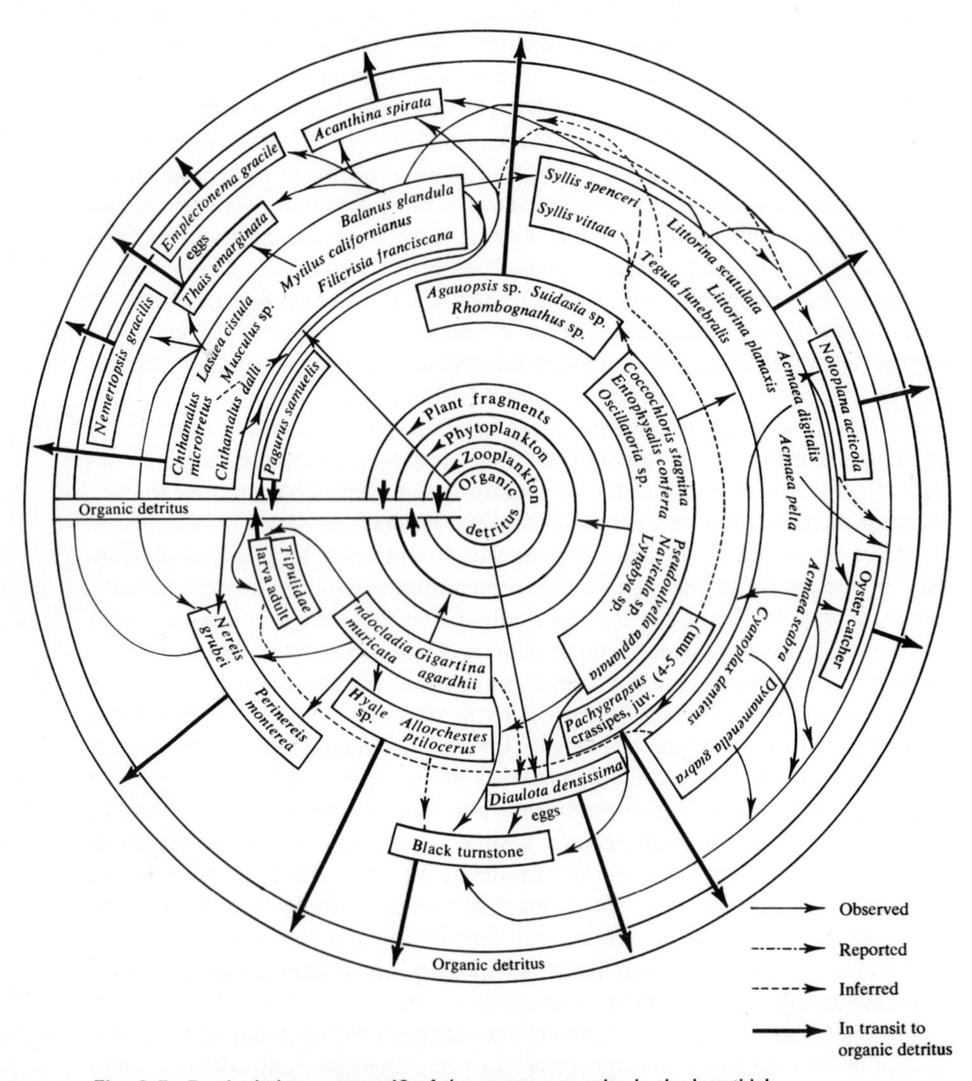

Fig. 3-5 Food relations among 43 of the common species in the intertidal *Endocladia-Balanus* association from Pacific Grove, California. The central circles indicates the most abundant food substances brought in at high water; the outer circle represents detritus, to which all food chains contribute, and which serves to supply some of the suspended detritus of high water (arrow from left). (After Glynn, 1965.)

Communities may be described, then, as composed of populations of certain species that are associated in space at any given time and that live within some characteristic type of habitat or within a characteristic range of habitats.

Ecosystems Are Associations of Niches

The functional aspect of a community is an ecosystem. As a community is an association of populations of different species, so a community ecosystem is an association of niches. Every ecosystem has characteristic properites, which depend in great measure upon the properties of the niches of the populations involved. Naturally, the ecosystem properties depend also upon the nature of the environment.

These relations can be readily visualized by reference to the hyperspace model of the environment. Within the biospace realized at any given time, there is a region occupied by the niches of all the populations that are associated within any given environment or environmental range to form a community. This region contains the dimensions representing all the parameters of environmental interaction that affect the populations at that time, including biological interaction. The hypervolume defined in this way represents the *realized* ecosystem.

When the environment fluctuates, the shape and position of biospace change within the environmental hyperspace lattice, and usually the region of overlap between biospace and the ecosystem also changes. The realized ecosystem is therefore altered. However, limits exist to the range of environments within which any given ecosystem can exist. When these limits are surpassed in any region, the biospace that can support the ecosystem has disappeared, and the ecosystem is extinct in that region. The environmental limits of existence of an ecosystem enclose a hypervolume within the environmental hyperspace lattice that represents the *prospective ecosystem.*

Ecosystem structure is a concept analogous to niche structure. It is defined as the way in which the hypervolume of the ecosystem is occupied by niches. In some cases, there are few niches and few trophic levels in an ecosystem, with the populations at the higher trophic levels feeding upon most populations at the next lower level or the next few lower levels. Energy flow is rather generalized in this case. On the other hand, there are sometimes a great many niches at each level and a great many levels in the trophic structure of an ecosystem. Populations at some levels may tend to feed upon only one other population. Energy flow in part of the ecosystem is therefore highly canalized into numerous streams and the structure is intricate. Each type of ecosystem structure is the result of adaptation (achieved on the population level) to different historical conditions.

Community Systems Are Altered as Niches Change

Communities change in a variety of ways, which may be characterized by the ways in which the collection of populations in the community changes. The populations may change in quality, proportions, or diversity. Changes in quality may stem from changes within the populations, chiefly based on changes in gene frequency (or sometimes on changes in genotype mix, in part), thus altering the niche of the populations, or on replacement of one species population by another that naturally has a different niche. Such changes cause the prospective ecosystem to change [Fig. 3-6(*a*)].

Changes in the proportions of populations involve the relative increase or decrease in the number of individuals of one population with respect to others.

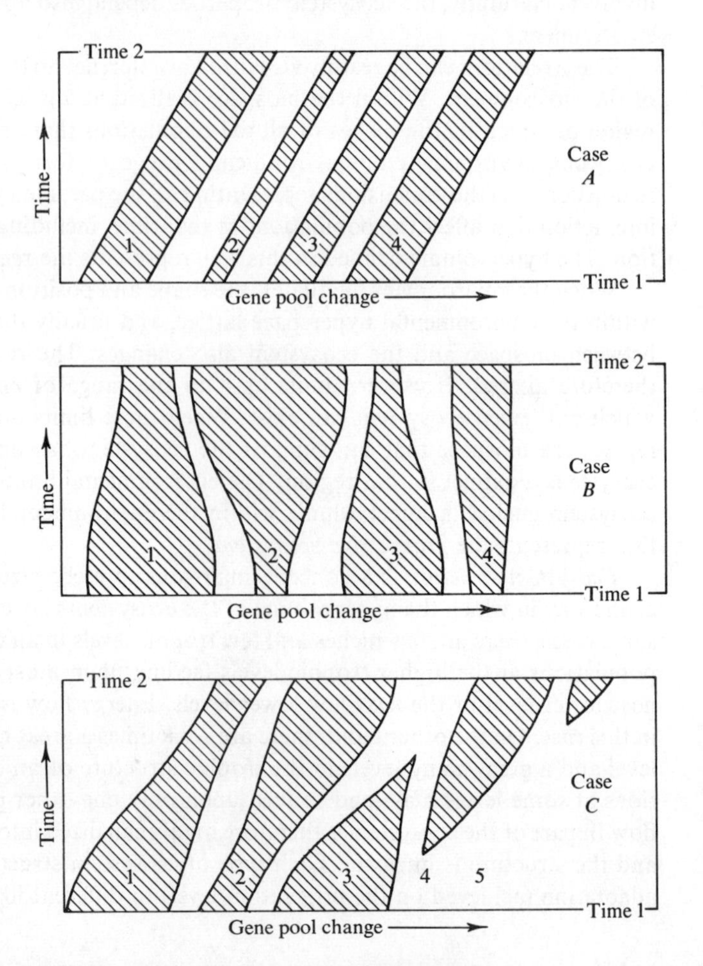

Fig. 3-6 Components of community change. The shaded areas represent populations; their widths are proportional to population size. Case *A:* Populations are evolving in concert, so that gene pools and population niches are changing, although population sizes remain constant. Case *B:* Populations are not evolving genically, but the environment shifts, altering the relative population sizes. Case *C:* Both of the previous sorts of changes are occurring, and in addition populations are disappearing and appearing so that the number of populations changes. This is a general case of community change. (After Valentine, 1968b.)

Such a change would alter the relative flow of energy (as food) through the community from some populations to others. The ecosystem structure would thus change [Fig. 3-6(*b*)], and the realized ecosystem would evolve. As changes in population size need not be at all associated with changes in gene quality within populations, but instead may be due to environmental changes that did not provoke genetic changes, the prospective niches, and therefore the prospective ecosystem, need not be altered; only the realized ecosystem need be changed.

Finally, changes in the number of populations present would alter the structure of the ecosystem and would constitute a change in both the prospective and realized ecosystem. Additions or subtractions (or, for that matter, simple replacement) of species populations need not be associated with genetic changes within any of the populations; therefore, change in the prospective ecosystem need not involve internal change within the prospective niches of the populations, but merely change in the populations that are present. In Fig. 3-6(*c*), a general case of community evolution is depicted, in which all the kinds of changes previously mentioned are occurring.

Provincial Biotas Are Associations of Communities

The associations of organisms inhabiting widely separate parts of the earth are usually very different from each other in taxonomic composition. As an example of a community we have used the recurrent associations of organisms inhabiting rocky shores along a ten-mile stretch of coastline. Suppose that stretch of coastline is along western Europe in the temperate zone, perhaps along the Atlantic coast of France. A similar stretch of coastline in the temperate zone of eastern North America along Virginia will also be inhabited by rocky-shore communities, but relatively few species are represented by populations in both France and Virginia. Similarly, a ten-mile stretch of rocky shores along the Oregon coast of western North America supports a rich rocky-shore biota, but this biota is quite unlike either of the others in species composition. Today, the barriers between France and Virginia are strong, and those between Virginia and Oregon are nearly impenetrable for shallow-water marine invertebrates, for they do not permit gene migration from one shore to another on any regular basis.

If a species were to succeed in establishing populations in both Virginia and Oregon, each population would ordinarily be quite isolated from the other, except for species with truly unusual powers of dispersal. For the great majority of species, the barriers would effectively block gene migration. The genetic isolation would lead in time to the development of separate species along each coast, owing to the normal processes of evolution.

The most obvious barrier preventing migration of rocky-shore marine species between France and Virginia is the deep Atlantic Ocean basin; between Virginia and Oregon the obvious barrier is the North American continent. These barriers are so gross that the fact that they are ecological barriers almost goes without saying. The marine species that compose the rocky-shore communities are not adapted to life on land or in the deep sea.

If the rocky-shore community from the Oregon coast is contrasted with the community found in similar habitats along a ten-mile coastal stretch in the Gulf of Alaska, it is found that a large number of species is not represented by populations in both regions. Precise figures are not available, but we can estimate from data on the mollusks and a few other groups that about half the species appear to be restricted to communities in only one of these regions. In the same way, if we contrast an average community from a ten-mile coastal stretch along southern California with an analogous Oregon community, we shall find that about half the species are different. Obviously, there are barriers to distribution along the west North American coast that are fairly, although not completely, effective. Yet there are no gross barriers, such as ocean basins or land masses. The major differences between the environments of these separate coastal stretches seems to be temperature, which suggests that temperature is a fairly effective ecological barrier in the sea.

This suggestion is strengthened if we contrast the communities found in the tropics with communities in analogous habitats in the temperate climatic regions to the north or south along the same coasts. Along western North America, for example, figures are available for mollusks, and it appears that only about 20 percent of the tropical species live also in temperate communities. Although the percentages vary from taxonomic group to group and from region to region, the same general situation exists everywhere.

There are discontinuities in the distribution of species. These discontinuities form the boundaries between regions having different biotas. Within one of these regions, communities are rather similar in species composition from place to place; between these regions, analogous communities are less similar and may even be entirely different. The boundaries are created by barriers to species distribution. The barriers may be gross, such as ocean basins, or more subtle, such as a change in a marine climatic regime. The regions that constitute distributional units of organisms are called biogeographical regions or *provinces* (biotic provinces, faunal provinces, floral provinces, and so on).

Thus, provinces differ from one another in kinds of species and, of course, in kinds of higher taxonomic categories as well, such as genera and families. They differ in still other ways, such as in sheer numbers of species, for there is a well-known latitudinal gradient in species diversity. Equatorial regions are extremely rich in species whereas polar regions are relatively impoverished. The changes in species diversity with latitude are fairly gradual, but seem most pronounced at provincial boundaries. There are also longitudinal differences in species diversity. For example, the shallow-water provinces of the western Atlantic contain fewer species than those of the western Pacific, even when climatically similar provinces are compared.

Another way in which provinces differ is in the numbers of communities.

In general, there are fewer in high-latitude than in low-latitude shelf provinces. There are also fewer in regions that have relatively homogeneous environments (fewer habitats are present) than in regions that are environmentally diverse. There are other significant differences between provincial biotas as well, and these will be discussed in Chapters 8 and 9. At present, let us examine the functional implications of just those few differences mentioned here. They are extremely important and demonstrate the significant place that the provincial level holds in the ecological organization of the world's organisms.

Provincial Systems Are Composed of Ecosystems

There is no generally accepted term for the functional aspect of a biotic province. The word "biome" has been used in this sense by some authors, but it is also commonly used in another way—to signify a group of communities which are more similar to each other than to other communities, so that they form a special unit larger than a community. There is not necessarily any implication that the biome as so defined is coextensive with the sort of biogeographic region which is termed a province herein. Therefore, to designate the functional aspect of a province, we shall use the simple term *provincial system*.

A provincial system may be conveniently visualized in terms of the environmental hyperspace model. At any given time, there is a certain hypervolume within the totality of environmental hyperspace that represents the realized environmental interactions of all the ecosystems composing any provincial system. This hypervolume represents the realized provincial system. Enclosing this hypervolume is a larger one that includes all the prospective functions (which are not incompatible) of all the ecosystems, and this represents the prospective provincial system. Throughout this book, the unqualified term "provincial system" refers to the realized provincial system.

The different hypervolumes that represent provincial systems may be occupied by ecosystems in different ways; the patterns among ecosystems comprise the *provincial system structures*. For example, contrast the provincial structure in two provinces alike in climate and in area but different in that one has a relatively narrow range of habitat conditions and contains few ecosystems, whereas the other has a wide range of habitats and contains many ecosystems. The province with fewer ecosystems will be inhabited by fewer different species than the other province, but species will have broader areas of occupation, on the average. Assuming everything else to be equal, with identical amounts of energy entering the provincial system in each province, the average number of individuals of each species will be larger at any time in the province with fewer species.

Now contrast two provinces having equal amounts of production and equally rich in physical habitats, but having greatly different species diversities (a situation that may arise due to climatic differences). Ordinarily, the province with more species will have more communities as well. Whether it does or not, the species will be represented by smaller average populations than in the province with fewer species. Thus, in the province with more species, populations will have to be less densely distributed or will have to occupy more restricted areas than in the province with fewer species.

It is obvious that the shapes, sizes, and packing of the ecosystem hypervolumes differ from province to province—that provincial structure varies greatly. Furthermore, the structure of the ecosystems obviously varies as well from one provincial system to another. In fact, the frequency of niche structural types varies from ecosystem to ecosystem and from province to province also. The salient features of these variations will be examined in subsequent chapters. It suffices here to point out that populations in different provincial systems may face quite different problems in maintaining adaptation. Clearly, the provincial level is an important one in ecology and evolution.

Provincial Systems Are Altered as Community Systems Change

The biotic changes which occur in provinces may be conveniently discussed in terms of changes within the communities or changes in the relations among the communities that make up a province.

The changes that occur *within* community ecosystems have already been outlined (Fig. 3-6 and explanation in text). These changes range all the way from mutations and recombinations that affect the functional ranges of individuals up to population size-frequency changes that alter the niche relations within ecosystems.

The changes *between* ecosystems chiefly involve changes in their relative sizes (Fig. 3-7). These usually accompany changes in the habitats within the provincial region. For example, there might be a change in the supply of sediment that is reaching a segment of continental shelf. At one time, the sediment there may be chiefly fine, with only a minor sand supply that comes to rest near or on shore. If the supply of coarse material increases owing to shift in drainage, to tectonic activity, or to climatic changes inland, the muddy shelf substrate may be replaced by sand. The deposition of sand sheets will provide opportunity for the spread of sandy-bottom communities and restrict the area occupied by muddy-bottom communities. Even if no species or communities appear that are new to the province, the relative sizes of muddy- and sandy-bottom communities have changed, and the provincial system has been altered. On the other hand, the deposition of coarse sediments in deeper subtidal shelf zones may result in the appearance of associations of species that were not formerly present, at least as associations; a new community has appeared. Or perhaps one or more communities living on shallow subtidal muddy bottoms entirely disappear as the muddy habitat is eliminated. These events cause important changes in the provincial system (Fig. 3-7).

If the number of communities within a province increases or decreases, the

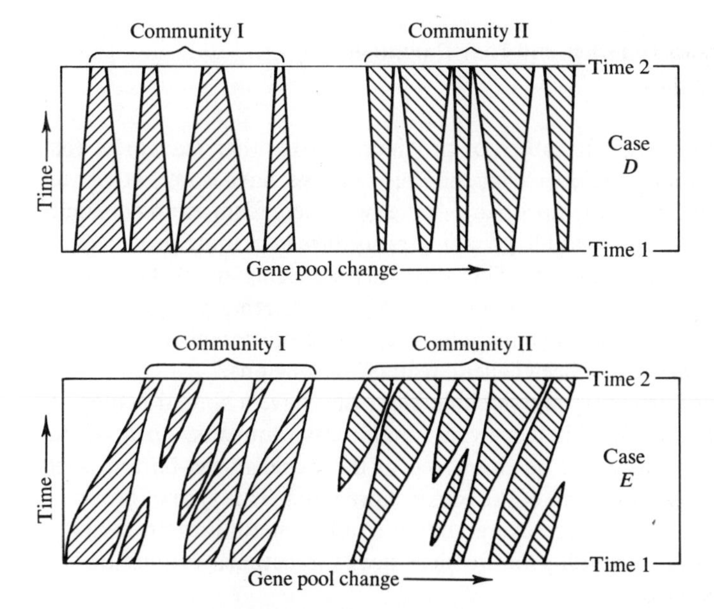

Fig. 3-7 Components of provincial change. (See explanation of Fig. 3-6.) Case *D:* Two communities, I and II, change their relative proportions but no other type of change is occurring. Case *E:* The communities are changing in quality (see Case *C*, Fig. 3-6) as well as in proportions. If new communities were appearing and/or old communities disappearing, a case not shown, then the types of provincial change would be complete. (After Valentine, 1968b.)

structure of the provincial system is altered. The processes that change community diversity are, naturally, those which allow introductions or cause extinction of communities so that a simple one-to-one replacement of communities does not occur. Often this involves an increase or reduction in the diversity of habitats available within the province.

It is easy to visualize changes in provincial systems in terms of the hypervolume models of prospective and realized provincial systems. Most genetic changes within any of the populations involved alter the *prospective* provincial system; the shape of the hypervolume that represents the system will be changed. The genetic changes usually are accompanied by an alteration in the *realized* provincial system also, but this is not inevitable since the genetic changes may not be realized, at least not immediately, as functional changes. The changes in relations among populations or among communities alter the shape of the realized provincial system hypervolume. The prospective provincial system is not affected unless the changes involve qualitative genetic changes that are expressed functionally, which is usually the case. Indeed, most or all of the various processes of provincial change will ordinarily occur at the same time and will go hand in hand, representing responses of different sorts to a given environmental change.

Add All the Provinces Together and They Comprise the Total Planetary Biota

It is possible to recognize groups of provinces or of other categories that form units larger than single provinces. These groups have some attributes which set them apart, but it is improbable that any of them form ecological units of fundamental sorts. One possible grouping procedure is to unite provinces on the basis of taxonomic similarity. Thus, provinces sharing many families or genera or species would be grouped, and separated from other groups of provinces with different but internally characteristic taxonomic compositions. Such grouping results in what have been called "realms" or "regions."

A second type of grouping may include biota that have a similar ecological or adaptive aspect and inhabit somewhat similar environments, on a world-wide basis. Thus, groups could be composed of the world's level-bottom shelf biota, of the coral reef biota, and of the rocky-shore biota. This type of grouping commonly involves assembling several but not all communities from many provinces, rather than grouping entire provinces. Sometimes groups of this sort are called "biomes."

Finally, groupings are sometimes made on the basis of climate—tropical biota are grouped, temperate biota (or north-temperate or south-temperate biota) are grouped, and so on. Because there is a relation between taxonomic composition, climate, geologic history, and major ecological aspect, all these groupings have features that allow them to be considered as ecological groupings to some extent. Yet none of them constitutes a unique category of ecological unit that forms a level of organization above the provincial level.

A series of ecological units similar to that within the marine environment occurs also within the terrestrial environment, building up from the functional ranges of individuals to the terrestrial biosphere. The ecological interchange between marine and terrestrial organisms has been relatively limited, consisting in such interactions as predation of sea life by coastal birds. The interchange of environmental factors, such as heat, gasses, nutrients, and salt, is more significant and links these two major environments into a single planetary system. Nevertheless, it is often considered that the total planetary biota is made up of marine and terrestrial divisions, forming separate subsystems of the planetary biosphere.

The Biosphere Is the Largest of All Ecological Systems and Is Altered by Changes in Provincial Systems

The functional aspect of the total planetary biota is called the *biosphere*, and is composed of simply all life on earth together with that part of the planet with which living things interact. In terms of the hyperspace model, the realized biosphere comprises all the environmental interactions of all living things.

Provinces may be regarded as the major components of the biosphere, and by their number and relative sizes and interrelations form the *biosphere structure*. When any of the provinces changes internally, owing to the sorts of causes discussed in the preceding sections, the biosphere is naturally affected (Fig. 3-7), as it is also whenever the relations among the provinces alter. For example, the increase in size of one province relative to another (perhaps a polar province expands and an equatorial province contracts during a cooling trend) constitutes a change in the structure of the biosphere even if no other changes occur (Fig. 3-8). New provinces may appear or old ones disappear in response to changing climates or geographic relations, and thus provincial numbers may change. Biosphere changes may be classed as changes in provincial quality, proportions, or numbers.

The Architecture of the Biosphere Is Hierarchical

The preceding sections have traced the ecological units from level to level of organization. At the highest level is the biosphere, the system of all life, and it is composed of subsystems (provincial systems), each of which, in turn, is composed of smaller subsystems (community ecosystems), and these of even smaller subsystems (niches). The hierarchy continues down to levels of organization below those that are ordinarily considered to represent ecological levels—right down to the genetic levels and below. The precise nature of organization at genetic levels is not yet clear, although vague outlines of molecular systems loom obscurely ahead. A tentative outline of the ecological hierarchy is presented in Table 3-1.

Two sets of terms are employed to describe the ecological hierarchy. One set simply describes the collections of organisms that form ecological units within the various hierarchical levels, such as populations and communities. The other set suggests the functional aspects of the units at each level, such as niche and ecosystem. The two sets of terms thus describe the same organisms, but the functional set includes the organism-environment interactions and parts of the environment involved in the system; it is therefore truly ecological in its concepts.

Naturally, the organisms in each unit have a genetic basis; individuals have genotypes, populations have gene pools, and the units belonging to higher levels are underpinned by appropriate collections of gene pools. These are indicated in Table 3-1. A level below the individual, the level of the functional genetic unit, is included in the table to indicate that genetic organization exists below the level of the genotype, even though all the details at this level (there may be more than one level) are not known.

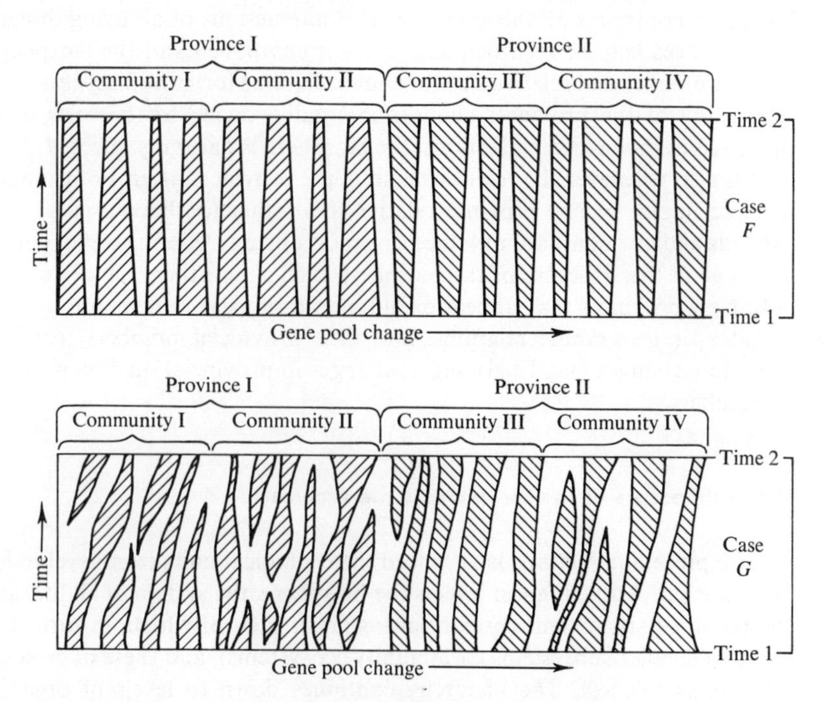

Fig. 3-8 Components of biosphere change. Case *F:* The relative proportions of two provinces are changing, but no other type of change is occurring. Case *G:* The qualities of the provinces are changing. If new provinces were appearing and/or old provinces disappearing, a case not shown, then the types of biosphere change would be complete.

From this point on, it will prove useful to have a general term to refer to a part of environmental hyperspace, realized or prospective, that may be occupied by a specific biological unit. The term *ecospace* is appropriate. The ecospace of an individual is its functional range: of a population, its niche; of a community, its ecosystem; of a province, its provincial system; and of the total planetary biota, the entire biosphere. The relation of the ecospace of a biological unit to the entire environmental hyperspace lattice is indicated in Fig. 3-9.

If we examine the ecospace that represents a biosphere—for an example, let's take a realized biosphere similar to today's—we find it contains a set of smaller ecospaces representing the provincial systems, some of which are overlapping to a degree along many dimensions but distinct along many others, so that each provincial system is a unique combination of parameters. Within each provincial ecospace is found a smaller set of ecospaces representing communities, and inside each of these is a set of still smaller ecospaces representing populations. Finally, inside each population ecospace is a set of ecospaces re-

Table 3-1 A Hierarchy of Ecological Units

GENETIC TERM		ECOLOGICAL TERM	
Unit of organization	*Organization*	*Functional*	*Descriptive*
"Gene"	Genotype	Functional range of individual	Individual (phenotype)
Genotype	Gene pool	Niche (population system)	Population (deme, cline, sp.)
Gene pool	Note 1	Ecosystem (community system)	Community
Note 1	Note 2	Provinical system	Province
Note 2	Note 3	Marine biosphere (world ocean system)	Total marine biota

Note 1. There seems to be no single term for an organized collection of gene pools.
Note 2. There seems to be no single term for an organized collection of gene pool collections.
Note 3. This is the collection of all the nucleotides in the world.

presenting the functional ranges of the individuals—sort of "individual systems." No doubt there are still smaller hypervolumes within the individuals that represent the phenotypic aspect of organized combinations of nucleotides, but we shall not be much concerned with these little-known systems.

The hypervolume model we shall be chiefly concerned with, then, is formed architecturally of a hierarchy of nested ecospaces, ranging in scale from the part of the environmental hyperspace lattice occupied by an individual to the entire biosphere. Hierarchies of this sort have certain general properties, two of which are especially significant for the ecological hierarchy. The first of these properties is a function of the simple fact that the changes at any given level appear at all higher levels. Thus, a change in gene frequencies may occur causing a niche to evolve, and therefore any community, province, or biosphere containing that niche changes also. The size of the unit being affected obviously increases greatly at each higher level, so that a given change affects a much larger proportion of the smaller units at lower levels than of larger units at higher levels. For any unit of time, the average change in systems at the low hierarchical levels is relatively high, whereas proceeding up the hierarchy to larger systems causes the average of the proportional change within each system to become less and less. A community ecosystem, for example, will tend to evolve more slowly than the average of its component niches, although as a limiting case it may evolve as fast (but no faster). The same situation applies to each succeeding hierarchical level. To contrast the lowest and highest level with which we are much concerned, we see that the individual genotype is recombined each generation, whereas the biosphere usually changes from one structural configuration to another only on the order of tens of millions of years.

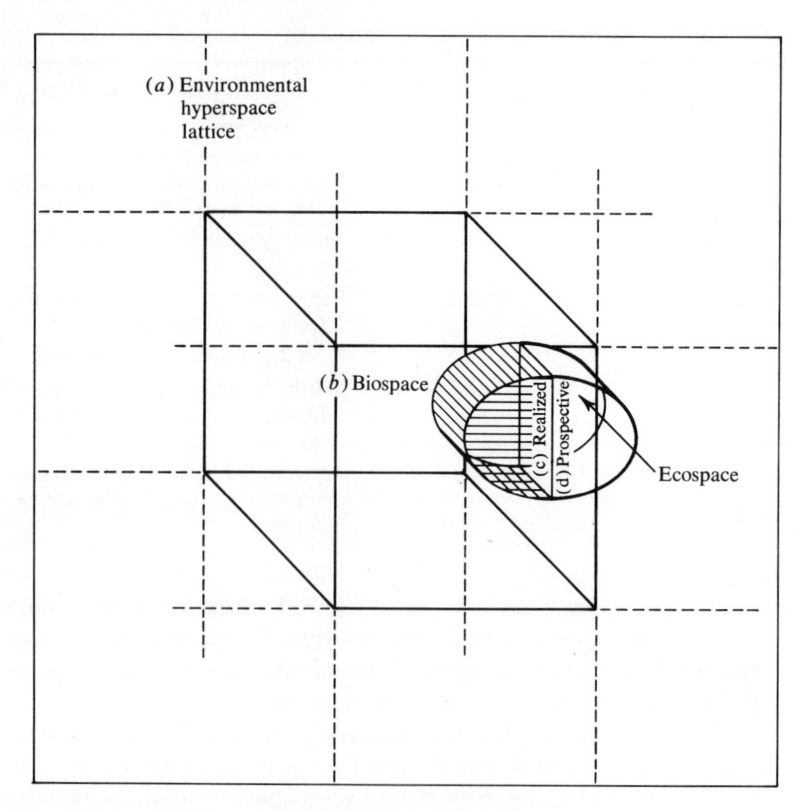

Fig. 3-9 A highly diagrammatic representation of the multidimensional space composing the environmental hyperspace lattice (*a*), wherein each point represents a unique combination of environmental factors. The realized environment is represented by *biospace* (*b*), which is available to organisms. Parts (*c*) and (*d*) represent the region of environmental space which coincides with factors tolerated by a biological unit and which is bounded by the limits of tolerance of that unit—the *ecospace* of that unit. Only a portion of the ecospace is realized (*c*); the remainder is prospective ecospace (*d*), which may become realized if the environment changes so as to include more of that portion of the lattice. (After Valentine, 1969, by permission of the Palaeontological Association.)

The second special property of hierarchies which has special interest for paleoecology is a function of the fact that changes in the structure of units at each level can be specified in terms of the interrelations of the units at the next lower level alone. No other levels need be considered. For example, evolution of provinces may be evaluated in terms of the interrelations among their communities. It is not necessary to know what goes on within the communities; they may be treated as sorts of "black boxes." Neither is it necessary to be concerned with the effects of provincial change on the biosphere structure. The evolution of a province may be "satisfactorily" explained solely in terms of the ebb and flow of change in communities, if there is no special interest in other levels. This property of hierarchies has been discussed by Simon (1962), who calls it the property of *near decomposability*.

"Decomposable" is a term in mathematics as well as biology, and Simon is referring to the mathematical usage. When it is asserted that the hierarchy of the biosphere is nearly decomposable, it follows that each level in the ecologic hierarchy can be studied nearly independently. In fact, each system can be studied in terms of its major subsystems alone. This will yield perfectly valid and significant scientific results, although it will not yield the "whole story." However, we shall not know all about everything for a very long time.

An example will help clarify this property, and Darwin provides us with the most famous example of all. It happens that he did not understand heredity, and, indeed, he proposed an hypothesis of heredity, called "pangenesis," which was about as wide of the truth as it was possible to be. Darwin worked chiefly on the population level, and explained the evolution of populations in terms of changes in types of individuals that are present. He also assumed that individuals had certain properties, such as heritable variation, which he could observe and establish empirically, but he did not at all understand the internal workings of the "individual systems." Despite this drawback, it is generally conceded that he formulated an explanation of evolution that ranks as one of the major scientific contributions of all times. That Darwin didn't present the whole story of the evolution of populations by solving genetic systems also is no reason to belittle the magnitude of his contribution.

Thus we may conclude with Simon that it is possible, with care, conceptually to strip off a level from the hierarchy and to study the major subsystems of the level independently of other levels. We must make appropriate assumptions about the properties of the major subsystems of our chosen system, but need not completely understand the internal workings of these subsystems in order to arrive at scientifically valuable conclusions. It is certainly a good thing that we do not have to wait for the lowest level to be solved before we can proceed to study the next level. On the contrary, all levels are available for study at once. Some levels are, in fact, rather neglected today and provide special opportunities for scientific contributions.

The Fossil Record of Each Ecological Unit Has Special Properties Related to the Size and Internal Dynamics of the Unit

The intermittent character of the fossil record has been described in Chapter 1. The biotic samples that it contains do not appear to be randomly distributed in space or time; rather their distribution is related to the stratigraphic framework resulting from the tectonic framework and from the interrelated patterns of erosion, sedimentation, diagenesis, and metamorphism. Perhaps the record can be visualized as having a random component and also a component of clumping—clumping of samples in space and time by tectonic-sedimentary events. As there are many classes of such events there are many sorts of clumping, and no single coefficient of clumping is adequate to describe the pattern. The clumps are probably clumped themselves. Although the pattern of the record is so complicated, it is possible to generalize somewhat upon the differences between the records of each of the ecological units.

Figure 3-10 depicts an interval of geological time for an hypothetical region.

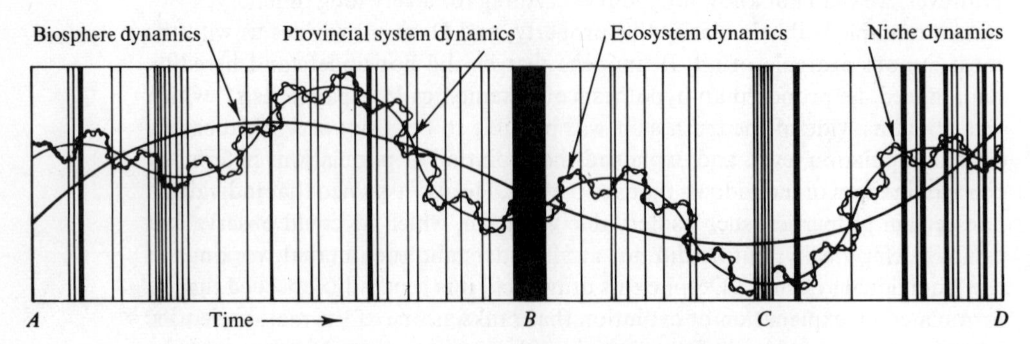

Fig. 3-10 The hierarchy of frequencies of the internal dynamics of change associated with the interactions of the largest subsystems at each level in the ecological hierarchy, and the character of their preservation in the fossil record. Vertical bars depict times or places represented by a fossil record, with unrecorded intervals between them. (After Valentine, 1968b.)

The parts of this interval of time that are represented by a fossil record are indicated by black vertical bars; the white intervals between the bars represent times for which no record is present. The sinusoidal lines represent levels of the ecological hierarchy; the different wave frequencies employed for each level indicate the relative frequency of a given amount of structural change at that level. Since the structure of each higher level of the ecological hierarchy tends to evolve at a slower rate than lower levels, niches have the highest frequency and the biosphere the lowest.

The geological record of the changes within the structures of these ecological units has a different quality for each unit. For the population niche, a great deal of change is common during unrecorded intervals of time. This makes it difficult or impossible to reconstruct precisely the pathway of niche evolution between samples. On the other hand, there are some intervals, as at (*B*) and (*C*) in Fig. 3-10, wherein the record is far more complete than the average, and from these the realized niche changes can be followed to some extent. At least some of the changes in prospective (and therefore realized) niches are represented by morphologic changes in the fossil populations. Changes in other aspects of realized niches, such as population age structures, have not been much studied but are sometimes preserved.

On the average, realized ecosystems do not evolve as rapidly as do population niches. Therefore, a fossil record that is continuous enough to include a given percentage change in an average niche will include less percentage change in the ecosystem with which it is associated. There is not so likely to be a reasonably continuous record of community evolution as there is of population evolution. On the other hand, samples from an evolving community, spaced some time apart, will record proportionately less change in the community than in the average population represented. Thus, there is a better chance of obtaining multiple samples of evolutionary trends of a given magnitude within a community than within a population (Fig. 3-10).

The same principles apply to provincial systems when compared with the average ecosystem. There is little chance of sampling a complete record of a major change in provincial structure. There is, however, a better chance to obtain multiple samples of a trend in provincial evolution than of a trend of proportionate magnitude in community (or population) evolution (Fig. 3-10).

Similarly, the entire biosphere tends to evolve slowest of all these units, but it is commonly possible to obtain so many samples of it (all of our samples represent it) that the course of its evolution may be in no doubt. This is certainly in sharp contrast with the evolution of populations; we do not now know and never will know much of the course of evolution of any population lineage.

Time is not the only dimension along which sample quality varies with ecological level, for it varies in space also. The situation depicted in Fig. 3-10 almost fits the geographic situation as well as the temporal one. If the horizontal scale is considered to represent space at a given time, then the hierarchy of frequencies can be read as a hierarchy of geographic ranges. The biosphere has the greatest range—indeed, it is as great as possible, and in this respect the figure errs, for it should show but one biosphere curve for space. Next in geographic range comes the province, several of which may compose the biosphere; next is the community, and finally the local population. It is possible that some communities will range farther than a province and that some populations will range through more than one community. The average community will, however, occupy only a portion of the province in which it lies, and the average population area will be only a fraction of that of the community. Geographically dispersed samples will miss many populations, fewer communities (which may be represented in several samples), and probably no provinces (which will be heavily sampled), and all samples will represent the biosphere.

Ecological Units, Living and Fossil, Are Polythetic

Another general property of ecological units is that they cannot be defined precisely on the basis of any particular combination of components or even of any single one of their components. Instead, a more flexible approach is required. An important insight into the nature of units of this type can be derived from Beckner's analysis (1959) of a class of biological entities, which we shall call *polythetic* (following Sokal and Sneath, 1963).

Let us imagine a set C of properties $p_1, p_2, \ldots, p_n$. Now, we have a number of samples, each of which contains a large number (unspecified) of the properties in C. Furthermore, each property in C is present in a large number (unspecified) of the samples; however, no property in C is present in all the samples. Nevertheless, each sample will resemble many of the others, since they have many properties in common, although no two samples need be exactly alike insofar as possession of the properties in C is concerned. The samples will have a "family resemblance." Such a class of samples may be called polythetic.

In contrast to polythetic classes are *monothetic* ones, which have special properties necessary for membership. For example, a chemical element may be defined as having atoms whose nuclei contain a given number of protons. Oxygen nuclei contain eight. If a nucleus contains more or fewer than eight protons, it is not an oxygen nucleus. The class of nuclei that comprises oxygen is monothetic.

The lineages of ecological units are either fully polythetic or are parts of polythetic lineages. The nature of each unit will be discussed in later chapters, for each level of the hierarchy has special properties and special problems. If ecological lineages are polythetic as claimed, then to define each unit we must identify a class of characteristic properties, which we shall call the *definitive set*. The occurrence of a certain number of these properties is then required to identify any particular fossil association as belonging to the unit. For most ecological units, the characteristic properties will consist in a number of the major components. For example, a community may be identified on the basis of the presence of a certain number of species populations out of a definitive set of species, but no particular species need be present in order to identify the community.

Some polythetic units are more homogeneous than others; in highly homogeneous units, there will be a high proportion of recurrence of definitive components in multiple samples. In very heterogeneous units, there may be only a few components that recur at all commonly. The fossil records of ecological units include a range from highly homogeneous to highly heterogeneous units. For example, samples from the fossil record of a community at different localities may include numerous characteristic populations in common or they may include few, and unusual samples of some communities may include no populations in common at all, although ordinarily they would both include (different) species from the definitive set. When the composition of a fossil community is not previously known, it is necessary to take numerous samples to establish the composition of the definitive set for units represented by heterogeneous records.

The Ecospace Model May Be Extended to the Taxonomic Hierarchy

The geometric model of environment-organism interrelationships pursued in this chapter has certain advantages that justify the effort spent in its development. One is that the geometric model can be formalized and at least partially quantified so that the ecospace patterns may be analyzed numerically (see Hutchinson, 1957a). We shall not attempt to capitalize on this property here. Another advantage is that it clearly separates the properties of the environment from the properties of organisms, and permits their interrelations to be displayed. Finally, it is a model in which the properties of taxa as well as of ecological units can be conceptualized. These latter advantages are especially significant for our purposes. For these reasons, we shall employ the biospace-ecospace model in preference to the adaptive zone model of Simpson (1944, 1953).

For example, the ecospace of a population has already been defined as the biospace occupied by all its component individuals; it is a geometric representation of the population's environmental interactions, that is, of its niche. The ecospace of a species, then, is simply the combined ecospaces of all its populations. We may examine the species ecospace in an instant of time, or we may look along the time dimension of biospace and examine the temporal ecospace changes displayed by the lineage. Similarly, the ecospace of a genus is composed of the combined ecospaces of all its species. At any time, these ecospaces will usually appear as a partially interpenetrating cluster of hypervolumes. Examined through time, generic ecospace will begin with the niche of the root or founder species and then partially segregate into separate but interpenetrating pathways whenever cladogenetic processes create new lineages. In theory, the ecospaces of families, classes, phyla, and even higher phylogenetic groupings may also be defined in biospace. In practice, of course, we do not actually know enough to do this with precision. If we define all the ecospaces of all the marine taxa at any one time, the resulting structure will be identical with the marine biosphere of that time, and the nested hypervolumes of the ecological hierarchy will appear.

The ecospace of a phylum throughout its existence will somewhat resemble the family tree of that phylum, branching according to its evolutionary history. The fundamental significance of adaptive divergence and radiation is thus verified in ecospace—that is, the morphological divergences reflected in taxonomy represent functional divergences and radiations. It is when dealing with convergent and parallel evolution that the ecospace model proves particularly instructive.

Convergence, it turns out, is nearly as universal as divergence, but this is not easily discovered from taxonomic or morphological work. Except in the case where a lineage is evolving so as to enter a previously unoccupied region of biospace, divergence is always accompanied by some convergence. That is, a given lineage may be radiating from its own root stock as it invades an environment where it was not previously present, but it is simultaneously converging functionally with the lineages that already occupy this environment. The convergence may result in the ecospace of a lineage of one phylum coming to overlap strongly with the ecospace of a lineage in another phylum. Such overlaps are certainly common. Within a shallow marine rocky-bottom community today, for example, representatives of several (six, eight, or more) different phyla are commonly associated as benthic suspension feeders. They share out the habitat space and food supply between them in some way (Chapter 7). The original stocks of many of these phyla were not benthic suspension feeders; the evolutionary pathways that have led these lineages into their present niches represent divergences from ancestral ecospace, but they also form convergences with each other. However, the morphological modifications undergone by different lineages during this convergence may have little in common. Take, for example, mussels, barnacles, and tunicates.

In the same way, functional parallelism may be present within many lineages that are only distantly related, but because the morphological changes are so different in the different taxa it may not be identified. Imagine a gastropod and a crustacean lineage inhabiting the same community and both evolving toward greater trophic specialization, owing to a general environmental change. Their morphological changes may not be very similar; the snail may alter its radular apparatus and perhaps evolve specialized apertural characters that aid in securing some special prey; the crustacean may modify its mandibles and claws. Functionally, these changes represent parallel trends, but this is not easily recognizable.

For that matter, divergences also occur that involve modifications of morphology in very distantly related lineages, and they may go unnoticed, since the correlation of changes in different directions among lineages in different phyla may not seem especially significant, although it may represent an important event ecologically. An example can be based on an invasion of a newly available portion of biospace. Such an invasion could occur by a single lineage, which then radiates to produce a variety of niches. Or it might occur through the invasion of a number of separate lineages, each of which represents a single split in its evolving stock, which produce the same variety of niches. The result of these alternate invasions is much the same so far as the structure of the ecosystems is concerned, but it is not at all similar taxonomically.

It seems likely that at certain times in geological history general ecospace divergence became very common, and at other times it was rare. Also, at times, ecospace partitioning seems to have been common, whereas at other times ecospace appears to have been consolidated by favored lineages. The waxing and waning of these trends have at times involved nearly all the marine taxa simultaneously, whereas at other times the response has been very selective. All these sorts of events, which bring about major changes in the structure of the biosphere both ecologically and taxonomically, most probably correspond to times when biospace is changing, so that evolutionary success requires a new set of strategies. At this point, then, it is suitable to examine the present marine environment, to attempt to understand the sort of biospace with which marine invertebrates must deal.

CHAPTER FOUR

MAJOR FEATURES OF THE MARINE ENVIRONMENT

It is usual to speak of an animal as living in a certain physical and chemical environment, but it should always be remembered that strictly speaking we cannot say exactly where the animal ends and the environment begins. . . .

—Charles Elton, *Animal Ecology,* Sidgwick and Jackson, Ltd., London, 1927.

The environmental diversity of the world is so great that many organisms have little in common. Animals living in the abyssal zone and those inhabiting subtropical deserts possess very different ecospaces, even though they share some basic requirements, such as oxygen. Their limits of tolerance for most environmental parameters do not overlap, and some of their general requirements are qualitatively different. Even if we restrict ourselves to the sea, the environment encompasses so many different conditions and combinations of factors that it is not even possible to describe what is known of the marine environment in any detail.

Fortunately, there are a few physical parameters in the sea that are of outstanding importance to the majority of marine invertebrates. Accounts of these important parameters not only will serve as examples of the ways in which organisms can react to factors, but also will describe a large proportion of the factors that actually limit individual organisms in the seas at present. Furthermore, these factors also play major roles in determining the structures of niches, ecosystems, provincial systems, and the entire biosphere.

The Major Circulation Pattern of the Oceans Is Governed by the Principles of Fluid Motion

The global patterns of many marine environmental parameters are controlled chiefly by the circulation pattern of the ocean, and even many local environmental patterns are strongly affected thereby. The circulation pattern and the principles governing it are therefore of fundamental ecological importance. The major currents are partly due to the action of wind, which drags and pushes the surface waters along, and partly due to density differences in seawater, which chiefly result from temperature (denser at lower temperatures) and salinity (denser at higher salinities) variations.

The primary temperature effect is due to the latitudinal temperature gradient that arises from the unequal reception of solar radiation on the earth's surface (Fig. 4-1). More solar radiation is received in low latitudes than is reradiated, whereas more is radiated at high latitudes than is received. The total

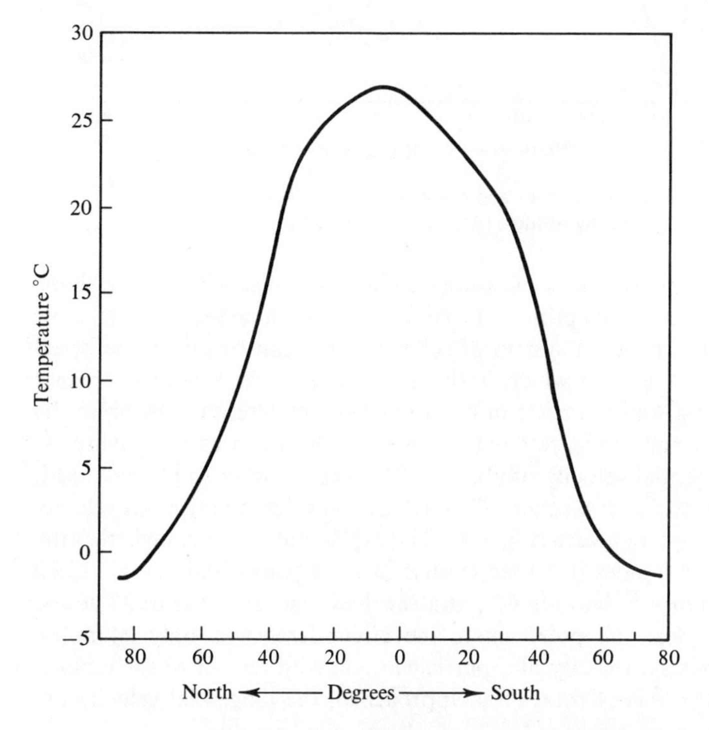

Fig. 4-1 Sea-surface temperatures in the Atlantic, by latitude. (After Wust *et al.*, 1954.)

energy reradiated must equal the amount received if the average temperature of the earth is to remain steady, and heat must flow poleward if the average temperature at each latitude is to remain steady. Variation in salinity is chiefly a function of evaporation (causing high salinities) and precipitation and runoff (causing low salinities) (Fig. 4-2). The combined effects of temperature and salinity produce latitudinal density distributions, such as those in Fig. 4-3.

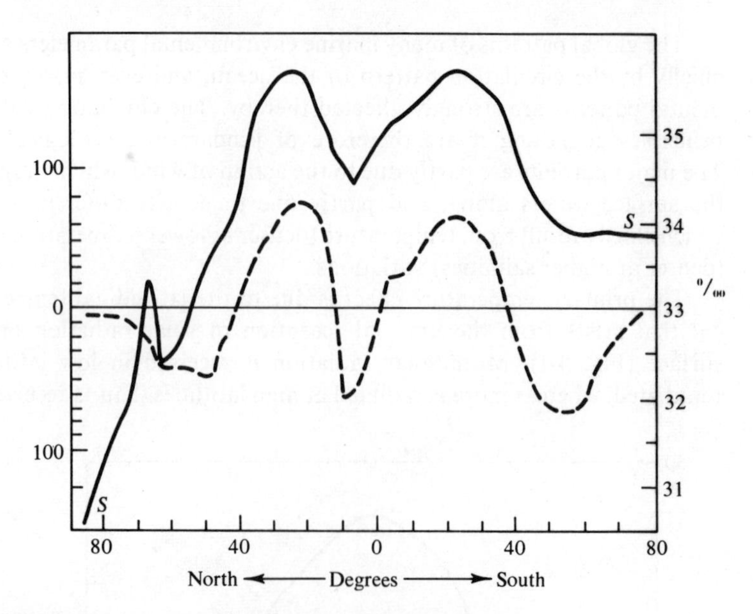

Fig. 4-2 Sea-surface salinity (solid line) and evaporation minus precipitation (dashed line) in the Atlantic, by latitude. (After Wust *et al.*, 1954.)

To understand the dynamics of ocean currents, we must consider the shape and motion of the earth. The primary inertial system in which the currents move is not the earth, but rather a system of reference that can be said to be based upon the fixed stars, and to which both the motion of the earth as a planet and the motion of a water particle in a current must be referred. Owing to the earth's rotation, a particle at rest with respect to the earth at the equator is traveling at a tangential velocity of about 1000 miles per hour (mph) eastward, whereas a particle at rest at latitude 60° north or south has a tangential velocity of only about 500 mph eastward (Fig. 4-4). The explanation is, of course, that the earth is about 24,000 miles in circumference at the equator and about 12,000 miles in circumference at latitude 60°, and revolves eastward once in 24 hours. At some latitude near the pole, where the circumference is only 24 miles, the eastward tangential velocity of a particle at rest with respect to the earth is only 1 mph. As the axis of rotation is approached, the tangential velocity decreases.

In accordance with the law of conservation of momentum, a particle on the

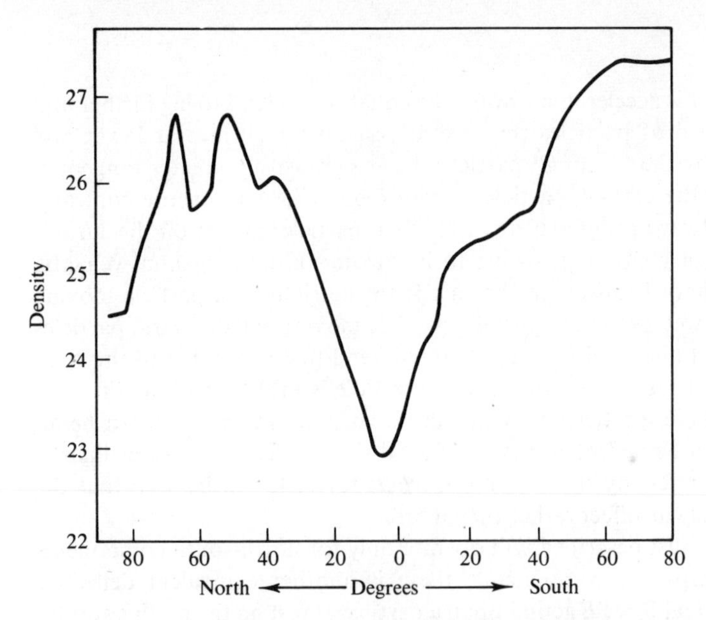

Fig. 4-3 Sea-surface density in the Atlantic, by latitude. (After Wust *et al.*, 1954.)

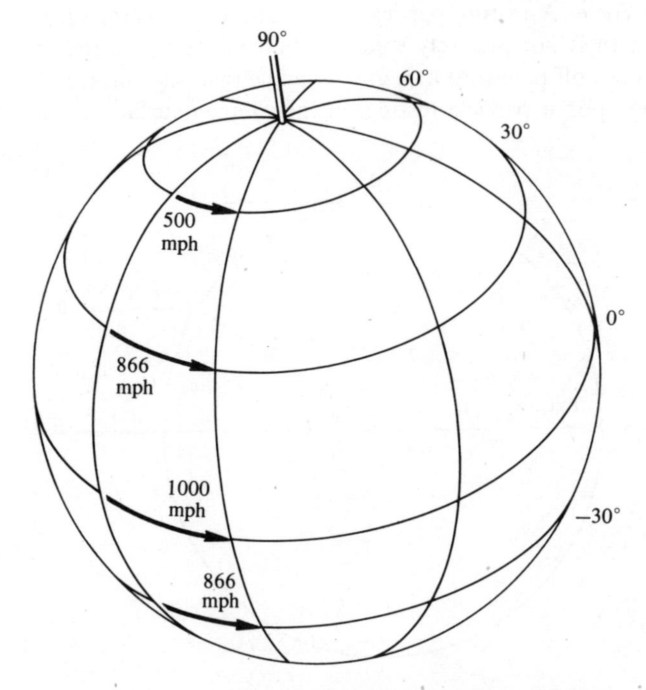

Fig. 4-4 Tangential velocities of particles at rest on the earth's surface. (After Von Arx, *An Introduction to Physical Oceanography,* 1962, Addison-Wesley, Reading, Mass.)

earth's surface that is accelerated toward the equator will tend to lag farther and farther to westward of particles that are at rest on the surface at lower and lower latitudes, for the "resting" particles have successively greater tangential velocities toward the east. A particle moving poleward will, on the contrary, tend to move eastward progressively faster than particles at rest on the surface in successively higher latitudes, owing to its greater initial tangential velocity. Since in the northern hemisphere the east is to the right of a particle moving poleward and the west is to the right of a particle moving equatorward, particles moving across latitudes in either direction will tend to be deflected to the right relative to the earth's surface. Since the South Pole is in the opposite direction from the North Pole, particles moving across latitudes in the southern hemisphere will tend to be deflected to the left. All these deflections are ordinarily accounted for by referring them to the *Coriolis* force. It can be seen that the deflections are really an effect rather than a force.

The deflections just described will operate only for north-south components of motion; for east-west components, there is another equivalent deflective effect. The centrifugal force F acting upon a particle at rest on the earth's surface is directed perpendicularly to the axis of rotation (AB, Fig. 4-5). Therefore, there is an equatorward tangential component of this force (AC, Fig. 4-5), except at the equator itself, the magnitude of which is a function of the angular velocity of a particle. A resting particle does not move equatorward, however, because the earth is not precisely spherical but bulges equatorially; therefore, the surface slopes off poleward at a greater angle than the surface of a sphere. Gravity acting upon a particle resting on this sloping surface has a

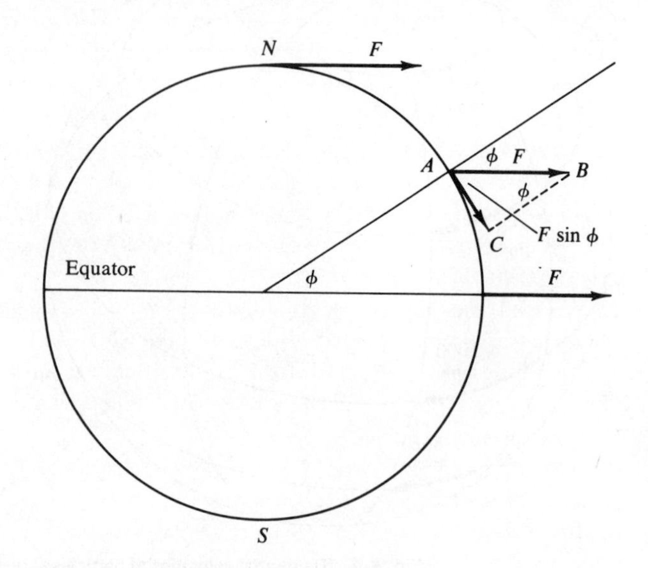

Fig. 4-5 Components of forces acting on a surface current. (After Raymont, 1963, from Harvey, 1928.)

poleward component that balances the equatorward component of the centrifugal force, and the particle can remain at rest. However, when a particle is set in motion across the surface to eastward, its tangential velocity is increased, the centrifugal component is increased, and the particle therefore tends to move toward the equator—to the right in the northern hemisphere and to the left in the southern hemisphere. When a particle moves westward across the surface, its tangential velocity is decreased, the centrifugal component is reduced, and gravity tends to move the particle poleward (right in the north, left in the south).

Thus, the Coriolis effect operates in all directions; the force per unit mass is given by the expression,

$$(2\Omega \sin \phi)u, \quad \text{for east-west motion}$$

$$(2\Omega \sin \phi)v, \quad \text{for north-south motion}$$

where Ω is the angular velocity of the earth's rotation at latitude ϕ, u is the east-west component, and v is the north-south component of the horizontal velocity of a fluid particle. The acceleration of the Coriolis force is very small, being about 0.00015 times the horizontal velocity of a fluid particle on the earth's surface at the poles, and falling to zero at the equator. Nevertheless, the effect is sufficiently great to be a major determinant of the direction of flow of ocean currents, as we shall see.

A particle of water at rest within the ocean is overlain by a column of water and air, which exerts a downward pressure. Beneath the particle is an upward-directed pressure which approaches the downward pressure as the particle decreases in size to a point, so that the pressures are balanced. If surfaces of equal pressures within the water are horizontal, then points on opposite sides of the particle are under equal pressure also, and the entire particle is under a hydrostatic confining pressure that permits it to remain at rest. If, however, the surfaces of equal pressure—*isobaric* surfaces—are inclined, then although vertically directed pressure may keep a particle from sinking directly downward, there is now a component of gravity down the inclined isobaric surface, and the particle will tend to move down the surface. In Fig. 4-6, the *isobars* (lines of equal pressure) are tilted down to the left.

Within the oceans, pressure differences arise chiefly because of differences in water densities. For example, in Fig. 4-6, the water on the right is the less dense, and therefore the overlying column of water and air weighs less and the pressure is less, at any given level, than on the left. The horizontal pressure differences cause an acceleration of water particles.

As soon as the particles move, however, they are influenced by the Coriolis effect and are deflected (toward the right in the northern hemisphere). Instead of moving down the tilted isobaric surface, the particles tend to move horizontally along it, at right angles to the direction of maximum inclination. The deflecting effect is in balance with the acceleration provided by gravity. The greater is the angle of inclination of the isobaric surface, the greater is the accelerating force

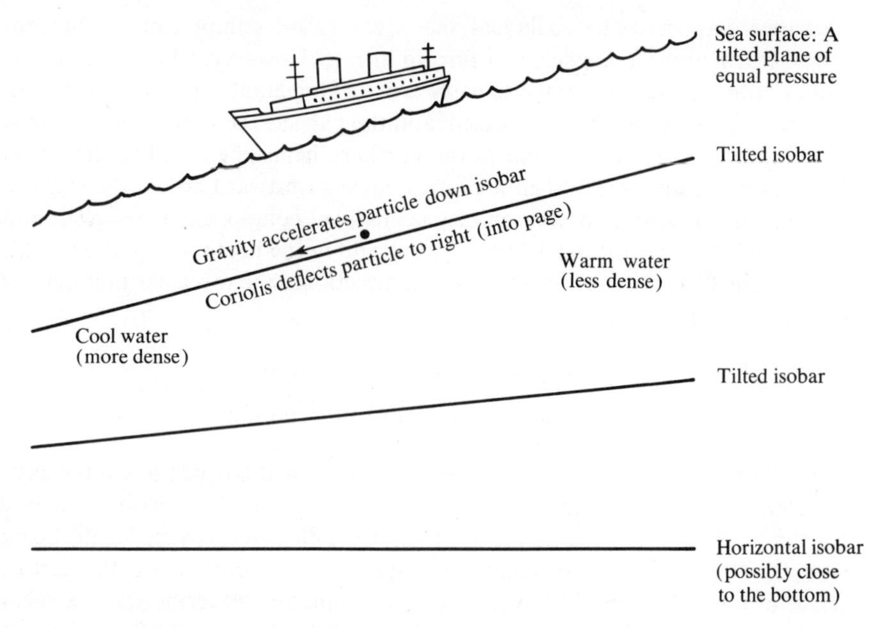

Fig. 4-6 Some aspects of geostrophic flow.

of gravity down the incline, and the greater is the deflective effect of Coriolis, so that the net effect remains a horizontal flow along the isobaric surfaces. In Fig. 4-6, a current would be flowing into the page in the northern hemisphere and out of the page in the southern. Flow of this sort is called *geostrophic flow*.

Perfect geostrophic currents probably never exist in the real ocean, owing to the interference of other forces that we have not considered. Forces of friction, for example, will tend to slow the speed of water particles; thus currents will not be deflected far enough by Coriolis to achieve a perfectly horizontal flow, but will move down an isobaric surface at some low angle to the horizontal. The tendency toward geostrophic flow is nonetheless real enough, and has a major effect in determining the distribution of important ecological factors. For example, the general case for currents that tend to be geostrophic is that light water lies on the right-hand side in the northern hemisphere. Since density differences are usually due to temperature, water on the right of currents is usually warmer than that on the left.

In addition to flowing because of density differences, currents are driven by wind, which pushes water when it blows upon the backs of ripples and drags water along by friction. However, oceanic water is not usually driven directly before the wind. V. W. Ekman calculated that over an idealized ocean (unbounded and of infinite depth and with certain properties constant), a steady wind would drive a surface layer at an angle of 45° to the wind direction; as usual, the deflection is to the right in the northern hemisphere and to the left in the southern. At increasing depths beneath the surface, the velocity of the current decreases logarithmically and the angle of deflection increases, until at some depth the deflection reaches 225° and the current is flowing in a direction opposite to the surface direction (180° difference), and is moving about $\frac{1}{23}$ as fast as on the surface (Fig. 4-7). The depth of the point of current reversal varies with latitude owing to variation in the Coriolis effect, being deepest at the equator and rising toward the poles. In midlatitudes it is near 100 m.

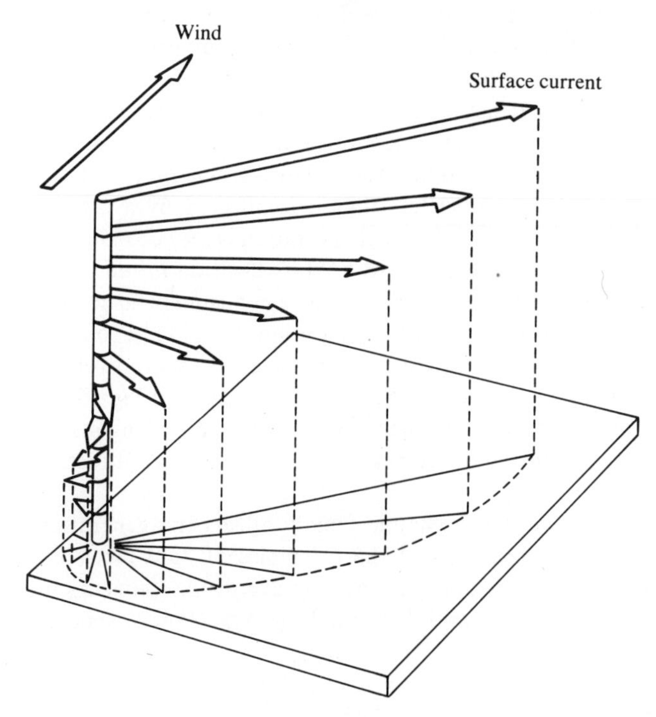

Fig. 4-7 Theoretical course and strength of a wind-driven current with depth; the Ekman spiral. (After Sverdrup, Johnson, and Fleming, © 1942, renewed 1970.)

As the ideal conditions specified in Ekman's calculations do not hold in a real ocean, the actual deflections may depart widely from those of the Ekman spiral. The effects of ocean boundaries, boundaries of other currents, unequal distribution of eddies, changes in water depth, and other factors, commonly cause the deflection to be far less than in the ideal case. A tendency for the deflection to occur has, however, been verified by observations.

Vertical circulation in the ocean is caused chiefly by density differences but is also influenced by winds. The major vertical movements result from the sinking of water of high density and the rising of water of low density. Low-density water is created in low latitudes by heating (and in low latitudes the general vertical circulation is rising) and in all latitudes by rainfall or runoff of water of low salinity. High-density water is created chiefly in high latitudes by cooling, but also results from high evaporation rates and the subsequent increase in salinity at places in middle and low latitudes. Water of high salinity from low latitudes that flows poleward and is cooled at high latitudes is among the densest in the oceans. Sinking in high latitudes, it flows poleward again at depth.

The density, temperature, and salinity of ocean water are thus properties that result from surface processes; water is said to be "conditioned" at the surface. Mixing of different waters in various combiantions may produce water with intermediate characteristics. When relatively dense water is created, it will sink to depths where equivalent densities are found, and will thereafter move laterally. These movements will displace other water at depth; at the surface, there will be an inflow to replace the sinking water. Thus, a general circulation pattern is created. Incidentally, when water sinks it is affected by a Coriolis "force," for its distance from the axis of the earth's rotation is reduced and thus it need not have as high a tangential velocity to hold its position in relation to the earth. It will be deflected, to the right in the northern hemisphere, as it sinks, so that when we look down on a descending water column, it will tend to rotate clockwise (the "anticyclonic" direction). Rising water has less tangential velocity than is required to hold its place relative to the earth's surface, so it lags, being again deflected to the right in the northern hemisphere; since it is rising instead of falling, it will tend to rotate counterclockwise (the "cyclonic" direction) when viewed from above.

A number of less important hydrodynamic factors affects the general circulation pattern of the oceans. Accounts of the principles of fluid mechanics that are appropriate to paleoecology are presented by Sverdrup, Johnson, and Fleming (1942, especially Chaps. 12 and 13) and Von Arx (1962, especially Chap. 4).

The Currents and Water Masses of the Oceans Provide a Framework for the Distribution of Environmental Parameters

The same principles of fluid motion that affect the circulation of the oceans govern the circulation of the atmosphere. Most of the energy in the form of solar radiation that penetrates the atmosphere reaches the ocean or land surface; some is absorbed there, but most is radiated back into the atmosphere, which is thus chiefly heated from below. There is a vertical temperature gradient as well as a latitudinal gradient, and, in general, the warmest air is found at the surface near the equator and the coolest at high altitudes near the poles.

Warm equatorial air rises and then flows poleward, owing to vertical and latitudinal pressure gradients, respectively. Some of this air descends toward the surface near latitude 30° and flows equatorward to replace the rising air, forming in each hemisphere a cell of *meridional* (north-south) *circulation*, termed the *Hadley cell.* Naturally, the surface winds beneath the Hadley cell are deflected by the Coriolis effect and are easterlies (flowing in a westward direction), the trade winds of the sailor (Fig. 4-8). Poleward of the Hadley cell are zones of generally westerly winds (flowing eastward), stretching roughly from

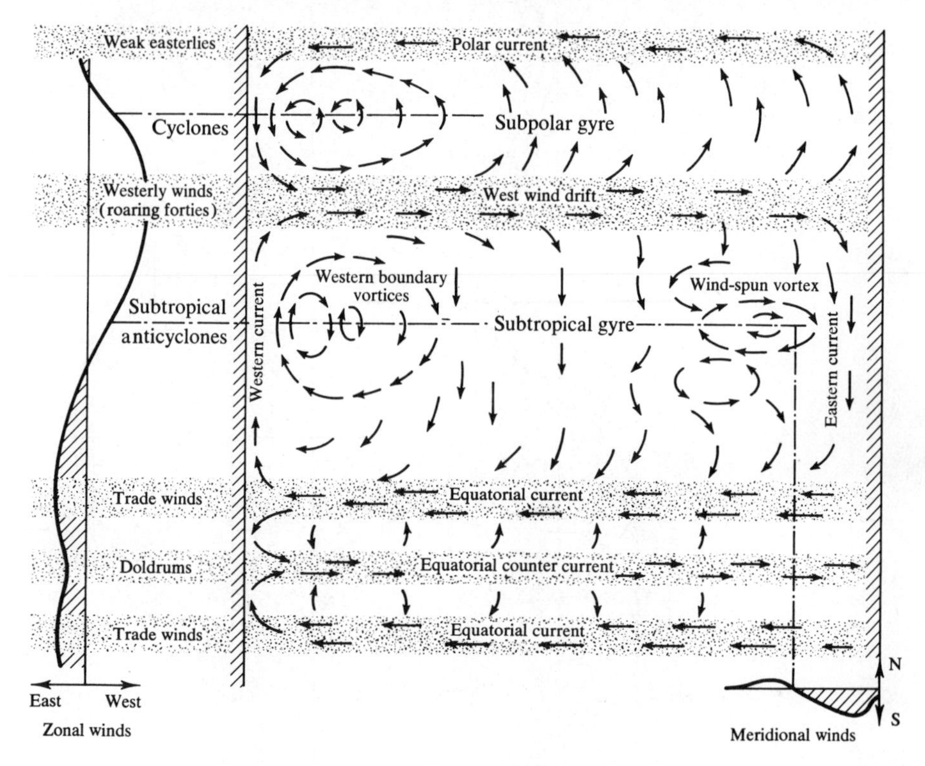

Fig. 4-8 Model of surface circulation in a rectangular ocean, with winds as indicated. (After Munk, 1950.)

30 to 60° latitude in each hemisphere (Fig. 4-8). These westerlies have been assumed to be related to meridional cells, called the *Ferrel cells,* which were postulated to contain poleward air rising near 60° latitude and equatorward air descending near 30° latitude, producing the appropriate surface wind direction when the Coriolis effect is considered. The existence of this cell has not yet been verified, however, and it is possible that the westerlies are associated with a zone wherein winds at all altitudes tend to flow poleward due to a strong temperature gradient between latitude 30° and the polar front, and are deflected to the east by the Coriolis force. Poleward of 60° latitude, prevailing surface winds are easterly (Fig. 4-8), but the form of the meridional movements is not clear.

Major unsolved questions still exist about the general circulation of the atmosphere, but the effects of the observed surface winds on ocean currents are well known (Fig. 4-8). The *zonal* (east-west) *circulations* of the trades, the wester-

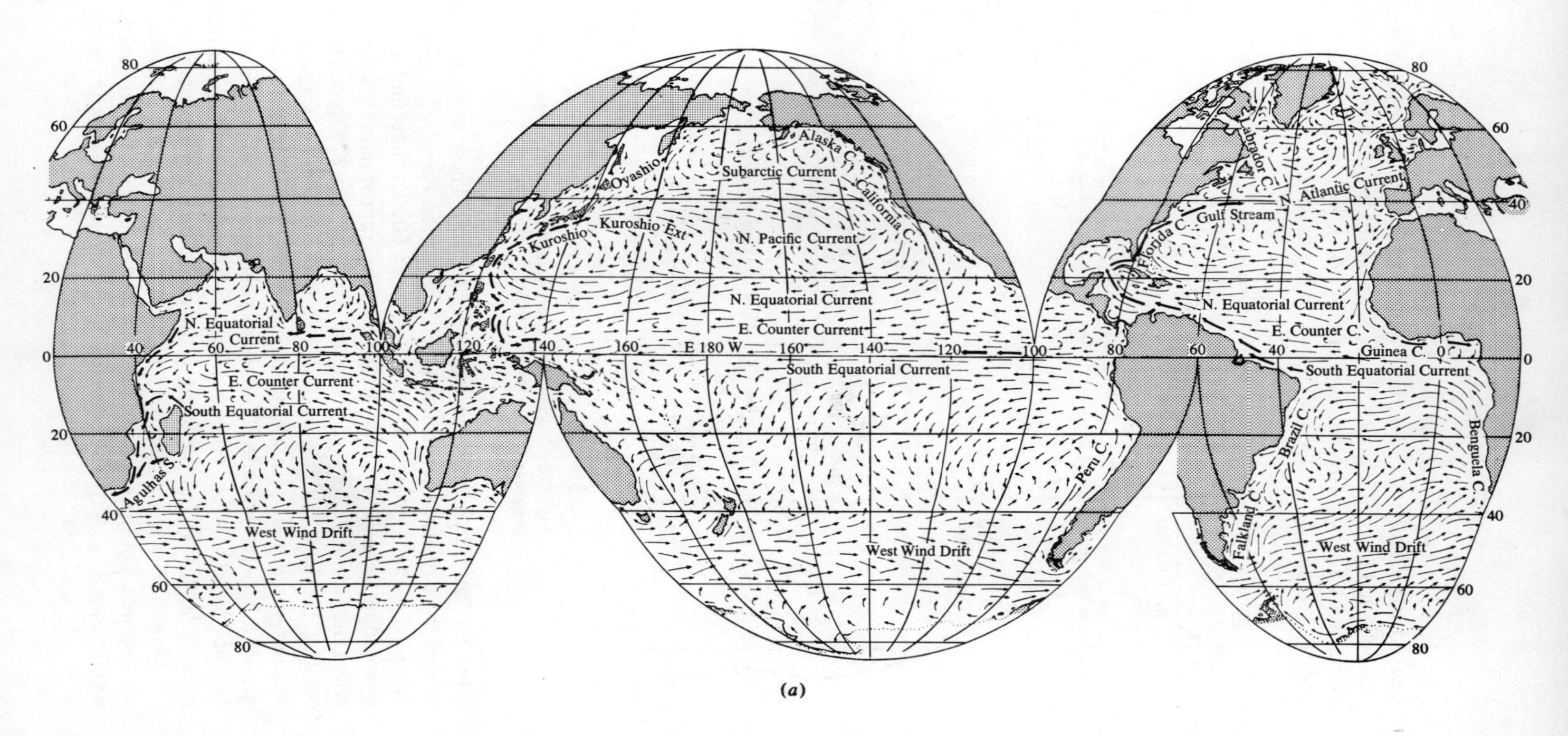
N. Equatorial Current
E. Counter Current
South Equatorial Current
Agulhas S.
West Wind Drift
Oyashio
Kuroshio
Kuroshio Ext
Subarctic Current
Alaska C.
California C.
N. Pacific Current
N. Equatorial Current
E. Counter Current
E 180 W
South Equatorial Current
West Wind Drift
Peru C.
Labrador C.
N. Atlantic Current
Gulf Stream
Florida C.
N. Equatorial Current
E. Counter C.
Guinea C.
South Equatorial Current
Brazil C.
Benguela C.
Falkland C.
West Wind Drift

(a)

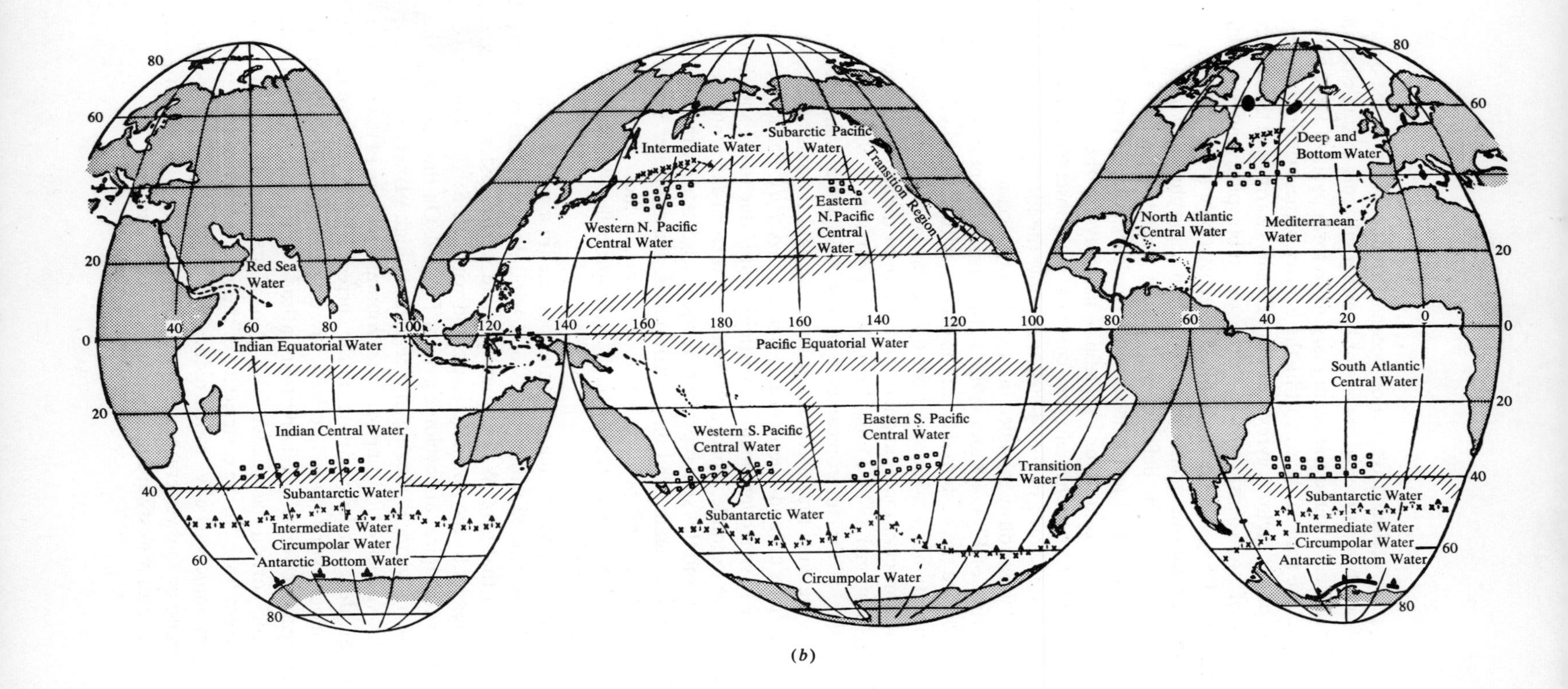

Fig. 4-9 Surface waters of the oceans. (*a*) The general circulation pattern in February and March. Note the similarities to the theoretical pattern in Fig. 4-8, with modifications related to the shapes and sizes of the ocean basins. (*b*) Water masses, boundaries approximate. Arctic or Antarctic Intermediate Waters sink along the lines of crosses; Central Waters are formed at regions marked by squares. Compare with the current pattern. (After Sverdrup, Johnson, and Fleming, © 1942, renewed 1970; base map copyright by the University of Chicago, Department of Geography.)

lies, and the polar easterlies drive major ocean currents. Most of the ocean currents are confined by continental masses or by other currents to single ocean basins, and are deflected around the basin margins to form parts of large gyres [Figs. 4-8, 4-9(*a*)]. The spin of the earth tilts the density planes within the gyres so that the centers of anticyclonic gyres are of low density (warm) and the centers of cyclonic gyres of high density (cool)—as in the subtropical and subpolar gyres, respectively, of Fig. 4-8—according to the geostrophic principles. Water tends to drift to the right of currents (in the northern hemisphere) owing to the principle of Ekman, and to be deflected with depth from the path of surface motion in accordance with the Ekman spiral. However, the current flow is limited and channeled into more or less definite pathways by the geometry of the ocean basins and the presence of other currents. The observed currents of the present oceans are shown in Fig. 4-9(*a*); note how the ideal scheme of circulation in Fig. 4-8 is variously modified in each separate ocean.

A striking feature of Figs. 4-8 and 4-9(*a*) is the displacement of the centers of gyral motion toward the western sides of the oceans, with strong narrow western boundary currents but weaker and more diffuse eastern boundary currents. This is due to the conservation of angular momentum of the water on opposite sides of the gyres (Stommel, 1948; Munk, 1950). Water particles moving poleward on the western side of the subtropical gyre in Fig. 4-8 have a component of velocity owing to the wind that drives the particles around the axis of the gyre. Circulation of particles around a vertical axis is termed *vorticity*. Particles on the earth's surface may have a component of vorticity that is due to the spin of the earth. At the pole, for example, a propeller fixed on a vertical axis would revolve once every 24 hours, and this spin represents vorticity associated with the earth's spin. At the equator, if again fixed on a vertical axis, the propeller would not spin at all around the earth's pole, to which its axis is, in fact, perpendicular. The axis does revolve around the earth, but the propeller has no vorticity associated with the earth's axis.

Now, instead of a propeller, imagine water particles in a gyre with a vertical axis. The vorticity associated with the axis of a wind-driven gyre is symmetrically disposed about that gyre. Water on eastern and western sides of the gyre, however, will differ as to the vorticity associated with the earth's spin. Take, for example, the subtropical anticyclonic gyre of Fig. 4-8. On the west side, the the poleward-moving water is moving into regions of high planetary vorticity, where the cyclonic vorticity of resting particles (counterclockwise in the northern hemisphere) is greater; vorticity of water in the current lags farther and farther "behind" the cyclonic vorticity of the local "resting particle." The current is picking up anticyclonic vorticity, relative to its latitude, as it proceeds. This adds to the anticyclonic vorticity associated with spin of the anticyclonic gyre and enhances the current flow. On the eastern side of the gyre, water moving equatorward picks up cyclonic vorticity, which reduces the equatorward anticyclonic flow. Strengthening of the western boundary currents narrows the western side of the gyre and causes a westward displacement of its center.

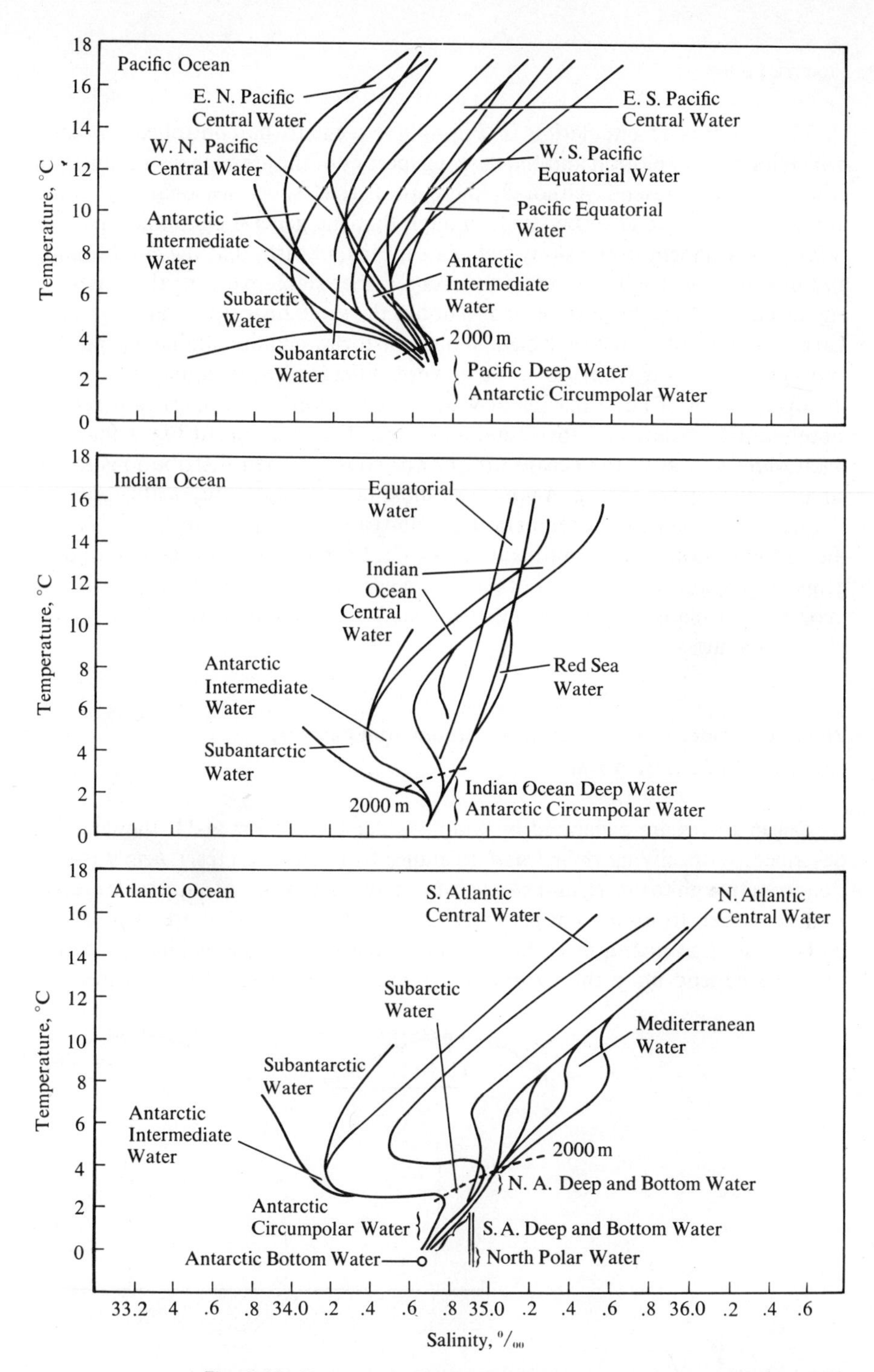

Fig. 4-10 Temperature-salinity relations of the principal water masses of the Pacific, Indian, and Atlantic Oceans. (After Sverdrup, Johnson, and Fleming, © 1942, renewed 1970.)

The patterns of circulation in Fig. 4-9(*a*) seem to be controlled by basic principles of fluid motion and the basic properties of the lithosphere, oceans, and atmosphere at present. Although these properties show seasonal and other fluctuations, they tend to be stable enough that the general circulation pattern is rather permanent in the short run. This means, in turn, that the conditioning and mixing of water in the ocean follows a rather stable pattern; the waters in any given gyre have been subjected to about the same histories, year after year. Large volumes of water that have similar histories acquire similar properties, and are marked off from other water with different histories and properties; the distribution of large masses of water of characteristic properties is related closely to the patterns of horizontal and vertical circulation. In Fig. 4-9(*b*), the main *water masses* of the oceans are shown; compare their distributions to the circulation patterns in Fig. 4-9(*a*). The most characteristic physical properties of each water mass are the temperature-salinity relations (Fig. 4-10). In addition, the concentration of certain dissolved gases, salts, and other substances, of oxygen isotopes, and the values of other parameters are characteristic in each water mass so that many environmental parameters may change markedly at water-mass boundaries.

Waves and Tides Are Significant Environmental Factors, Especially in Coastal Water

Ocean waves are generated by wind. As depicted in Fig. 4-11, they may be described by specifying their *length* (distance from crest to crest), *height* (elevation from trough to crest), and *period* (time between passage of successive crests). In nature, their forms are not perfectly sinusoidal, however, but are variable and complex so that such a description is incomplete. Both the height and period of waves depend upon the nature of the wind that generated them, especially

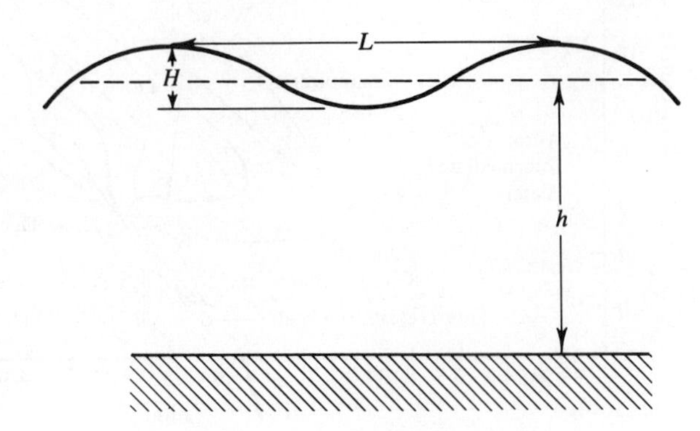

Fig. 4-11 A wave of height H and length L in water of depth h. (After Inman in F. P. Shepard, *Submarine Geology*, 2nd ed., 1963, Harper & Row.)

its velocity, the distance over which it blew (the fetch), and the length of time that it blew.

In regions where wind traction is generating waves, the ocean surface usually has a choppy appearance called *sea*. Upon running out of the area of generation, the longer, faster waves segregate as *swell*. The wave form is propagated through the water but is accompanied by only a small net water movement, a *wave drift* in the direction of propagation. Water particles are chiefly displaced through a circular orbit as each wave passes, returning nearly to their starting points (Fig. 4-12). Waves commonly travel thousands of miles from their generating

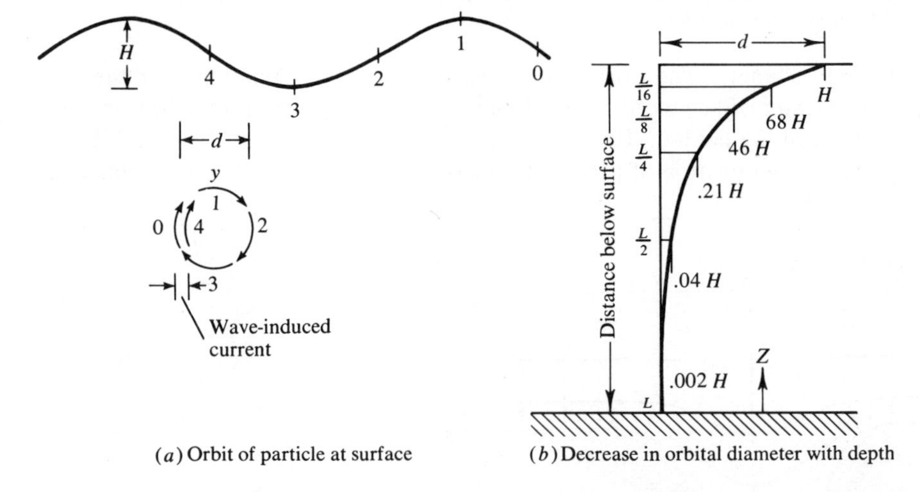

(*a*) Orbit of particle at surface (*b*) Decrease in orbital diameter with depth

Fig. 4-12 Theoretical orbital velocity of a wave traveling over deep water. This type of wave is called an *Airy wave*. Note (*a*) that the diameter of an orbiting particle at the surface equals the wave height, but that the diameter decreases with depth (*b*). (After Inman in F. P. Shepard, *Submarine Geology*, 2nd ed., 1963, Harper & Row.)

areas in deep water. Upon approaching shallow water, the wave velocity is checked by drag as the waves "feel bottom" at a depth of one-quarter the wave length. As the waves slow down, a number of processes occur. The most spectacular involves a decrease in wave length as the crests of following waves, still moving faster than preceding ones, narrow the distance to the crests of the waves in front. This compresses the wave energy formerly distributed over a longer wave length into a smaller region; as a result, wave height increases. The slower, shorter, higher waves resulting from this transformation retain the same period as in their deep-water states, however. Finally, waves break at or near the shore; their final form depends upon their original form, upon the steepness of the shoaling, and upon local winds (Fig. 4-13).

When wave velocity is altered in shallow water, wave fronts will be refracted in the same manner that, in optical physics, a light wave is refracted when its

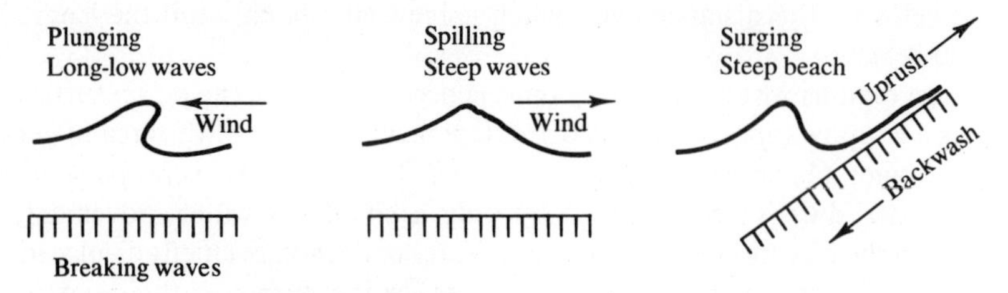

Fig. 4-13 Variation in the forms of breaking waves according to wind direction and beach steepness. (After Inman in F. P. Shepard, *Submarine Geology,* 2nd ed., 1963, Harper & Row.)

velocity changes. The portion of a wave front that is running over the shallowest water will be more impeded than neighboring portions over deeper water, and it will lag behind. This causes a deflection in the wave front, and, consequently, the direction of propagation along the deflection is altered (Fig. 4-14).

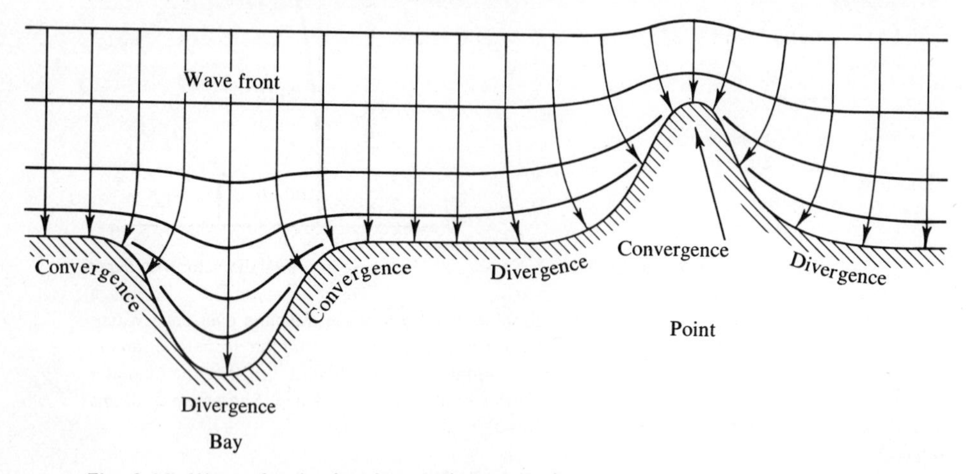

Fig. 4-14 Wave refraction into bays and around points.

When waves approach points or headlands that extend seaward as submarine ridges, they are refracted inward so that wave energy is concentrated upon the points. In embayments that extend seaward as submarine troughs, waves are refracted outward so as to distribute their energy around a longer shoreline, and thus there is less energy expended per unit length of shore. The pattern of wave energy received along a shoreline therefore depends in great measure upon the coastal configuration and bottom topography (Fig. 4-14).

As waves approach shore, the net amount of water that is translated increases so that an excess of water is brought inshore. The excess is moved in a *longshore current* that parallels the shore, flowing in the surf zone. The direction of the current is determined by the direction of approach of the wave front to the shore (Fig. 4-15). A longshore current cannot relieve the shoreward buildup of wave-drifted water, however, unless it were to accelerate indefinitely; instead, the water is discharged seaward at intervals along the shoreline in *rip currents* (Fig. 4-15). Inman gives a concise account of the major features of ocean waves and associated currents (in Shepard, 1963, Chap. 3).

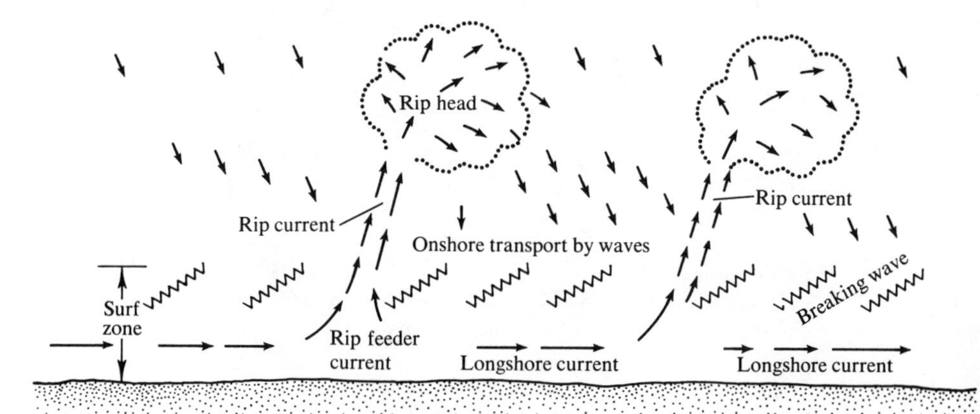

Fig. 4-15 Major features of the nearshore water transport system. (After Inman in F. P. Shepard, *Submarine Geology,* 2nd ed., 1963, Harper & Row.)

Ocean tides are generated by the gravitational attraction of the moon and sun. The moon, close as it is to the earth, produces over twice as much effect as the sun. The sun's attraction reinforces or opposes the moon's attraction as the positions of these bodies change relative to the earth. The tidal effects arise because the net attraction is strongest at the point on earth nearst the moon and is weakest at the point opposite. Only at points 90° from the moon's position is the attractive force balanced by the centrifugal force associated with the revolution of the earth-moon system. The differential attraction creates an elongation of the earth's surface along a line connecting such points, but the relatively rigid crust responds far less than the mobile oceans, which rise and fall relative to the crust. The surface is highest directly beneath the moon, where the attraction is greatest, and also directly opposite it, where the attraction is least; therefore, the surface at that point is least affected, and, under the influence of centrifugal force, it remains higher than points nearer the moon. The sun strengthens the tides most when it is directly over the moon or directly beneath it, and weakens tides most when it is 90° from the moon. The strengthened tides are called *spring tides* and the weakened ones *neap tides*.

Tidal rise in water level is accomplished by horizontal water motion. Water flows toward the points beneath or opposite the moon, which are constantly moving around the earth (Fig. 4-16). The greatest tractive force is exerted 45° from the points beneath and opposite the moon. Ordinarily two high tides would be expected for each lunar day, one associated with the point beneath the moon and the other with the point opposite. However, the irregular configurations of ocean basins and shorelines, of bays and estuaries and inlets, and the dis-

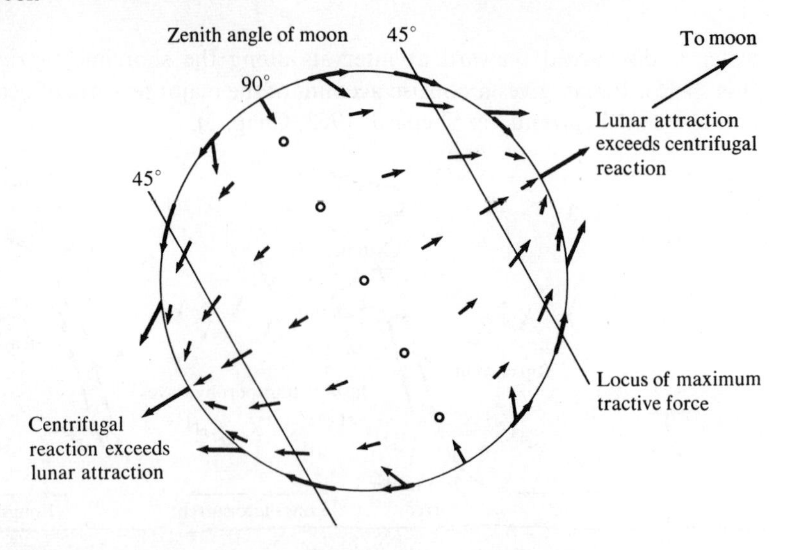

Fig. 4-16 Interrelations between the attractive force of the moon and the centrifugal force from the earth's rotation on water particles on the ocean surface. (After Von Arx, *An Introduction to Physical Oceanography,* 1962, Addison-Wesley, Reading, Mass.)

position of barriers to tidal motion, all create tidal effects that cannot be precisely accounted for as yet, owing to their complexity. They are predictable from empirical observations, of course. Some regions have only a single tidal cycle per lunar day, as at places in the Gulf of Mexico, in the Okohstk Sea, and in the Maylasian-Indonesian region (Doty, 1957, Pl. 1). The Mediterranean and Black Seas are almost tideless, whereas in the Bay of Fundy the tidal range reaches 50 feet. As the sun-moon-earth relations are constantly shifting, the tides follow very complicated cycles. Some tidal types are illustrated in Fig. 4-17. Causes of variations in tidal cycles are discussed by Doty (1957).

It is at the coastline that effects of wave and tide are most important ecologically. Both cause turbulence and local variation in sea level, and tend, in general, to enhance the variability of the environment. The tide especially causes periodic currents that run in alternatively opposite directions, often bringing waters of markedly different properties with them and covering and uncovering the littoral zone. Much of the fossil record is found in rocks bearing evidence of wave and tidal action.

Nearshore Circulation Patterns and the Distribution of Characteristic Local Waters Provide a Local Framework for Environmental Parameters

Over continental shelves, within semienclosed basins, and in general along the margins of continents, hydrographic effects are encountered that have a profound influence on the local or regional distribution of environmental param-

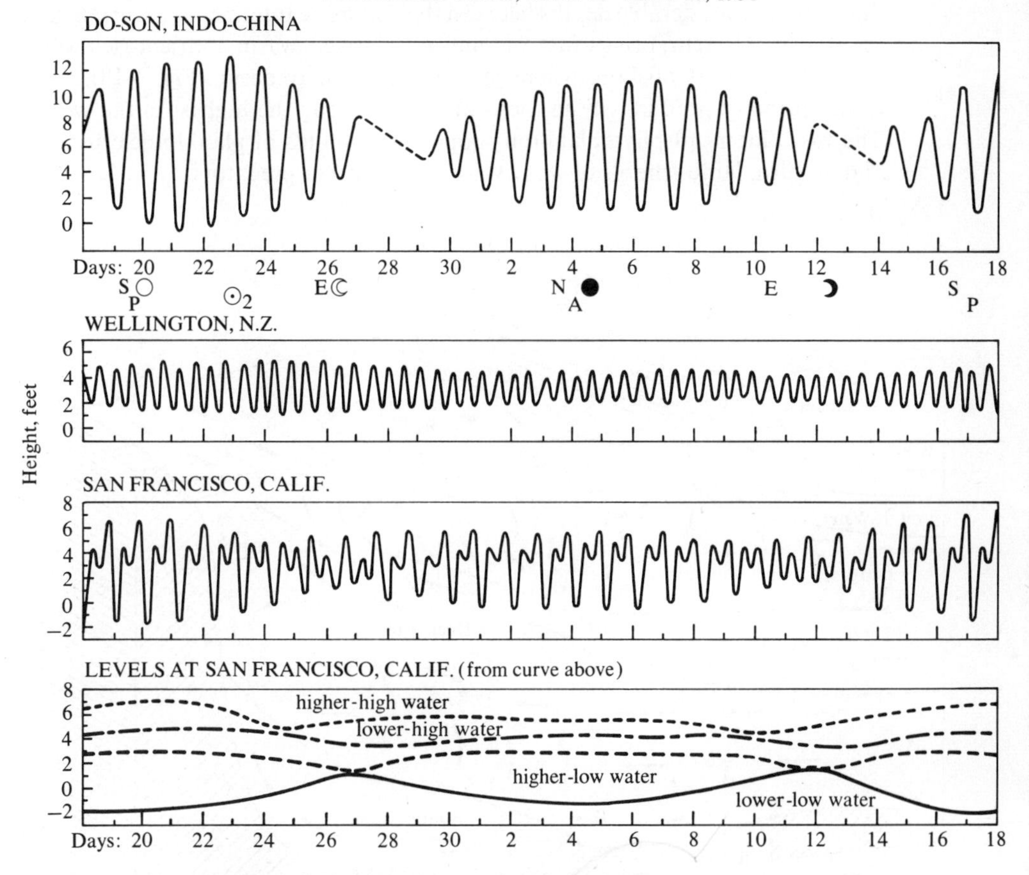

Fig. 4-17 Three different types of tides, plotted for 30-day periods. (Compiled by Doty, 1957, from U. S. C. & G. S. Tide Tables.)

eters. Among the broadest scale of these effects are those that involve runoff of fresh water from drainage basins or ice masses into the sea, thereby diluting the ocean water and altering the density patterns.

A striking case occurs around Antarctica, where surface water is diluted by meltwaters and thus, although cold, remains relatively light. The surface water drifts northward and eastward away from the continent, remaining on the surface until reaching about latitude 50° south, where it meets and descends beneath a lighter southward-drifting water layer at the Antarctic Convergence (Fig. 4-11); after the descent, it is known as the Antarctic Intermediate Current. Below the surface off Antarctica, water of higher salinity is cooled and becomes quite dense, sinking down the continental slope as the Antarctic Bottom Current, and moving eastward around the continent, emitting branches which flow northward into the major ocean basins. With both surface and bottom currents flowing away from the continent, there must be a replacement of water through a compensatory current flowing toward the continent. Such a current, which is drawn from considerable depths between the Antarctic Intermediate Current and the Bottom Current, brings vast amounts of relatively warm, nutrient-rich water to shallow depths offshore where photosynthesis may occur (Fig. 4-18). As a result, primary productivity in these waters is among the highest in the world. The vast blooms of phytoplankton support rich stands of planktonic crustacea and, in turn, large schools of whales, fishes, and other predators.

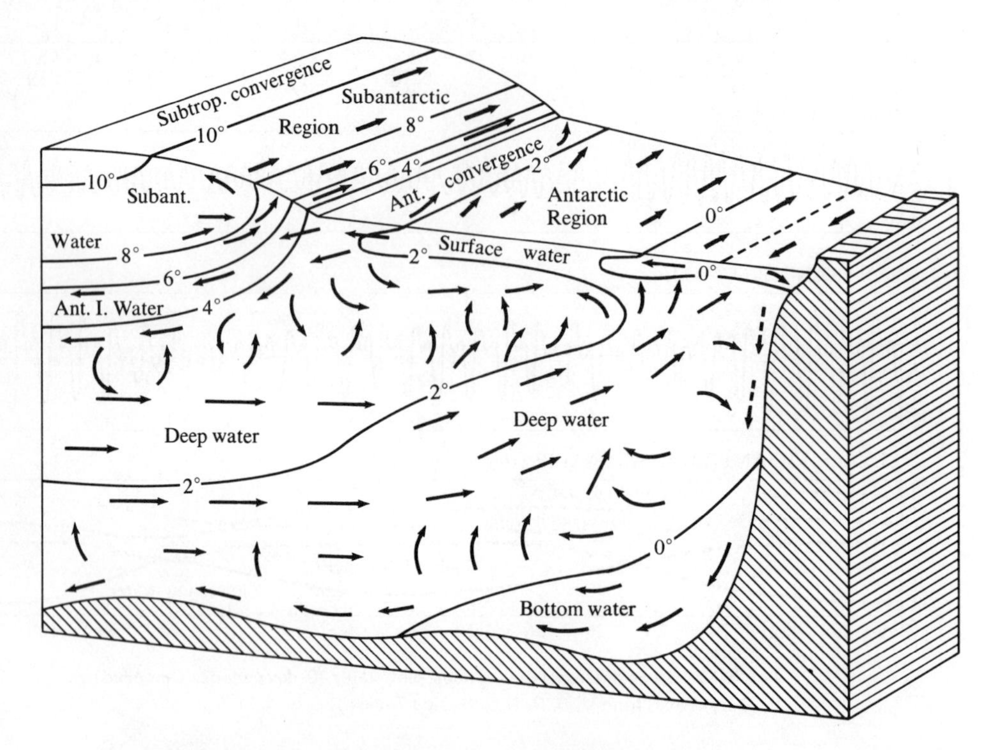

Fig. 4-18 Circulation along Antarctica. (After Sverdrup, Johnson, and Fleming, © 1942, renewed 1970.)

Salinity effects are especially strong in partially enclosed basins, such as the Arctic Ocean and the Baltic Sea, where there is a large drainage from adjoining land masses so that surface waters of characteristically low salinity are found. In the Mediterranean and Black Seas the effects of both excessive runoff and excessive evaporation are displayed. On its western margin, the Mediterranean opens into the Atlantic Ocean through the narrow and shallow Strait of Gibraltar; in the east, it connects with the Black Sea through the narrow and shallow Dardanelles and Bosporus (Fig. 4-19). Within the Mediterranean, evaporation greatly exceeds precipitation and runoff, so that there is a net loss of water, and the salinity of Mediterranean water is raised above that of nearby Atlantic

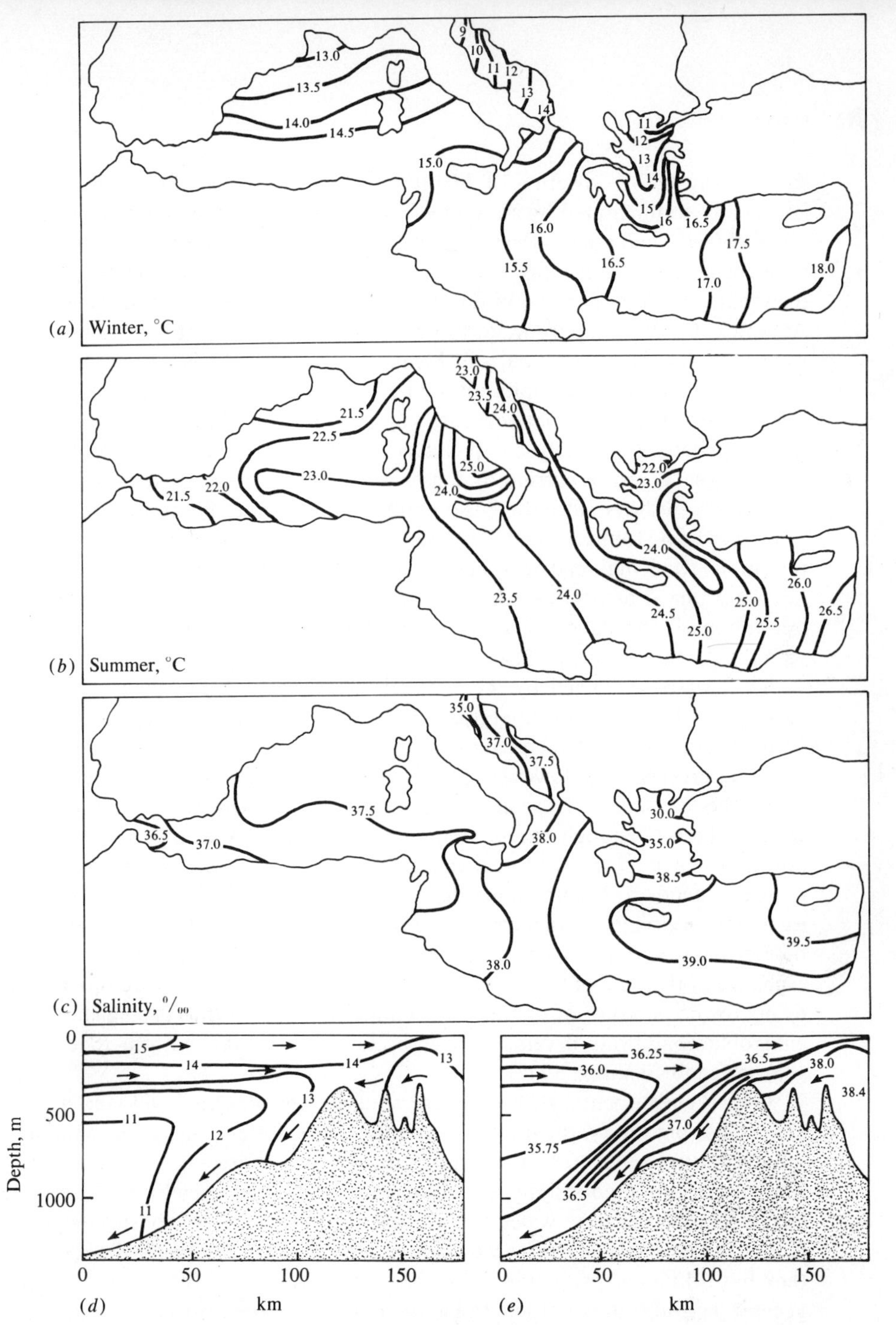

Fig. 4-19 Some aspects of Mediterranean oceanography. (*a*) Winter surface temperatures. (*b*) Summer surface temperatures. (*c*) Surface salinities. (*d*) Temperature profile across the Strait of Gibraltar. (*e*) Salinity profile across the Strait of Gibraltar. [(*a*), (*b*), and (*c*) are from Bruns, 1958, after Raymont, 1963; (*d*) and (*e*) are after Sverdrup, Johnson, and Fleming, © 1942, renewed 1970.]

surface water. The lowering of Mediterranean sea level through evaporation is partially compensated by an inflow from the Atlantic; the inflowing water is relatively light and crosses the straits as a surface current. Saltier Mediterranean water beneath this current is denser than Atlantic water at equal depths outside the straits, and therefore flows out over the sill and spreads out in the eastern Atlantic, at a depth appropriate to its density, a great fan of warm but saline water (see Sverdrup, Johnson, and Fleming, 1942, pp. 642-49).

In the Black Sea, precipitation and runoff exceed evaporation, so there is a buildup of water of relatively low salinity that flows out into the Mediterranean, where it partially compensates for the loss of water there through evaporation. A return current of denser water flows into the Black Sea beneath this surface outflow, since there is a gradient of lesser density toward the Black Sea (Sverdrup, Johnson, and Fleming, 1942, pp. 649-51). Thus, the Mediterranean is receiving relatively light water in surface currents and returning heavier water at depth at both ends; from the exchange, it makes a net gain in water mass, to balance the excessive evaporation (Fig. 4-19). The interchange with the Atlantic is by far the most important.

The water exchanged by the Mediterranean Sea is more than enough to balance the loss due to evaporation. The excess is due to the replacement required for the outflow generated by the horizontal gradients of density (higher in the Mediterranean and lower beyond the entrance straits) that are found at each end of the sea. This high exchange provides for a replacement of Mediterranean waters; there is enough excess at present to completely change the Mediterranean water in 75 years. Moreover, Mediterranean surface water, when cooled in winter, tends to become quite dense where saline and sinks, thus providing for a vertical circulation that aerates the bottom. On the other hand, the exchange by the Black Sea just balances the excess precipitation and runoff. There is no exchange of the bulk of the water, which furthermore has a stable density profile owing to the presence of lighter fresher water above and thus is stagnant, with anaerobic conditions developed over broad areas of the bottom (Sverdrup, Johnson, and Fleming, 1942).

Over the continental shelves, where water depths are small and land masses are in close proximity, the properties of the oceanic water masses are commonly modified into local waters of unique characteristics. Waters of some characteristic temperature and salinity are called *water types*, and sometimes these local waters are referred to as water types. These shallow wedges of coastal water are collectively termed the *neritic zone*, not because they are all similar, but because they have different properties from the shallow waters of the open ocean. In general, neritic waters are less saline than oceanic waters owing to runoff, but locally excessive evaporation raises the salinity; even the drainage may be from saline reservoirs, and therefore the runoff may raise salinity (as in parts of Florida Bay; Lloyd, 1964). Temperatures in neritic waters are sometimes significantly affected by runoff, such as when there are massive infusions of cold meltwaters. Temperatures are also raised by solar radiation in shallow areas where the substratum reradiates heat; thus, the water column may be heated from below as well as from above. The ionic composition of neritic water is occasionally unusual owing to peculiarities in the composition of water in adjoining drainage. Many of these effects are seasonal or at least episodic, and thus neritic waters are markedly more variable than oceanic waters; variability may itself be the most important single characteristic of the neritic zone.

The waters over the shelf are subjected to a large number of mixing processes, many of which are episodic and enhance the variability of the neritic environment. Rising waters in the oceans are termed *upwelling*; we have noted an especially large-scale upwelling around Antarctica. Upwelling occurs at divergences, where currents flow away from other currents or from land. Upwelling is especially notable along the west coasts of continents—the eastern boundaries of oceans. There, water is flowing equatorward with the major ocean currents, and migrates seaward owing perhaps to the effect of the Ekman spiral and to the Coriolis parameter. Coastal upwelling is required to compensate for this loss. The upwelling water is denser and usually cooler than surface water, thus adding to the temperature and density difference across the currents. This generalized sort of upwelling does not seem common along east coasts, perhaps because of the high velocity of western boundary currents.

Upwelling also occurs locally where surface waters are blown offshore by winds. This is very common along western coasts and is also known off eastern ones. Even short intervals of high offshore winds, such as the "Santa Ana" of southern California, will drive surface water off to be replaced by upwelling water. Where points of land lie athwart the prevailing winds, upwelling may be of more or less regular occurrence to leeward. Such is the case near the Channel Islands off southern California (Fig. 4-20), where persistent northerly winds blow surface waters to the south, and a patch of cool upwelling water is located semipermanently in the lee of the islands and of Point Conception. Since this patch of cool water is dense, isobars in the region are tilted toward this dense water and a counterclockwise current flows around the dense center in accordance with the principles of geostrophic flow. This current brings warm surface waters from the right-hand side of the ocean current [the California Current, Fig. 4-9(*a*)] inshore, where they create a high temperature gradient, from cool north of Point Conception to warm to the south (Fig. 4-20). This temperature change forms an important barrier to the distribution of shelf biota. A more local but spectacular situation occurs at Punt Banda, Baja California, Mexico, an east-west point that protrudes into a southerly wind drift. On the north side of the point, warm surface waters pile up; on the south side, where moderate depths occur close to shore, the southerly drift of surface water results in upwelling of cold water. The result is a surface temperature contrast of over 12°C at times in summer in less than 2 miles. Upwelling is certainly of major biological importance, not only because of the temperature effects but especially because the rising waters commonly bring nutrients into shallow lighted waters to form the basis of high organic productivity.

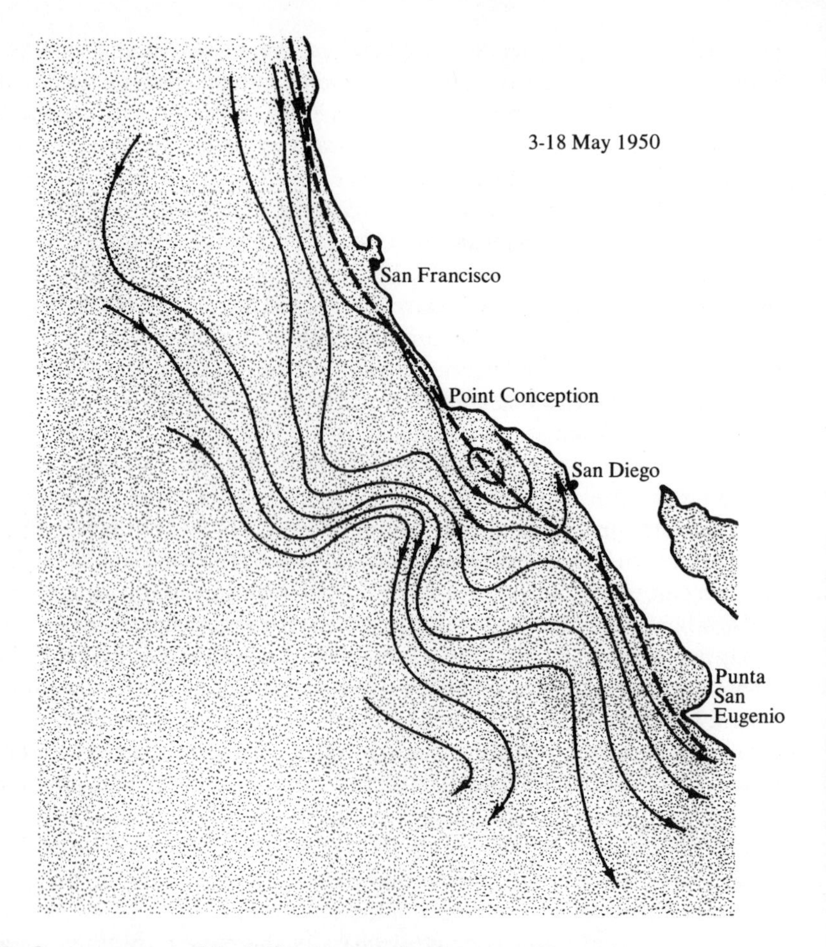

Fig. 4-20 Current patterns off the California coast, showing the semipermanent gyre south of Point Conception with associated drift of offshore waters to inshore regions of southern California. Geostrophic principles indicate that the water in the center of the gyre is more dense (cold) and the water along shore is less dense (warm). (From Marine Research Committee, 1953.)

Downwelling occurs at convergences in the open ocean or over the shelf; downwelling is powered from the momentum of the currents and from gravity. It also occurs where surface waters become denser. In shallow waters, the turbulence that is brought about by sinking surface waters tends to cause vertical mixing with the establishment of a homogeneous water column and may cause nutrients to be introduced to surface waters. However, when water depths are moderate or great, downwelling regions usually support only limited organic production, for nutrients and even plankters are swept out of the euphotic zone by the sinking waters.

As the continental slope and shelf become generally shallower toward shore, the density of bottom water tends to decrease with depth. When vertical mixing processes form fairly homogeneous water columns over the shelf and bordering continental slope, the average density of water columns will decrease shoreward, and a lateral density gradient will appear. This might give rise to a longshore current. If the vertical mixing is accompanied by falling temperatures, however, the shallow water columns near shore will become denser than those further offshore, where the increasing density of surface layers is distributed through a deeper column. In this case, a strong density gradient develops along the bottom, this time denser inshore, and water flows off the shelf and down the slope (Nansen, 1913; Cooper and Vaux, 1949). Presumably, it will spread out into the oceanic water mass at depths appropriate to its density, and will be replaced by shoreward-drifting surface water. This seaward flow of dense bottom waters is called *cascading*, and may seriously deplete the shelf of nutrients by washing them down the continental slope. Cascading is known to occur in the Celtic Sea and elsewhere.

Vertical movements of water on the shelf also result from the deflection of currents that run over irregular bottoms. Tidal currents in the Gulf of Maine are mixed vertically by turbulence from this cause (Gran and Braarud, 1935), and bottom currents on the shelf off southern California are sometimes so deflected over irregular bottoms that relatively dense cool water currents break out at the surface, a process called *projection* (Stevenson and Gorsline, 1956).

Finally, in many respects the intertidal zone represents the ultimate in environmental variability in the sea, for it is swept by waves and tides, it is alternately terrestrial and marine, and it is influenced by ecosystems of both these great ecological realms. Nevertheless, it is populated chiefly by marine organisms.

Ocean Temperature Has a Vast Range of Effects on Marine Organisms, Involving Their Form, Function, and Distribution

Temperature affects organisms directly according to their adaptations to thermal regimes, and it also affects many other ecologically significant characteristics of ocean water, such as density, viscosity, and the solubility of dissolved substances.

Water has a very high *specific heat*, which is defined as the number of calories required to increase the temperature of 1 gram (g) by 1°C. Specific heat varies with temperature, salinity, and water density; at 17.5°C, and under atmospheric pressure, pure water has a specific heat of 1.000, and ocean water (salinity of 35 percent) has a specific heat of 0.932 (Sverdrup, Johnson, and Fleming, 1942). Among reasonably common substances, only liquid ammonia has a higher specific heat than water. This means that water has the ability to store much heat—a great *heat capacity*—without raising its temperature very far, and likewise can give off heat in large amounts without reducing its temperature much. Fluctuations of the temperature of the atmosphere due to daily and seasonal fluctuations in incoming heat are damped down in the ocean. Marine

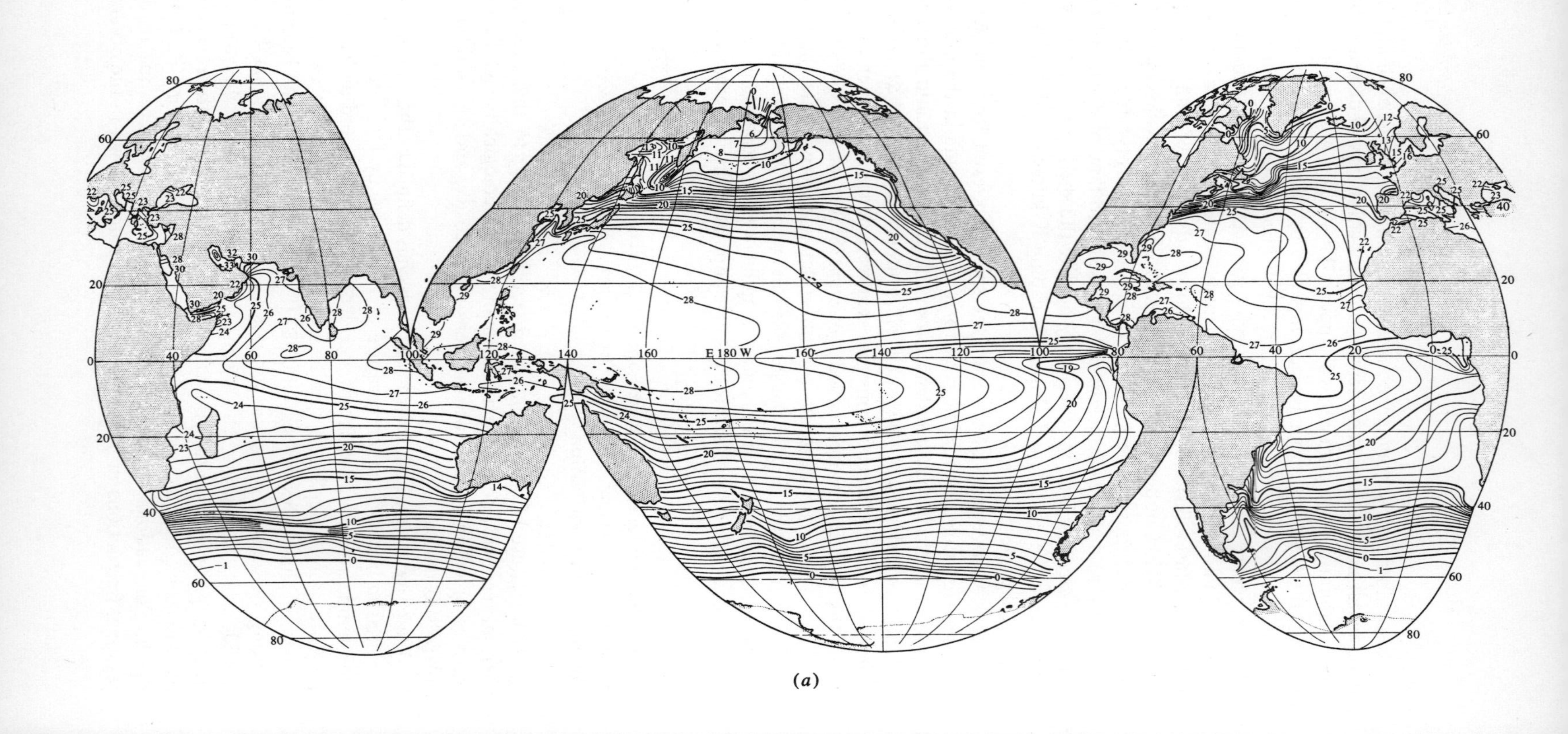

(*a*)

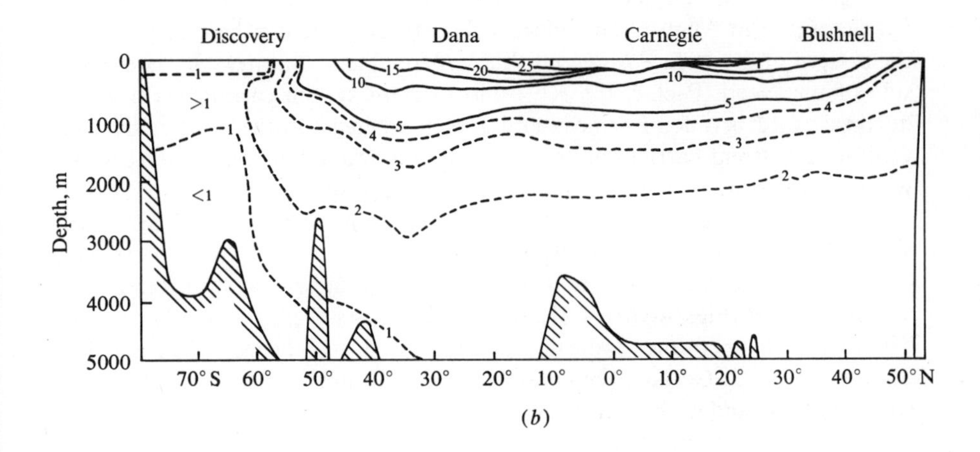

(b)

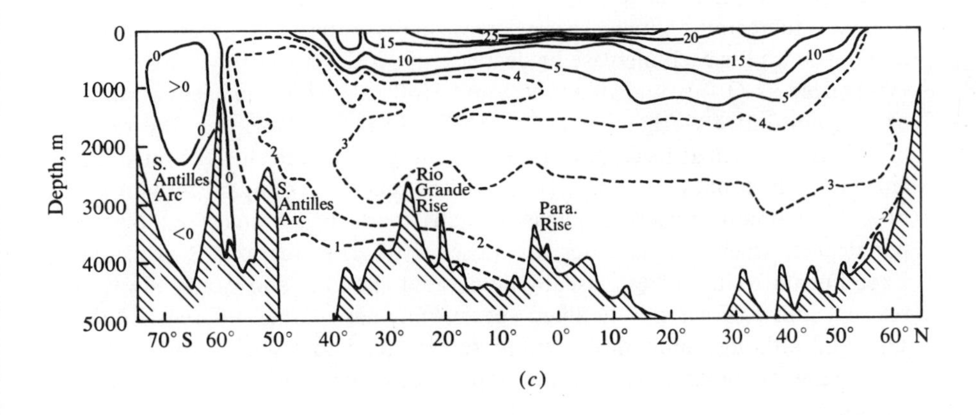

(c)

Fig. 4-21 Temperature in the oceans. (*a*) Surface temperatures during the northern summer. (*b*) Temperature profile in the Pacific Ocean along longitude 170° west. (*c*) Temperature profile in the Atlantic Ocean. Note the low temperature of the great masses of intermediate and deep water and the restriction of regions of rapid temperature change to shallow depths. (After Sverdrup, Johnson, and Fleming, © 1942, renewed 1970; (*a*) base map copyright by the University of Chicago, Department of Geography.)

climates are usually mild (not too hot or cold) and equable (rather similar the year round). It is in the interior regions of continents, far from the oceans, that the great extremes of temperature are found.

Figure 4-21 depicts the broad temperature pattern of the ocean surface in February and in August. The effects of ocean currents can be clearly seen in the pattern of the surface *isotherms* (lines of equal temperature). On the western side of the North Pacific, the Kuroshio Current system carries warm waters northeastward between Formosa and Japan, whereas on the eastern side, the California Current carries cool subarctic Pacific water southward along northwestern America; as it flows, the water is mixed and conditioned to form a "Transitional" water mass [Fig. 4-9(*a*)]. The effects of the Gulf Stream are clearly seen in isotherms on both sides of the North Atlantic, where a warm northeasterly trend can be found from off Cape Hatteras to the sea between Iceland and the British Isles [Figs. 4-9(*a*), 4-21]. Similar situations occur in the other oceans. The highest temperatures indicated in Fig. 4-21 in the Persian Gulf and Red Sea exceed 30°C. Oceanic temperatures this high are also found at times off southeast Asia and in the Caribbean during northern summer, and off northeast Australia during southern summer. Within shallow embayments, on reef flats, in tidal pools, and at other favorable sites, temperatures sometimes exceed 40°C. Temperatures in high latitudes fall to nearly −2°C. The range of temperatures in the sea is indicated in Fig. 4-22.

The basic physiological reaction of organisms to temperature change is part of the general phenomenon of change in the rate of chemical reactions with change in temperature. Increase or decrease in reaction rates is generally between 2 and 3 times for each increase or decrease of 10°C, respectively. This relation, symbolized as Q_{10}, is known as the Van't Hoff rule, after the chemist who did early quantitative work on it (Van't Hoff, 1884). In organisms, the rates of life activities (such as feeding, respiration, and growth) are chiefly controlled by enzymes, which obey the Van't Hoff rule, in general. An organism with a Q_{10} of 2.5, a common value, will increase its activity rate by 2.5 times, or 9.6 percent per degree, when temperatures are raised 10°C (Prosser and Brown, 1961). Experimental values of Q_{10} usually lie between 2 and 3, and nearly always between 1 and 5. Even closely allied species may have rather different values of Q_{10}, and although individuals within species tend to resemble each other in this character, separate conspecific populations sometimes have different Q_{10} values.

Most organisms in the sea, including all protists and invertebrates, do not actively regulate their body temperatures, but approximate the temperature of the ocean water in which they are immersed; such organisms may be termed *conformers* (such as most of the "cold-blooded" or *ectothermic* animals). A few animals maintain body temperature at some given level or levels and are called *regulators* (such as the warm-blooded or *endothermic* animals). The regulators can control temperature-related reaction rates even in the face of environmental temperature changes. Organisms that cannot regulate their temperatures respond to changes in environmental temperature either by altering the rates

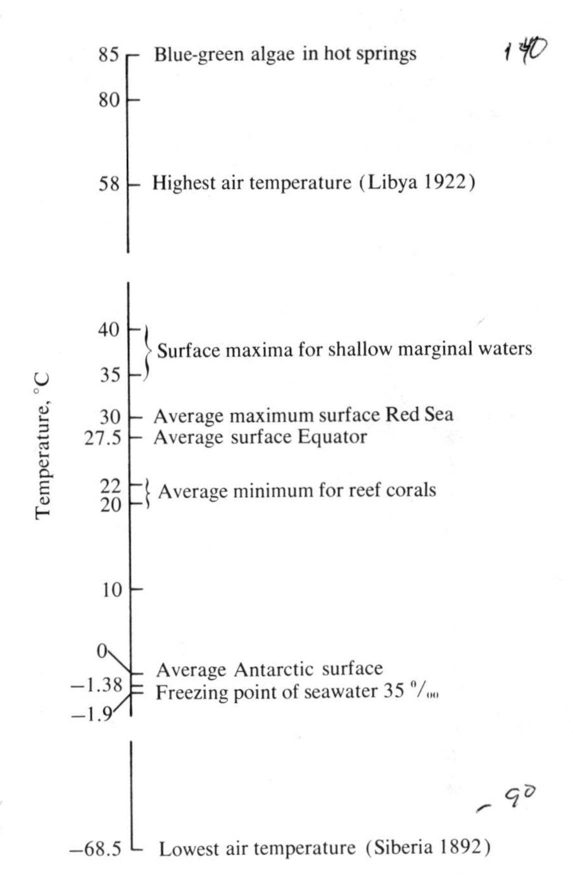

Fig. 4-22 The temperature range found in air and water. (After Raymont, 1963, from Gunter, 1957.)

of their biological processes in accordance with Q_{10}, or by adjusting the level of their temperature-related reaction rates.

In the simplest case, a given ectotherm can be thought of as having a fixed activity rate at a given temperature, and a fixed Q_{10} that will determine the activity rate at any other temperature. The fixed activity rate may be considered to be governed by an *activity coefficient*. In general, species living in warm water have lower activity coefficients than species living in cold water. This difference commonly more or less balances the effects of Q_{10}. According to Van't Hoff's rule, assuming a Q_{10} of 2.5, ectotherms that live at 25°C would be expected to have activity rates five times higher than other ectothermic species that live at 5°C. However, it commonly happens that the coefficients of activity of such organisms differ by a factor near five, with the warmer-water species having the lower coefficients, so that the activity rates of both warm- and cold-water groups are similar. In a number of cases, adjustments of the activity coefficients for low temperature seem to compensate only partially for activity differences due to Q_{10} (Scholander *et al.*, 1953). Most investigators believe that on balance the activity level in high latitudes is somewhat below that in low latitudes, since the compensation for Q_{10} is sometimes incomplete. It is not yet certain that this is true, but if it is, it may be that the lack of complete compensation represents a lag in cold adaptation by members of the high-latitude biotas that will eventually evolve higher activity coefficients, or it may be that adaptation to the difficult environments of high latitudes is sometimes achieved by a strategy involving a lowering of activity rates.

An excellent example of a balance between activity coefficients and Q_{10} has been presented by Thorson (1952), who has measured the rate of oxygen consumption of populations of three species of mussels, from the Persian Gulf, from Denmark, and from Greenland (Fig. 4-23). Their rates are closely com-

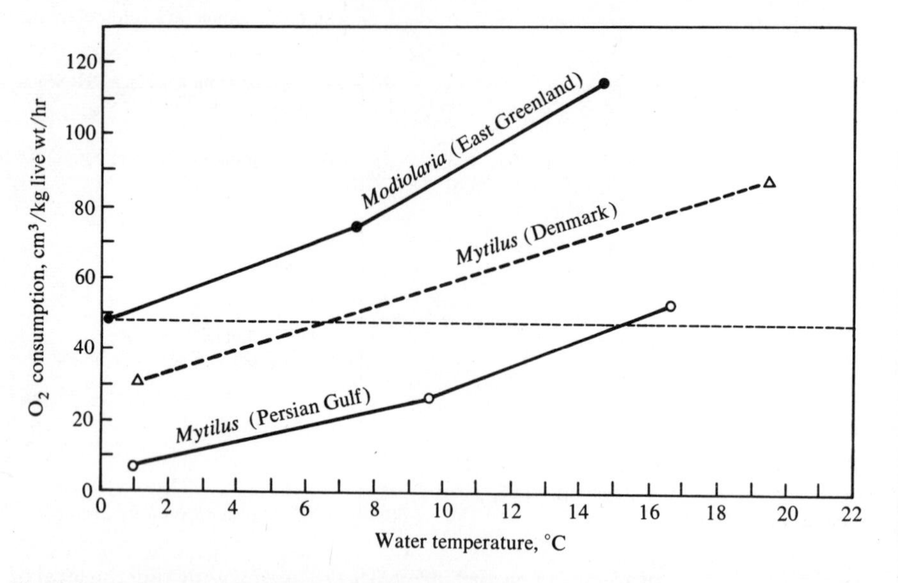

Fig. 4-23 Temperature dependence of rates of oxygen consumption of mussels living in three different temperature regimes. The horizontal line crosses the temperature-rate curves at approximately the temperature at which the mussels live naturally. (After Thorson, 1952.)

parable at their home temperatures, indicating that differences in their activity coefficients compensate for rate differences that would ordinarily arise from Q_{10}. Individuals from each population display the normal Q_{10} reaction, increasing their oxygen consumption with increasing temperature by two or more times for each 10°C. If the activity coefficients of all populations were equal to those of the Danish one, then individuals from Greenland would have extremely low, and those from the Persian Gulf extremely high, oxygen consumption in home waters.

Organisms may be adjusted to a temperature regime by adaptation or by acclimation. *Adaptation,* as employed herein, refers to all the genotypically fixed environmental relations of an organism, such as the activity coefficients of Thorson's mussels. *Acclimation* refers to the ability of an individual to adjust physiologically to different conditions. Sometimes a sudden prolonged rise or fall in water temperature of a given amount proves fatal, whereas if the change is gradual, or if the organism is subjected to the change for only short periods at first, a conditioning process occurs. The organism becomes more and more fit for life at the new temperature until it is hardy enough to survive indefinitely. It has become acclimated. This process often requires changes in the phenotype, changes that are certainly within the temperature limits fixed genotypically. The ability to acclimate is an adaptation.

Temperature acclimation is also associated with the ability of an organism to alter the activity coefficient and thus to alter activity rates at any given temperature. Presumably, the rates to which an animal acclimates tend to be optimal. Acclimation in the mussel *Mytilus californianus* has been investigated by Segal, Rao, and James (1953). (See Fig. 4-24). The pumping rate—that is, the rate at

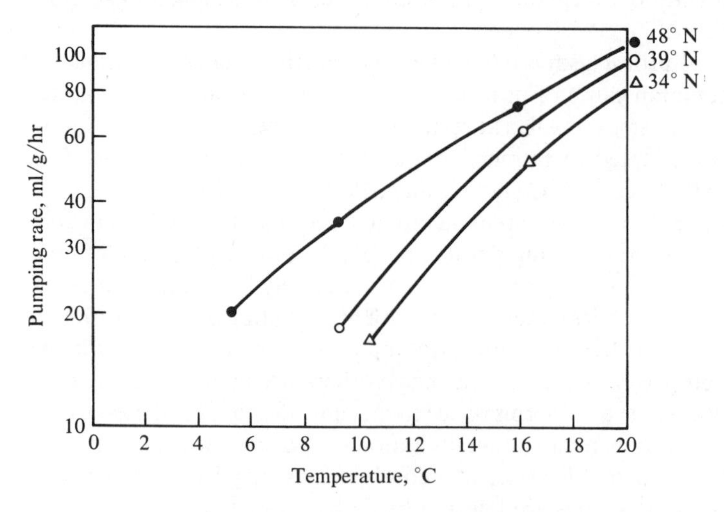

Fig. 4-24 Temperature dependence of pumping rate in the mussel *Mytilus californianus* at three latitudinally separated localities; the rate is measured under standardized conditions. (After Bullock, 1955, from Rao, 1953.)

which water is circulated through the mantle cavity by ciliary activity—was studied in three populations living adjacent to each other. One was subtidal and lived at relatively cool temperatures, a second was from the low intertidal zone and lived at a higher mean temperature, and the third was from higher in the intertidal zone and lived at a still higher mean temperature. At a single given temperature, the lowest population pumped the fastest and the highest pumped the slowest. At the different temperatures of their respective habitats, then, all the populations would tend to pump at rates more equal than if their activity coefficients were equal, and Q_{10} alone governed their rates. *Mytilus* spawns into the water where fertilization occurs, so that the three populations are probably not isolated genetically and the acclimation is truly phenotypic.

Several terms are in common use to describe the temperature relations of organisms. Forms requiring warm environments are called *thermophiles*, and those requiring cold environments *cryophiles* (or sometimes *frigophiles*). Tolerant forms that have broad temperature ranges are called *eurytherms*, whereas those with narrow ranges are *stenotherms*. *Homoiotherms* are forms that live at constant body temperatures, and *poikilotherms* have variable body temperatures.

The precise physiological mechanisms that limit the temperature ranges of organisms are not well known, and the causes of death of organisms killed by temperature extremes are often uncertain. Among the more frequent suggestions for causes of heat death are coagulation of protoplasm; melting of lipids; denaturation of certain proteins; asphyxiation based upon increased oxygen requirements; failure of enzyme systems; and lowering of adjustment to other environmental factors, which then prove fatal (see Kinne, 1963, 1964, for a more exhaustive list).

Coagulation of protoplasm accompanies heat death in many organisms and is probably a common cause of heat death at the higher oceanic temperatures. However, may organisms are killed at relatively low temperatures. Some Antarctic fishes that are adapted to extreme cold and that normally live at -1°C or less display "heat" effects at 0°C (Wohlschlag, 1964).

The protoplasm of some warm- and cold-adapted organisms differs in composition. Lipids with low melting points are found in the protoplasm of cryophiles, which therefore tend to be destroyed at relatively low temperatures. On the other hand, thermophiles contain lipids with high melting points (see Gunter, 1957). Despite the fact that heat limits are imposed on many organisms by destruction of protoplasm, it seems that many others are limited in some other way. The ability of some ectotherms to acclimate indicates that for these forms at least a significant factor in temperature limitation is associated with activity rates; apparently, the reason that heat-acclimated organisms may live at higher temperatures is that activity rates are lowered by a change in activity coefficients.

The limiting mechanisms at the low-temperature extreme are also puzzling. For organisms that are cold-adapted, freezing is usually fatal, although there are a few cases of tolerance to freezing (Kinne, 1963). But the vast majority of marine ectotherms are limited at temperatures well above freezing. As the low-temperature limit is characterized by low activity rates, it is reasonable to suspect that some critical functions, such as respiration, may finally be reduced at some low temperature to a rate so low that life cannot be sustained. The general lowering of activity certainly makes cold-affected organisms much easier prey to the dangers of their environment. General numbing from cold may reduce or eliminate escape responses from predators, and shallow-water marine invertebrates are sometimes so inactivated by cold as to be washed from their normal habitats and stranded ashore by waves (Gunter, 1957).

Bullock (1955) has presented an hypothesis which may account for some puzzling cases of temperature limitation. For the sake of simplicity, it has been assumed in the preceding paragraphs that all the manifold activities of an organism could be described by a single rate coefficient. This is certainly not the case. Furthermore, the value to Q_{10} is somewhat different for different activities. In general, within the range of optimum temperatures, reactions occur at speeds that are consistent with an optimal balance of functions. As temperatures change more and more, the ratios of different reaction rates depart more and more from their optimum, and an imbalance in rates occurs. This imbalance, when carried to extremes, results in a breakdown in related functions and leads to death. Thus, the imbalance hypothesis can account for temperature limits that are well short of causing direct, deleterious physical effects.

The rates of growth and differentiation of ectotherms are affected by environmental temperature, which, in turn, affects their forms. In addition, modifications of form that correlate with temperature and that may be temperature adaptations are also found. Growth tends to be faster, the onset of reproduction earlier, and death sooner in warmer water than in colder water. Despite slower growth, cold-water individuals commonly attain a larger size than conspecific warm-water individuals, presumably because growth continues longer owing to the delay in reproductive activity and in death. There are numerous exceptions to this trend, however (Moore, 1958; Kinne, 1963), and the differences in growth rates and age of reproduction between warm- and cold-water individuals are usually not as great as expectable from Q_{10}, suggesting that growth and reproduction are somewhat acclimated or otherwise adapted to temperature.

In invertebrates and protists, temperature-correlated morphologic modifications tend toward elaboration in warm water and simplicity in cold. Individuals of planktonic species from warm southern waters frequently have larger spines than individuals from cold northern waters, and summer generations are more spinose than winter generations at the same locality. Furthermore, warm-water species tend to be more spinose than cold-water species (Gunter, 1957).

The spines are thought to be adaptive for buoyancy in the less viscous warm water. However, benthic invertebrates also display this sort of tendency. For example, species of Bivalvia that bear long slender spines only inhabit waters above 25°C, species with shorter spines that are over 10 mm only inhabit waters above 20°C, whereas species with very short spines only inhabit waters above 10°C; no spinose bivalves live in shallow waters below about 10°C, and probably not in abyssal depths either (Nicol, 1967). For very cold-water bivalves (5°C and less), ornamentation is subdued or absent, consisting chiefly of striae when present, and shells are on the average thin and colorless (Nicol, 1967). It appears that the trend toward elaboration of skeletons in warm water may have other causes than selection for buoyancy. It may be a very important trend indeed, especially at the community and provincial levels, and we shall return to this point in later chapters.

Reproduction is commonly affected by temperature; not only is the age of reproduction sensitive to temperature, but in invertebrates, which chiefly have seasonal reproduction, the precise time of year when spawning occurs is often adjusted to local temperatures. Many marine invertebrates are reproductively stenothermal, even though they may be rather eurythermal for vegetative functions. For most, reproduction occurs in the warmer part of the year. Even in the tropics, where seasonal temperatures differ only a little, most spawning seems to occur in the summer, although winter spawning is more prevalent than in higher latitudes (Kinne, 1963). Spawning is often elicited by the rise of temperatures to a key level in the spring or summer. Some species spawn at different temperatures in different regions, each temperature appropriate to the local climate, so that individuals living in cool water spawn at lower temperatures. The bivalve *Crassostrea virginica* spawns at about the same time at a series of localities along the American East Coast, although the temperatures differ from one locality to the next (Stauber, 1950).

The adaptive strategy of reproductive stenothermy appears to be to permit fertilization and early stages of development to proceed in the most favorable possible environment. Temperature tolerances of early ontogenetic stages are frequently quite narrow; usually, the cleavage stages require a narrower temperature range than do the later larval stages, which, in turn, require a narrower range than do stages after metamorphosis (Kinne, 1963). Selection presumably occurs for spawning at temperatures that normally presage favorable developmental conditions, including food. The dangers of life as a larva, planktonic or benthic, are great, and mortality due to predation alone is extremely high. Therefore, the sooner metamorphosis can occur the better, other things being equal. The upper range of local temperature, favoring rapid development to metamorphosis, would seem to be advantageous for early ontogeny. Furthermore, the rate of change in the organization of a phenotype is far greater in early ontogeny, and the complex biochemical pathways that control development must surely require close coordination of reaction rates, for which a narrow temperature range is advantageous.

It is useful to summarize briefly the salient ecological features of temperature in the oceans. For ectotherms, the primary effects are owing to the influence of temperature on reaction rates. For each, there is usually an optimum temperature range within which activity rates are well coordinated and no deleterious effects on biological materials are observed. Beyond this optimum, effects from rate extremes or rate imbalances or from the breakdown of biological materials occur, and the functioning of the organism is impaired, until, at some distance from the optimum range, death occurs. The rate of impairment may be gradual as temperatures depart from the optimum range, with the probability of death from immediate causes other than temperature rising as impairment increases; or the rate of impairment may be rapid, with death occurring immediately as some key function is destroyed.

Temperature tolerances vary during ontogeny. Usually the younger stages are the least tolerant and the older stages the most tolerant, except that reproduction is commonly possible only within a narrow temperature range. The temperature of reproduction is adjusted to the requirements of early ontogenetic stages, which often lie near the warmer end of the temperature range of the adult. However, reproduction during winter at relatively low temperatures is not uncommon in the ocean, especially toward the warmer end of the geographic range of an ectotherm. Most adults live nearer the upper end than the lower end of their temperature ranges, presumably owing to reproductive thermophily and to the advantages to be derived from rapid growth and development at higher reaction rates. Many ectotherms are able to escape from the confines of temperature imposed by Q_{10} by acclimating, that is, by changing their activity coefficients so as to maintain optimal reaction rates over a broad temperature range.

Ocean Salinity Is Normally Less Variable Than Temperature but Has a Wide Range of Effects

Salinity as a major physical attribute of ocean water is one of the more important marine ecological parameters. The water properties most affected by salinity changes, apart from the actual concentration of the ions themselves, include density and the *colligative properties* of osmotic pressure, vapor pressure, freezing point, and boiling point. The colligative properties are so called because they are tied together; if one is known, the remainder may be calculated. Of these, it is the osmotic pressure that is most important biologically.

Salinity may range from pure to fully saturated water, from 0 ‰ to 260–280 ‰. The chemical analyses required to measure salinity directly are exceedingly laborious; however, it happens that the proportions of ions in "normal" seawater are nearly invariant, so that it is possible to estimate the salinity accurately by measuring the concentration of only one of them. Usually the chlorinity of seawater has been measured and converted to salinity. In recent years, the development of accurate measures of the electrical conductivity of seawater, which varies with density at a given temperature, has provided a more accurate and rapid method of salinity measurement.

Like temperature, salinity displays rather characteristic oceanic patterns that result from large-scale conditioning and mixing processes, associated in this case with global evaporation, precipitation, and ice melt patterns (Fig. 4-10). Surface salinities of the ocean and vertical salinity profiles of the Atlantic and Pacific Oceans are depicted in Fig. 4-25. As the figure indicates, the normal range of salinities in the open ocean is quite small, usually from about 33 ‰ to

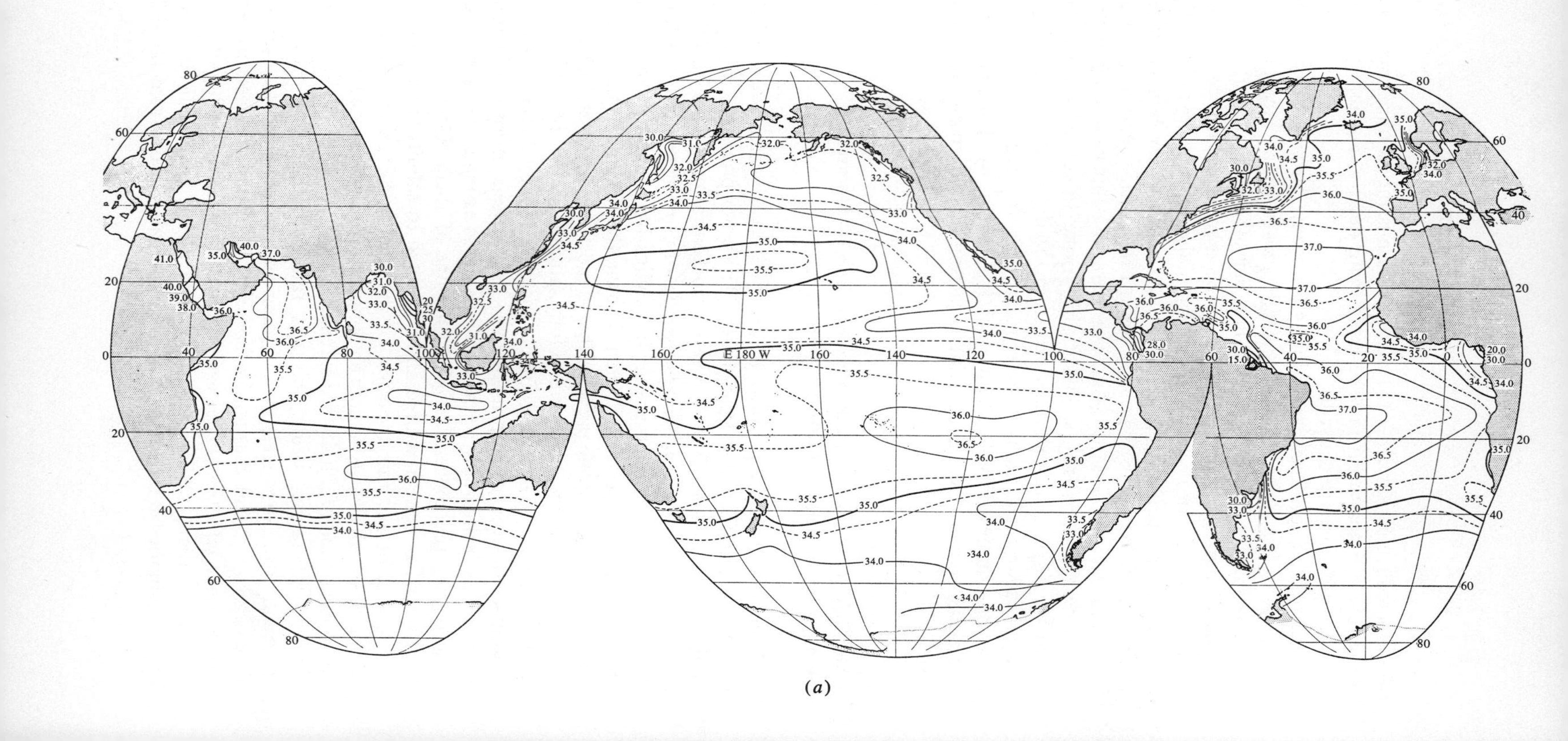

(*a*)

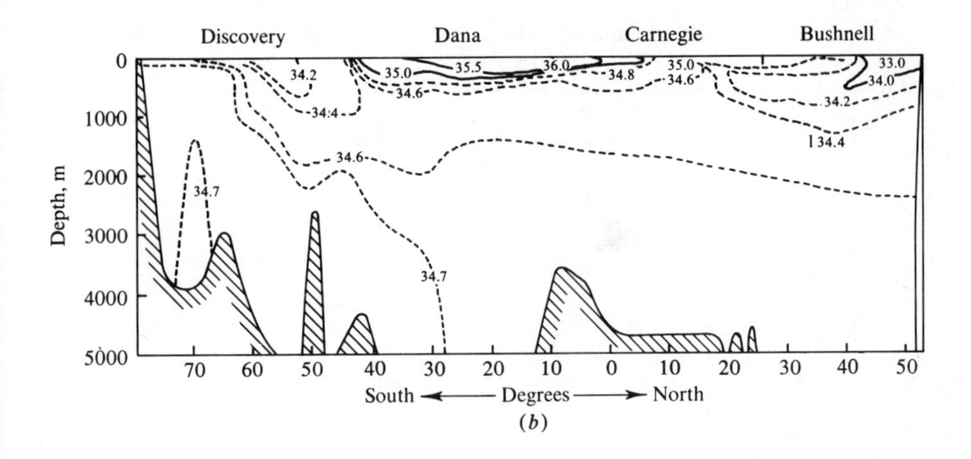

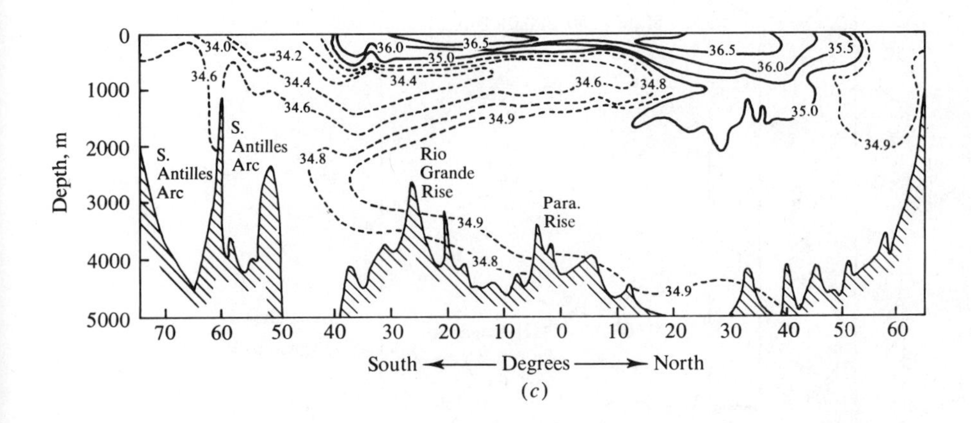

Fig. 4-25 Salinity in the oceans. (*a*) Surface salinities during the northern summer. (*b*) Salinity profile in the Pacific Ocean along longitude 170° west. (*c*) Salinity profile in the Atlantic Ocean. Note the high salinity of low latitudes, the extensive horizontal isohaline tongues at intermediate depths, and the generally higher salinities in the Atlantic. (After Sverdrup, Johnson, and Fleming, © 1942, renewed 1970; (*a*) base map copyright by the University of Chicago, Department of Geography.)

37 ‰ and rarely beyond 30 ‰ to 40 ‰. In regions where circulation is restricted, however, and precipitation and runoff are high, salinity frequently falls far below this range, grading into fresh water. Where evaporation is excessive, salinities frequently exceed 50 ‰, and even 100 ‰ in some lagoons. A variety of schemes is available to classify water according to salinity; the terms

‰	This book	Other terms
	Saturated	
250		
200		
150	80:00–260/280.0: Brine	
100		
	40:00–80:00: Hypersaline	Metahaline, Ultrahaline } over 30*
50		
	30.0–40.0: Normal marine	Polyhaline 15/20–30*
	0.5–30.0: Brackish	Mesohaline 2/3–15/20*
0	0–0.5: Fresh	Oligohaline 0.2/0.5–2.0/3.0*

*Class limits vary among authors.

Fig. 4-26 Some terms describing water salinity.

to be employed herein follow Kinne (1964), and are presented in Fig. 4-26, along with some terms that are especially common in the literature. For reviews and references to classifications, see Hedgpeth (1957) and Remane and Schlieper (1958).

The physiological reaction of organisms to salinity as such is chiefly associated with their osmotic properties. *Osmotic pressure* is one of the colligative properties of a solution, all of which behave in concert (Fig. 4-27). In any solution, an increase in the number of solute particles results in a lowering of vapor pressure, depression of the freezing point, elevation of the boiling point, and raising of osmotic pressure. To understand how osmotic pressure is thus raised,

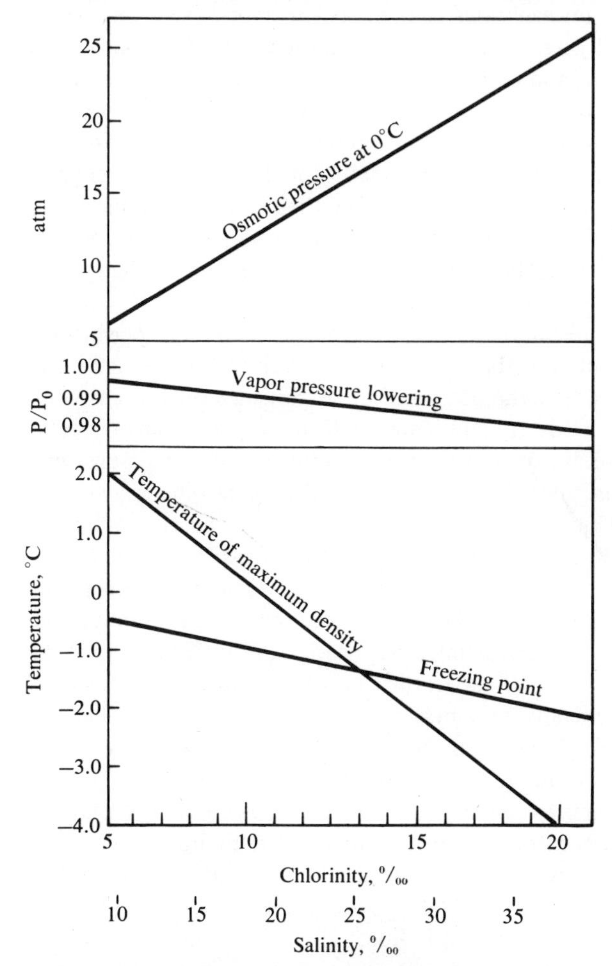

Fig. 4-27 The colligative properties and their relations to chlorinity and salinity variations. (After Sverdrup, Johnson, and Fleming, © 1942, renewed 1970.)

consider two aqueous solutions of different salinities, separated by a membrane that is permeable to the solvent but not to the solutes. The vapor pressure is greater in the solution with fewer solute particles, and therefore a vapor pressure gradient occurs across the membrane; solvent molecules flow from the less saline, higher pressure side of the membrane to the more saline, lower pressure side. The pressure necessary to prevent such flow is called osmotic pressure. Cell and tissue membranes of organisms are not perfectly semipermeable but are sufficiently so that osmotic flow may occur.

The extracellular tissue fluids of organisms are *isosmotic* with their cellular fluids; that is, no vapor pressure gradient occurs between them. Furthermore, the body fluids of marine invertebrates tend to be isosmotic with the ocean water in which they normally live. In this case, when the water salinity changes, the osmotic pressure changes and osmotic flow tends to occur across the membranes of the organisms. More strictly, it is the concentration of *all* solute particles in the water, not just those contributing to salinity, that governs the osmotic concentration; however, this difference is normally small and we shall neglect it. Some organisms are osmoconformers, permitting their body-fluid concentrations to conform with the water medium. Others are osmoregulators, which maintain a nearly constant body-fluid concentration in the face of changes in the salinity of the medium.

Some osmoconformers respond to salinity changes simply by permitting water intake or expulsion until they become isosmotic again. Whenever environmental salinity falls, water enters the body and the organism swells; if environmental salinity rises, body fluids are eliminated and the organism shrinks. The limitations to this response are obvious: upon swelling, the body tissues and cells eventually burst; upon shrinking, the tissues and cells become dehydrated and eventually collapse. This simple response is common in invertebrate eggs, such as echinoderm eggs, and among parasitic marine protistans. Other osmoconformers control their volumes, taking salts into, or removing them from, body fluids to maintain an isosmotic relation. Usually there is an initial period of bodily swelling or shrinking when salinity first changes, followed by a gradual return to the original body proportions. In the case of swelling, water is usually excreted as urine or, in protistans, in contractile vacuoles; in exceptional cases, some water may be stored (Prosser and Brown, 1961). Physiologically, these volume adjustments are not simple but require a great chain of responses. Many mollusks and echinoderms can adjust in a limited way to salinity changes through controlling the concentrations and volumes of their body fluids (for examples, see Prosser and Brown, 1961; Pearse and Gunter, 1957; Kinne, 1964; and references therein).

Osmoregulators can maintain their body fluids in a *hyposmotic* state (that is, they can be *hypotonic*, with a lower concentration than the medium) or in a *hyperosmotic* state (they can be *hypertonic*, with a higher concentration than the medium). This is accomplished in various ways. Often, most of the membranes in contact with the water are relatively impermeable, and respiratory membranes are reduced in area. Water or salts are actively taken up or excreted to maintain the required concentrations. Marine invertebrates that are somewhat osmoregulatory are chiefly estuarine and thus hyperosmotic. They include arthropods, such as estuarine crabs and crayfish, and worms, such as the polychaete *Nereis*. A few invertebrates that inhabit salt marshes and ponds are hyposmotic, such as the brine shrimp *Artemia*. Marine bony fishes and marine mammals are osmoregulators.

Most marine invertebrates are stenohaline (that is, tolerant of only a narrow salinity range), which does not limit them much due to the narrow range of salinities found in the ocean. Among those that are not stenohaline, the ability to tolerate osmotic changes varies enormously. Both osmoconforming and osmoregulating organisms may be either stenosmotic or euryosmotic. Many organisms are more stenosmotic for reproduction than for vegetative existence, but there does not seem to be such a widespread ontogenetic increase in tolerance for salinity changes as there is for temperature changes, although eggs tend to be more stenohaline than adults (Moore, 1958), perhaps owing to their complete or nearly complete conformity to environmental salinity. Smith (1964) has described the salinity tolerances of early ontogenetic stages of the polychaete *Nereis diversicolor* in the Baltic. For a population in southern Finland, fertilization may occur in salinities from about 1.8 ‰ to 33.5 ‰, but cleavage of the eggs occurs only between about 5 ‰ and 15 ‰. Near the limits of tolerance, development is retarded. Once early postlarval stages are passed, however, the animal becomes euryhaline. *N. diversicolor* living in other regions also have lowered tolerances during cleavage and larval stages, although the tolerance limits may be different. However, the young of many species appear to tolerate or even prefer lower salinities than adults, so that there is sometimes a size gradient, based on an age gradient, associated with salinity gradients. Smaller, younger organisms dominate in waters of lower salinity (Pearse and Gunter, 1957).

Much less is known of the effects of salinity on growth and form than is known of the effects of temperature. Marine organisms tend to be smaller in less saline water, presumably because the growth rate is slower there. Marine invertebrates with calcareous skeletons usually attain a smaller terminal size and have thinner shells in brackish water (see Moore, 1958, and Kinne, 1964, for numerous references). However, a few species that seem particularly well adapted to brackish water grow larger there (as the bivalve *Rangia*), and also secrete thicker shells (as some oysters; Pearse and Gunter, 1957). A few invertebrates display marked variation in morphology that can be correlated with variations in salinity. *Artemia* not only is reduced in size at higher salinities, but its appendages are reduced in number and body proportions are altered (Hesse, Allee, and Schmidt, 1951).

Salinity limitations may be due to salinity effects other than osmosis. A relatively euryosmotic organism may be rather stenohaline if other salinity effects, such as vapor pressure changes, are limiting. This and other factors sensitive to salinity are discussed in sections that follow.

Many Ecologically Significant Substances Are Dissolved in Seawater

Quite apart from the effects of salinity, the concentration patterns of the substances dissolved in seawater are of major ecological significance, for many of them are incorporated into the bodies of organisms and, indeed, are required to support life. Others are toxic at certain concentrations, but such concentrations are reached only locally and total effects are minor. Dissolved substances are conveniently classed as inorganic salts, gases, and organic compounds.

Elements composing the dissolved salts fall into two somewhat arbitrary classes. First are the *conservative elements*, which vary in concentration with the salinity. Although we have seen that runoff of waters unusually rich in certain ions may cause local variations in relative ionic abundance, this effect is usually restricted to small semienclosed basins or lagoons or to local regions of the continental shelf, and is usually minor on a planetary scale. In general, the pattern of concentration of conservative elements is regulated by the conditioning and mixing processes that we have reviewed. There may, however, be very long-term trends of compositional change even among the conservative elements, measured in hundreds of millions of years and related to the rates of supply of elements to the sea from the lithosphere and to the trapping of elements in sediments (Garrels and Mackenzie, 1971). The pattern of distribution of conservative elements, then, can be accounted for chiefly by the principles of physical oceanography and geochemistry.

Second, and in contrast to the conservative elements, are *nonconservative elements*, which have patterns that cannot be explained by these principles. These elements are present in the seas in substances that are taken up by marine organisms in significantly large amounts relative to their concentrations, and employed in constructing bodies and skeletons. As they are extracted from the general circulation pattern, there is a fractionation of water composition, and the extracted substances enter into a *biochemical circulation* pattern (Redfield, Ketchum, and Richards, 1963). Their distribution then depends upon the properties and activities of organisms until they are returned to the general circulation through excretion, respiration, or death and decomposition of the organism.

Even the conservative elements have important ecological effects. Table 4-1 lists the concentrations of the major inorganic salts in seawater. Although most invertebrates conform to seawater in osmotic concentration, and some have similar ionic concentrations in their body fluids, none conform ionically with seawater at the cellular level. Thus, ionic regulation occurs within all cells, and organisms must concentrate some elements in amounts exceeding their con-

Table 4-1 The Twenty-two Most Abundant Elements in Seawater, in Weight Percent

Element	Weight percent
Oxygen	85.89
Hydrogen	10.82
Chlorine	1.90
Sodium	1.06
Magnesium	0.13
Sulfur	0.088
Calcium	0.040
Potassium	0.038
Bromine	6.5×10^{-3}
Carbon	2.8×10^{-3}
Strontium	1.3×10^{-3}
Boron	4.8×10^{-4}
Silicon	2×10^{-4}
Fluorine	1.4×10^{-4}
Nitrogen	$0.3\text{–}7 \times 10^{-5}$
Rubidium	2×10^{-5}
Lithium	1.2×10^{-5}
Aluminum	1×10^{-5}
Phosphorus	5×10^{-6}
Iodine	5×10^{-6}
Arsenic	1.5×10^{-6}
Barium	1×10^{-6}

SOURCE: After V. M. Goldschmidt, *Geochemistry*, Clarendon Press, Oxford, 1954.

centration in seawater and discriminate against others. Ionic regulation must be more primitive (and probably more fundamental) than osmoregulation (Prosser and Brown, 1961).

Among the ions that organisms concentrate are some that are much rarer than those listed in Table 4-1. For example, copper is concentrated in many autotrophs and mollusks, vanadium in ascidians, and beryllium in sponges. These ions are so rare that they may well be limiting at times when locally depleted by organisms. This possibility has been suggested for ions of molybdenum and gallium, which are required in minute amounts by diatoms; it may prove that diatom population growth is sometimes checked by their depletion during rich blooms (Harvey, 1945). Perhaps some of these elements should be considered nonconservative. Their distribution cycles and ecological significance are still little known.

An interesting possible effect of limiting concentrations of conservative elements has been pointed out by Pora (1962), who notes that the Black Sea, which supports relatively few species, may be impoverished not only owing to salinities somewhat lower than normal, but to the unusual ionic concentrations of the water, which is relatively enriched in K^+, Ca^{++}, and Mg^{++}. In the Baltic, where ionic proportions are similar to seawater, the biota is richer than the Black Sea when areas of similar salinities and habitats are compared.

Nonconservative elements are usually called *nutrients* and *vitamins*, and include phosphorus, nitrogen, silicon, and iron. They are taken from seawater in various chemical forms and then *regenerated*—returned to the water in forms that make them available again for uptake—so as to have a cyclic biochemical pattern. Their general concentrations in the oceans are indicated in Fig. 4-28. Phosphorus is utilized chiefly as phosphate (PO_4). It is synthesized as a phosphoric acid residue into proteins and lipids of high molecular weight, and is involved in basic metabolic processes. The acid residue is easily detached, and therefore PO_4 is quickly regenerated, both by decay of phosphate-bearing excretory products and more importantly by the decomposition of protoplasm.

In the ocean, four vertical layers are generally associated with PO_4 (and other nutrient) concentration: a surface layer of low concentration; a zone below this of rapid increase in nutrient concentration; a zone of maximum concentration, often between 100 and 500 m; and a lower zone extending to the bottom, in which rather high concentrations are found. The surface waters are certainly depleted owing to uptake by planktonic populations, and with increasing depth photosynthesis, and therefore uptake, fall off whereas nutrients are regenerated, thus accounting for the swift rise in the second zone. The maximum zone has sometimes been attributed in part to the accumulation of nutrients after sinking from the surface; perhaps the density of the nutrient materials is nearly as low as the water there and their sinking rate retarded so as to concentrate them more highly. However, much PO_4 (and other nutrients) is also introduced by intermediate, nutrient-rich waters from higher latitudes, reinforcing the concentration levels (Redfield, Ketchum, and Richards, 1963). The deeper waters of the oceans act as a great nutrient sink and reservoir.

These vertical zones are developed differently in different regions. Where upwelling is on a large scale, as off Antarctica, surface waters are, so to speak, drawn from the deep reservoir and may have high nutrient concentrations. The upper two zones therefore become indistinct, especially during the winter "night." In summer, when photosynthesis proceeds, even the high nutrient concentration in such surface waters is lowered and the zonation reappears; but even so, the concentrations tend to remain high. Over continental shelves, PO_4 (and other nutrients) is introduced inshore by upwelling and other vertical mixing processes, and is often patchily distributed, as are the upwellings. In low latitudes, blooms of plankton occupy these nutrient-rich waters year around.

Nitrogen is utilized chiefly as nitrate (NO_3); its uptake and regeneration is somewhat more complicated chemically than phosphorus. First, nitrogen participates in the backbone structures of organic compounds, and is therefore more complicated to regenerate, so that there may be a lag in the production of nutrient nitrogen, when compared to nutrient phosphorus, from decomposing organic material. Second, there is sometimes biological uptake of other nitrogen compounds, especially ammonia (NH_3) and nitrite (NO_2); these compounds

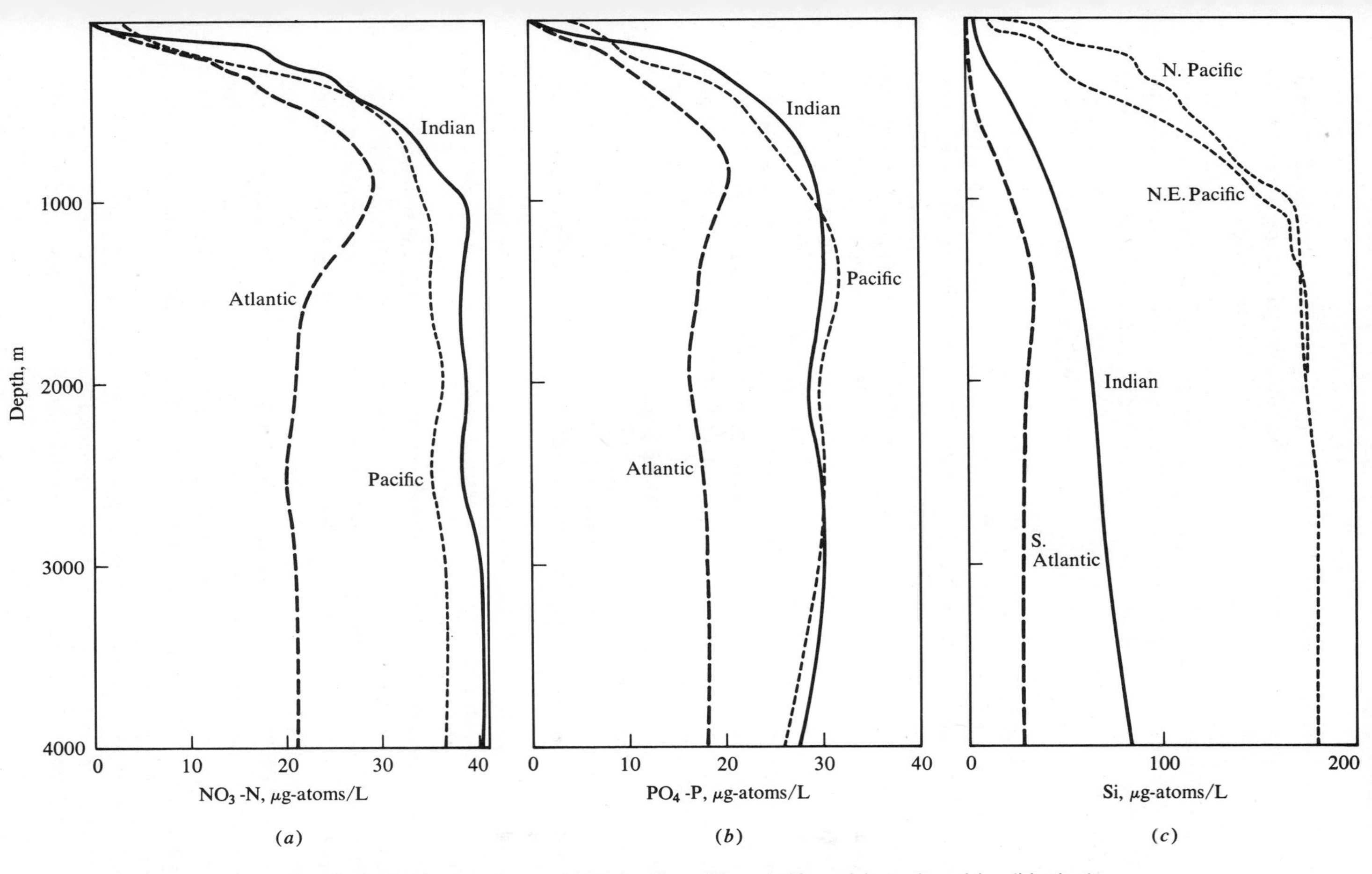

Fig. 4-28 Vertical distribution of nitrate (*a*), phosphate (*b*), and silicate (*c*) at selected localities in the Atlantic, Pacific, and Indian Oceans. The localities represented by the silicate curves are different from those represented by the nitrate and phosphate curves. (After Sverdrup, Johnson, and Fleming, © 1942, renewed 1970.)

are consecutive intermediate products in the degradation of large nitrogenous organic molecules to nitrate. In general, the amounts of the intermediate products are highest during times of great decomposition, that is, in spring and summer, and during these times ammonia may become the principal source of nitrogen in some areas. During fall and winter, oxidation of nitrogenous compounds proceeds and nitrate becomes the more common of them. As Fig. 4-28 illustrates, the vertical distribution of nitrate is similar to that of phosphate, although it is an order of magnitude more abundant. Surface concentrations of nitrogenous nutrients tend to follow the same principles as those of phosphorus, and thus the vertical zonations are modified strongly in areas of upwelling or unusual physical circulation.

Silicon is utilized as silica (SiO_3), which is not an essential requirement of all plants; among the phytoplankton, it is utilized chiefly by the diatoms in their skeletons (tests). Figure 4-28 shows that the vertical distribution of silicon is different from that of phosphorus and nitrogen nutrients. This is due to several causes. First, diatoms vary in their relative abundance from one part of the ocean to another, so that the uptake and regeneration of silicon does not correlate with total phytoplankton. Second, the regeneration of silicon is not related to the decomposition of organic materials, but merely to the dissolution of siliceous tests. Finally, most animals do not utilize silica, and therefore tend to exclude it from the nutrient cycle at higher trophic levels (Redfield, Ketchum, and Richards, 1963). Silica is, however, most abundant in deeper waters and richest in surface waters where upwelling occurs, or where there is local runoff from silica-rich streams.

Iron is an additional element required in minute amounts by all organisms; in the sea, it is present only in minute quantities, very seasonally in most waters. Uptake appears to be chiefly by absorption of colloidal (or larger) particles of ferric hydroxide, which is essentially insoluble. Iron may become limiting in open oceanic areas that have little vertical mixing, such as in the Sargasso Sea (Menzel and Ryther, 1961).

Gases dissolved in the ocean that are of major ecological significance are oxygen and carbon dioxide. The oxygen content of the atmosphere is nearly exactly 21 percent or 21 cubic centimeters per liter (cc/l). At saturation, pure water dissolves 10 cc/l of oxygen at 0°C, 6.6 cc/l at 20°C, and 5.6 cc/l at 30°C. Seawater of a salinity of 35 ‰ contains even less oxygen at saturation: 8 cc/l at 0°C and 4.5 cc/l at 30°C, for example. Thus, the amount of oxygen is much lower in sea water than in air. Nevertheless, most marine organisms have relatively small oxygen concentration requirements, with many of the more active zooplankton able to utilize as little as 3 cc/l without evident impairment of functions. Shallow-water benthic forms, such as corals, seem to function normally at levels of 50 percent oxygen saturation (Yonge, Yonge, and Nicholls, 1932).

Oceanic oxygen is derived from atmospheric exchange (which is greatly speeded by turbulence) and from photosynthesis. Therefore, oxygen supplies originate in shallow water. Figure 4-29 shows the vertical distribution of oxygen in subtropical latitudes. Note the high surface concentrations, which fall to a marked *oxygen minimum zone* just above 1000 m. Oxygen depletion correlates well with the decomposition of organic matter; comparison of nutrient maxima (Fig. 4-28) shows that nutrients have maximal concentrations near the oxygen minimum layer. The trends of nutrient concentration are, in general, opposite to those of oxygen, except in deeper water, where oxygen and nutrient levels both remain rather high.

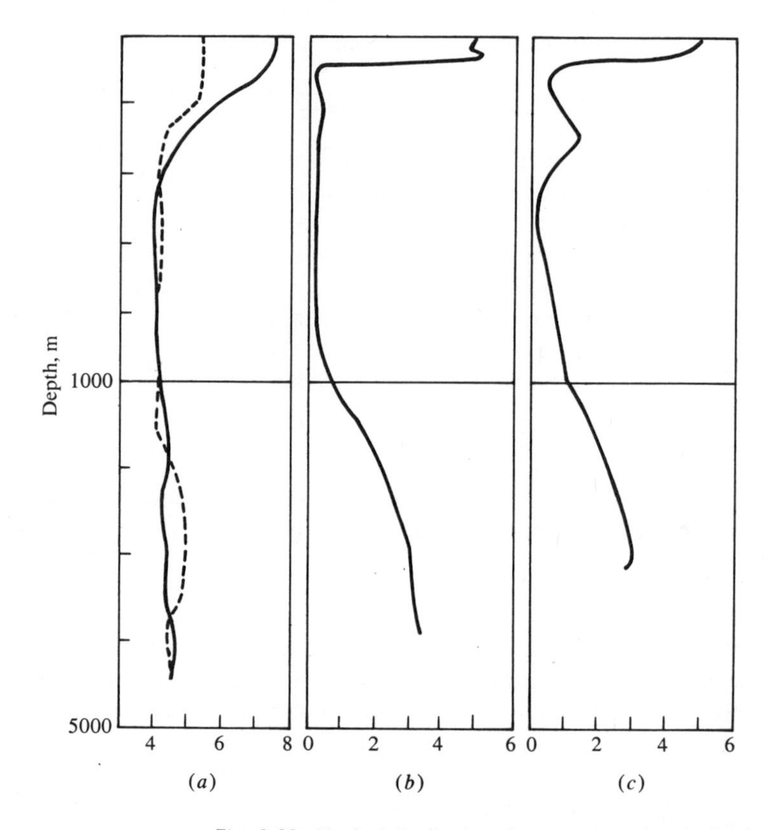

Fig. 4-29 Vertical distribution of oxygen at stations in the Antarctic Convergence (*a*) and in the eastern tropical Pacific [(*b*) and (*c*)]. Note that the curves are in general the opposite of those in Fig. 4-28(*a*) and (*b*) for nitrate and phosphate. (After Richards, 1957.)

In general, oxygen occurs at levels considerably below saturation, except near the surface where oxygen concentration is stabilized by atmospheric exchange. Even though deeper waters may be conditioned at the surface, oxygen is employed for respiration and for the decomposition of organic matter, and the supply is reduced. Nevertheless, the levels are high enough to support life even over abyssal bottoms, except for localized regions of restricted circulation where water actually becomes stagnant. Such areas cannot support aerobic life and they lack plants and animals.

Carbon dioxide is present in the atmosphere in very small amounts, about 0.03 percent, and the amount of free CO_2 present in seawater is very small indeed. However, CO_2 enters into a variety of compounds, which permit a vast increase in the total CO_2 content. First, it forms carbonic acid, which ionizes somewhat. Furthermore, CO_2 may be "bound" into bicarbonate and carbonates, and the quantity of bound CO_2, chiefly as HCO_3^-, is considerable (see Raymont, 1963). These reactions may be expressed as follows:

$$CO_2 + H_2O \rightleftarrows H_2CO_3 \rightleftarrows H^+ + HCO_3^- \rightleftarrows 2H^+ + CO_3^=$$

Addition of CO_2 to the system moves the equilibrium to the right, and results in CO_2 becoming associated with cations to form highly soluble bicarbonates. The secretion of carbonate skeletons is a matter of removing the H^+ from the bicarbonate and forming calcium carbonate, which is nearly insoluble; there is only a trace amount of $CO_3^=$ ions in seawater. These relations buffer the pH of seawater so that it does not much depart from the range of 8.1 to 8.3 in the shallow waters of the open ocean. If H^+ is added to the system, it drives the equilibrium toward the left in the preceding equation, so that

$$H^+ + HCO_3^- \rightleftarrows H_2CO_3$$

but the CO_2 concentration is not altered and the pH remains unchanged at equilibrium. Conversely, if OH is added, the equilibrium shift is

$$OH^- + H_2CO_3 \rightleftarrows HCO_3^- + H_2O$$

which leaves the amount of H^+ unaffected and, again, the pH unchanged. Local variation in oxygen, in the concentration of decomposition products, and other factors do cause local variations in pH. The pH associated with the oxygen minimum zone, for example, ranges from 7.5 to 7.7 or so, whereas in surface layers of embayments and restricted areas of the sea, the pH may approach 9 during high photosynthetic activity (Raymont, 1963).

It is interesting to relate the patterns of the nonconservative elements to the general circulation. Table 4-2 presents analyses of the major elementary composition of plankton together with those of seawater samples from the northwest Atlantic. In the plankton, for each atom of P used in building protoplasm, 6 atoms of N and 106 of C are required (line 2, Table 4-2), and about 276 atoms of 0 are set free during the synthesis. The change in concentration of these elements with depth should be in the ratios of their abundance in plankton, if the increase in nutrient levels is due entirely to the oxidation of sinking organic debris, and

Table 4-2 Ratios of the Elements Present in Plankton of Average Composition, and Those Involved in the Oxidation of Organic Matter in Seawater at Depth by Atoms

	ΔO	ΔC	ΔN	ΔP
1. Northwest Atlantic	−180	105	15	1[a]
	O	C	N	P
2. Plankton analyses	−276[a]	106	16	1

SOURCE: From Redfield, Ketchum, and Richards, 1963, after Richards and Vaccaro, 1956.

[a] Estimated assuming 2 atoms 0 ⇔ 1 atom C and 4 atoms 0 ⇔ 1 atom N.

the loss of oxygen should also be in the indicated ratio ($\Delta O/\Delta P = -276$). As exemplified by the analyses summarized in line 1, Table 4-2, the ratio of change in C, N, and P suggests derivation from a planktonic source, but oxygen change (depletion) falls far short. This is interpreted as indicating that many of the nutrients are so-called *preformed nutrients*, regenerated from plankton in previous times and now circulating with the water currents; the oxygen required to degrade these has long since been expended. Probably only the fraction of the C, N, and P that can be associated with the $\Delta 0$ of -180 is being regenerated in situ. This implies that below the light surface layers, seawater owes much of its nutrient content to the biochemical regime in which it is conditioned; subsequently, nutrient concentrations may be modified by mixing with other waters or by the sinking of locally regenerated nutrients from above.

The excretion and decomposition of marine organisms produce a varied supply of dissolved organic substances, which commonly contain a large fraction of the total nutrient elements P and N present in seawater. Whether significant amounts of these dissolved substances are taken up by animals as food has been a disputed point for over half a century, and the answer is still uncertain. It appears unlikely at present that a large proportion of the energy available to animals in marine ecosystems is routed directly through dissolved nutrients. Bacterial decomposition of the substances and their subsequent assimilation by autotrophs seems the usual pathway for return of these nutrients to the biosphere. However, some organisms base or supplement their diets on this source. Furthermore, it is certain that some dissolved organic substances play the role of "micronutrients" or vitamins in the sea, and provide to some important species of phytoplankton a supply of substances which they cannot synthesize. On the other hand, the dissolved organic metabolites of some phytoplankton species are known to be toxic to other phytoplankton and to zooplankton as well. Seawater is thus biologically conditioned, and this conditioning may possibly play an important role in the distribution, association, and abundance of marine species, especially of the plankton. Much remains to be learned of these factors.

Solar Radiation Is the Indispensable Primary Source of Energy for the Biosphere

In many respects, this book concerns the utilization of solar radiation for the creation and elaboration of biological systems. Solar radiation must be the most fundamental of all ecological factors, for this energy source supports photosynthetic activity and therefore the metabolism of nearly the entire biosphere, and also drives the circulation of the atmospheric and oceanic systems. In this section, we are concerned only with the gross aspects of light as a factor in photosynthesis, and as it affects the behavior of some invertebrates.

The amount of light intercepted by the earth may be considered as constant for practical purposes, but the amount reaching the earth's surface varies with the angle of incidence, and therefore with latitude. It also varies with scattering, absorption, and reflection by the atmosphere; cloud cover is a major source of this variability. Upon reaching the sea, light penetration depends upon the angle of incidence and upon the reflectivity of the sea surface. The light entering the water is further absorbed and scattered as it proceeds, so that the light intensity is reduced with depth. The quality of the submarine light also varies with depth, because the shorter wavelengths of light tend to be attenuated more rapidly than the longer, owing chiefly to the presence on the water of absorptive substances, presumably of organic origin. The amounts of these substances vary from place to place, being especially high in coastal waters. For a good review of these factors, see Holmes (1957). In addition to the variations in light arising from combinations of the previous causes, there is the strong latitudinal variation in seasonality of illumination. It grades from nearly similar day-night alternations throughout the year in low latitudes through progressively stronger differences between summer and winter days at midlatitudes to the high-latitude situation where the day-night changes become seasonal.

As a result of these factors, photosynthesizing organisms are restricted to the upper layers of the water where light is sufficiently strong. In general, photosynthesis and growth are proportional to light intensity except at higher intensities. At some level of illumination, photosynthesis does not increase further with increasing light intensity, and this level is approached asymptotically by the growth and photosynthesis curves.

As light decreases with depth, so does oxygen production by photosynthesis. Therefore, there is a level at which oxygen consumed by the respiration of plants precisely balances the oxygen they produce. This is the *compensation level;* above it, a net oxygen surplus is produced by plants and is available for animal respiration; below it, animals and plants as well are net consumers of oxygen, which must be imported. The depth of the compensation level varies with light intensity and therefore with all the other factors previously discussed, and also with the composition of the plankton, for different species have different photosynthetic reactions to light. In midlatitudes, it probably lies near 50 m in the open ocean, it is between 25 to 45 m in normal coastal waters, and it falls to 1 to 12 m in turbulent situations.

Photosynthesizers are often vertically zoned within the water according to their tolerances for strength of light and also to their adaptation to light of certain wavelengths; the photosynthesizing pigments of different plants vary and achieve optimum response at different wavelengths. This effect is most noticeable among macroscopic algae, for green algae tend to be most common in shallow water, brown in intermediate water, and red in deeper water.

The distribution and abundance of animals are commonly regulated by the distribution and abundance of plants, and thus are strongly affected by light variations. Furthermore, animals often respond directly to light, often so as to regulate behavior through the employment of reflex responses keyed partly to light intensity. An example has been described by Evans (1951) for the chiton *Lepidochiton cinereus*, which retreats under stones at low tide. In bright light, these chitons are stimulated to move, and will do so until they reach a shaded locale, when they stop. They are also stimulated to move on wet surfaces (as opposed to submerged or dry ones), and, when exposed to air, move downward in response to gravity. These three simple responses therefore have the result that when the tide falls exposing a *Lepidochiton*, it moves over the wet surface downward until reaching a shady spot beneath a rock, where it stops. As Barrington (1967) has pointed out, these three simple reflex responses are thus integrated into a highly adaptive behavior pattern.

Another use of light as a stimulus for an adaptive response is by the zooplankton. Since light intensities vary with depth (and opacity) of water, many zooplankters have evolved responses to light that help maintain their optimal position in the water column, which usually means the position where their food is most likely to be. They sink unless swimming, and thus by swimming toward the light at low intensities and stopping at high intensities they can remain within a certain range of light intensity; at night, different responses are present. The given response to light varies from species to species, and from one ontogenetic stage to another. For example, in some forms the stage in which they pass the winter does not respond, or responds only weakly, to light intensity and remains in relatively deep water, perhaps to conserve energy since little food is present in shallow layers at this time (Raymont, 1963).

Marine Substrates Are Important Ecological Factors, Especially on the Community Level

Many benthic and nektobenthic animals are adapted for life upon (*epifaunal*) or within (*infaunal*) a relatively narrow range of substrate types. Some investigators believe that substrate is the single most important factor in the ecology of the benthos. Certainly the obvious correlation between the animal and substrate when similar substrates are examined in different regions, and the distinctiveness of animal associations in different substrates in the same region, indicate that the substrates (or parameters highly correlated with them) exert a major influence on the composition of the biota.

In general, rock substrates and coarse sediment substrates occur in regions of high turbulence, whereas progressively finer-grained substrates are found in progressively quieter water, where decreasing competence to transport particles permits finer particles to settle out. Particulate organic matter is of low density and usually settles out with the finer sediments. Interstitial water circulation is best in coarse sediments and can be practically nil in clays. All these circumstances combine to create distinctive ecological situations associated with each major type of substrate.

For example, on rocky substrates in turbulent water, animals will ordinarily be provided with abundant oxygen and suspended food and with benthic plants. Animal associations in such environments will be composed principally of epifaunal species with efficient adaptations for anchoring (such as streamline shapes and clinging or attachment mechanisms) and will be chiefly grazers or suspension feeders and predators upon these. Some infaunal rock-borers may be present. In nearshore sands, the sediment will still contain little organic debris and may be disturbed by waves and thus have rather a high mobility. Animal associations will tend to be composed of rapid burrowers capable of maintaining their optimal position in moving sand; they will consist mostly of suspension feeders and their predators. As sediments become finer, the organic content rises and water turbidity and internal circulation decrease. The fauna tends to include more deposit feeders (see Sanders, 1968). As the sediment mobility declines, the number of epifaunal organisms may rise. Finally, the increasing fineness of the sediment renders it soft, and special stabilizing mechanisms are required of the epifauna to keep from sinking. Infaunal animals must contend with decreasing oxygen supply in interstitial water as lowered circulation and rising organic content eventually lead to reducing conditions beneath the sediment surface; of course, many animals use water from above the sediment surface. Deposit feeders are commonly dominant. Purdy (1964) has discussed sedimentary substrates.

The mineral composition of the sediments is not ordinarily a factor in itself, although it does sometimes give rise to ecological parameters. Granite boulders may be smoother than volcanic ones and support fewer large algae or attached forms; unusual sediment types, such as large oolites and pisolites, are chiefly calcium carbonate, and thus there is a correlation between the environment associated with these spheroids and their composition. In general, however, the composition is secondary in ecological importance to grain size and correlated factors.

Environment Can Be Regarded as a Hierarchic System of Regimes

From these brief reviews of only a few of the many factors that are of considerable ecological importance in the sea, it is obvious that the factors are by no means independent. To change one factor, such as water temperature or salinity, involves a great train of concomitant changes in water properties, and in discussing even the most common effects of such factors, such as light or substrates, it is necessary to consider many associated environmental parameters, such as turbidity and turbulence. Although for experimental purposes it can be essential to isolate the effects of each parameter, in nature the factors are interrelated in complex ways to form one environmental system.

Organisms are also systems, highly integrated assemblages of characters with harmoniously coadapted functions. Even if an organism responds to the variation of only a single environmental factor, the response is likely to be complex, and to involve the organism's response to other factors which have not even changed. Take, for example, a change in water temperature; this alone does not alter water salinity, yet it commonly affects the salinity tolerance of organisms as well as their tolerances for other factors, such as oxygen (Fig. 4-30; Kinne, 1964).

The effects of a persistent change in any given environmental factor are thus difficult to guess, as they commonly depend upon the state of many other factors; ultimately, the organism is responding to the environment as a whole. Furthermore, organisms frequently employ environmental clues to trigger their behavioral responses, which further expand and complicate the ecological effects. Some oysters prepare to spawn as the waters warm in the spring; others may require some weeks to manufacture eggs and associated tissues or materials. Therefore, they require a certain lead time, a period of anticipation of optimal or at least tolerable spawning conditions. When stimuli for reproductive preparations and for spawning are received as temperature, the ecological significance of the temperature is increased far beyond its mere association with activity rates. In the same way, the use of light as a stimulus for movement that is part of a highly adaptive behavior pattern extends the significance of light as an ecological factor.

Even a single organism and its environment, then, can be regarded as a sort of ecosystem. The boundaries of the system are formed by the limits of tolerance of the individual. For a given environmental situation, individuals with broad tolerances will have wide ranges, whereas individuals with narrow tolerances will be restricted to regions where the appropriate environments are localized.

Partly because environmental factors are interrelated, and partly because

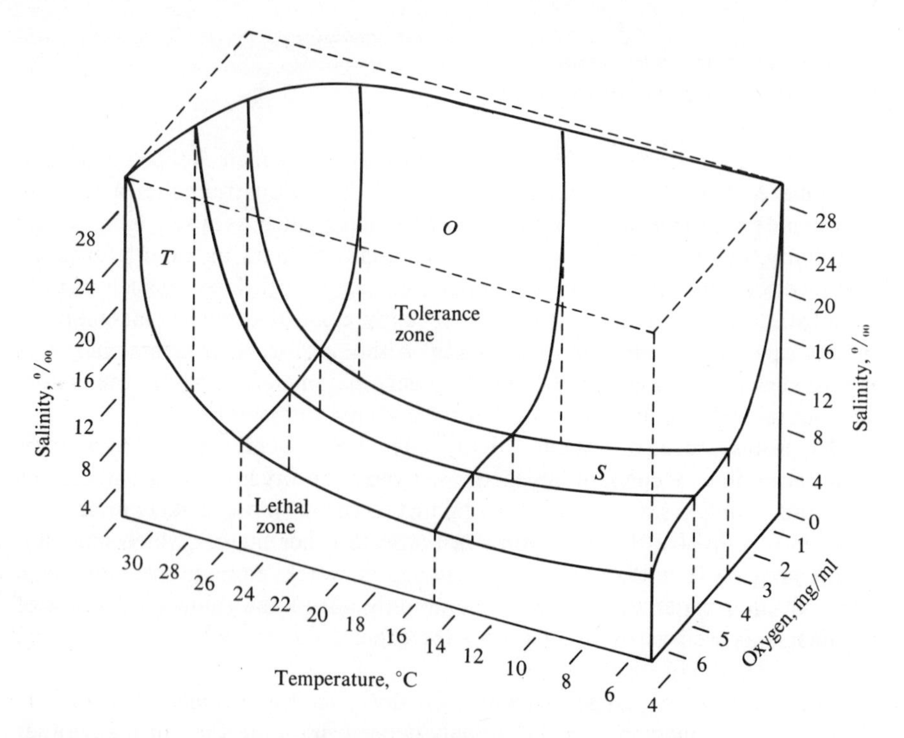

Fig. 4-30 Tolerance limits of lobsters for temperature, salinity, and oxygen, which vary in various combinations of these parameters. The letters *T*, *S*, and *O* represent the regions in which temperature, salinity, and oxygen, respectively, are the lone limiting factors. (After McLeese, 1956.)

the planet is heterogeneous, the environment does not vary in an even manner along linear gradients; in general, concentrations tend to change rapidly in some regions and slowly in others, which produces a mosaic pattern, a patchy environment with regions of reasonably similar habitat conditions (which may be broken into a mosaic of microenvironments) separated by regions of rapid change. The regions of relatively similar conditions are called *biotopes*, and may support the characteristic biotas called communities. The systems formed by these entities are, of course, ecosystems.

It is particularly important to regard heterogeneity as a character in its own right, since so many environmental factors are closely interrelated by either physical law or normal circumstance so that variations in one or a few parameters nearly always entail a great array of concomitant variations in other parameters.

These considerations make it useful to develop a general concept of environmental heterogeneity, which can be thought of as a description of the structure of realized biospace. Heterogeneity is expressed both in space, as geographic variation in environmental conditions, and in time, as temporal fluctuations of the environment. For either case, there is a problem of classifying heterogeneity because it occurs on a great range of scales. It may be expressed spatially as a very local minute change or as one that involves a gradient across a planetary hemisphere, and temporally as a nearly instantaneous small variation or as a long-term fluctuation that lasts many millions of years. Heterogeneities of all these sorts have ecological effects. It is convenient to classify the heterogeneities in a hierarchic manner and to bring them into approximate scale with the environmental aspects of the hierarchy of ecological units.

For spatial heterogeneites, then, we may recognize a class that affects the population level. These are the variations that form population boundaries, and, as we shall see, help to regulate the geographic patterns of individuals within the population and have considerable effects on adaptive strategies. In a spatially monotonous environment, even narrowly tolerant forms are able to range widely, whereas in highly heterogeneous environments, even very tolerant organisms may be restricted to localities with favorable conditions. The scale of significance of spatial heterogeneity is, naturally, different for different types of organisms; microgeographic variations that greatly affect population structure and that are limiting to minute, specialized forms often are without effect on larger and wide-ranging animals.

At the community level, the heterogeneity present within the biotope is a major determinant of the number of different species' populations that may be present. A high degree of heterogeneity implies a large number of potential habitat types, which we may say means that there is a large realized biospace, in which the ecospaces of a relatively large number of populations may be accommodated. Therefore, although the biotope is a region of relatively uniform conditions, these conditions may be highly heterogeneous within the region but similar throughout. At the provincial level, high heterogeneity is expressed by a larger number of biotopes, which permits more communities to be present. At the level of the entire marine biosphere, heterogeneity is expressed by the major variations that form provincial boundaries; the higher is the heterogeneity at this level, the more provinces may be present.

Temporal heterogeneity is expressed as fluctuations in environmental parameters. As there are a large number of rather regular environmental fluctuations arising from various periodic motions within the solar system—day and night, tidal cycles, and seasonal cycles, for example—much temporal heterogeneity is periodic. However, numerous sources of aperiodic environmental fluctuations exist, such as effects of storms and weather in general and of the many minor perturbations of the general hydrologic framework. Temporal environmental fluctuations may be described in terms of their amplitude, frequency, and regularity. Slight and regular changes with infrequent periodic perturbation are typical of the present inner tropics, and probably of many of the shelf seas of the Paleozoic. Large and regular changes, together with numerous aperiodic events, are typical of some present high-latitude environments.

Populations inhabiting environments that vary in a regular manner (curves 1 and 3) may evolve life cycles that are closely geared to the normal succession of environmental changes. If generation lengths are significantly shorter than a year, for example, a population may display a succession of two or more phenotypic modes from generation to generation and repeat the succession each year—a phenomenon called *cyclomorphosis*. If generation lengths are near a year, then reproduction may be arranged to occur at the most favorable time for incubation and larval food supply (usually when temperatures are high and development rapid), and "resting" stages may occur during inclement periods. When the fluctuations are of very low amplitude, the homeostatic mechanisms of the population may be able to compensate for the entire variation, so that no effects on population habit, size, or structure are evident. This mode of population reaction to environmental change may be called *regulatory*. When fluctuations are of high amplitude, most populations will find it necessary to alter some aspects of their environmental interactions to adjust to this change; if they are well adapted to the environmental regime, however, they should be able to approximate a given ecological state each time the environment returns to a given condition—or, at least, to do so after a reasonable lapse of time to allow establishment of some equilibrium condition. This mode of population reaction may be called *conformational*. If the fluctuations are very irregular, populations must be highly flexible in their responses and must be opportunistic in order to take advantage of the occasionally favorable conditions.

CHAPTER FIVE

THE MODE OF LIFE AND FUNCTIONAL RANGE OF FOSSIL ORGANISMS

It is an indispensable principle that structure must be considered in relation to function; in isolation it is meaningless.

—R. B. Clark, *Dynamics in Metazoan Evolution,* Clarendon Press, Oxford, 1964.

The individual organism (or, sometimes, the colonial "superorganism") is the ecological unit on which natural selection actually functions, for the relative success of individuals in propagating a sample of their genotypes determines the course of evolution. The extent to which one individual proves to be superior to another in genotype propagation is a measure of the relative "fitness" of individuals. Fitness will vary with the environment, of course—it will vary with changes in any factor that affects relative genotype propagation.

Although they are selected in individuals, many of the factors that contribute to fitness are actually adaptations at higher levels of the ecological hierarchy, such as the population or community levels. For example, a high reproductive potential might be required of a given population for survival under given environmental conditions. It is achieved through selection for fecundity on the part of individuals; those individuals that produce the most young propagate more of their genotypes, which tend to be programmed for high fecundity, and selection for other differences may become relatively less important. Thus, although the population parameter of "high reproductive potential" is achieved by selection on the individual level, the increased fecundity does nothing special for the individual, except in the sense that it permits the population to survive, and thus the individual to occur. In fact, it involves an expenditure of energy that cannot be employed for adaptations on the individual level.

In this chapter, we are interested in those adaptations that benefit individual organisms as such. It is convenient to consider these in three parts: the adaptations for very early life, which are commonly not preserved in the fossil record; the adaptations to accommodate growth; and the integration of adaptations for feeding, protection, maintaining a place in the habitat, and all the other factors that together form a characteristic mode of life.

Paleoecologists must reconstruct these adaptations to explain the data of the fossil record. On the one hand, they must reconstruct the environment itself, the ancient biospace. On the other hand, they must reconstruct the organism from its fossil remains. Finally, they must reconstruct the former organism-environment interactions, the functional range of an individual—its ecospace. Because the fossil record is fragmentary, paleoecologists must make extrapolations from such data as they possess, partly by constructing functional models (usually conceptual models) of the organisms. In Chapter 1 we reviewed the possible sources of ideas for such models. The richest source is simply the present biosphere.

The First Formative Stages of Ontogeny Are Very Important for Understanding the Functional History of the Individual, Although They Are Rarely Preserved in the Fossil Record

The growth of a multicellular organism from a zygote involves not only increase in size but also development of form, which may be accomplished in several ways. One is simply by increase in the size of the cells. Actually, this is an uncommon method among multicellular organisms; usually cell size is independent of organism size. The other methods involve cell multiplication. One is through the multiplication and then the differential movements of cells, which migrate from the site of their formation. This can provide more cells to some regions than to others, resulting in a differential growth that generates a form. Cell movements of this sort—*morphogenetic movements*—are very common at early growth stages in animals. Another method involves control of the direction of division of multiplying cells that do not migrate. When all or most cell divisions occur in a given direction, growth ensues at right angles to the plane of the cell divisions. Form is thus controlled by the direction and amount of divisions. This is an especially common sort of morphogenesis in plants, and is not uncommon in animals. Additionally, growth may be accomplished by the localized multiplication of cells in certain growing regions that retain their structural or morphological positions. This is a common morphogenetic mode in plants, for example in the increase in stem length due to the operation of a growing region, the *meristem*, at the tip. This mode of growth is also well known in animals; the expansion of the mantle and accompanying shell in brachiopods and mollusks involves a generative zone in the mantle lobes. Finally, growth is sometimes achieved by rather random, helter-skelter cell divisions, the geometry of which cannot be related in any simple way to the geometry of the form that results.

The development of every multicellular organism involves differentiation as well as growth. Differentiation is the creation of new types of cells with different properties from the mother cells. This is absolutely necessary for truly multicellular organisms, for simply multiplying a single-celled organism does not generate many special advantages; the advantage of multicellularity lies in the specialization of functions among different parts. The differentiation of cell lineages into numbers of special types gives rise to a variety of tissues and eventually to organs with separate functions.

Inasmuch as most adult multicellular organisms are orders of magnitude larger than their zygotes and contain differentiated cells (sponges), tissues (diploblastic metazoa), and organs (triploblastic metazoa), they can be expected to have different ecologies during their life histories; in particular, their very young stages would have different requirements from their adult forms. Plans of development are commonly rather distinctive among different phyla, and, of course, there are as many specific developmental pathways as there are species. However, the general problems of early growth are similar anong most marine invertebrates. There are two major strategies for coping with early growth, and sometimes closely allied species follow different strategies. In one, zygotes develop directly into adult form by relatively even or gradual differentiation and growth, and the early environment is an egg. In the other, zygotes develop first into larvae, which leave the egg and which are morphologically and ecologically different from adults and change into adults by a relatively abrupt process of metamorphosis.

Most marine invertebrates have larval stages, and most of the larvae are planktonic. About 95 percent of marine species in tropical waters have planktonic larvae, although this proportion drops to between 55 and 65 percent in boreal provinces (Thorson, 1950). Although larvae are rarely found as fossils, their properties are of considerable importance in determining adult living patterns. It is therefore valuable to consider the more important phenomena of larval life.

The major contributions of the larval stage appear to be three: (1) simply as a stage of growth and differentiation that leads from the zygote to the adult; (2) as an instrument of dispersal of the population; and (3) as an instrument of habitat selection for the adult. The first contribution, that of growth and differentiation, must be made in any ontogeny and may be by a process of direct development rather than larval development. Mortality among planktonic larvae is high, so there is a great wastage of zygotes and consequently an associated selective pressure for great fecundity, which requires much energy. Nevertheless, the advantages of maintaining a planktonic larval stage are evidently great enough to offset the difficulties it entails.

A great part of the advantage must be the ability for dispersal of young in the planktonic stage. Although some larvae remain in the plankton only a few hours, most remain from several days to a few weeks, and in some cases the planktonic stage is prolonged for several months. During this time, the larvae may be transported passively far from their parental population and settle in a new place. This avoids competition of young with their own parents and promotes hybridization between offspring of separated adult populations that could not otherwise exchange genes. Furthermore, it provides a population with the potential for spreading out into new geographic regions whenever they become available for occupation, and prevents permanent range restrictions and undue reduction of species population size that can result from locally disastrous mortality, from whatever source.

The importance of dispersal may be often related to the third major contribution of the larvae to an ontogeny—the selection of a suitable habitat for the adult. When the planktonic larvae of sessile or weakly locomotory animals settle to begin a benthic life, they commit their adult stages to an existence at or near the point of settling. Therefore, it is a great advantage if the larvae can exercise some selectivity in settling, and choose a habitat that is propitious for the oncoming adult stages. The larvae of the barnacle *Balanus balanoides* are adapted to seek rock surfaces in shallow water, suitable for adult inhabitation. If they do not find a suitable settling surface, they may delay metamorphosis and continue to search for up to two weeks. Not all larvae exercise very much habitat selection, but a number of cases are now well known, and it is likely that the function is widespread.

Most planktonic larvae feed on the smaller plankters, a habit termed *planktotrophic.* Others do not feed on the plankton at all, but depend upon the yolk of their eggs for food, even though they may spend some time in the plankton after hatching; these are called *lecithotrophic.* Perhaps 10 percent of tropical and temperate species are lecithotrophic. In the arctic, a great many species do not have larvae at all, but develop directly in large yolky eggs. This is probably an adaptation to the paucity of planktonic food during much of the arctic year (Thorson, 1950). Furthermore, direct development is evidently considerably more econimical; in two allied species living in similar environments, one with larval development produced far more egg cytoplasm than one with direct development (Chia, 1970). Under arctic conditions, the hazards of planktonic life have come to outweigh the advantages for many species. Even among species with indirect development, some have larvae that are bottom dwelling, some of which feed and some of which do not (Pearse, 1969).

Benthic species with planktonic larvae lead a sort of double life and have bimodal ecospaces. Selective pressures act upon the larval portion of their ontogenies to adapt it to a planktonic habitat, and act upon the adult portion to adapt it to a benthic habitat. Each habitat has a somewhat different set of physical parameters and is certainly associated with different organic communities. The morphology and functions of the two life stages are therefore quite distinct and may evolve independently. However, the two modes must be coupled into an integrated life cycle. The coupling places some limits upon the extent of specialization permitted to the larvae, for it is necessary that the basic structure of the adult be prepared if not actually laid down. The coupling process, then, consists chiefly of structural constraints upon larval morphology, habitat selection if it occurs, and the metamorphosis itself.

Since different stages in the life cycle may have radically different morphologies and functions, the total functional repertoire of an individual is thereby greatly increased over that at any one stage. Should a function that is present at one stage subsequently prove valuable at another, the shifting or extension of this function in ontogeny is sometimes possible, bringing it into association with different ontogenetic stages and thereby creating a new functional assemblage. Ontogenetic shifts of this sort may result in major evolutionary steps. The classic example of this sort of process is the shift of the onset of reproduction into early ontogenetic stages (through a speeding up of sexual maturation or a retarding of the development of other adult features). This brings adult status to larval forms, which become self-reproducing and can complete their life cycle without metamorphosis. In the case of functionally bimodal life cycles, this process can eliminate the adult mode, and further evolution tends to be toward increasing adaptation and diversification of the former larval mode. The embryologist Garstang suggested in 1894 and later that the phylum Chordata may have arisen from the planktonic larval stage of a benthic adult which acquired reproductive functions; this remains a leading hypothesis of chordate origins. Similarly, the first benthic worms may have evolved from the larvae of coelenterates.

A large number of terms have been coined to describe the introduction of new characters at particular ontogenetic stages and the changes in character associations brought about by extending the time range of characters within ontogeny. Most of these terms originally served the theory of recapitulation, which postulated in its extreme form that the succession of morphologies in the ontogeny of an organism recapitulated the succession of adult stages of the ancestral forms in the phylogenetic history of the organism. This view is no longer tenable (De Beer, 1958). Some of the terms are still useful, however, and these include *caenogenesis*, the introduction into a juvenile stage of a novel character that does not affect the adult, and *neoteny*, a process of retardation of development that brings juvenile characters into association with adult characters. The retention of larval characters into an adult stage as visualized for chordate origins by Garstang is an example of neoteny. Since characters that arise early in ontogeny tend to be more generalized than those that arise later, neoteny may provide a means of escape from the path of incessant specialization (Hardy, 1954) and may form an important evolutionary mechanism.

The larval stages of marine invertebrates are rarely known as fossils. They are unskeletonized or their skeletons are so small and fragile, and often constructed of metastable minerals such as aragonite, that they are not commonly preserved. It is usually necessary to infer the developmental strategy that a fossil lineage has had from indirect evidence. Some fossils appear to have had brood pouches, suggesting that they began independent life in an adult form. Our speculations on the early lives of fossils can at least be qualified and guided by a knowledge of the early ontogeny of their close living allies, and of the larval adaptations usually displayed by organisms that are ecologically analogous in some respects. It is certainly worth the trouble to infer early life stages, for it is necessary to consider the entire ontogeny of an organism in order to understand its full functional range.

Growth Presents a Number of Functional Problems, Often Resulting in Changes in Form

Assuming that an organism has a constant density, growth implies an increase in volume and therefore an increase in some linear dimensions. Often, growth follows some basic pattern, such as is graphed in Fig. 5-1, but many different growth patterns are known that commonly involve abrupt changes in growth rates at certain ontogenetic stages. The growth rate changes are sometimes correlated with a change in form, as in metamorphosis, or with the onset of reproduction. A few generalities about growth are evident from the curves in Fig. 5-1. Organisms characteristically have an S-shaped growth curve when their weights are plotted against their ages [Fig. 5-1(*a*)]. Although the *growth rate* may reach a peak early in life [Fig. 5-1(*b*)], the *specific growth* of an organism —that is, the measure of the increase in living substance per unit of living substance present—declines throughout life [Fig. 5-1(*c*)], indicating that living matter progressively loses the power to multiply itself at the rate at which it was formed (Medawar, 1945). This curve is not concerned merely with the *addition* but with the *multiplication* of living matter. Although the specific growth rate always declines with age [Fig. 5-1(*d*)], it falls more and more slowly as age advances. As Minot (1908) has said, organisms age faster when they are young.

There are wide differences in intrinsic growth rates among different organisms, and marked environmental growth controls as well. Figure 5-2 shows shell growth data from 12 species of bivalves; even within this single class there is much variation, as between *Mytilus californianus* (curve 12), a rapid grower, and *Cardium edule* (curves 1 and 2), a slow grower. Species in the lower, warmer latitudes tend to grow rapidly and to be short-lived, whereas in higher, cooler latitudes, they tend to grow slower and live longer; contrast *Siliqua patula* from

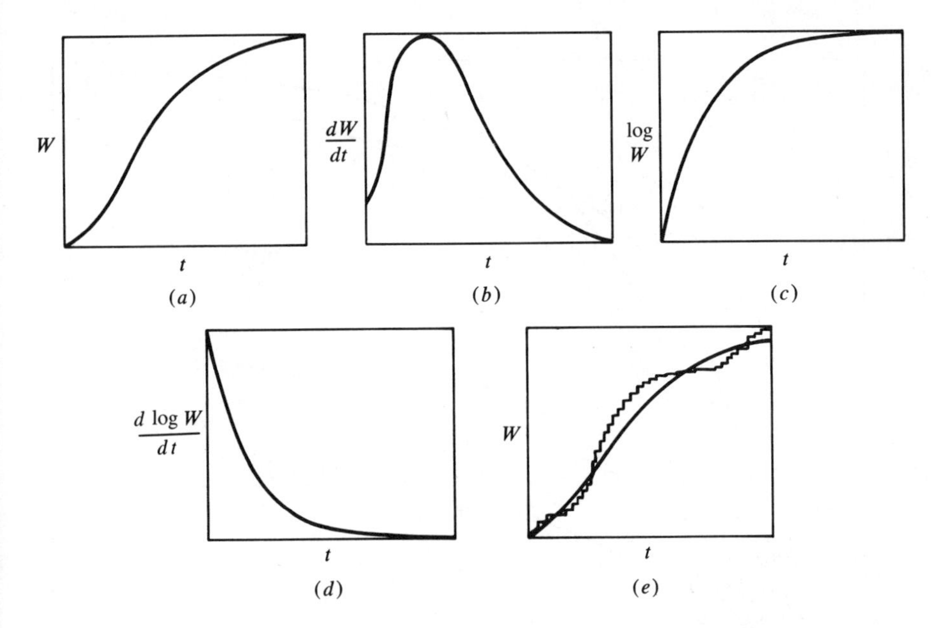

Fig. 5-1 Several aspects of growth; W is the weight of an organism, t is time. (*a*) Simple *growth* curve of weight against time. (*b*) Curve showing variation in the *growth rate*, based on the growth curve in (*a*). (*c*) Curve showing variation in the *specific growth*, based on the growth curve in (*a*). (*d*) Curve showing the *specific growth rate*, based on the growth curve in (*a*). (*e*) Curves indicating diagrammatically the actual course of growth, allowing for seasonal and other variations (tidal, diurnal) in growth rate. [Parts (*a*) through (*d*) after Medawar, 1945.]

California (curve 11) and Alaska (curve 8) in Fig. 5-2. Where seasons are relatively similar, as in parts of the inner tropics, growth may be more or less continuous, whereas where seasonality is high, growth may be concentrated in certain productive months and cease altogether in unproductive periods (usually winter and early spring). Growth curves that included seasonal effects would have a staircase appearance. Indeed, since skeletal growth commonly occurs only at particular times of day, precise growth curves would have minute diurnal steps imposed on the larger seasonal steps, all following the general trend [Fig. 5-1(*e*)]. Most growth curves are, obviously, smoothed to eliminate these irregularities.

One of the more significant and pervading effects of growth follows from the simple geometric fact that, for an increase in linear dimensions, the area of a body of a given shape increases as a square, but its volume increases as a cube (Fig. 5-3). The classic account of the biological problems posed by this fact is by Thompson (1917), and a review by Gould (1966) provides recent information; much of the discussion here is drawn from these works.

The surface area/volume (A/V) problems of growth are important partly

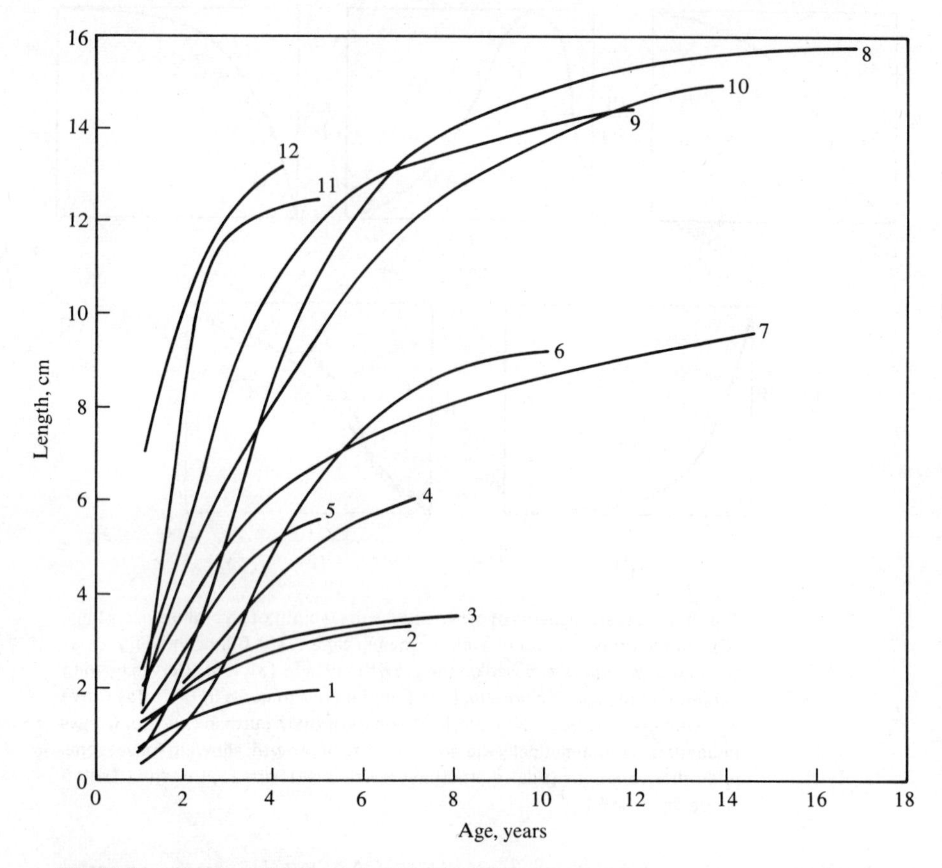

Fig. 5-2 Growth curves for various species of bivalves: 1 and 2, *Cardium edule;* 3, *Venus gallina;* 4, *Mya arenaria;* 5, *Mytilus edulis;* 6, *Cardium corbis;* 7, *Mercenaria mercenaria;* 8, *Siliqua patula* from Alaska; 9, *Pecten maximus;* 10, *Tivela stultorum;* 11, *Siliqua patula* from California; 12, *Mytilus californianus.* (After Hallam, 1967, who lists authorities. By permission of the Palaeontological Association.)

because environmental exchanges occur through surfaces, and less exchange per unit volume can be accomplished as size increases. Thus, for example, to maintain a constant metabolic regime, respiratory oxygen exchange must increase apace with body volume, but unless there are changes in shape the area available for respiration increases far more slowly than the body volume. This problem is often solved by convoluting or branching the respiratory surfaces or otherwise enlarging them disproportionately to general body area so as to maintain a constant proportion to a unit of volume. A similar solution occurs commonly with feeding and digestive surfaces and to other surfaces that must enlarge with volume rather than area for physiological reasons.

As volume increases, the interior of an organism becomes more remote from

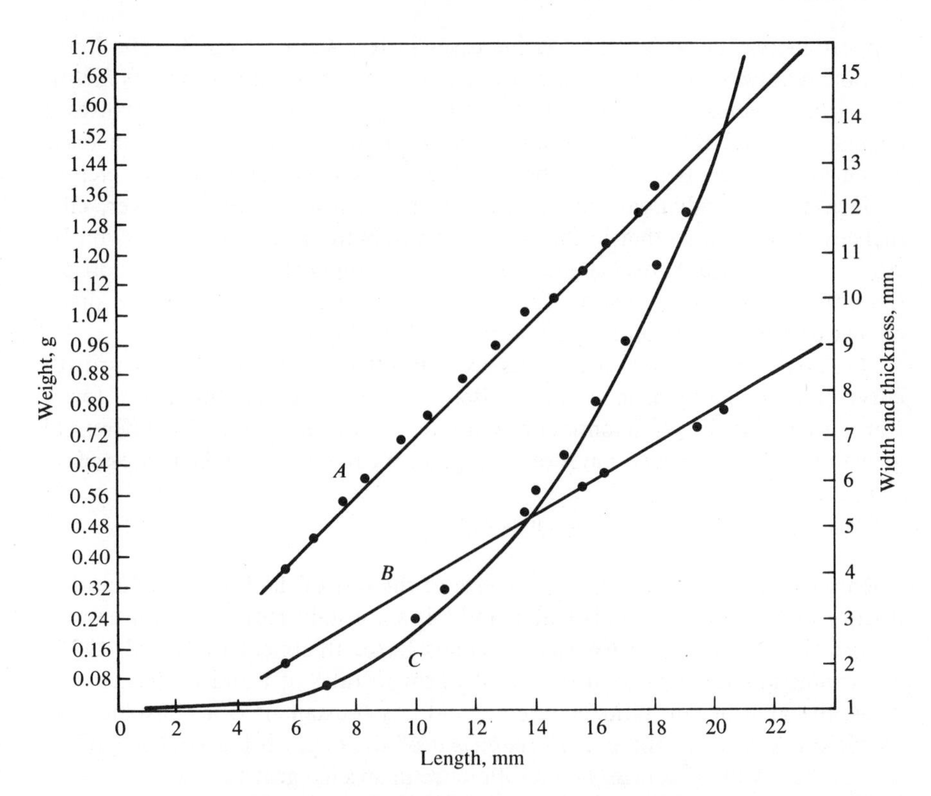

Fig. 5-3 Growth in several dimensions of *Donax* (*Latona*) *cuneatus*. Line *A* represents length and width, which are isometric; line *B*, length and thickness, which are simply allometric; line *C*, length and weight, the ratio between which constantly changes with size. (After Nayar, 1955; and after Wilbur and Owen, 1964.)

the environment. This makes internal regulation easier and promotes homeostasis, but requires more elaborate systems of communication and supply from the exterior. Nerve networks and circulatory pathways must be greatly lengthened if the same service per unit volume is to be continued. Of course, many functions are accommodated to the need to serve larger volumes of tissue by an increase in efficiency that does not necessarily involve size changes. Nevertheless, the A/V ratio is clearly responsible for many such innovations. A/V problems are a constant accompaniment of growth, it is common for an organism to lay the morphological groundwork early in ontogeny for structural complexities that are to be required in later, larger stages of the life cycle.

Since weight increases as the volume rather than as the area, area/weight problems occur during growth. These are more serious to terrestrial than to marine animals, since in the sea the weight of organisms is partly supported by the water. The cross-sectional areas of limb bones, for example, must be increased disproportionately to their lengths to remain strong enough to support increasing weight as size increases in terrestrial tetrapods. The limbs become relatively more bulky and stumpy, so that size is limited by this factor. In marine organisms, size is not so limited, but size change often involves a change in the relative energy expended in locomotion. In fish, muscles increase as the volume increases and thus provide the necessary power with size increase; in fact, for hydrodynamic reasons, the fastest fish are large. However, cilia increase with area, so that in the case of invertebrates and larvae that swim by ciliary action, surface area becomes inadequate to support enough cilia to propel the organism as size increases. Either the area must be elaborated by shape change or a new locomotory method must be employed.

Different parts of most organisms grow at different rates, so that the growth curve of a whole organism is not precisely related to the development of form. Form is generated by differential growth rates in different parts. The differential growth ratio between two parts of an organism can commonly be expressed as

$$y = bx^k$$

when y is the size of an organ, x the size of the rest of the body or of another organ to which comparison is made, and b is a constant required in the expression so that y does equal bx when k equals 1 (see treatment by Gould, 1971). The exponent k is the ratio of the specific growth rates of y and x, respectively. Thus, the growth-rate ratio between x and y is constant if k is constant. The coefficient k is called the *growth coefficient* of the organ being studied. When $k = 1$, there will, of course, be no difference in specific growth rates between x and y (assuming that they are dimensionally equivalent), although there may be a difference in size as expressed by b. As growth continues, the addition to both x and y will be equal. This is called *isometric growth.*

When k is not 1 but remains constant over a certain size range, there will be a progressive change in proportions between x and y, which is called *allometric growth.* When x is total body size and k remains constant, the organ represented by y will grow disproportionately larger when k is greater than 1, and smaller when k is less than 1. This is simple allometry, and it has been found that the growth patterns of many organs in a wide variety of animals can be described by this expression. In Fig. 5-3, curves *A* and *B* show isometric relations between dimensions of the living bivalve *Donax cuneatus.* In Fig. 5-4, inferred growth-rate ratios between width and length and between width and thickness of the Carboniferous brachiopod *Brachythyris ovalis* are plotted as straight lines on an arithmetic graph, implying that k was constant for both ratios. The data were based upon measurements of a number of individuals of different sizes. A number of examples were presented by Huxley (1932); especially striking is the case of the large claw or chela of the fiddler crab, *Uca pugnax* (Fig. 5-5). The chela increases in weight at a greater rate than the body, and k is constant from a very young stage until sexual maturity, when the value of k changes abruptly but then again remains

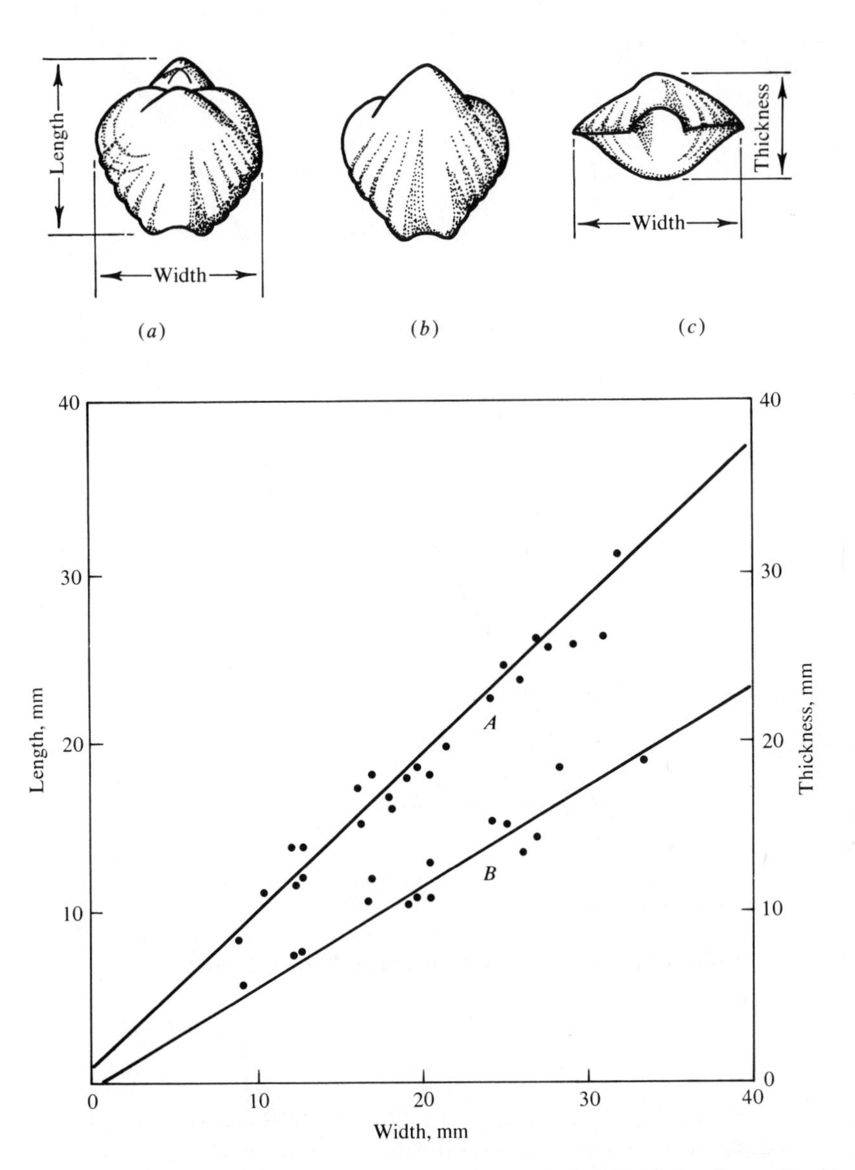

Fig. 5-4 Growth of several dimensions of a Carboniferous brachiopod, *Brachythyris ovalis*, shown at top in dorsal view (*a*), ventral view (*b*), and anterior view (*c*). Line *A* represents length and width; line *B*, thickness and width. (After Parkinson, 1960.)

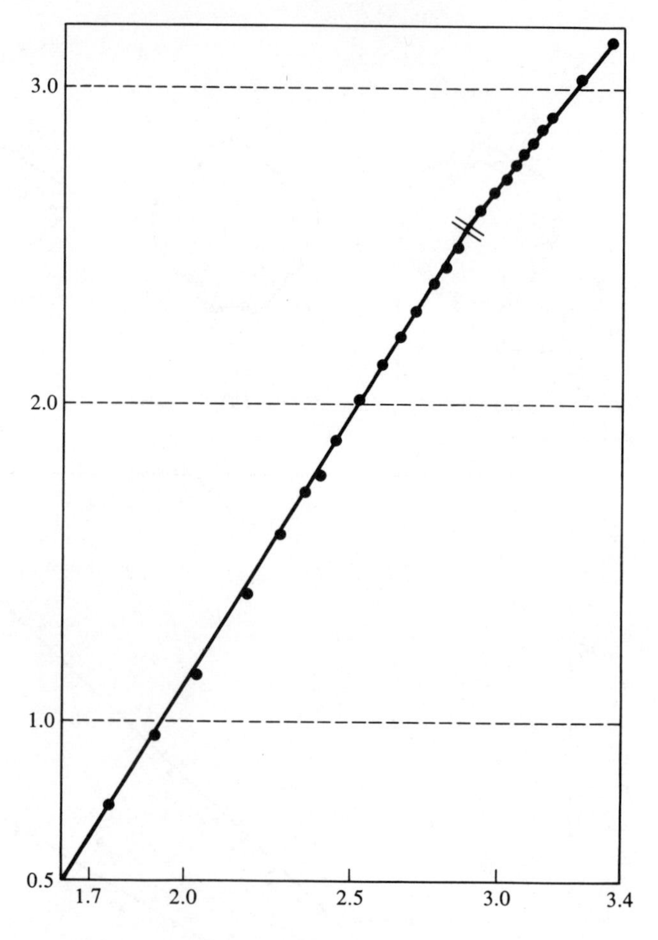

Fig. 5-5 The chela weight of male fiddler crabs plotted against total body weight; log plot. (After Huxley, 1932.)

constant at a smaller value, evidently until death. Now the equation

$$y = bx^k$$

can be written

$$\log y = \log b + k \log x$$

and in this form the function will plot as a straight line if k is constant. Figure 5-5 is based on such a logarithmic transformation, and the change in k (from about 1.62 to about 1.25) is evident. A sharp ontogenetic change in k is a fairly common phenomenon, occuring frequently at metamorphosis, when there may be a general reorganization of growth patterns, and at sexual maturity.

When k remains constant over a size range, it means that the ratio of specific growth rates [Fig. 5-1(c)] of x and y is remaining constant. There are also many cases when k is gradually altered and the logarithmic plot of the data forms a smooth curve. Rhythmic ontogenetic fluctuations of k and irregular changes that cannot be easily quantified are also recorded.

A *growth gradient* results when there is a gradient in k along one or more dimensions of a body or organ. Imagine the diameter of a long cylindrical organ with a high diametrical k at its distal end grading to a low k at its proximal end. During growth the distal end would expand disproportionately and the cylindrical organ would become a truncated cone. Form is obviously generated in a great many cases by just this sort of growth gradient.

The Constellation of Ecological Functions Performed by an Individual Is Coadapted and Integrated into a Particular Mode of Life

In addition to special functional problems of very early life and of growth, the individual is at all times faced with a great variety of physical and biological conditions and problems, and its responses to this constellation of conditions must be coordinated. This means that a compromise must usually be reached between the conflicting demands of different functions. A simple example is the timing of metamorphosis; ordinarily, the later it occurs the better is the larval dispersal, whereas the earlier it occurs, the better is the larval survival, because predation on larvae is particularly high. Obviously, the precise compromises reached between competing functions will favor those for which selection pressure is stronger.

In a way, these tendencies will favor specialization, for each adaptive response may place constraints upon the range of other possible responses. And, in a way, they will favor generality, since the adaptive potentials of a lineage will rarely or never be focused entirely upon a single factor, and thus the ultimate response to any given factor may not be realized. Instead, we find a harmoniously balanced, coadapted suite of responses. This is not to say that certain functions cannot evolve fairly independently under appropriate conditions. This is the functional aspect of *mosaic evolution*, usually noticed when one character or character complex undergoes evolution which changes its morphology while the remaining characters are stable—or when some characters change more rapidly than others.

The major ecological functions of an organism are easily grouped into such categories as feeding, locomotion or attachment, protection, respiration, reproduction, and so on. It is possible to erect a classification for each of these major functions, and to employ it as the basis of an ecological classification of organisms. For example, organisms may be grouped according to their food types and feeding habits, which will provide a very basic sort of classification, since food relations are fundamental aspects of community structures. Table 5-1 summarizes the major feeding habits of some important marine organisms. A feeding-type classification must consider several aspects of feeding, such as food

Table 5-1 Major Feeding Habits, by Food Type and Feeding Method, of Marine Invertebrates

FOOD ORIGIN, SIZE, AND LOCATION				FEEDING METHOD									REMARKS ON CHIEF FEEDING TYPES
				Biters & raspers	*Scrapers*	*Swallowers*	*Ciliary currents*	*Mucous nets*	*Setae & cirri*	*Adhesive tentacles*	*Pseudopods*	*Suckers*	
Plant	Living	Small	Water column				X	X	X	X	X		Suspension feeders
			Atop substrate	X	X								Browsers on sessile diatoms, etc.
			In substrate										
		Large or masses	Water column										
			Atop substrate	X									Browers on larger algae & phanerogams
			In substrate										
		Fluids	Water column										
			Atop substrate									X	Plant-juice feeders, as Saccoglossans
			In substrate										
	Dead	Small-particle detritus	Water column				X	X	X	X	X		Suspension feeders
			Atop substrate		X	X					X		Surface detritus feeders[a]
			In substrate			X							Sediment feeders[a]
		Large or masses	Water column										
			Atop substrate	X									Detritus feeders on algae & phanerogams
			In substrate										
		Fluids											

Animal	Living	Small	Water column				X	X	X	X	X		Suspension-feeding predators
			Atop substrate	X	X								Small predators
			In substrate	X									Small predators
		Large	Water column										
			Atop substrate	X									Large predators
			In substrate	X									Large predators
		Fluids	Water column										
			Atop substrate									X	Parasites
			In substrate									X	Parasites
	Dead	Small	Water column										
			Atop substrate	X	X								Small scavengers
			In substrate	X	X	X							Small scavengers
		Large	Water column										
			Atop substrate	X									Large scavengers
			In substrate	X									Large scavengers
		Fluids											

[a] Food commonly includes or consists of bacteria.

Table 5-2 Major Methods of Locomotion or Attachment Among Benthic Marine Invertebrates

INFAUNAL			EPIFAUNAL			LOCATION AND MOBILITY
Sessile	*Semivagrant*	*Vagrant*	*Sessile*	*Semivagrant*	*Vagrant*	*Locomotion or attachment method*
					X	Cilia
					X	Pedal waves
		X			X	Limbs—tentacles
	X	X			X	Peristalsis
	X	X			X	Hydraulic system
	X		X	X		Pedunculate
			X	X		Byssate
			X			Hold fast
			X			Cemented
X			X			Resting free
X						Boring—burrowing

origin and quality, food size, food location, and the food-gathering and ingesting mechanisms of the feeders. These aspects are somewhat intermingled in Table 5-1; a true multifactorial classification would take many tables to present. The column labeled Remarks on Chief Feeding Types on the right side of the table summarizes the feeding relations shown in the body of the table, which are already rather generalized, into a few general classes that are employed in Table 5-3. The complexity that is possible and that is perfectly legitimate within an ecological classification is suggested by Table 5-1, since the divisions in the body of the table could easily be subdivided much further, and the subdivided classes employed in Table 5-3.

Table 5-2 presents a simplified classification for another major ecological function, locomotion or attachment, for benthic forms. The categories may be multiplied or reduced to serve the purposes of the classifier. Now it is easy to make a classification based on both the feeding and the locomotion and attachment functions (Imbrie, 1959; McAlester, 1968). Such a classification, based on generalized categories that are not subdivided by feeding or locomotory methods, is depicted in Table 5-3. Some combinations of the categories of these functions are common whereas others are unknown. The common combinations seem to be wholly reasonable associations of functions, such as high mobility with predatory feeding or sedentary habits with suspension feeding. The combinations that are not known to exist seem to be unlikely functional associations, such as rock-boring with predation. Nevertheless, unusual combinations of functions may be advantageous in unusual circumstances, and we should not decide that

Table 5-3 Classification of Some Invertebrate Taxa by a Combination of Feeding and of Locomotion or Attachment Type

LOCOMOTION OR ATTACHMENT		HERBIVORES				CARNIVORES	
		Consumers		Recuperators		Consumers	
		Browsers	Suspension feeders	Detritus and sediment feeders	Scavengers	Parasites	Predators
Epifaunal	Vagrant	Gastropoda Echinoidea Amphineura	Crustacea	Crustacea	Gastropoda Annelida Crustacea	Gastropoda Crustacea	Gastropoda Annelida Crustacea Asteroidea Crinoidea Ophiuroidea
	Capable of site change		Gastropoda Bivalvia Porifera				Coelenterata
	Immobile		Ectoprocta Articulata Bivalvia Gastropoda Crustacea Crinoidea				Coelenterata
Infaunal	Vagrant		Bivalvia Annelida(?)	Bivalvia Gastropoda Annelida Ophiuroidea Echinoidea	Crustacea		Gastropoda Crustacea
	Capable of site change		Bivalvia Annelida Inarticulata				
	Immobile		Porifera Bivalvia				

any combination is impossible biologically unless it is actually impossible physically.

It is a simple matter to expand further the two-function classification by adding yet another function—for example, protection. A classification of protection would include boring (which is actually already in Table 5-3, since it serves two functions) and high locomotory speed. Boring would not ordinarily be adopted as a protective device by predators, whereas locomotory speed is physically impossible for cemented forms and would be unlikely for sediment feeders. Thus, many of the combinations of the three functions will never have occurred in real organisms. We may continue to add functions to the classification. As it gets larger, the graph representing it becomes a multidimensional lattice, and the functions may be arranged in the same manner as in the biospace lattice. In other words, the functional classifications in Tables 5-1, 5-2, and 5-3 are ways of representing various biospace regions by means of fairly large classes, so that the structure is simplified and becomes clearer and more easily comprehensible.

Bivalve Functions Are Harmoniously Integrated

Any taxon could be used as an example of the manner in which functions are coadapted to form a mode of life. It is convenient to use the Bivalvia, a relatively homogeneous taxon that nevertheless displays a variety of form and function which has been appreciated widely only in the last few decades. Figure 5-6 depicts a few of the more common adaptive types of Bivalvia; there are three major feeding methods: by labial palps [Fig. 5-6(*g*) and (*h*), by septal pumping [Fig. 5-6(*o*)], and by ctenidial filtering (the remaining parts in Fig. 5-6). Bivalves may swim, may be cemented to hard substrates, or may be attached by special fibers called byssal threads; they may bore into hard substrates or nestle in crevices or the burrows of other organisms; or they may creep or dig through soft substrates, wherein they may live shallowly or deeply, in coarse or fine sediment. Some bivalves take in water well above the sea bottom, some take in water (which may include resuspended particles) just at the surface, and some take in water below the surface from tubes or simply through sediment interstices. Some siphonate forms take in surface sediment and feed hardly at all on suspended material. There are many variations in habits besides these few. In each case, the bivalves are well integrated to carry on their particular mode of life. Let us examine one family in more detail, as an example of the sorts of functionally harmonious assemblages of characters that serve to form each adaptive type.

The family Lucinidae is composed of eulamellibranch suspension feeders that burrow to moderate depths by bivalve standards, depths often several times the length of the shell. Major features of the functional anatomy of the family and of its close relatives have been described by Allen (1958). Unlike most moderately deep-burrowing bivalves, lucinids do not utilize posterior siphons for the intake of feeding and respiratory currents (Fig. 5-6). Instead, they are provided with water currents via an anterior tube, which communicates with the surface and leads to an anterior mantle gap just in front of the anterior adductor muscle. This tube is built by the foot, which is modified into a long vermiform organ anteriorly (Fig. 5-7); the vermiform portion of the foot is capable of extension to several times the shell length. The tip of the vermiform foot is encircled by a ring of mucous glands. A more conventionally shaped posterior part of the foot is sometimes well developed; it is called the *heel*, and is chiefly employed for burrowing. The foot is attached to the shell by two pairs of pedal muscles, which insert just above and slightly central from the adductor muscle attachments.

The animal burrows into the substrate primarily by using the heel and employing a seesaw rocking motion of the shell. Most lucinids burrow only slowly, descending in the sediment with the plane of commissure more or less vertical and the hinge upward; when settled, they usually retain this upright position [Fig. 5-6(*k*)], although a few species have other orientations. The usual upright position is unlike that of most bivalve burrowers, which orient themselves with anatomically posterior margins upward, extending their posterior siphons thence to the surface, as in Fig. 5-6(*l*) and(*m*). After a lucinid has finished burrowing, the vermiform foot is extended gradually to the surface by alternate expansions and contractions. As it progresses, the sides of the tube that it is creating are impregnated with rings of mucus, which harden and preserve the form of the tube. The foot may then be withdrawn from the completed tube, which maintains its shape and serves as an inhalant passage. The foot may be extended within the tube periodically to clean out debris or make repairs.

Water currents that are drawn down the tube by ciliary action enter the mantle cavity above and in front of the anterior adductor muscle, which is elongated in a unique manner for bivalves (Fig. 5-7); the elongated faces of the muscle are ciliated, and food particles are sorted thereon. Particles that are accepted are passed onto the palps and into the mouth. The gills are aerated by respiratory currents that are directed by gill cilia. Water currents then exit posteriorly via an exhalant siphon. "Mantle gills," respiratory folds in the inner mantle wall, lie in the inhalant current around and beneath the anterior adductor, near where the current first enters the mantle cavity. The posterior exhalant siphon may be extended beyond the shell to several times the shell length. The siphon is retractable, but does not retract into a marginal mantle space to create a pallial sinus. Instead, it turns inside out when contracted and lies in a cavity behind and above the gills in the *suprabranchial chamber* (Fig. 5-7).

The special features that characterize lucinaceans are believed to be a response to life in rather poor habitats, wherein food is scarce and the oxygen content of water in the substrate is relatively low (Allen, 1958). A burrowing existence in such habitats is made practicable by the anterior inhalant tube and mantle gills. The anterior sorting system, which is more crude than the elaborate gill-sorting systems of bivalves with posterior inhalant feeding currents, leads to

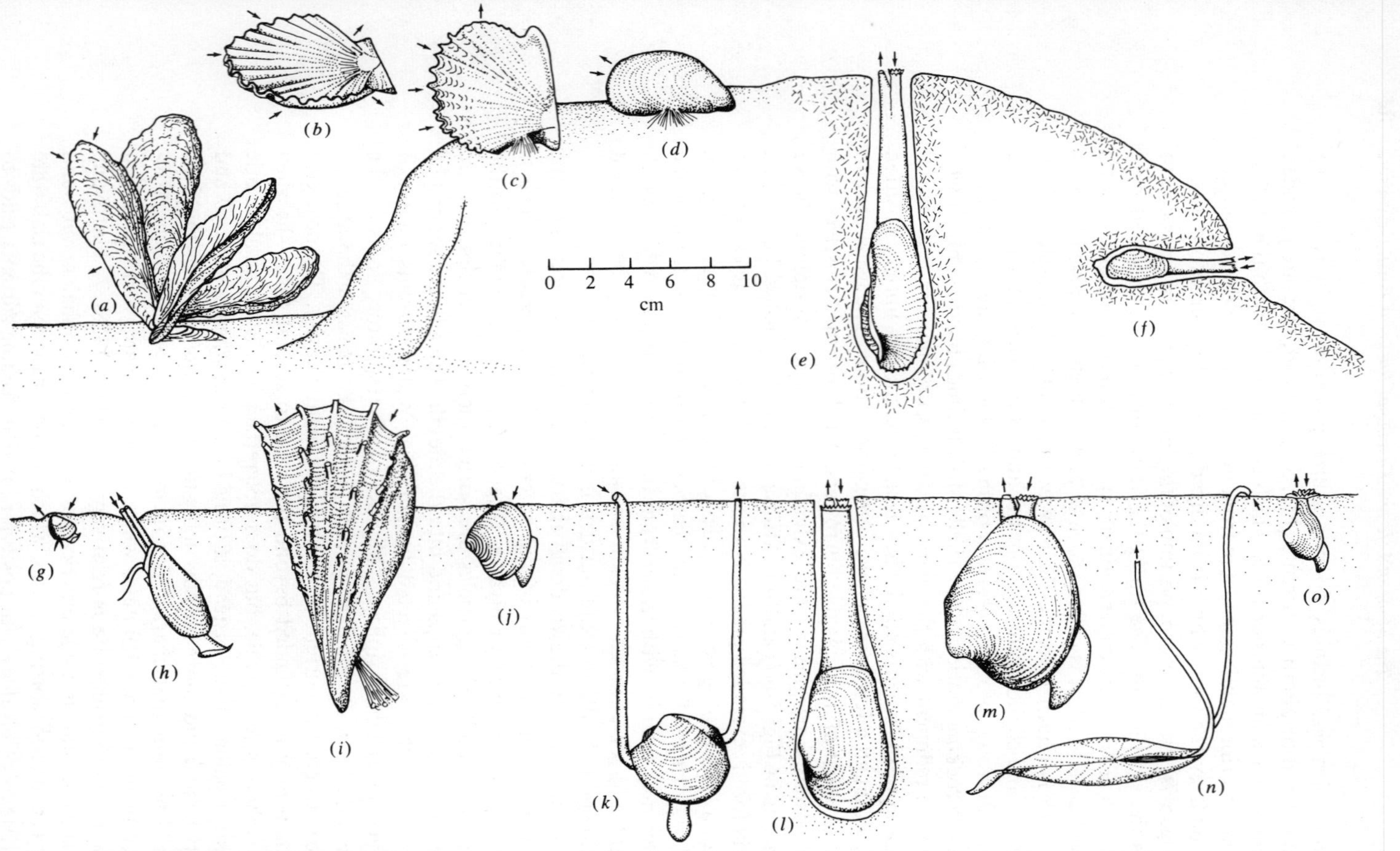

Fig. 5-6 Common modes of life of the Bivalvia. (*a*)–(*d*) Epifaunal bivalves, all suspension feeders: (*a*) *Crassostrea*, cemented; (*b*) *Pecten*, which can swim; (*c*) and (*d*) *Pinctada* and *Mytilus*, attached by byssal threads. (*e*)–(*o*) Infaunal bivalves: (*e*) and (*f*) *Pholas* and *Hiatella*, siphonate suspension feeders living in rock, the former boring, the latter nestling; (*g*) *Nucula*, nonsiphonate labial palp deposit feeder; (*h*) *Yoldia*, siphonate labial palp deposit feeder; (*i*) and (*j*) *Atrina* and *Astarte*, nonsiphonate suspension feeders; (*k*) *Phacoides*, infaunal mucus tube feeder; (*l*) and (*m*) *Mya* and *Mercenaria*, siphonate suspension feeders living in sediments; (*n*) *Tellina*, siphonate deposit feeder; (*o*) *Cuspidaria* (septibranch), siphonate carnivore. (After Stanley, 1968.)

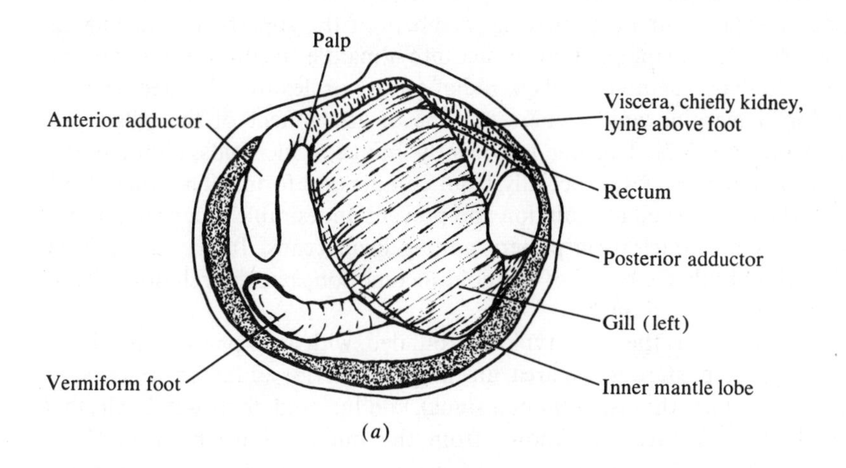

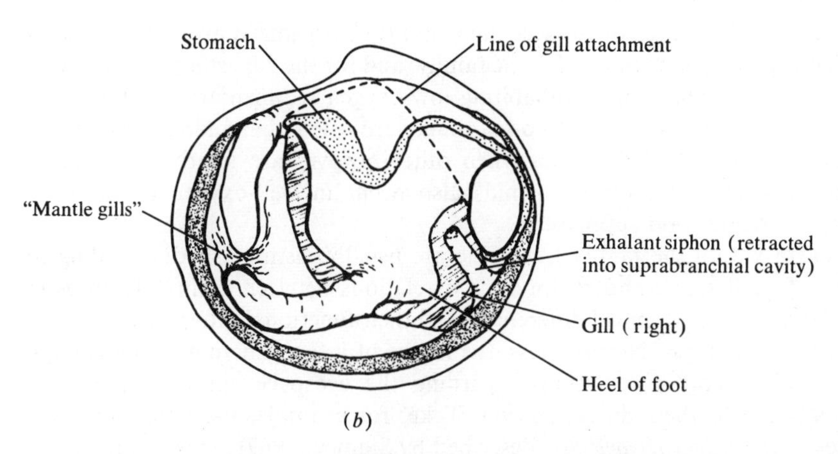

Fig. 5-7 Generalized lucinoid anatomy. (*a*) With left valve, mantle and right hinge plate removed. (*b*) As (*a*), but with left gill also removed and digestive tract shown inside visceral mass; note the "mantle gill" folds beneath the anterior adductor muscle. (After Allen, 1958.)

the acceptance of large food particles and therefore acceptance of a greater percentage of all incoming particles—an obvious advantage where food is not abundant. Modifications of the digestive and other systems found among lucinids are consistent with this interpretation, and are described in some detail by Allen (1958).

Licinoid shells are subcircular in lateral view. This shape is obviously highly functional in burrowing with the seesaw technique. In addition, the shell is somewhat expanded anteriorly, presumably to accommodate the enlarged muscle and organs associated with the anterior inhalant tube. An elongate anterior adductor scar reflects the elongated anterior adductor muscle, and is a characteristic of great value in identifying members of the superfamily Lucinacea. Because there are no siphons that retract into a mantle sheath, there is no pallial sinus. On many lucinoid shells, a radial groove or flexure is located near the posterodorsal margin, which marks the line of attachment of the gills to the visceral mass (Fig. 5-7). This line is evidently unique to lucinoids, although it is absent from some species, especially from compressed forms. The hinge teeth that form the dorsal shell articulation and prevent undesirable independent shell rotations have a characteristic pattern among lucinaceans. It may possibly be related to the peculiar stresses of lucinoid locomotion, but the relation, if any, has not yet been worked out.

A lucinoid valve is therefore typically rounded, with an expanded anterior, a flexed or grooved posterodorsal area, an elongate anterior adductor muscle scar, an entire pallial line (that is, without a sinus), and lucinoid dentition. Shells that include all these features are known from the middle Silurian of Gotland, Sweden, and continue to appear in the fossil record until the present. Clearly, these shell features reflect the special lucinoid mode of life, which was developed at least by Middle Silurian time.

The typical features of the Lucinidae are thus organized around a mode of life that is probably primitive for the family, and for the superfamily Lucinacea. Today, marine communities inhabiting low-oxygen sedimentary habitats, such as are found in lagoonal or deep-sea muds, are often characterized by one to several species of lucinids, even when other bivalves are relatively scarce or absent altogether. However, lucinids also occur in well-oxygenated habitats having abundant food supplies.

Genera within the family Lucinidae are usually distinguished according to details of shell shape and sculpture. Correlations between their differences in shell form, their anatomical differences, and their functional differences have not yet been worked out. Nevertheless, the habits of a few genera are well enough known to support speculations regarding the ecospace specializations that originally led to their differentiation. Take, for example, the habits of some species of the genus *Divaricella*, described by Stanley (1969). This lucinid genus is characterized morphologically by the possession of divaricate ornamentation (Fig. 5-8), consisting in ridges that begin at anterior and posterior margins as if they were to form a concentric pattern, but instead bend upward to form a broad inverted "V" along an axis (the "line of demarcation") just anterior to the midline of the shell (Fig. 5-8). There is a narrow shelf atop each ridge, and the lateral face of each ridge slopes down to the inner margin of the shelf atop the next lower ridge. Species of *Divaricella* that have been observed burrowing move downward nearly vertically, their hinges more or less upward, aided by an anterior-posterior rocking motion through an angle of about 45° (Stanley, 1969). When *Divaricella* is burrowing in this manner, the beveled lateral slopes of the ridges of the descending part of the valves cut down through the sediment, meeting relatively little resistance. On the part of the valves that is rocking

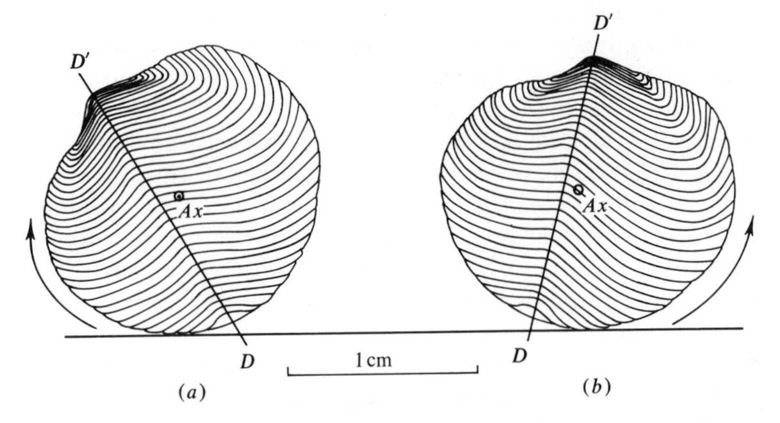

Fig. 5-8 Divaricate sculpture of *Divaricella quadrisulcata,* which aids in burrowing. (*a*) Shell at maximum forward rotation. (*b*) Shell in position from which forward rotation begins. Rotation occurs around *Ax.* (After Stanley, 1969. Copyright 1969 by the American Association for the Advancement of Science.)

upward, however, the ridge-top shelves provide more resistance and, in fact, carry sedimentary particles upward on the shelves. Thus, the animal excavates its way down into the sediment. *Divaricella* is rather a slow burrower when compared with some of the elongate heterodont bivalves, but it is nevertheless the fastest burrower known among the lucinids. It lives in very fine- to medium-grained sands that are relatively mudfree, which is the size range in which the sawlike divaricate mechanism appears to work especially well. Here, then, is a lucinoid genus with a specialization for inhabiting clean sands that are usually well aerated. Features of soft-part anatomy that may correlate with this habit are not known, but the shell features indicate the probable ancestral specializations which have led to the rise of this morphologically distinctive genus with its distinctive adaptation.

In sum, many living lucinoid species appear to inhabit the poor environments to which the earliest lucinaceans were probably adapted. Some lucinid lineages have re-entered richer marine environments and are presumably well adapted thereto; these forms retain the assemblage of characters that makes up the lucinid adaptive type.

Many Skeletons Function as if They Were Mineralized Tissues or Organs

The fossil record of ecological functions can be surprisingly good, although, of course, it is never complete in detail. The precise habitat of an organism can often be reconstructed from direct taphonomic evidence. Shells may be found still in their living positions, in burrows or associated with surfaces that formed their substrate, and an environmental reconstruction can then be made from sedimentological and stratigraphic evidence. But interpretations of the particular adaptive responses that permitted organisms to live in their habitats must ultimately rest on evidence from the fossils themselves.

The fossils are chiefly of skeletons, or molds, casts, replacements, or fragments of skeletal material, or, occasionally, of cuticular material. The only other type of fossil that is at all common is the trace fossil, the trails, burrows, and other sedimentary features formed by organisms. These certainly provide direct evidence of the activities of organisms, but they have usually not been created by evolution in order to perform or aid in ecological functions, as skeletons have. Here we are interested in the extent to which the ecological functions of individuals are recorded in their skeletons. From the example of the lucinoid skeleton, we can learn a very great deal about the mode of life of the Bivalvia from their shells. The fossil bivalve skeleton is clearly coadapted with a variety of anatomical organs and organ systems and may aid in locomotion, protection, support, and as a rigid architectural frame to help preserve the spatial relations of the soft parts. Furthermore, the skeletal shape and markings are related to the feeding mechanism, which can be inferred by drawing analogies with living allies. Skeletons are clearly useful for functional interpretation. It is therefore worthwhile to examine the ways in which skeletons are formed.

Some broad similarities exist in the secretion of mineralized skeletons by invertebrates of different phyla. The minerals are precipitated from a solution, and many properties of the solution, such as the ionic concentrations, are controlled to some extent by the organism. Extracellular body fluids of most marine animals conform in osmotic concentration to seawater. However, these body fluids, even the mesoglea of coelenterates, the water-vascular fluids of echinoderms, and cellular fluids of all organisms so far as is known, have their ionic concentrations regulated so that some ions are more, and some less, concentrated than those in seawater (Chapter 4). Ion exchange is thus a very basic physiological reaction (Prosser and Brown, 1961).

To secrete a skeleton, ions must be selected from the environment, transported to the depositional site, and appropriately concentrated for precipitation (although since precipitation may be mediated by enzymes, concentrations may be far below the requirements for strictly inorganic precipitation). Sometimes ions are stored before final utilization. Evidently there are many mechanisms of ionic concentration and transport, none of which is understood in detail; most seem due to enzyme activity or appear to involve exchange molecules contained in membranes (Danielli, 1954).

It is not necessary to review here all the known details of shell formation and their variations among invertebrates; our knowledge is very incomplete at any rate. Instead, general aspects of shell formation in brachiopods and bivalve mollusks will be used as examples.

An articulate brachiopod is enclosed within a sheetlike organ, the mantle, which drapes over the internal organs and secretes a calcite shell (Fig. 5-9). The course of skeletal deposition in brachiopods has been described by Williams (1968a). The form and structure of the shell are under the control of the mantle. Because the mantle cannot be extended significantly beyond the shell margins, shell growth takes place at the margins themselves or on the inner shell surface. Growth in shell size occurs as the mantle grows in size, by expanding peripherally. The peripheral mantle expansion is accommodated in the epithelial (surface) layer chiefly through the generation of new epithelial cells in a zone along the mantle groove (Fig. 5-9). Cells change their position relative to fixed morphologic points around the outer or inner lobes because newer cells continuously form behind them.

The epithelial cells around the outer lobe and outer mantle surface secrete the shell. In most articulates, the shell has three layers. The outermost is a non-mineralized layer chiefly of protein (Jope, 1967) called the *periostracum*. The epithelial cells first secrete a coating of (probably) mucopolysaccharide as they "migrate" from the mantle groove toward the tip of the outer lobe (Williams, 1968b). Beneath this coating is secreted the periostracum, which varies in detailed structure among living brachiopods but which may usually consist in a vacuolated mucoprotein layer bounded by one or two triple-layered membranes; the inner membrane is discontinuously underlain by a protein cement (Williams, 1968b). At the mantle lobe tip, the cells begin to deposit a calcite layer, the *primary layer* (Williams, 1968a), which rests on the periostracum but is essentially free of organic material. The primary layer is nucleated by calcite seeds secreted upon the protein cement of the periostracum (Fig. 5-9). Membranous extensions of epithelial cells, termed *microvilli*, are at first attached to the periostracum, but as shell deposition proceeds, the calcite seeds grow and coalesce to form a calcite sheet. Although the microvilli penetrate the primary layer and anchor the epithelium thereto, they are soon separated from the periostracum. The primary layer continues to thicken by epitaxial growth.

The cells continue to change their morphological positions, becoming progressively more distant from the tip of the outer lobe, owing to outward mantle growth. When the shell edge has been extended a certain distance, rather constant within each species, the cells switch from deposition of primary layer and begin to deposit a new, *secondary layer*. The thickness of the primary layer is thus chiefly a function of the rates of calcite secretion and cell generation. The secondary layer consists in calcite fibers sheathed in protein that is deposited contemporaneously. Each fiber, with a portion of its protein covering, is secreted by a single epithelial cell.

In the Mollusca also, a mantle secretes materials that form a mineralized exoskeleton pervaded by an organic matrix (Fig. 5-10). The shell mineral may not be only calcite, but aragonite, or vaterite, or combinations of these. Vaterite is rare and seems to be a product of skeletal repair rather than an original shell constituent. In bivalve mollusks as in brachiopods, the mantle cannot be much

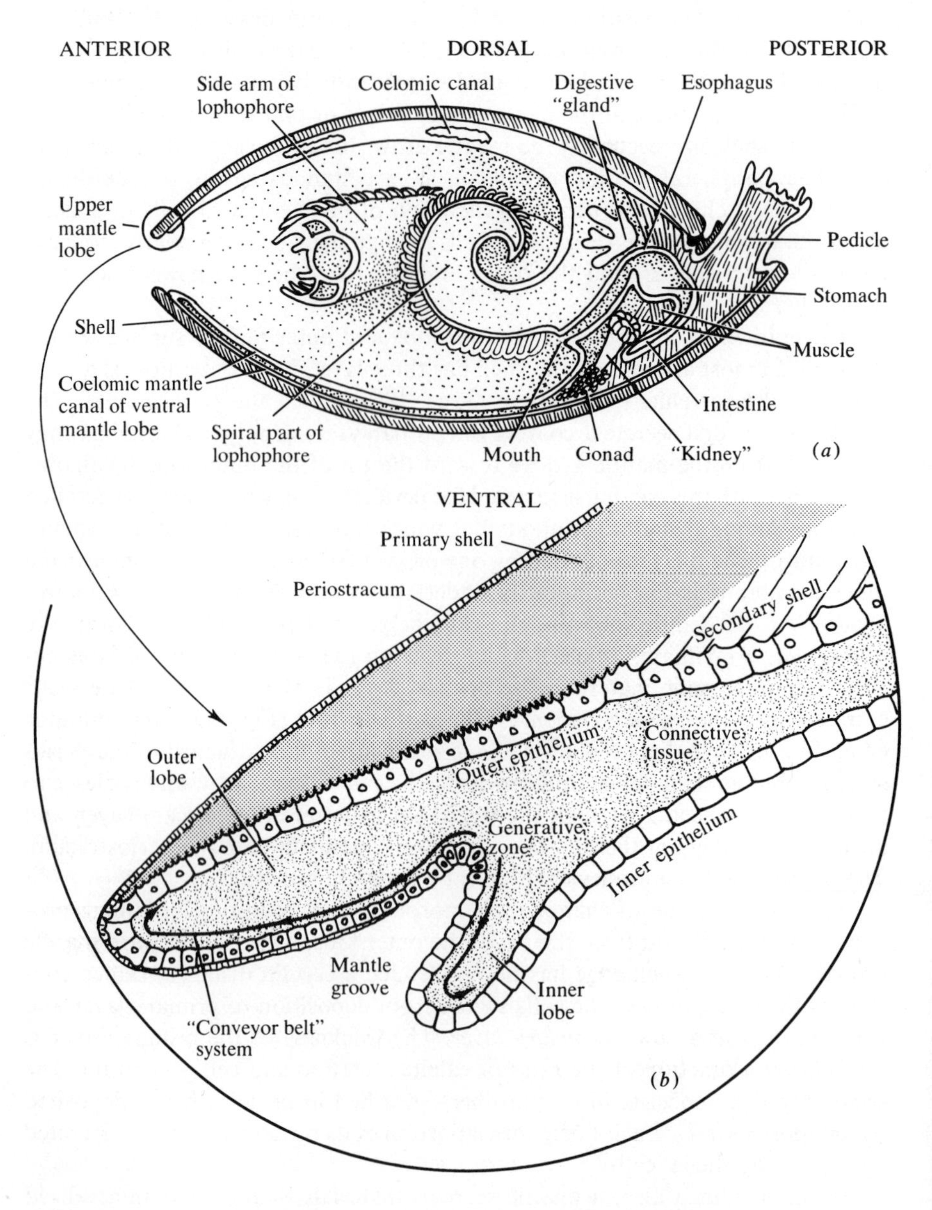

Fig. 5-9 (*a*) Generalized cross section of a brachiopod. (*b*) Detailed cross section of edge of mantle and shell in the brachiopod *Notosaria*, showing outer cell layer in mantle. [(*a*) After Beerbower, 1968; (*b*) after Williams, 1968a.]

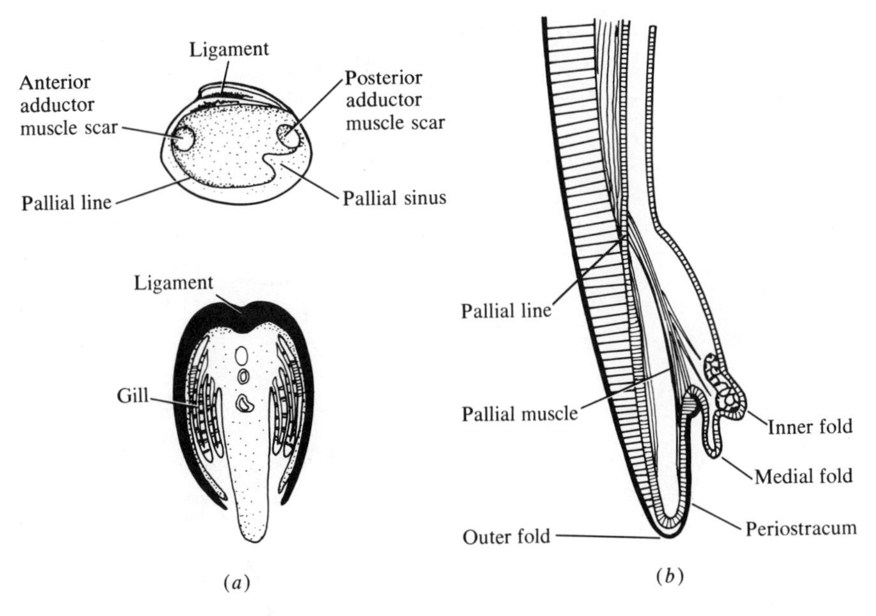

Fig. 5-10 (*a*) Generalized cross section of a bivalve. (*b*) Detailed cross-section of edge of mantle and shell. [(*a*) After Beerbower, © 1968; (*b*) after Wilbur, 1964.]

extended over the exterior shell surface, so that once the outer shell is secreted it cannot be built up or dissolved from the outside by the bivalve; shell thickening or thinning must be done interiorly or at the shell margins.

The outermost shell layer, the *periostracum*, is not mineralized and is composed of protein (Beedham, 1958). The mineralized portions of most bivalve shells contain two or three distinctive layers, different parts of the mantle being responsible for different layers. Layers differ in texture and sometimes in mineralogy.

During shell formation, the mantle is separated from the growing shell surface by a narrow space containing a fluid, the *extrapallial fluid* (Fig. 5-11), from which shell minerals crystallize and the organic shell matrix is derived. Thus, the mantle surface is not in direct contact with the growing shell surfaces as in brachiopods. Usually (perhaps always during shell deposition), the extrapallial space is cut off from the exterior by a mantle edge pressing against the shell or periostracum. Materials are secreted into the extrapallial fluid by the mantle epithelium. Calcium may be derived directly from seawater, or from body fluids. Carbonate is formed from CO_2 derived from a pool in the mantle, part of the general body pool that has come from seawater or food. In either event, it may have been metabolized; Wilbur (1964) notes that normal respiratory activity of the mantle alone could easily supply the shell carbonate in the common eastern American oyster, *Crassostrea virginica*, which has a rather massive

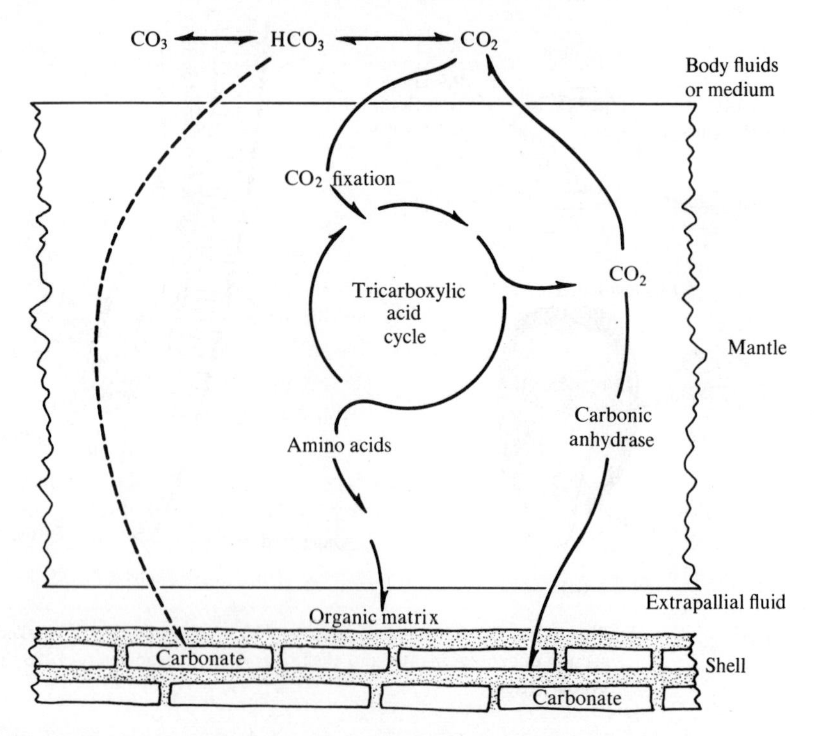

Fig. 5-11 Diagrammatic representation of the relations suggested between CO_2, organic matrix, and shell minerals in bivalves. (After Wilbur, 1964.)

shell. Carbonate deposition is promoted by an enzyme, carbonic anhydrase, inhibition of which retards shell formation (see Wilbur, 1964).

Mantle growth in bivalves as in brachiopods involves the formation of new epithelial cells in a generative zone near the mantle periphery, with change in morphologic position of the cells as new cells are formed after them. Each epithelial cell, in the words of Taylor, Kennedy, and Hall (1969), "... in turn secretes periostracum, outer shell layer and middle shell layer in succession, becomes an attachment area at the parallel line, and secretes inner shell layer, perhaps finally becoming adductor muscle attachment." The morphology of the cells changes, at least in some cases, when changing from secretion of one shell layer to another. The composition of the extrapallial fluid is known to be different in calcitic species (one protein is present) than in aragonitic or bimineralic species (three proteins are present). In addition to proteins, the extrapallial fluid contains other organic substances and inorganic ions.

It has been suspected that the composition of the extrapallial fluid influences the mineral phase directly. On the other hand, there is evidence that the organic matrix acts as a structural template for carbonate secretion. The question of the control of the mineral phase is still an open one.

Portions of the shell that are deposited beneath muscle attachments have a distinctive texture and are collectively termed *myostracum*. The valves are joined by a ligament of organic material that may be partially mineralized. So far as is known, all ligamentary mineral and all myostracum are aragonite. The larval bivalve shell, the *prodissoconch*, may have a mineralogy different from the adult shell, and is often aragonite.

Rates of skeletal secretion vary in all organisms during different stages of ontogeny and during certain different environmental conditions; some details of these variations have been described in bivalve mollusks. During most of shell growth, the shell layers are deposited contemporaneously; that is, secretion of outer, middle, inner, and myostracal shell parts occurs simultaneously, and fluctuations in rate of shell secretion are generally in phase in all layers. Secretion rate fluctuations involve acceleration, deceleration, or stoppage in calcium carbonate deposition. Shell color usually varies with rate of deposition, being darker at slower rates and lighter at faster ones, so that rate fluctuations appear as color bands within the shell; the bands cut across shell layer boundaries. The darker color is probably caused by an increased ratio of organic material to mineral. Shell material deposited during a certain period of time, such as a single color band or several distinctive bands, is called a *growth increment* (Fig. 5-12).

By studying the growth increment patterns in bivalves that lived in a region where conditions were known, it has been possible to distinguish several sorts of depositional rhythms which can be detected in thin sections of shells, living or fossil, whenever the growth increments are preserved (Table 5-4; Barker, 1964; Pannella and MacClintock, 1968). The basic growth increment is the daily increment; shell secretion rates vary during 24-hr periods in a rhythmic manner that generates an increment consisting of a lighter color band (itself composed of thin layers) bounded by sharp, dark surfaces representing times of little or no deposition (Fig. 5-12). The daily increment is sometimes complex, with internal surfaces that subdivide it. Daily increments vary in width in a periodical manner, being wider during times of favorable depositional conditions. Thus, summer increments are thick and light and winter increments are thin and crowded. The alternation of lighter summer and darker winter bands is often visible to the naked eye; these seasonal "growth rings" have long been known.

Tidal cycles are well expressed by growth increments, appearing as clusters of thicker and thinner daily increments. Tidal periods of approximately 14 days are displayed by some *Mercenaria mercenaria* (Fig. 5-12). Synodic month periods of approximately 29 days are formed by the grouping of tidal cluster into pairs, each pair consisting of one strongly and one weakly expressed tidal cluster. A bidaily pattern consisting of one thicker and one thinner daily increment has been detected in *Mercenaria*, but its cause is not yet known.

In addition to the rhythms based upon periodical environmental fluctuations, biological rhythms, such as breeding periods, are also reflected in growth increment clusters, as shell deposition is reduced during spawning. The breeding periodicity may vary from about monthly to about yearly. Furthermore, episodic

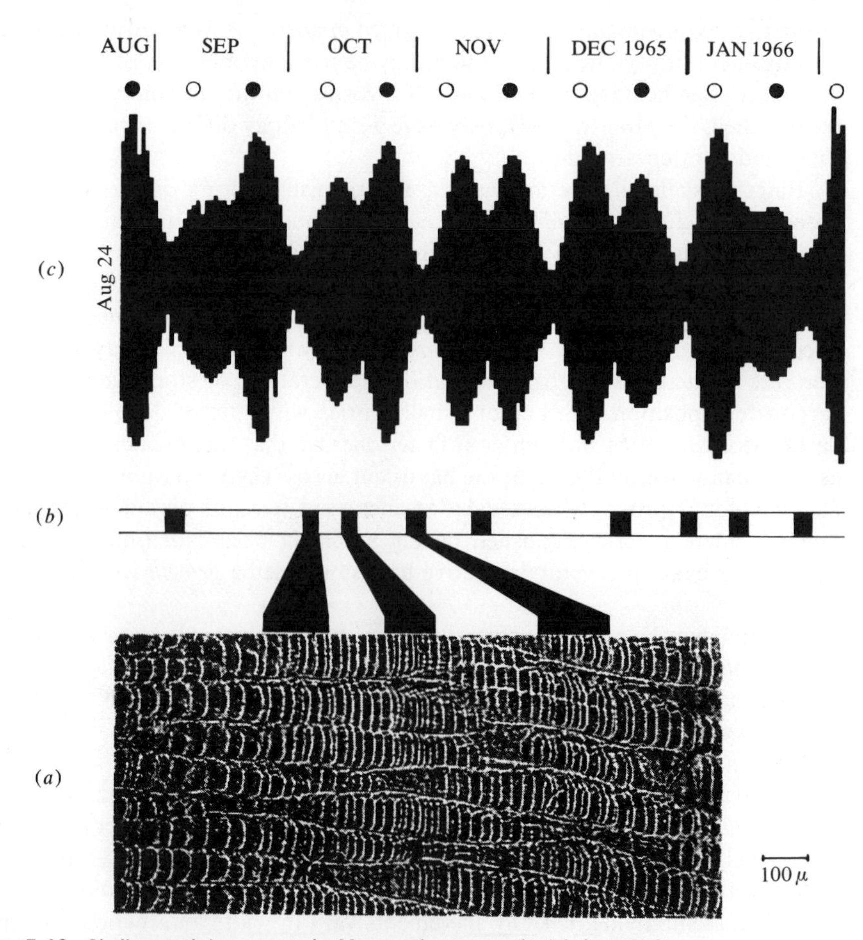

Fig. 5-12 Shell-growth increments in *Mercenaria mercenaria*. (*a*) A peel of the shell, × 200, showing three clusters of growth increments. (*b*) Shell-growth increment clusters of this shell arranged on a time scale. (*c*) Maximum daily tidal ranges at the locality (Barnstable Harbor, Massachusetts) and during the time that the shell was deposited. Growth was clearly less during periods of low tides and greatest during periods of high tides. (After Panella and MacClintock, 1968.)

events, such as storms, unseasonable temperatures, or attacks by predators, may cause bivalves to keep their valves closed and thus prevent normal shell growth. The irregular nature of such events makes them easily separable from the periodic shell-secretion rhythms. The common growth-increment rhythms are summarized in Table 5-4. It is evident that all these events, recorded in the structure of the shell, may be investigated in fossils and thus give evidence of growth rates, environmental and biological rhythms, and other episodes affecting shell secretion in the distant past. The bivalve shell may, in effect, be considered

Table 5-4 Physiologic and Environmental Events Recorded in Molluscan Growth Patterns

Events	*Circadian rhythms*	*Spawning*	*Winter*	*Summer*	*Tides*	*Storms and individual accidents*
Characteristics	Shell layers consisting of dark and light bands	Sudden depositional break, followed by a slow recovery	Gradual slowing down of deposition rate to a minimum, followed by gradual increase of deposition rate	Thickest daily increments, often with many subdaily bandings	Regularly recurrent patterns with 15- to 30-day periodicity	Sudden depositional break, followed by fast recovery (or death)
Time and place of occurrence	Daily worldwide	Variable with species and geographical position	Best shown in mid- to high-latitude shallow water areas		Best developed in intertidal and shallow subtidal species	Storm frequency highest during equinoctial time
Some paleoecological applications	Geochronometry Paleoecology	Paleobiology	Paleoclimate Paleolatitude Geochronometry Paleogeography	Paleoclimate Paleolatitude Geochronometry Paleogeography	Geochronometry	Paleoweather

SOURCE: D. C. Rhoads and G. Pannella, "The Use of Molluscan Shell Growth Patterns in Ecology and Paleoecology," *Lethaia* 3: 143–161, 1970.

as a long-term continuous environmental recorder (Rhoads and Pannella, 1970). In addition to bivalves, corals are known to be useful for such studies (Wells, 1963; Scrutton, 1964), and some gastropods and echinoids are probably suitable.

From what is known of the processes of skeletonization in other groups of invertebrates, they all elaborate crystals (or at least mineral substances) from a fluid, the contents of which are chiefly governed by the organism. There are certainly a number of variations in secretory mode; in echinoids, for example, skeletal secretion begins within cells (Okazaki, 1960) and continues within tissues. However, in all cases it appears that skeletonization involves the carefully controlled activity of tissues and is well integrated with the general metabolic system of the organism. Skeletons are not simple accessory armor plates, but are more like tissues and organs that happen to be mineralized to enhance their particular functions.

Skeletonized Higher Taxa Have Characteristic Skeletal Ground Plans

Skeletons can be regarded as functional organs that are coadapted with some of the other organ systems. Inasmuch as they furnish the bulk of the physical evidence of the morphology of fossil organisms, and because we are interested in interpreting the ecological functions of those organisms, it is clearly necessary to inquire into the functions for which skeletons are themselves adaptive, and into the roles they play in limiting or expanding the adaptive potentials of other organ systems with which they may become coadapted. The multiplicity of functions for which skeletons are employed can nearly all be classed as for protection, for tissue support, and for locomotion and maintenance of orientation and position in the habitat. The various functions associated with these categories sometimes favor conflicting skeletal modifications, and compromise solutions are evolved.

The great majority of invertebrate skeletons are external. Even echinoderm skeletons, which are endoskeletons physiologically, are exoskeletons mechanically. External skeletons are, of course, highly advantageous for protection in that the entire organism may be enclosed within an all-encompassing armor. Furthermore, external skeletons are more resistant to most kinds of mechanical failure, weight-for-weight, than are internal skeletons. An exoskeleton of bone could provide seven times the resistance to bending provided by an internal skeleton of the same weight (Curry, 1967). This fact is owing to the same principle that makes a hollow tube more resistant to distortion than a solid rod of the same weight. However, internal skeletons are more resistant to the effects of impact, chiefly because soft tissues can absorb much energy without being damaged; when damage does occur, it is localized and does not spread (Curry, 1967).

It is evident from the most cursory knowledge of invertebrates that the major taxa, such as phyla or sometimes classes, have distinctive skeletal plans. Indeed, nearly every possible plan of construction that will approximate the specifications for an efficient skeleton has been employed by one taxon or another. Each plan has certain advantages and disadvantages; these have been explored for most skeletal plans by Vermeij (1970), whose terminology is followed here. Skeletons are classed first as permanent (retained throughout postlarval life) or transient, and secondly as immutable (do not change subsequent to formation) or modifiable. Perhaps the only workable approach that has not been used is the single secretion of an immutable, permanent skeleton in final form. If an organism is to be extensively encased within a skeleton, there must be some provision for growth. It is conceivable that a skeleton could be made so capacious and of such proportions that it would encompass the largest body size reached by the organism, but no invertebrates have followed this plan, perhaps because it would require too much energy of the young organism.

Another possibility is that the skeleton can be immutable but transient. It can be secreted just a little too large; then, when the organism grows into a very tight fit, it can be molted and discarded and a new skeleton formed. This is the method of the Arthropoda, and it has proven to be a very successful one. Exoskeletons that are to be molted may be secreted in almost any shape, depending upon the shape of the generating surface, which is simply the surface of the organism; thus, the skeleton may encase long spindly appendages or cylindrical body segments. Growth in the diameter of these body parts is accommodated simply by sloughing off the skeleton when it becomes too tight; many arthropod skeletons are somewhat elastic, also. Each new skeleton can be tailor-made to the growth rate variations among the encased soft parts, and differential growth rates present no special problem. Growth curves based upon arthropod skeletons must be extrapolated from molts of discrete sizes. Skeletons of this sort seem ideally suited for casing segmented organisms. The tubular skeletal elements of each body or limb segment may be articulated and provided with muscles enabling the segments to be flexed in appropriate ways at these joints. Appendages for walking, running, or swimming, or for sensory purposes, may be developed, each complete with skeletal covering. A disadvantage to this plan is that the periodic molting is most successful if the skeleton is thin, so that massively mineralized arthropod exoskeletons are practically absent. Indeed, the acorn barnacles have evolved a secondary permanent skeleton (Vermeij, 1970).

Still another way of constructing a protective skeleton is to build it up of numerous small articulated elements, like links in mail armor; growth may then be accommodated either, or both, by multiplying the number of elements or by enlarging them, and the multiplication or enlargement may be differential to allow for differential growth rates in soft parts. Like the system of multiple skeletons in successive sizes, this type of skeleton may be of complicated shapes, which may be changed with relative ease, and it may possess long, armored appendages. This approach has been successfully employed by the echinoderms.

Because the echinoderm skeletons are physiologically endoskeletons, they are permanent but modifiable. Actually the echinoderms have not much exploited the possibilities of shape, being mostly ovoid, sack-, or disk-shaped, although stalks and appendages have developed very commonly. Probably this is because the secretion of echinoderm plate usually begins at a generative zone; then the plates migrate in morphological position as new plates appear. Radical changes in plate shape that would constantly be required to fit them to re-entrants and complex surfaces would pose many difficulties; instead, the skeletal shapes tend to be gently curved, so that remaking of plates is minimized.

Another approach to skeletal construction is to secrete a permanent skeleton that is of such a geometrical shape, as a cone, that growth may be accommodated simply by marginal additions, thus enlarging the internal volume. A tube will do as well as a cone, providing the organism grows only along the axis and not in any diametrical direction; evidently this is a serious restriction, however, since the organism must get relatively thinner with growth, and permanent tubular exoskeletons are practically unknown among invertebrates. Variously modified cones are common, such as in corals, bryozoans, brachiopods, some worms, and most classes of mollusks. Most of these accretionary skeletons have immutable exterior surfaces; bivalves and brachiopods, for example, cannot extend their mantles over their outer shell surfaces to modify them. Some conical skeletons can be modified, however, as in cypraeid gastropods, which have their shell covered with mantle tissue. Because conical skeletons are in such wide use, especially immutable ones, we shall examine them at some length.

The conical shell shape presents certain problems of its own. One is that the open end of the cone, although ideal for egress of body parts and for environmental exchange, is unprotected and gets larger during growth. A variety of solutions to this problem have been employed. The body frequently possesses a protective plate, such as the operculum of gastropods or the aptychus of ammonites, which fits the shell aperture more or less snugly when the animal is fully retracted. Or the aperture may be furnished with a hinged lid, as in bryozoa. Or the skeleton may be composed of two cones, closely fitted at their open ends and provided with a hingement so that they may swing apart to allow environmental exchanges. This is the sort of skeleton employed by the Brachiopoda and Bivalvia.

Another problem with the conical shell form is that, for an organism that undergoes a significant size increase during growth, a single straight cone must be either short with a broad aperture, and thus fail badly in protecting the soft parts, or tall with a narrow aperture, and thus be an extremely unwieldy skeleton in many habitats. The common solution to this problem has been to secrete a tall shell with a narrow aperture but to coil it in order to have a more compact shape, either planispirally, as in some gastropods and most ammonites, or helicospirally ("trochospirally"), as in most gastropods. Organisms that use two hinged shells may have them shaped as short, broad cones and still attain protection. Most of the Brachiopoda and Bivalvia are not only short and broad but also coiled—the former planispirally, the latter more or less helicospirally, presumably to permit the formation of a hinge structure at a margin which is as stable as possible with respect to shell enlargement and to attain compact shapes that are coadaptive with general functional architecture and mode of life.

Because the spiral shell form is employed by a number of important groups of invertebrates and is geometrically tractable, it has received much attention. A standard review is by Thompson (1917); Raup (1966b) has given the subject a new treatement. Much spiral coiling approximates a logarithmic spiral, the chief advantage of which is that size increase does not involve any shape change, and thus growth may occur with form constant. The descriptive scheme for logarithmically coiled shells given by Raup and Michelson (1965) is used here.

The dynamics of shell generation are conceived in a framework of cylindrical coordinates containing an axis of coiling (y), about which a *generating curve* (symbolized S) revolves (Fig. 5-13). The generating curve is usually formed from the outline of the growing edge of the shell; a circle is the simplest generating curve. As it revolves, three basic things may happen. The size of the whorl increases at a steady rate, the *whorl expansion rate* (W). The generating curve may become increasingly distant from the axis (distance is symbolized by D). In helicospiral coiling, the generating curve migrates along the axis, at a certain *translation rate* (T). If the shell is planispiral, T is 0, and coiling occurs in a plane.

In Fig. 5-13, r_o is the initial distance of an arbitrary point on the generating curve from the axis of rotation y. After θ revolutions, the position of the point, $r\theta$, is given by

$$r\theta = r_o \, W^{\theta/2\pi}$$

where W is the rate of whorl expansion. The distance (D) of the generating curve from the axis is defined as the ratio between the r values of the axial and outer margins, respectively. Thus, $D = 0$ when the axial margin of the whorl is at the axis, a common situation in "tightly coiled" gastropods.

All possible variations among the values of the parameters W, D, and T may be expressed in a three-dimensional bloc, in which each parameter is represented by one of the dimensions, as in Fig. 5-14. The effects of changing parameters are seen in Fig. 5-15. With increasing T, shells depart farther from a plane spiral and become more and more high spired. The effect of increasing W is to increase the proportions of the last whorl at the aperture at the expense of the proportions of the early whorls. The effect of increasing D is that shells become more and more "loosely coiled" (evolute) as the generating curve moves farther from the axis of coiling. Note the dashed line labeled $W = 1/D$ on the right-hand side, W-D face of the bloc in Fig. 5-14. This represents the locus of points where the increasing distance of the generating curve and the increasing rate of whorl expansion are so balanced that each succeeding whorl is precisely in contact with the preceding one; there is neither whorl overlap (involution), as at a lower W or D, nor any space between whorls (evolution), as at higher W or D.

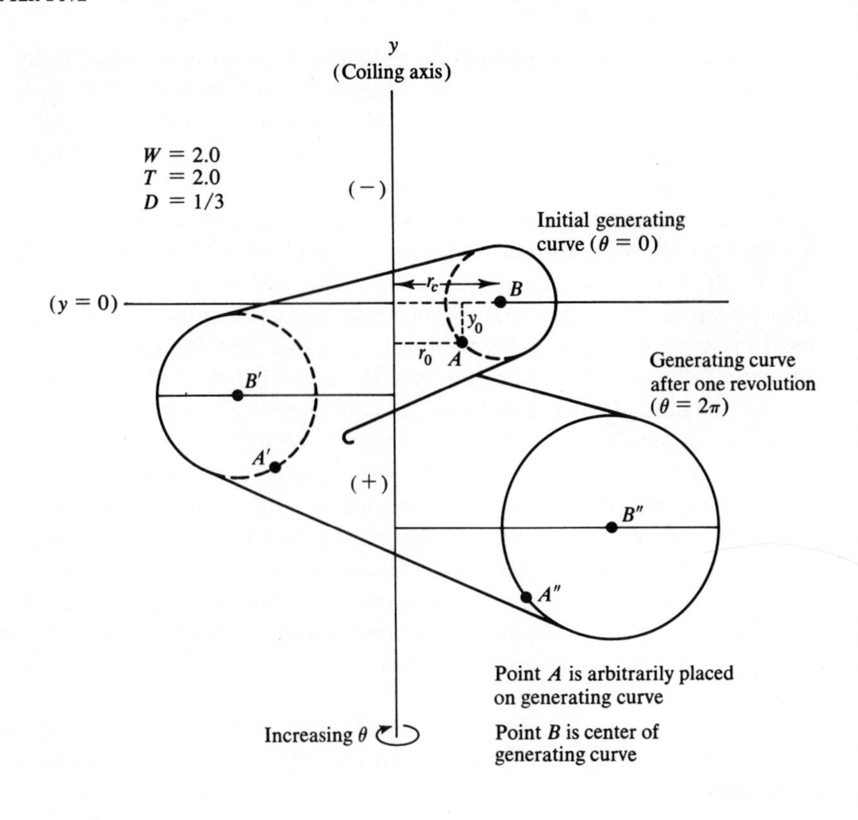

Point	Coordinate θ	Coordinate r	Coordinate y
A	0	r_0	y_0
A′	π	$r_0W^{1/2}$	$y_0W^{1/2} + r_cT\ (W^{1/2}-1)$
A″	2π	r_0W	$y_0W + r_cT\ (W-1)$
B	0	r_c	0
B′	π	$r_cW^{1/2}$	$r_cT\ (W^{1/2}-1)$
B″	2π	r_cW	$r_cT\ (W-1)$

Fig. 5-13 Cylindrical coordinates as applied to describe coiled shells. (After Raup, 1966.)

Although we are considering only the simplest cases of coiling, neither planispiral nor trochispiral shells can be formed without pronounced apertural growth gradients. Only a straight cone could be produced by equal growth increments around the aperture, for once the cone is curved, the outer diameter becomes larger than the inner around the axis of curvature or coiling, and therefore

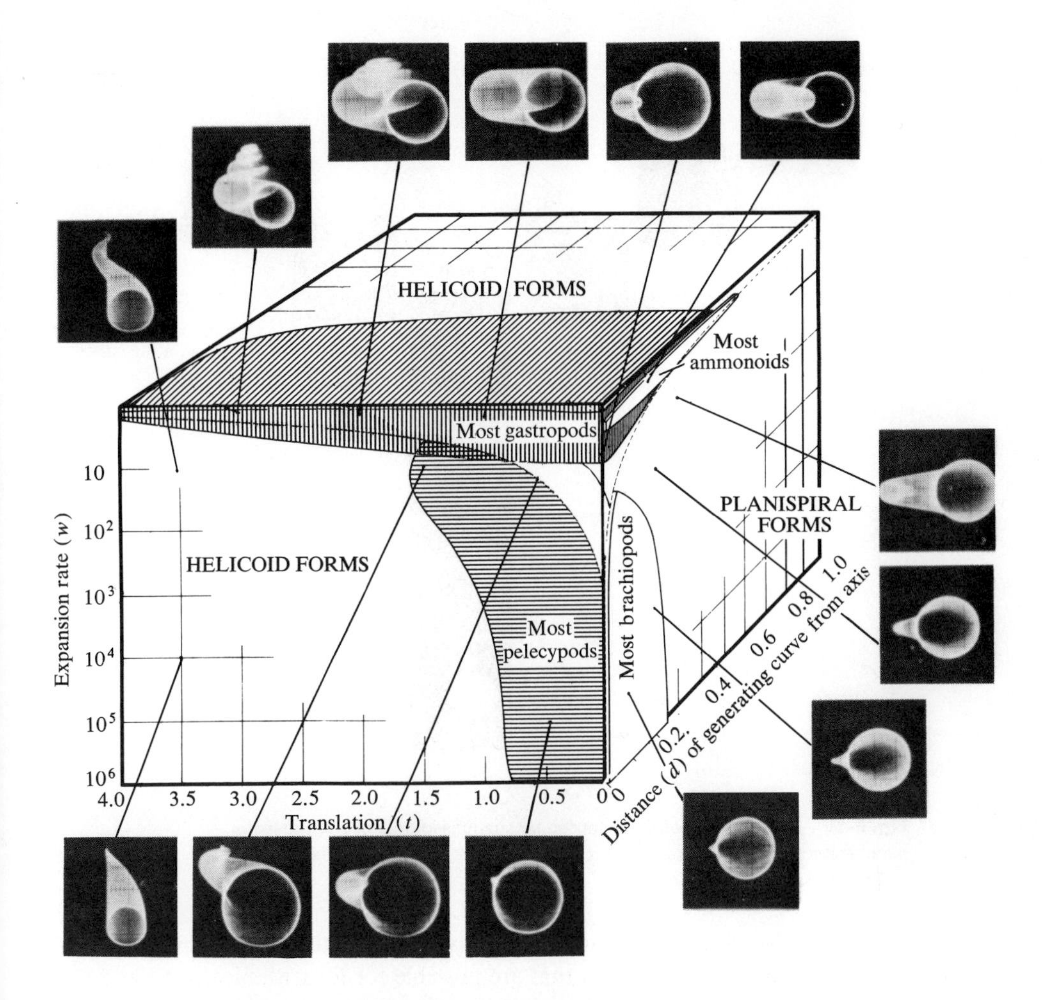

Fig. 5-14 Variations in coiling geometry that result from changing the major parameters of coiling—the expansion rate, the distance of the generating curve from the axis, and the translation of the generating curve along the axis. The combinations of these parameters as represented in the block are not all realized in nature; some of the regions that characterize the coiling geometry of various higher taxa are indicated. (After Raup, 1966.)

shell growth must be faster opposite the axial position. In planispiral shells, the growth gradients are symmetrical on each side of the plane joining the proximal and distal points on the aperture. For a circular generating curve, they grade evenly in either direction from the low point nearest the coiling axis around the shell to the high point farthest from the axis [Fig. 5-16(*a*)]; in helicospiral shells, the growth gradients are not symmetrical on either side, but there must be a second gradient, related to T [Fig. 5-16(*b*)]. The portions of the bloc in Fig. 5-14 that are chiefly occupied by brachiopods, bivalves, gastropods, and ammo-

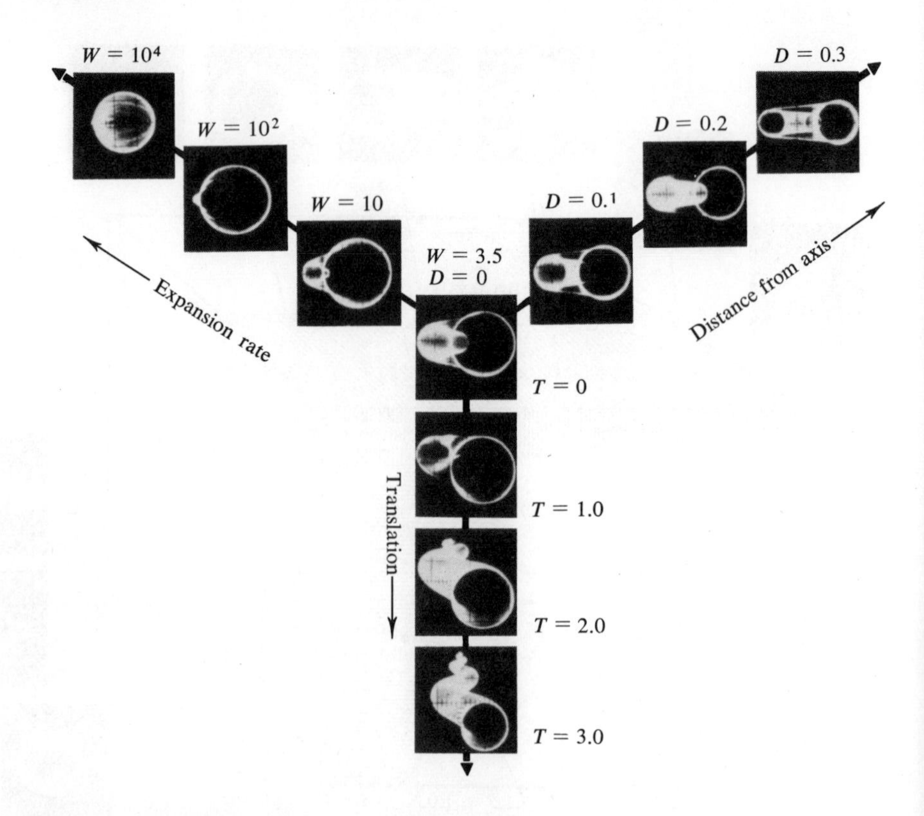

Fig. 5-15 Changes in shell form caused by changes in only one of the major parameters of coiling, using $T = 0$, $D = 0$, and $W = 3.5$ as a standard. (After Raup, 1966.)

noids are labeled. It is striking that great volumes of the bloc are unoccupied—most of the geometrical forms that are possible for logarithmically coiled shells have not been utilized. Also striking is that there is relatively little overlap between the forms employed by different groups of animals. Some of the reasons for this distribution of coiling geometries are apparent from the general specifications for external skeletons, and some from special specifications for bivalve skeletons. Notice that the univalve forms are chiefly found in the low-W portion of the bloc, near the top. This restriction must arise from a requirement for as complete an enclosure as possible, which involves as small an aperture as possible for a shell of given volume, other things being equal. A few coiled univalve shells are found in higher-W portions of the bloc—for example, the cap-shaped limpets and abalones among the gastropods (not shown in the diagram). These forms live on solid substrates, such as rocks, shells, or algal thalli. They gain protection by clamping their shells down firmly against their substrates, which they are in effect using as a protective cover. Other departures from the main

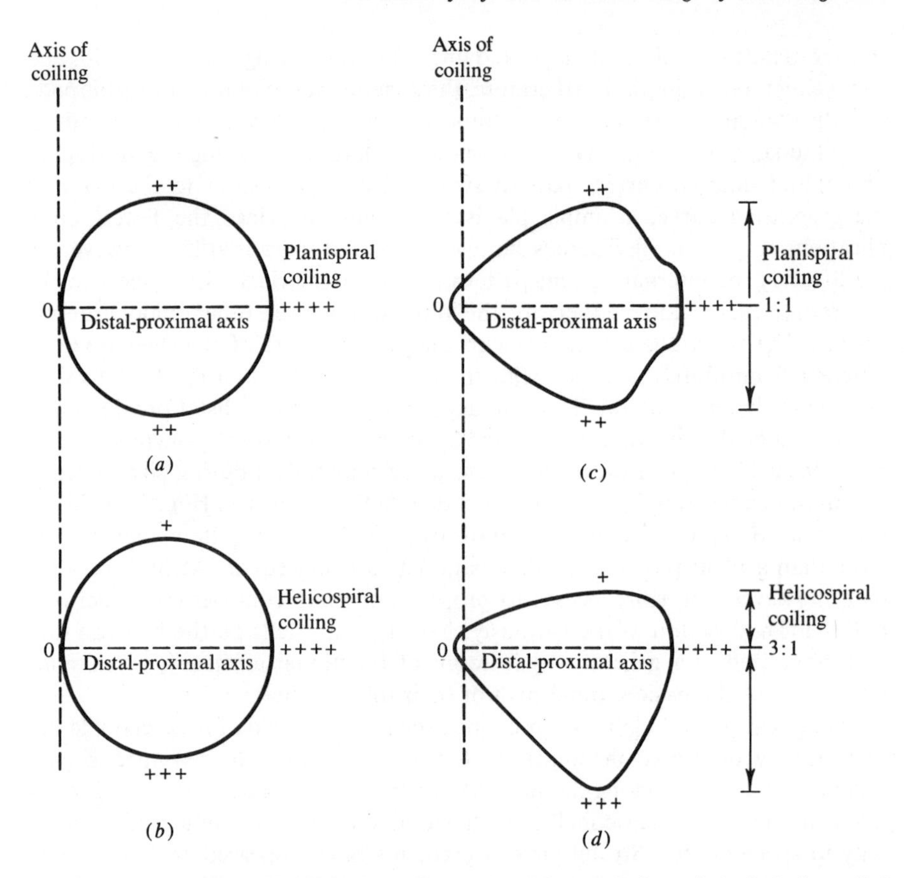

Fig. 5-16 Growth gradients around helicospirally and planispirally coiled generating curves.

region of shell coiling are found among gastropods as well as other groups, and are all owing to special habits or conditions.

The two coiled bivalve groups are narrowly restricted in shell geometry; Brachiopoda lie on the planispiral (W-D) face in the region of low D; Bivalvia lie in a helicospiral region of low T and also of low D. The brachiopods do not extend into the involute part of the W-D face, above the $W = 1/D$ line. This condition is doubtless owing to the necessity of providing for a hingement along the shell margin nearest the axis of coiling; the shells cannot overlap. Bivalvia cannot be involute either, for the same reason, and thus they are not found above the dashed line on the W-T face, which represents the change from involute to evolute forms on that face.

The restriction of bivalve forms to the low-D region is probably due to the advantages of a rather broad flat shell over a high peaked one. Raup has pointed out that muscles that cross the inside of the shell from valve to valve will be shorter in flattish shells than in peaked ones, thus improving their A/V ratios and enhancing their strength. Furthermore, the general modes of life of Brachiopoda and Bivalvia alike certainly favor compact rather than peaked biconical valves. In epifaunal or semi-infaunal Brachiopoda, with food-gathering organs (lophophore) and internal current pattern symmetrical with respect to the plane of the generating curve, a simple planispiral coiling provides the fewest complications [Fig. 5-16(*c*)]. There is simply no reason for translation. However, in the Bivalvia the internal current pattern is not symmetrical with respect to the generating curve, and neither are the foot, siphons, or other major internal organs. The functions associated with the anterior part of the shell are quite different from those associated with the posterior parts; as a result, the valves are markedly inequilateral. Thus, an asymmetry about the distal-proximal axis is required of the generating curve and strongly asymmetrical growth gradients must occur if the shell proportions are to be maintained during size increase. The asymmetric growth gradients generate a helicospiral shell [Fig. 5-16(*d*)]. At low values of T, a helicospiral, inequivalved shell will be quite compact, even more than a planispiral shell, other parameters being equal. At higher values of T, however, the spire begins to project, and thus compactness is actually reduced. The location of the variously shaped spiral shells on the bloc can thus be rationalized in a general way in terms of functional efficiency, given some knowledge of the basic ground plan of their inhabitants.

Modification and elaboration of the skeleton to provide for special adaptations is a widespread phenomenon, for most well-skeletonized species have unique shell forms. Among the Bivalvia, diversity of form depends considerably upon modification of basic shell-growth patterns, so that the simple model applies only to special cases. Strong growth gradients have appeared that are asymmetrical with respect to the proximal-distal axis and often to all principal planes and axes related to the coiling geometry. The parameter T frequently varies at virtually every point along the generating curve. Because of this complexity, it is useful to analyze the growth direction of a point on the generating curve of a bivalve shell in terms of *growth components*, vectors of growth represented along three normal axes (Fig. 5-17). The plane of the generating curve is commonly used as a reference plane for growth components. Growth owing to the rate of translation (T) is called the *tangential* component [Fig. 5-17(*a*)], whereas growth owing to the rate of whorl increase is resolved into two components, *radial* and *transverse* [Fig. 5-17(*b*)]. Many modifications of basic coiling geometry may be evaluated as gradients in these growth components around the generating curve (Owen, 1953; Wilbur and Owen, 1964).

A few examples will indicate the sort of complex patterns displayed by Bivalvia. First, there are shells, such as *Glycymeris*, that do approximate a simple logarithmic spiral, with a low T so that they are nearly planispiral [Fig. 5-18(*a*)]. The *beaks*, early portions of the shell, are only slightly "twisted" relative to the plane of commissure; the twist demonstrates that they are helically coiled. Then there are shells, such as *Exogyra*, that have a higher T and a correspond-

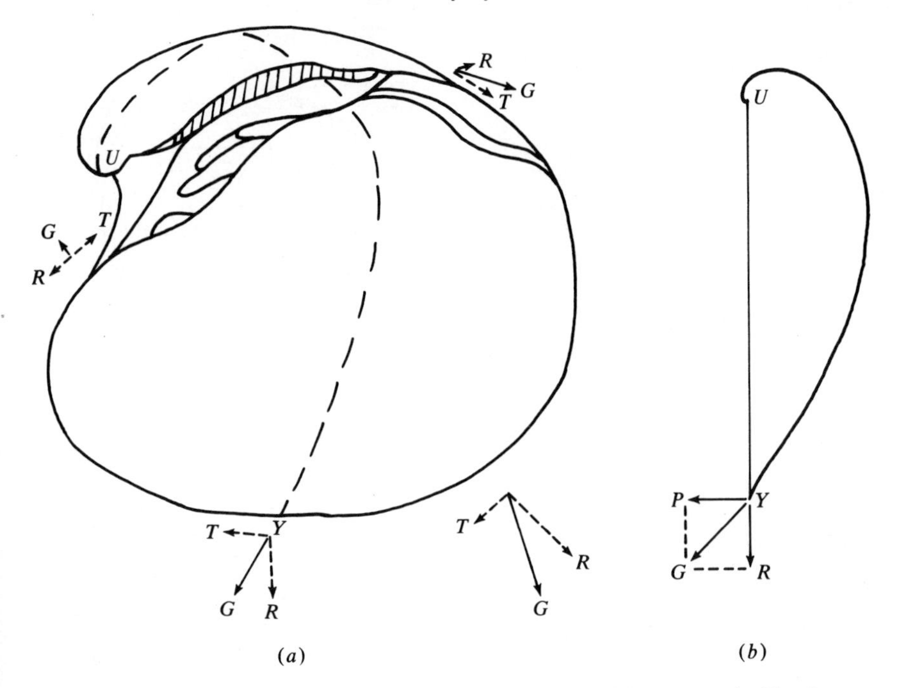

Fig. 5-17 Growth components of the Bivalvia, measured with reference to the plane of commissure. (*a*) The *tangential* component (*T*) owing to translation, and the *radial* component (*R*), the radially directed vector owing to whorl increase. (*b*) The radial component (*R*) and the *transverse* component (*P*), the vector owing to whorl increase that is normal to the commissure. The direction of growth in each view is indicated by *G*. (After Wilbur and Owen, 1964.)

ingly greater twist to the beaks and early shell [Fig. 5-18(*b*)]. More complex are shells such as some of the pectinids [Fig. 5-18(*c*)], which along their medial plane are coiled planispirally, without any translation; however, on either side of the plane they are helicospiral, positively so on one side and negatively so on the other. There is a gradient of *T* around the ventral shell margin.

A gradient of *T* also characterizes such a shell as *Solen* [Fig. 5-18(*d*)]. As in the pectinids, the plane that bisects the beaks is a plane of no translation, and the beaks do not "twist"; but in *Solen*, this plane is near the anterior end of the shell instead of being in a median position. Proceeding from this plane toward the anterior around the ventral margin is a gradient of *T* that is (arbitrarily) positive, but *T* does not assume a very high value there. Proceeding from the planispiral plane in the other direction, *T* is (arbitrarily) negative, and reaches a rather high value posteriorly. This is an asymmetric example of the sort of pattern displayed by the pecten. In one of the most common growth patterns among the Bivalvia, common especially among such families as the Veneridae, a plane of planispiral coiling lies near one end of the valve but the beaks lie to one side of this plane, so that they are helically twisted. The planispiral plane is

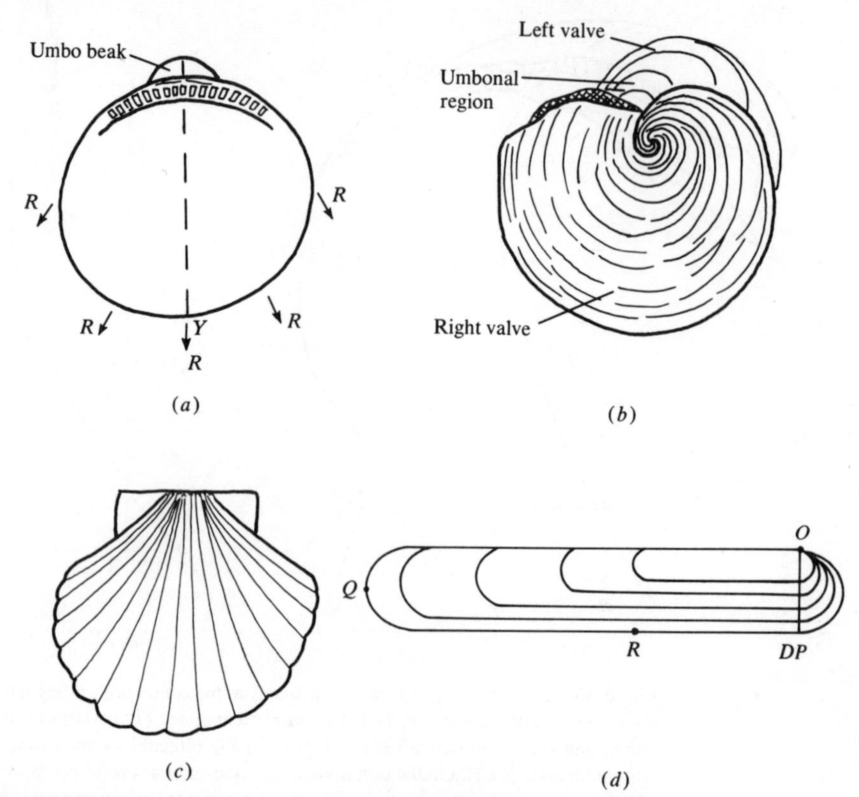

Fig. 5-18 Bivalves of different shapes generated by different growth-gradient patterns. (*a*) *Glycymeris*; essentially planispiral. (*b*) *Exogyra*; strongly helicospiral, with a translation along the axis in one direction only. (*c*) *Pecten*; center of valve is planispiral; anterior side (to right in figure) is helicospiral, with a progressively greater translation toward the anterior margin; posterior side (to left in figure) is helicospiral, but with translation in the opposite direction, with the greatest translation toward the posterior margin. Thus T varies along the entire ventral margin. (*d*) *Solen;* planispiral along the line *O—DP*, with translation grading in one direction toward the posterior (to the right in the figure), and with much greater translation gradient in the opposite direction toward the anterior (to the left in the figure). (In part after Carter, 1967, and Wilbur and Owen, 1964.)

obviously useful for descriptive purposes; some authors have employed it as a basic plane of reference (Lison, 1949; Carter, 1967).

Another type of modification of basic coiling geometry is found among the family Arcidae, which has rather inflated umbos that would ordinarily interfere with each other. The plane of the generating curve is altered from its usual position in a simple logarithmic spiral to form a "biological" rather than a "geometric" generating curve (Fig. 5-19). In this sort of shell, the growth components do not lie in the plane of the actual generating curve but in the plane of the geometric generating curve instead (Stasek, 1963).

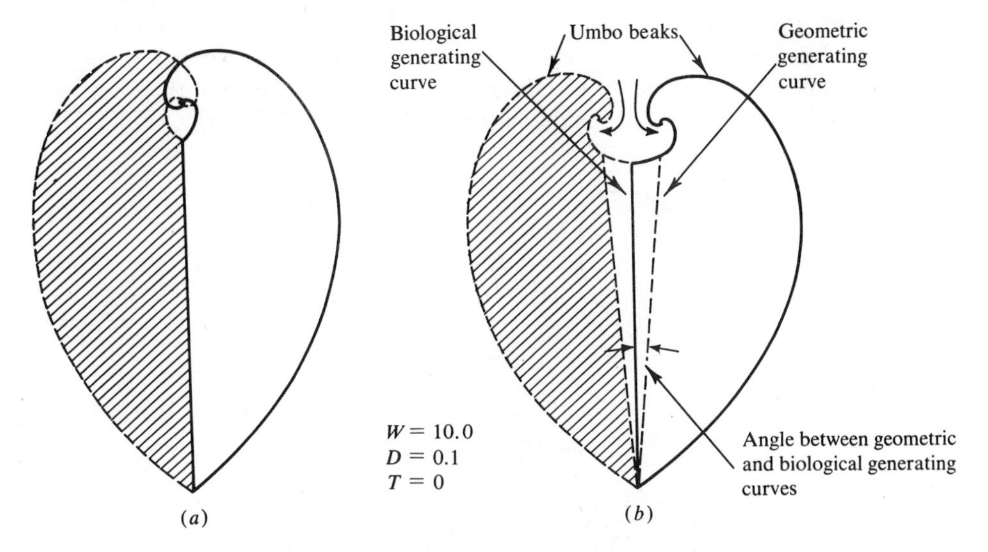

Fig. 5-19 (*a*) Bivalve with inflated umbos that would overlap if the commissure were formed from the generating curve. (*b*) Solution to the overlap problem adopted by the Arcidae, with interumbonal growth producing a different "biological" generating curve that separates the umbos. (From Raup, 1966, after Stasek, 1963.)

Even though alteration in coiling geometry can provide a great variety of shell form, the adaptive versatility of permanent skeletal features generated as primary fixed structures through the expansion of a generating curve (S) is limited, even if the shape of S is modified through time, and the pattern of such structures must be spiral (or radial). These limitations have been overcome to a considerable extent by the introduction of *time-line-structures* (Vermeij, 1970). These are features that appear in the generating curve at a time after the onset of shell formation; they may be restricted to a single position (in which case they are concentric) or they may be continued as radial features (in which case they are secondary fixed structures). Examples are spines, concentric ribs, flanges, and varices (concentric structures) and intercalated or split radial ribbing (secondary fixation).

Among brachiopods, gastropods, and ammonoids, variety of shell form is enhanced chiefly by elaboration of such features as ribs and spines and by altering the shape of S, but large departures of shells of these groups from simple logarithmic coiling models or large variations in T are exceptional. Among gastropods, great variety of shape is achieved by variations in the shape of S combined with a correlated variation in the angle between the plane of S and the axis of coiling, termed the angle E (Vermeij, 1971); usually, large E is associated with simple circular or ovoid generating curves and small E with elongate generating curves, such as that possessed by the siphonate Neogastropoda (Fig. 5-20).

Carter (1967) has strongly emphasized that these schemes of representing skeletal geometry are only descriptive of the form and do not necessarily imply that each reference point, vector, plane, or surface has special biological significance. The geometric descriptions can be applied because the form with which they deal is an inevitable outcome of a mode of growth.

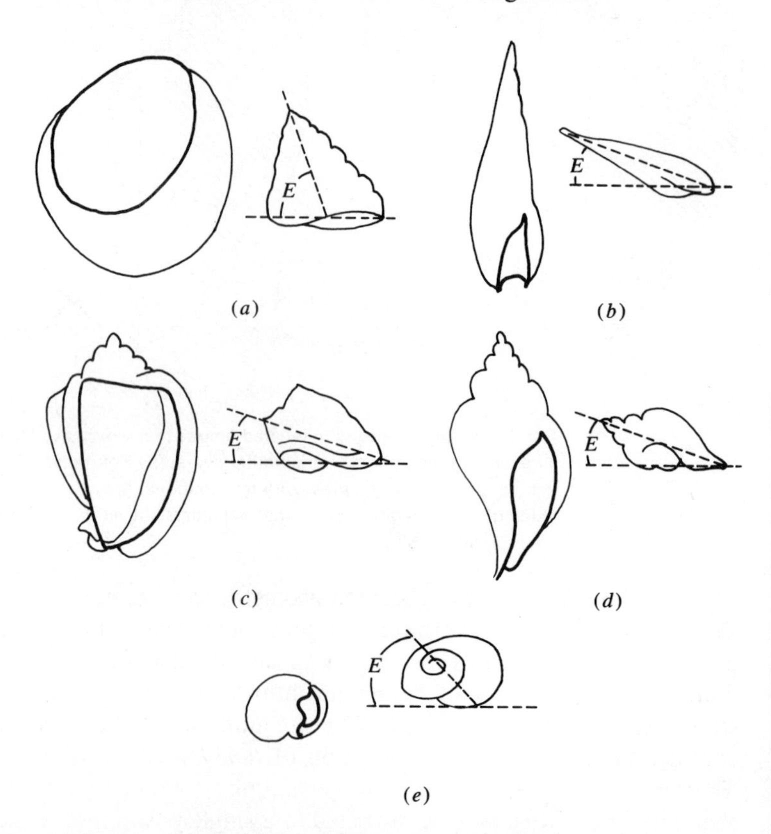

Fig. 5-20 *E*, the angle between the plane of the aperture and the axis of coiling, in various prosobranch Gastropoda: (*a*) *Tegula pellisserpentis*, (*b*) *Terebra felina*, (*c*) *Cassis tuberosa*, (*d*) *Fasciolaria hunteria*, (*e*) *Cyclope neritea*. Note that species with elongate generating curves (shown in heavy line) have a low *E*. (After Vermeij, 1971.)

In summary, the distinctive skeletal plan of each major taxon harmonizes well with its ground plan and promotes the adaptation of the taxon. The contrast between the transient, immutable jointed carapace of the segmented Arthropoda and the permanent, partially modifiable shell of the unsegmented and pseudosegmented Mollusca, for example, is very striking. Thus we may speak of a *skeletal ground plan*. The skeletal ground plan must itself exercise considerable influence upon the evolutionary potentialities of taxa. A given skeletal plan is presumably selected because it is coadaptive with the general anatomical ground plan of the lineage. Once adopted, the skeletal plan may provide the potential for enhancing certain functions and at the same time constrain the development of others. For example, the bivalve shell has evolved into an extremely effective burrowing device, supporting and enhancing the occupation of infaunal habitats and permitting the exploitation of deposit and suspension feeding modes. It has also proven amenable to certain epifaunal modes of life. However, the adoption of these modes of life has precluded bivalves from invading nonaquatic habitats, where their methods of locomotion and feeding cannot operate. There is clearly a feedback between the skeletal and anatomical ground plans, which operates to bring them into harmony, and which constrains the elaboration of either plan.

The adaptive range of taxa partially depends upon the extent to which the potentials inherent in their ground plans are realized; for skeletons, this commonly means the extent to which functional modifications and elaborations of the basic plan have occurred. The modifications presumably have direct adaptive significance, as indeed the skeletal ground plan itself must have had, and usually function ecologically. If the modifications work well, they are commonly diversified and consequently form a common morphological theme for intermediate or lower taxonomic categories, such as orders, families, or genera. Of course, similar modifications are commonly evolved independently in separate lineages; thus, morphology and phylogeny are not directly correlative.

The Mode of Life of Extinct Organisms Can Be Pieced Together by Integrated Functional Analyses of Skeletal Characters

The ecological-evolutionary significance of the major anatomical and skeletal ground plans of invertebrates will be discussed further in Chapter 10. This will involve reconstructing the functions of preserved skeletal and inferred anatomical features of what are presumed to be the primitive lineages of higher taxa. At this point, it is more useful to pursue examples of the interpretation of the functions of relatively well-known extinct organisms from their skeletal morphologies to illustrate the methods and principles employed. The articulate Brachiopoda, for example, were a dominant class among the skeletonized benthos during much of the Paleozoic, and are still represented today by 18 families, although many of the groups important during the Paleozoic are extinct. Some of these extinct groups are very unlike any living forms and have been the subject of extensive functional analysis. Brachiopods are unsegmented, bilaterally symmetrical coelomates, commonly with unequal dorsal and ventral valves (Figs. 5-21 to 5-24). A visceral region lies posteriorly, crossed by muscles to open and close the valves. Anteriorly, the mantle encloses a cavity containing the *lophophore*, a ciliated feeding organ. Most, but not all, brachiopods are attached to the substrate at some time during their ontogeny by a fleshy stalk or *pedicle* that issues from an opening on the median line posteriorly, chiefly or

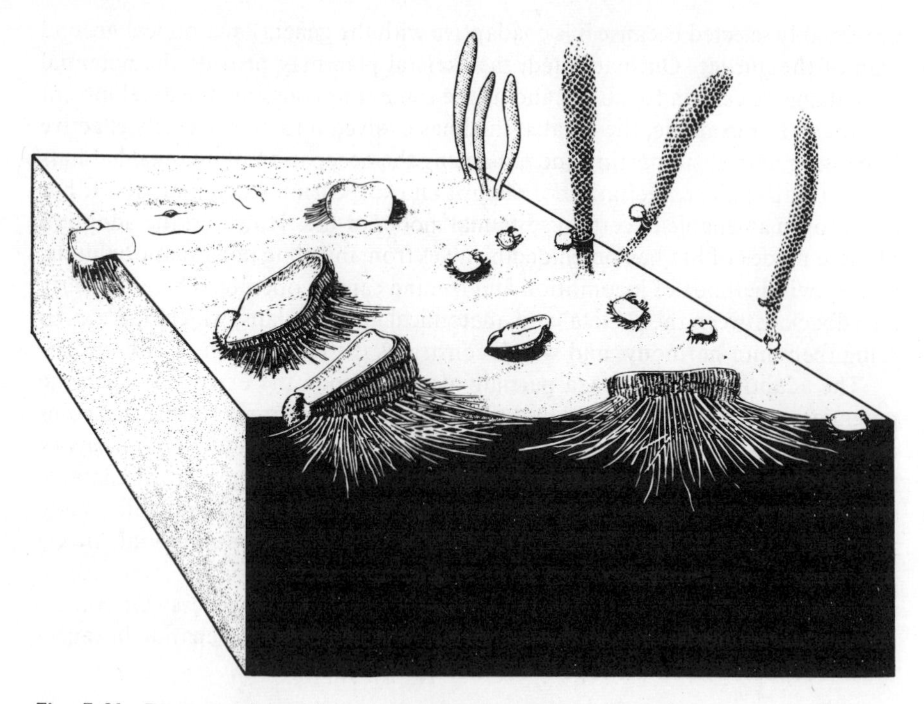

Fig. 5-21 Reconstruction of living positions of the Permian brachiopod, *Waagenoconcha abichi*; note very young attached as epizoonts by clasping spines, older juveniles settled onto muddy substrate, adults supported by mats of spines. (After Grant, 1966.)

entirely in the ventral valve. The pedicle may become obsolete in adults or may remain functional throughout life.

The brachiopods are a relatively homogeneous group; the diversity of adaptive types found in the Brachiopoda is on the same order and perhaps even less than the diversity found within the single molluscan class Bivalvia. The evolutionary patterns of the Brachiopoda have been summarized in the *Treatise on Invertebrate Paleontology* (Part H, Williams *et al.*, 1965). Many of the morphological variations among brachiopods seem to be associated with either attachment or feeding. The main organ of feeding is the lophophore, a looped or coiled ribbonlike structure that is bilaterally symmetrical about the mouth, which lies about in the median plane (Fig. 5-24). The lophophore bears rows of ciliated filaments. The cilia create currents that introduce particles, which are trapped in mucus on the lophophore and conveyed to the mouth along special grooves (see Atkins, 1960). Respiration may be partly accomplished by the lophophore but evidently occurs chiefly in tubular, branching *mantle canals*, which are extensions of the body cavity into the mantle.

After food and oxygen are procured from the current, the water is expelled from the mantle cavity. Most adult brachiopods have two lateral inhalant currents separated by a single median exhalant one. There is no mantle fusion, so a separation of inhalant and exhalant apertures is contrived by the form of the shell opening, by lophophore filaments, and sometimes by the aid of setae that help define these regions. One of the main evolutionary trends among brachiopod lineages has been the elaboration of the lophophore, presumably to increase feeding efficiency and, often, to keep pace with phyletic increases in body size. Presumably, the lophophore must increase its length allometrically compared with body size to provide the same food for a given amount of tissue. Lophophores are folded, looped, and coiled to pack a maximum functional lophophore length into the variously shaped mantle cavities; of course, an efficient filtering design must be maintained.

Most articulate brachiopods have been strictly epifaunal, resting upon the substrate and fixed thereto by pedicle, spines, or a cemented attachment area, or in some cases lying unattached upon the bottom. Suspension feeding obviously requires an environment with a reasonable amount of material suspended in the water, and in such environments, the suspension commonly includes much terrigenous clay. Columns of water containing the largest organic suspensions are frequently floored by soft muddy bottoms, and benthic suspension feeders are therefore commonly adapted to life on soft bottoms and in turbid waters. Many of the brachiopods seem to possess such adaptations.

One group that apparently has lived chiefly on soft bottoms is the Productoidina, a suborder which appeared in the Devonian and became extinct by the end of the Permian. The largest superfamily of productoidines, the Productacea, have more or less deeply convex ventral valves and flattened or concave dorsal valves, and usually numerous surface spines (Figs. 5-21, 5-22). Lophophores were coiled about dorsoventral axes [Fig. 5-24(*b*)]. Their shells are planispirally coiled with a low *D*, in keeping with their anatomical plan. Interpretations of the life modes of several members of the Productacea have been worked out.

An adaptation rather widespread among productaceans, is the infant attachment loop (Figs. 5-21, 5-22, 5-23). The larvae were clearly planktonic, for at about metamorphosis they settled onto invertebrate skeletons or vegetation that provided a relatively secure habitat above the soft bottom muds. A pair of spines growing posteriorly from the cardinal region encircled or clasped part of the host, which held the young brachiopod suspended in the water. As growth continued, the brachiopod eventually broke loose and came to rest on the mud, presumably at an ontogenetic stage when it was equipped to cope with the problems of a soft bottom.

One of the Permian productaceans, *Waagenoconcha abichi*, had a marginal aureole of surface spines that extended brushlike around the lateral and anterior margins (Fig. 5-21). These spines do not meet specifications for protection because they leave the commissure and its approaches unguarded. However, they would seem to have been rather efficient as supports, increasing the effective horizontal area of the shell and propping it up in soft ooze at the level where they spread out, much like snowshoes (Grant, 1966). When the larvae of this species first settled, they were suspended from invertebrate skeletons by attachment loops until adequate marginal spines were formed. Upon reaching the soft bottom, the convex ventral valve could sink down into the sediment relatively easily until the fan of spines was reached. The spines are so positioned that the commissure was kept just free of the sediment surface. Sediment seems to have become emponded upon the upper surface of the concave dorsal valve; in fact, small spines positioned dorsally seem designed to prevent the escape of such emponded sediment. The muddy cover may have served as camouflage and also to prevent the settling of larvae upon the dorsal shell surface. Subsequent sedimentary buildup above the spiny thicket was compensated by upward growth of the anterior shell margin, which curved up more and more strongly and sent out spines to reinforce the supporting mat. The anterior margin was slightly differentiated into median and lateral areas, probably the sites of median exhalant and lateral inhalant currents. When the dorsal valve gaped to permit mantle circulation, the effect must have been somewhat "reminiscent of the lair of a trapdoor spider" (Coleman, 1957).

A rather different pattern of spines is found in the Permian productacean genus *Marginifera* (Grant, 1968). These brachiopods also formed attachment loops upon first settling from the plankton, and while they were thus suspended, a few long surface spines grew out in an anterior direction from the sides and front of the ventral valve (Fig. 5-22). After breaking loose from their attachment site, the central portion of the ventral valve rested on the bottom, with the anterior margin propped up above the surface of the substrate by the spines. This living orientation is verified by the stratification of sedimentary accumulations within the fossil shells. As the animal grew, the anterior margin grew upward and became elevated, and the early spines sank into the substrate or were covered by sedimentation. Additional spines, heretofore growing free of the bottom from the exposed sides of the valve, were then brought into contact with the bottom and took over the function of propping up the anterior commissure. The buried spines served thereafter as anchoring devices. Sediment was evidently emponded within the concave dorsal valve, presumably for the same reasons as in *Waagenoconcha*. The commissure was kept just above the sediment, according to evidence from the geometry of the spines and from internal sediment accumulations, and in a position that permitted mantle circulation. The spines inhibited sinking or tipping of the shell. The direction in which the shell may be most easily tipped is posteriorly (Fig. 5-22), and the orientation after such tipping may have been permissible in any event. There is little doubt that the spines represent supporting and anchoring devices.

The generic name *Marginifera* refers to a shelly marginal ridge that runs along the inside of the valves. When the valve edges of *Marginifera* grew sharply upward during ontogeny, a bend or deflection of the shell in a dorsal direction was created; this bend is called a *geniculation*, and the flangelike shell band outside the geniculation is called a *trail*. The internal marginal ridge formed an internal baffle between the wedge-shaped mantle space in the trail and in the

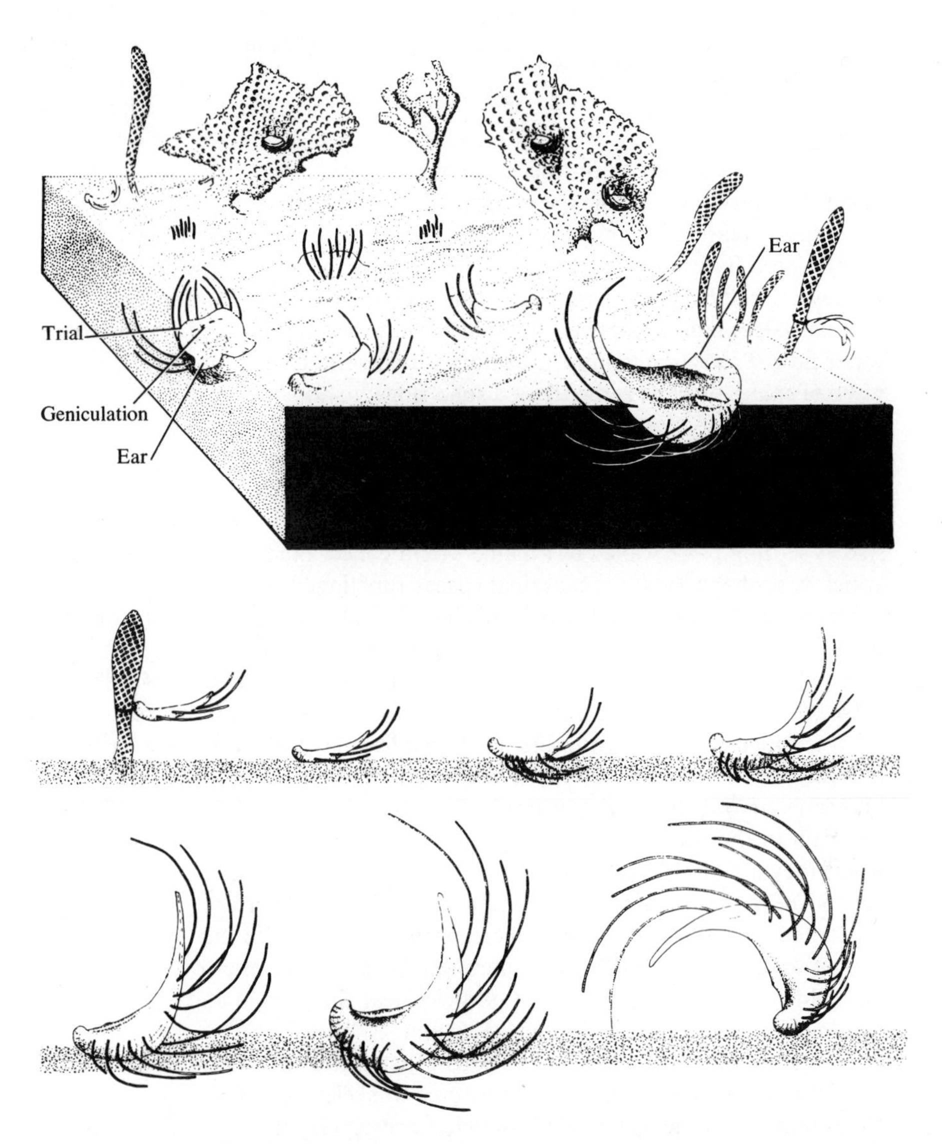

Fig. 5-22 Reconstruction of the living positions of the brachiopod *Marginifera*. Very young are attached as epizoonts above the substrate; older juveniles fall to the substrate; and adults are supported on ski-like spines. Upper figure represents a group of *Marginifera* in various stages of ontogeny; lower figure depicts an ontogenetic sequence, showing increased tilting of valves and lengthening of spines with age. The final position, at lower right is a hypothetical possibility if organism is disturbed and rotated backward; probably it could have survived in this orientation. (After Grant, 1968.)

rest of the mantle cavity containing the lophophore and viscera. Marginal ridges are also found in the posterolateral regions, where the shell is swollen into lateral extensions called ears. The form of the ears seems designed to inhibit sinking of the posterior margin of the shell into the substrate. The cavity within each of the ears is bordered by two ridges, one arising from each valve and crenulated so as to interlock with each other when the valves are closed. Thus, a double baffle is formed, cutting off the ear spaces from the visceral cavity.

It seems reasonable that these ridges were protective devices, for they certainly would have impeded access to the internal organs of the brachiopods. However a bafflelike, solid ridge prevents even the passage of water currents, so that the valves must be opened just that much wider to allow a given amount of water flow. Protection against predators would seem to call for a grid or meshwork of spines or bars, rather than a solid ridge, to allow water currents to flow through (see the following section). The baffle probably performed a biological role other than just protection from predators. Perhaps the ridges functioned as a protective device against the ingress of muddy sediment, which could not be sieved out mechanically by the skeletal devices owing to the small particle size. A settling basin walled by a baffle would collect mud particles very neatly, which is probably how the marginal spaces functioned (Grant, 1968). The ear spaces were pressed against or partially into the substrate, a dangerous source of mud. The rest of the commissure was guarded by a single marginal ridge. Internal spines arranged in regular rows were present in the interior of the dorsal valve; these may have had a function of protection against predators.

One other group of characters in *Marginifera* has been interpreted as adaptive for life on muddy bottoms. These are small conical tubercles of the sort known as *pseudopunctae* that are found on the interior of the valves. They are common in the ventral body cavity, where they are rather regularly spaced in rows; they are also found in the interior of the ear and are especially abundant on and near the marginal ridge and also on the inner surface of the trail. In short, they are most abundant where protection from fouling by mud seems most desirable. Grant (1968) has interpreted them as the sites of setae or bristles, which functioned to sieve out particles near the margin and to sense particles in the interior and aid in their rejection.

Probably productaceans had various other means of defending themselves against mud fouling that are not represented by skeletal characters. It is possible, for example, that the upper valve could have been "clapped" shut rather rapidly, creating a short-lived but relatively strong outgoing current all around the commissure to clear away marginal accumulations of rejected particles.

A number of other patterns of spine support and protection are known among productaceans, but the general productacean mode of life and the nature of its variations are well enough indicated by these examples. However, one additional and unusual adaptation is worth mentioning. *Linoproductus angustus*, which formed an infant attachment loop upon settling, solved the muddy bottom problem in a different but definitive way. It never descended to the sea floor at all normally, but continued to secrete cardinal spines that formed a series of loops and supported the adult brachiopod above the substrate throughout life (Fig. 5-23; R. E. Grant, 1963). However, this ingenious evolutionary solution did not prove to be a key to any enduring success.

Fig. 5-23 Reconstruction of the living position of *Linoproductus augustus;* both juvenile and mature individuals are epizoonts on crinoids. (After R. E. Grant, 1963.)

"Bizarre" Morphological Characters Commonly Indicate Functions That Are Unusual for the Taxon

The Richthofeniidae are an extinct family of productoidine Brachiopoda that in shell form fall well outside the modal range of the suborder and of the phylum. Richthofeniids are recorded in rocks that range in age from Lower to Upper Permian. Their unique features include an extremely deep, conical ventral valve, which in life was cemented to the substrate, and a flat platelike dorsal valve (Fig. 5-24). The dorsal valve is commonly recessed so that the margin of the ventral valve projects higher dorsally than the dorsal valve, and the shell interior of the ventral valve is thus partitioned into an outer and inner cavity by the dorsal valve. Both valves bear spines; the spines on the dorsal valve are

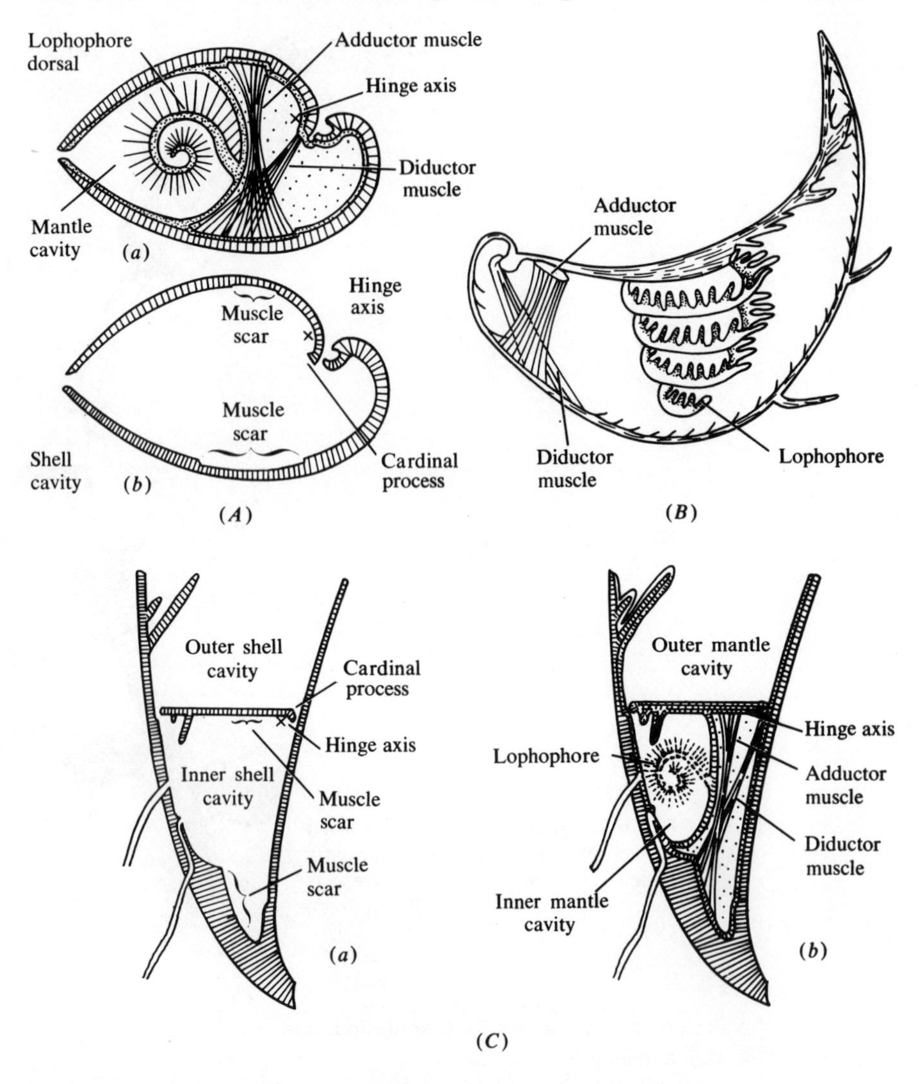

Fig. 5-24 (*A*) Cross section of a generalized living brachiopod, showing position of muscles and lophophore (*a*) and of muscle scars (*b*). (*B*) Cross section of a generalized productoid brachiopod, showing reconstruction of position of lophophore and muscles. (*C*) Cross section of a richthofeniid brachiopod, showing position of muscle scars and spines (*a*) and reconstructed positions of lophophore and muscles (*b*). [Parts (*A*) and (*C*) after Rudwick, 1961, by permission of the Palaeontological Association; (*B*) after Grant, 1968.]

internal and project into the inner shell cavity, whereas some spines of the ventral valve project into the outer shell cavity.

A "normal" productoid has valves that meet along their outer margins, such as in *Waagenoconcha* or *Marginifera*; Fig. 5-24(*B*) and (*C*) contrasts a normal type with a richthofeniid. In living brachiopods, the posterior part of the shell cavity (toward the hinge line and beak) contains the muscles that cause the valves to open (*diductor muscles*) and close (*adductor muscles*). The valves are opened on the lever principle. The dorsal valve swings on a hinge axis, and the diductors are attached to the dorsal valve where it projects behind the axis, so that when they contract the narrow rear part of the valve is pulled down and the broad forward part moves up and opens [Fig. 5-24(*A*)].

Richthofeniids were so bizarre morphologically that they present a special challenge to the functional morphologist. They have been the subject of functional studies by Rudwick (1961), which form the chief basis of the present account. Reasoning from the locations of muscle scars on fossil richthofeniids and on well-known "normal" brachiopods, he was able to reconstruct through homological argument the general distribution of viscera and lophophore within the inner richthofeniid shell cavity (Fig. 5-24). However, the richthofeniids also possessed an elaborate suite of shell characters, for which living homologues are unknown, and for which analogues are certainly not obvious if they exist. These shell characters may nevertheless furnish the morphological bases for functional models.

Consider the unusual dorsal valve. An obvious function for the shells of brachiopods is protection, for which a certain strength and enclosure of soft parts are required. The richthofeniid dorsal valve is very thin and waferlike, much more fragile than normal brachiopod valves, and does not protect the outer shell cavity. As we have seen, brachiopod shells are secreted by mantle cells in contact with the growing shell surface. Thus, at least during shell secretion, the mantle tissues must occupy the inner surface of the ventral valve that is exposed in the outer shell cavity. Since this surface is spinose, it is hard to see how the mantle could have been retracted even when shell growth was not occurring. The dorsal valve, then, is weak and does not enclose all the soft parts, suggesting that functions other than protection were involved.

The thin dorsal valve seems to have been operated by muscles of at least normal size for robust valves, judging from the scars, and is articulated by spindle-type structures that allow the valve to revolve on a sharp ridge. This is a hinge type that would involve little friction. Unlike normal brachiopod valves, the range of movement of many richthofeniid dorsal valves must have been great, from 60 to 90°; fossil specimens are known with the valves preserved in widely open positions. These facts suggested to Rudwick that the valves might have operated relatively rapidly, moving up and down through a wide arc. The requirements for such a function would be a light, flat, well-muscled valve with a rather precise hingement. The valve morphology appears to match this model closely.

What biological roles would such a function play? The most obvious suggestion is to bring water currents into and out of the shell cavities for feeding, respiration, and waste disposal. Normal sorts of brachiopods cause such currents chiefly by ciliary action; currents flow into the mantle cavity along the posterolateral margins and out near the midline posteriorly. Now if a mechanism of suspension feeding is to be efficient and dependable, it must meet certain basic requirements, especially that fresh water be introduced steadily to the food-gathering devices. The alternatives are that used and polluted water is recycled or that no water flows, and these could do the organism no good. Therefore, there must be an efficient separation of the inhalant and exhalant currents. Clearly, the plan of water circulation within the organism must prevent much mixture of fresh and used water. This requirement is usually met by an opening that functions as an inhalant aperture and a separate one that functions as an exhalant aperture. The apertures may be morphologically distinct, as in siphonate bivalves, or only functionally distinct. Associated with a suspension-feeding organism, then, is a region or chamber which is occupied by unused water and one which is occupied by used water, separated by the organ(s) that gathers food. In the case of suspension-feeding bivalves, and of most other molluscan suspension feeders, the gills separate inhalant and exhalant chambers; in living brachiopods, the lophophore does this job.

The presence of the outer shell cavity in richthofeniids poses a problem: How were inhalant and exhalant currents separated therein and prevented from mingling? There was probably no physical division by tissues, for the dorsal valve required a large free space in order to open within the outer cavity. Rudwick (1961) suggests that, in fact, the currents were not separated spatially but temporally: periodic water exchange promoted by the moving dorsal valve served to flush out the mantle cavity, removing used, and introducing fresh, water. This would be an alternative method of obtaining an unpolluted supply of fresh water.

To test the feasibility of his suggestion, Rudwick built working models of richthofeniids to natural scale and photographed the water movements that accompanied opening and closing of the dorsal valve. Results are diagrammatically presented in Fig. 5-25. As the valve opens, currents sweep over the anterior margins of the ventral and dorsal valves and down into the inner shell cavity. Water that had been standing in the outer shell cavity chiefly flows out of the cavity. When the valve closes, water escapes upward from the path of the descending valve and flows into the outer shell cavity from the posterior. Evidently, if the dorsal valve had operated as hypothesized, an exchange of water in the mantle cavity would have been accomplished.

Spines guard the aperture of the outer shell cavity in richthofeniids. In some species, the spines branch and anastomose to form a grillwork. A natural suggestion is that the mesh and grill serve to protect the organism from particles or from other organisms that are too large to pass through. Requirements for a protective mesh could include the following features: (1) the mesh bars must

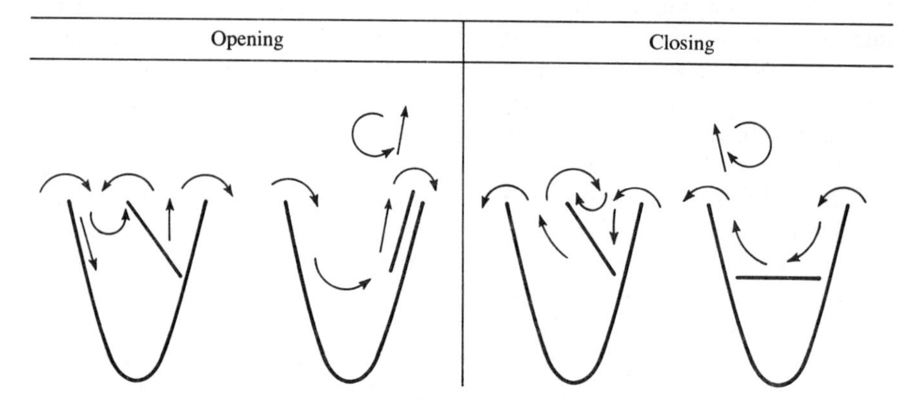

Fig. 5-25 Patterns of water currents during opening and closing of dorsal valve in a model of *Prorichthofenia*, in longitudinal section. (After Rudwick, 1961. By permission of the Palaeontological Association.)

impede the flow of water as little as possible; (2) the spaces between bars must be as uniform as possible; and (3) all the inhalant current must pass through the meshwork. The meshwork of one species investigated by Rudwick, *Prorichthofenia uddeni*, agrees well with these specifications, considering the geometric complications imposed by growth requirements. For example, the mesh bars are elongated at right angles to the aperture, thus blocking the aperture as little as possible while achieving a certain strength.

Requirements for a protective grill should include the same features: (1) the grill must not impede the current any more than possible; (2) the slots between the bars must be nearly the same width; and (3) all the inhalant current must pass through the grillwork. In this case also, the bars should be elongated at right angles to the aperture, and if there is more than one tier, they should lie in the same plane so as not to interfere unduly with water flow. The grillwork of a species investigated by Rudwick, *Prorichthofenia permiana*, did not agree very well with these specifications. The slots are fairly regular and the grill covers the aperture effectively, so that large particles would have been blocked out. On the other hand, the bars are stout and the slots between them relatively narrow; also, the bars are rounded, roughened by longitudinal grooves, and vertically offset from tier to tier to form irregular "thickets." The grillwork may have served a protective function, but it is so constructed as to intercept much of the water, causing currents to swirl through narrow spaces between offset bars. Indeed, the obstructive character of the bars suggests that they may have functioned as interceptors.

Rudwick suggested that the grill bars functioned to intercept and trap food particles, supplementing or perhaps even replacing the lophophore as a food-gathering device. Indeed, the spines seem to accord well with a model for food gathering of this sort, compromising with functions of protection and the necessity of admitting a current of water to the mantle cavity. Spines on the interior of the dorsal valve, extending into the mantle cavity, may have acted also as particle traps, for they stand in the path of strong water currents associated with valve movements. The bar tissues would have been ciliated and provided with mucous glands, with acceptance and perhaps rejection tracts developed in the longitudinal grooves.

The feeding mechanism proposed for these richthofeniids is unknown among living brachiopods. But, after all, their morphology is also unique in an extreme and aberrant way, so that an aberrant functional interpretation does not seem unwarranted. Also, there are organisms known that have evolved somewhat analogous feeding methods. Septibranch bivalves employ a rapidly moving membrane to suck food particles, often whole small animals, into an internal chamber, and lucinoid bivalves have developed a supplementary particle trap on their ciliate anterior adductor muscle that seems to have partly replaced the gill as a food-gathering device, as we have seen. A rythmic flushing method of feeding has been postulated for other fossil brachiopods, including the bizarre lyttoniaceans (Rudwick and Cowen, 1968) and the very flat plectambonitaceans (Cocks, 1970).

What selective pressures could have caused the development of the bizarre richthofeniids? Cowen (1970) has suggested that their exposed tissues were colonized by zooxanthellae, autotrophic protistans that live symbiotically in the tissues of present-day hermatypic corals and in some bivalves, most notably in the giant *Tridacna* of tropical reefs. Many features of the biology of *Tridacna* appear to be coadapted with zooxanthellae (Yonge, 1936; Stasek, 1962). Its larvae select an appropriately lighted site for permanent attachment. Mantle tissues are exposed to light, despite the danger of predation; few other organisms expose tissue on reefs, and those are generally toxic. The zooxanthellae evidently supplement the energy sources and metabolic processes of *Tridacna*, permitting rapid growth, large body size, and rapid calcification. *Tridacna* seems also to utilize a wider range of food items (even including small animals) than does the average bivalve. The analogy with the inferred paleoecology of the richthofeniids is very striking. They also lived on reefs, had exposed mantle tissue (some may have had translucent dorsal valves in addition), were relatively large, and may have ingested larger or more active prey than the average brachiopod, considering their postulated feeding mechanism.

The adaptive steps that may have led from an "average" productoid to the richthofeniids can now be provisionally reconstructed. The process probably began with the habitual exposure of some mantle tissue, perhaps in a lineage that commonly ingested at least part of its food by a rhythmic valve motion. Such exposure of mantle tissue, which may have served originally as a sensor to trigger an ingestion reflex, has been demonstrated in early lyttoniaceans (Rudwick, 1970). Once tissue was exposed, it could be turned to further advantage by supporting zooxanthellae; once established, this symbiosis would lead to the attainment of larger size and the exploitation of increasingly large and more active food items. Reef communities, developed in relatively clear, shallow waters and probably swarming with minute animals, would form an ideal biotic environment for lineages developing in this fashion.

Thus a picture can be built up involving progressive modification of a variety of coadapted characters that fulfill biological roles in a coordinated manner, and that lead logically to the mode of life postulated for richthofeniids. Of course, this is only a model, but it does render the evolution of these bizarre lineages comprehensible as an orderly modification of their ecospaces in adaptive directions that form adaptive pathways. Lineages cannot jump from one ecospace to another, but rather develop along continuous pathways in response to changes in selective pressures and to environmental opportunities.

Empirical Methods of Interpreting the Ecological Functions of Fossil Organisms Are in Wide Use

Paleoecologists do not often generalize upon the methodology of their functional interpretations, although they commonly describe and evaluate individual techniques. Table 5-5 is an attempt to categorize the main modes of interpretation that lead to knowledge of ecological functions at the level of the individual organism. Curiously, several of these modes require the investigation of patterns of populations rather than the occurrence of individuals; still, the ecological relations upon which they bear are on the individual level, for the most part.

We have already discussed *functional morphological interpretations*, most notably with examples from the productoid brachiopods. Other common methods are strictly empirical (Table 5-5, numbers 2-4) and do not seek to explain the basis of the environmental relations they invoke. One such method can be called *empirical morphological interpretation*. This involves the correlation of morphological characters with environmental parameters, without understanding (at least initially) the functional relationship that underlies the correlation, if any. Commonly, the empirical datum is based upon the living biota, but there is no reason in principle that a datum could not be established in the fossil record, providing only that the interpretation of paleoenvironmental parameters seemed sufficiently secure.

A good example is formed by the generalizations of Nicol (1967) on the shell morphology of cold-water bivalves (Chapter 4). He found that species living below about 5°C are generally small and have thin and chalky shells with subdued ornamentation and no spines. The reasons for these conditions are not well understood, yet assemblages of fossil bivalve species with these characters could be interpreted as possible cold-water associations. Among foraminifera, Bandy (1960) has noted that within a number of common genera there is a correlation between the morphology of certain species and their depth range. For example, in the eastern Pacific the genus *Uvigerina* is represented in outer shelf and upper bathyal depths chiefly by slender striate species, in upper to

Table 5-5 Some Methods of Interpretation of the Ecological Functions of Fossil Organisms

Interpretation	*Description or rationale*	*Chief pitfall*
1. Functional morphological	Inferring functions of morphological characters based on model of efficient operation.	Characters may represent a compromise between more than one function.
2. Empirical morphological	Correlation of morphological characters with paleoenvironmental parameters; adaptive relation assumed.	The correlation may not be causal.
3. Empirical taxonomic	Closely related taxa have similar functions.	Since taxa are not identical, their functions are probably not all identical either; in effect, the pitfall is evolution.
4. Empirical	Correlation of distributional pattern with paleoenvironmental parameters.	The correlation may not be causal.
5. Biogeochemical-structural	Skeletal composition or internal structure reflects environmental factors.	Skeletal properties differ among taxa and are commonly altered after death.

middle bathyal depths by costate forms, in deeper bathyal water by spinose forms, and, finally, in the lower bathyal by papillate forms. Depth trends in the morphology of species of the genus *Bolivina* include small smooth species on the inner shelves; medium-sized species with chambers that are striate, that terminate in short spines, or that are smooth, on the central and outer shelves; and costate species in outer shelf and bathyal zones. Other genera are also recorded as displaying morphological trends among their species with depth; sometimes similar trends are displayed within separate genera, and sometimes not. Although the reasons for these correlations are not understood, the morphological forms could be used as a tentative indication of depth range in fossils.

Waller (1969) has interpreted the habits of some extinct species of the *Argopecten gibbus* stock, a group of Middle and Late Cenozoic scallops, from a variety of evidence including empirical morphological evidence. Today, two species of scallop are common along the Atlantic coast of the United States: the bay scallop, *A. irradians*, which lives in semienclosed bays, sounds, or estuaries in water depths of a few inches to about 10 fathoms; and the calico scallop, *A. gibbus*, which lives offshore in open marine water in depths of from 5 to 200 fathoms. Along the coast of southern California and Baja California lives another *Argopecten*, *A. circularis*, which inhabits both bays and sounds and also open marine waters to depths of about 75 fathoms.

During the Miocene, a common *Argopecten* was *A. comparilis*, which is found in sediments interpreted as representing a variety of conditions including both protected bays and open oceans; it ranged from the western Atlantic through the Gulf of Mexico and the Caribbean to the Pacific. *Argopecten comparilis* evidently gave rise to both *A. gibbus* and *A. irradians*; it has some morphological features intermediate betwen them, but at the same time is more variable than either and has some characters that overlap the distinctive ranges of morphology which each displays. It appears that the ecological range of *A. comparilis* has been partitioned to permit two specialist species, each of which has distinctive morphological specializations, to occupy the ecospace it formerly inhabited alone. On the Pacific coast, *A. circularis* is descended also from *A. camparilis*, but has retained the primitive morphological characters of that species and is broadly adapted. By and large, the functional significance of the morphological characters involved has not been evaluated; it is their correlation with present and past environments that is interpreted. In fact, one fossil species, *A. anteamplicostatus*, from the Upper Miocene and Pliocene, resembles the living bay scallop in characters that correlate with bay specialization today, and is therefore considered to be a bay scallop itself (Waller, 1969). Whether these characters are directly adaptive to bay conditions is, however, unknown.

The usual rationale for empirical morphological interpretations is that the different species with similar characters in similar environments are products of convergent evolution for those characters, and that the environmental fidelity of the characters is therefore significant. The chief pitfall of this method is, clearly, that the correlation of character and environment may not be causal, and the morphological features may be related to some factor other than the environmental factors chosen for the correlation. Whenever this other factor itself does not happen to correlate highly with the chosen environmental trends, then the morphological correlation will break down and interpretations based on the morphology will be incorrect. This is particularly likely to be the case for depth interpretations, as depth itself is not an environmental factor any more than is latitude; it is the variation of a number of factors with depth that leads to biotic changes with depth, and in different regions or at different times most of these factors vary in different ways (as light in low and high latitudes or in turbid and clear waters; temperature with latitude and hydrography; and so on). Thus, trends of morphological change with depth may be expected to vary from region to region and time to time, depending on local conditions.

Another method of functional interpretation, on still a lower level of confidence, may be called *empirical taxonomic interpretation* (Table 5-5). This is a functional interpretation based upon the alliance of a fossil to some living taxon for which ecological data are available. The rationale is simply that because they are related, the fossil and living forms should have similar environmental responses. Therefore, there is no appeal to the functional interpretation of morphological characters at all, empirical or otherwise, except insofar as the forms involved share taxonomically significant characters that permit them to be considered as closely allied. Commonly, empirical taxonomic interpretations employ living species as a datum for the interpretation of fossils. Ancestral members of stocks now restricted to the tropics, for example, are employed as indicators of tropical conditions in the past, commonly without an understanding of the processes that operate to cause the restriction at present. Thus, because present colonial reef corals are essentially restricted to tropical climates, the presence of colonial corals as common constituents of Paleozoic reefs is sometimes considered to be evidence in favor of their tropicality, even though modern corals belong to orders separate from the Paleozoic reef-formers and are probably not descended from them, although, of course, they all share some remote common coelenterate ancestor.

On lower taxonomic levels, fossils are commonly interpreted as having ecological requirements similar to living conspecific, congeneric, or confamilial organisms. As an example of empirical taxonomic interpretation on a very low level, Pleistocene fossils that are conspecific with living forms are commonly inferred to have essentially the same ecological significance as their living allies. For some functions this inference is very likely, whereas for others it is much less so. Many marine mollusks lived during the Pleistocene in regions that they do not inhabit today. Often, there are no significant morphological differences between the fossils and living individuals. Using a strict empirical taxonomic interpretation of their functions, we must suppose that the factors limiting their ranges today had a different geographic pattern in the past, so that conditions at localities where they formerly lived have changed. Thus, northern species that are presumably limited at the southern ends of their ranges by warm temperatures must indicate cooler water in former times when they occur in southerly latitudes, and species that are southern today signify warm water when occurring well to the north during the Pleistocene. This empirical taxonomic interpretation clearly requires supporting evidence before being accepted as the most probable interpretation, for there are strong alternative possibilities. One is simply that the species have evolved so as to have different temperature tolerances at present than formerly; that is, the position and perhaps the extent of their ecospaces have changed along thermal axes.

However, if a Pleistocene lucinoid bivalve is inferred to have burrowed and to have fed on suspended particles drawn in through an anterior tube, the suggestion is not strictly based on empirical taxonomic grounds, but is actually far stronger. The fossil is identified as a lucinoid because it possesses taxonomic characters that happen to correlate highly with this mode of life. The inference is then at least as strong as an empirical morphological interpretation. In fact, the functional basis of these lucinoid characters is understood, and because they underly the activities and tolerances that permit the occurrence of the inferred habits, the basis of the interpretation is actually on the level of functional morphology. The interpretation may still be wrong; it is most likely to err in that the fossil, although *equipped* to live in the inferred manner, does not happen to have done so but functioned in a less specialized way.

The environmental preferences of fossils can also be studied by forming strictly *empirical interpretations* from the fossil record, without recourse to morphologic or taxonomic comparisons (Table 5-5). A good example is afforded by the work of Park (1969) on the bivalve genus *Venericardia* from the Cenozoic of the Atlantic and Gulf coasts of North America. This study was based on 198 samples, of which 85 contained specimens of *Venericardia.* The paleoenvironments represented at all the sample localities were inferred by empirical morphological analysis of several groups of microfossils and supplemented by the evidence and interpretations of previous investigators who had studied the fossil assemblages at these localities and who had used a wide variety of interpretative techniques. The difficulties inherent in empirical morphological analyses were minimized by the use of a large number of different taxonomic groups. The patterns of occurrence of subgenera of *Venericardia* were then compared with the inferred paleoenvironmental patterns. Each subgenus has a different pattern, and these were interpreted as indicating differences in the environmental preferences of species of each subgenus. That is, it was inferred that species within each subgenus has more similar environmental responses than species in different subgenera. The subgenera that are most closely similar morphologically seem to have had the most similar paleoenvironmental distributions. Three evironmental factors were inferred to have been chiefly responsible for the distinctive subgeneric ecospaces among *Venericardia*: sedimentation rate (presumably indicating variation in water turbidity); degree of normality of marine conditions ("open" versus "restricted," presumably related chiefly to salinity); and water depths (presumably related to a whole complex of factors including light, temperature, turbidity, and temporal heterogeneity). Park noted that there appear to be changes in the environmental patterns of some of the subgenera through time. *Pleuromeris*, a subgenus that was restricted to fossil samples interpreted as representing "shallow inner shelf" and "restricted marine" environments, lives today in greater water depths in open marine situations. *Cyclocardia* was restricted to "shallow inner shelf" environments in the fossil samples, yet ranges today to considerable depths; however, it occurs infrequently in the fossil samples. Furthermore, the use of subgenera rather than species as a basis of interpretation makes this partly an empirical taxonomic interpretation, and there is a question whether the differences between the fossil and recent habitats of certain species of these subgenera can be used as a basis for inferring that the ecospaces of the subgenera themselves have changed. Indeed, there is a question whether the total contemporaneous ranges of the ecospaces of these fossil taxa are adequately represented in the samples. Nevertheless, the matching of patterns of fossils with environments has permitted the erection of hypotheses concerning the habits and tolerances of the fossils, independent from their structure or taxonomic affinities.

Certainly, internal fossil evidence can be employed to interpret the local environmental ranges of higher taxa. Fossil coral reefs provide an especially good example, for the environmental framework of coral reef complexes may often be worked out because the physical structure of the reefs is commonly preserved or easily reconstructed. In particular, for any time, the position of sea

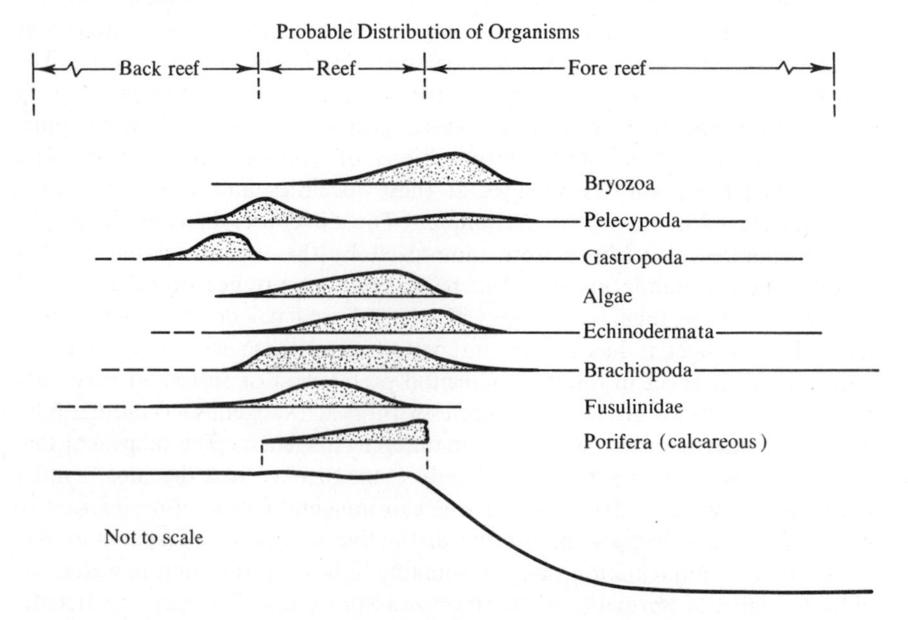

Fig. 5-26 The inferred living distributions of main groups of fossil reef organisms from the Permian Capitan Reef, shown in diagrammatic cross section. (From *The Permian Reef Complex* by Norman D. Newell, Rigby, Fischer, Whitman, Hickox, and Bradley. W. H. Freeman and Company. Copyright © 1953.)

level can often be closely estimated from the position of reef flats, and the depth of flanking rocks and basin rocks can sometimes be estimated when they can be mapped in the field. The general environmental setting of some west Texas Permian reefs has been worked out in this fashion. Fossils were collected from rocks representing the various biotopes within the complex of environments that formerly occurred in the reef region (Newell *et al.*, 1953). The distributions of various fossil groups were then compared with the habitats as inferred from the reconstructed environmental setting. Figure 5-26 depicts the inferred life distribution trends of several fossil groups in some of the major environments on the Capitan Reef. Fossil occurrences that appear to have been due to postmortem transport of shells from their normal living sites have been excluded. It is clear that the bryozoa, for example, which chiefly are cemented or encrusting in habit, were most common on the forereef and near the front reef edge, whereas algae occurred chiefly atop the living reef proper and gastropods lived chiefly behind it. Such an analysis may be made easily on the species level. Of course, it is possible that taxa restricted to certain habitats on this reef might be found in other habitats elsewhere.

Strictly speaking, the ranges of tolerances and functions inferred from the empirical methods that involve the paleoenvironmental correlation of large numbers of individuals cannot be regarded as representative of any given individual. They do indicate the general region of biospace that individuals of that species (or other taxon) can inhabit. However, the functional range of any one individual may be much less than the range of numerous individuals, even though they are all conspecific, or it may be much greater than is indicated by the preserved fossil record. Still, these methods produce general results that are useful in paleoecological interpretations. In fact, it is unavoidable that empirical methods play important roles in functional interpretations, for many functions are simply not reflected in skeletal morphology and thus can only be inferred empirically, chiefly by the pattern-matching of numerous individuals with paleoenvironments. Clearly, the methods of interpretation that entail the greatest confidence should be employed, and, whenever possible, multiple lines of inference should be used.

Skeletal Materials Sometimes Furnish Evidence of the Ecology of Individuals

Organisms that secrete or otherwise assemble skeletal materials which are preserved in fossils bequeath to us an actual sample of the materials in their former living environments, a sample unequivocally contemporaneous with at least part of their lives. Some aspects of the composition and structures of these materials are sensitive to environmental parameters, so that it is possible to learn something of the living conditions of an organism by investigating the chemical and physical properties of its skeleton (Table 5-5). A number of most ingenious attempts have been made to investigate this relationship, but various problems have so far prevented the utilization of such skeletal properties in a routine manner for paleoecological studies.

The most common skeletal material is $CaCO_3$, which is found both as calcite and as aragonite in invertebrate skeletons. Today, aragonite is the chief skeletal mineral, and many groups, such as the scleractinian corals, have entirely aragonitic skeletons. Other taxa contain some aragonitic and some calcitic species, whereas still others contain species that are bimineralic.

Some bimineralic taxa display a correlation between their shell mineralogy and certain environmental parameters. Temperature correlates very strongly with mineralogy in some species, which secrete more calcite in cooler water and more aragonite in warmer water. Thus, individuals from high latitudes tend to contain a higher percentage of skeletal calcite than conspecific individuals from lower latitudes, and individuals at given localities, record the passage of warm and cool seasons in their shell mineralogy (Lowenstam, 1954; Dodd, 1964).

Figure 5-27 depicts the shell structure of *Mytilus californianus*, which has three major mineralized shell layers beneath an organic periostracum: an outer prismatic layer of calcite; a nacreous layer of aragonite; and an inner prismatic layer of calcite. Also, blocky aragonite, the hypostracum, forms beneath muscle

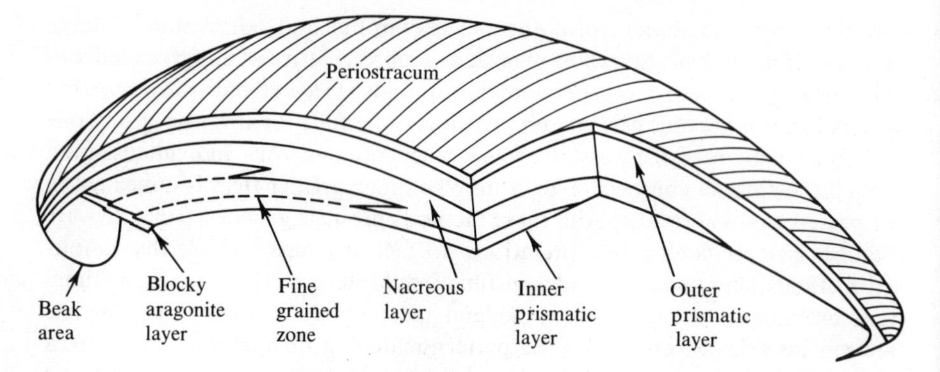

Fig. 5-27 Section of shell of *Mytilus californianus*. (After Dodd, 1964.)

scars. Note the intertonguing between the inner prismatic and nacreous layers. Dodd (1964) has shown that these tongues correlate with seasonal water temperatures. Summers are represented by extension of the aragonitic nacreous layer toward the beak, and winters by extension of the calcitic inner prismatic layer toward the beak. The results of such seasonal fluctuations in mineralogy are illustrated in Fig. 5-28, which depicts variation in the aragonite-calcite ratios

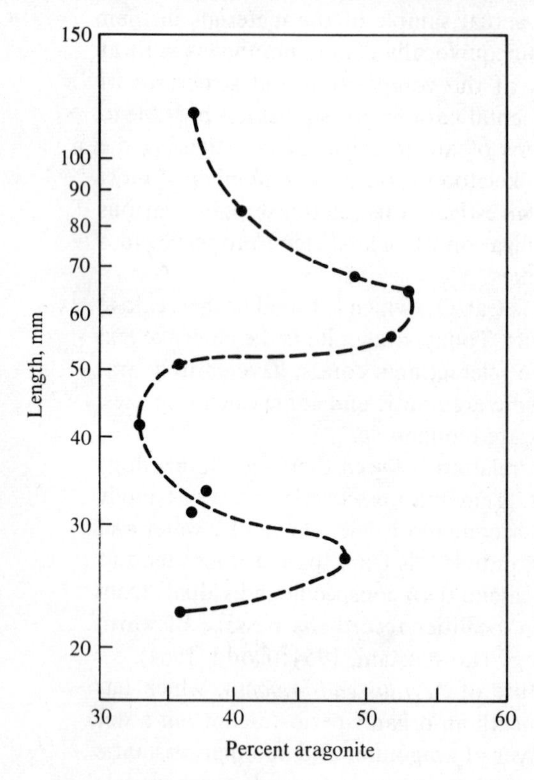

Fig. 5-28 Relationship between size (length, presumably related to age) and aragonite content in shells of *Mytilus californianus* that were collected alive in a single sample from La Jolla, California, in early winter. (After Lowenstam, 1954; from *The Journal of Geology*, the University of Chicago Press, copyright © 1954 by the University of Chicago.)

with shell size in a series of specimens of *M. californianus* collected alive on the same day in early winter (December 26). Very young shells of *M. californianus* do not have a temperature-correlated mineral ratio (Dodd, 1963), and probably the smallest shell plotted in the figure is in this category. The next largest shell, however, is probably considerably older and includes in its shell some growth from the previous summer: thus, it is more aragonitic. Shells between about 40 and 50 mm in length may be near a year old, and thus include temperature-sensitive winter growth from the preceding winter as well as summer growth. Accordingly, they contain relatively more calcite than younger shells with only summer temperature-sensitive growth. Alternating records of longer overall summer or longer overall winter growth in progressivley older shells produce the remaining oscillations in calcite and aragonite in the figure.

Different species have different mineralogical responses to temperature. Indeed, another species of *Mytilus*, *M. edulis*, exhibits only a slight change in its calcite-aragonite ratio with temperature. *Mytilus edulis* lacks the inner prismatic layer, so that the temperature-associated changes are quite unlike those in *M. californianus*. It is probable that even different populations of *M. edulis* may display different temperature effects (Dodd, 1963).

Environmental factors other than temperature may be correlated with aragonite-calcite ratios, and therefore with shell structures, at least in some species. In *M. edulis*, for example, low salinities may favor aragonite deposition (Lowenstam, 1954; Dodd, 1963), although this effect is evidently not found in all populations (Eisma, 1966). Shell thickness in *Mytilus* also correlates with aragonite-calcite ratios, which may, in turn, be related to such environmental parameters as effective food and oxygen supply, and thus to turbidity and microenvironmental effects.

In addition to the major elements that compose skeletal minerals, other elements appear in minor and trace amounts. Some of these are adventitious in the sense that they have been trapped within the growing mineral layers and represent mere "dirt," whereas some others form part of the mineral structure. Trace elements are also found (1) in separate trace mineral phases within the skeletons, (2) adsorbed on surfaces, or (3) in organic compounds (see Dodd, 1967, for a review).

Magnesium and strontium are the best known of the minor and trace elements in $CaCO_3$ skeletons. Commonly, they are in solid solution with calcium in aragonite or calcite structures, but sometimes they are present as separate mineral phases. They clearly vary in concentration with the mineralogy of the skeleton (Fig. 5-29). The calcite structure is more receptive to substitution of magnesium for calcium than is the aragonite structure, owing to the relatively small space occupied by Ca in the calcite structure (sixfold coordination), which more easily accommodates the small Mg ion than does the larger space in aragonite. As a consequence, high-magnesium skeletons (some with over 25 percent Mg) are always calcite, whereas skeletal aragonites usually contain less than 1 percent Mg. On the other hand, strontium (Sr) is favored in aragonite rather than in calcite

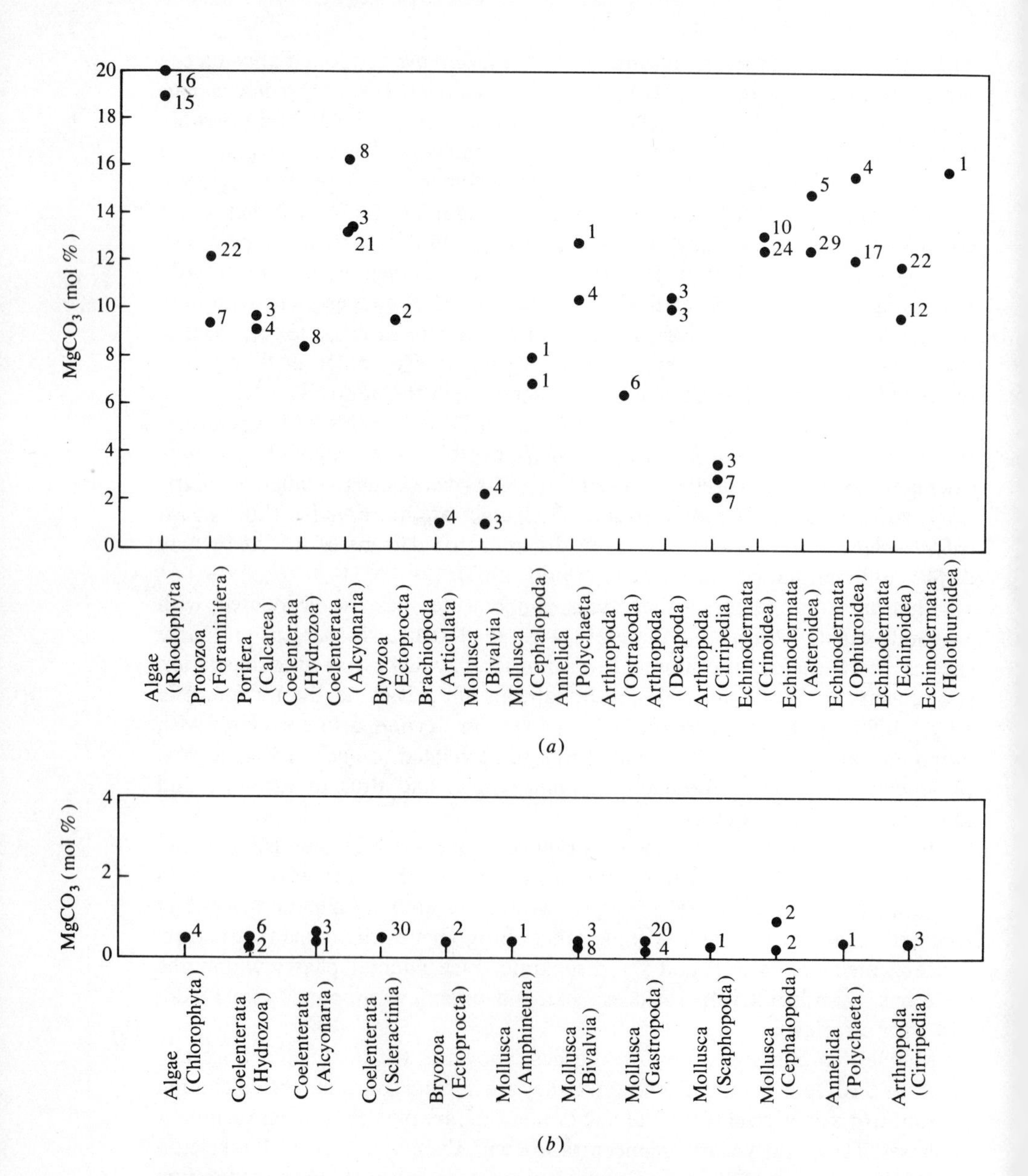

Fig. 5-29 (*a*) Magnesium in calcitic skeletons. Magnesium carbonate, $MgCO_3$ (mol percent); numbers by points indicate analyses used to establish the plot, which is the mean value. (*b*) Magnesium in aragonitic skeletons. Magnesium carbonate, $MgCO_3$ (mol percent); numbers and plots as previously. (*c*) Strontium in calcitic skeletons. Strontium/calcium atom ratio × 1000; numbers and plots as before. (*d*) Strontium in aragonitic skeletons. Strontium/calcium atom ratio × 1000; numbers and plots as before. (After Dodd, 1967.)

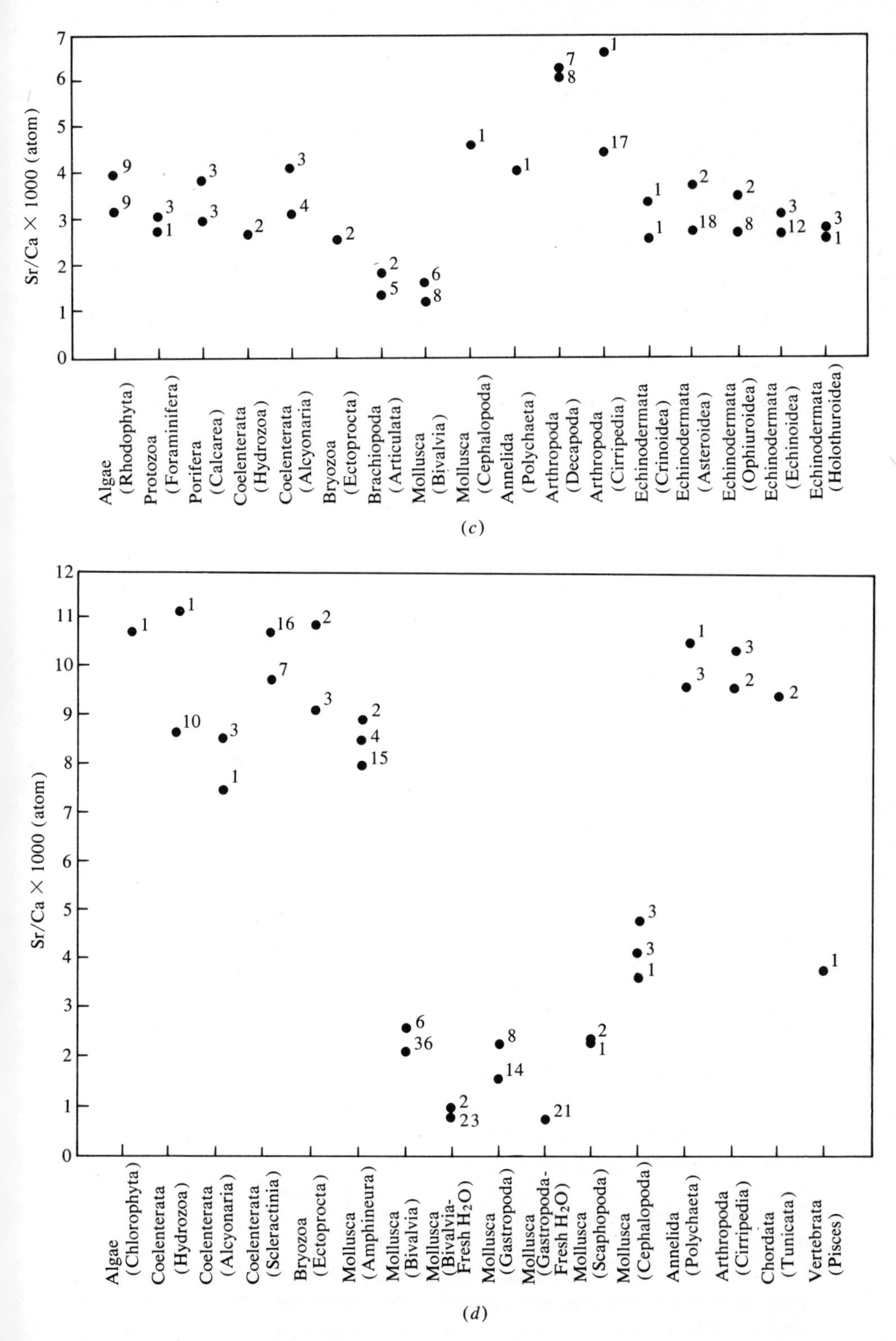

Sr/Ca × 1000 (atom)
Algae (Rhodophyta)
Protozoa (Foraminifera)
Porifera (Calcarea)
Coelenterata (Hydrozoa)
Coelenterata (Alcyonaria)
Bryozoa (Ectoprocta)
Brachiopoda (Articulata)
Mollusca (Bivalvia)
Mollusca (Cephalopoda)
Annelida (Polychaeta)
Arthropoda (Decapoda)
Arthropoda (Cirripedia)
Echinodermata (Crinoidea)
Echinodermata (Asteroidea)
Echinodermata (Ophiuroidea)
Echinodermata (Echinoidea)
Echinodermata (Holothuroidea)
Sr/Ca × 1000 (atom)
Algae (Chlorophyta)
Coelenterata (Hydrozoa)
Coelenterata (Alcyonaria)
Coelenterata (Scleractinia)
Bryozoa (Ectoprocta)
Mollusca (Amphineura)
Mollusca (Bivalvia)
Mollusca (Bivalvia-Fresh H_2O)
Mollusca (Gastropoda)
Mollusca (Gastropoda-Fresh H_2O)
Mollusca (Scaphopoda)
Mollusca (Cephalopoda)
Annelida (Polychaeta)
Arthropoda (Cirripedia)
Chordata (Tunicata)
Vertebrata (Pisces)

(*c*)

(*d*)

(Fig. 5-29), since the larger ionic radius of Sr is better accommodated to the larger (ninefold coordination) calcium site in aragonite.

Both Mg and Sr vary considerably in concentration among different higher taxa of invertebrates (Fig. 5-29), without respect to shell mineralogy. Presumably, this is due to differences in the physiology of skeletal deposition among different higher taxa. In general, the more primitive taxa deposit more Mg in both calcite (Chave, 1954) and aragonite (Lowenstam, 1963a). Even within higher taxa having a characteristic trace-element content, consistent variations in Mg and Sr concentration are found between different species. Thus, different species of *Mytilus* deposit different amounts of Mg in skeletal calcite, even under the same environmental conditions. Probably infraspecific differences in trace-element concentrations exist in some forms; there is evidence of such differences among semi-isolated populations of *Crassostrea virginica* (Lerman, 1965).

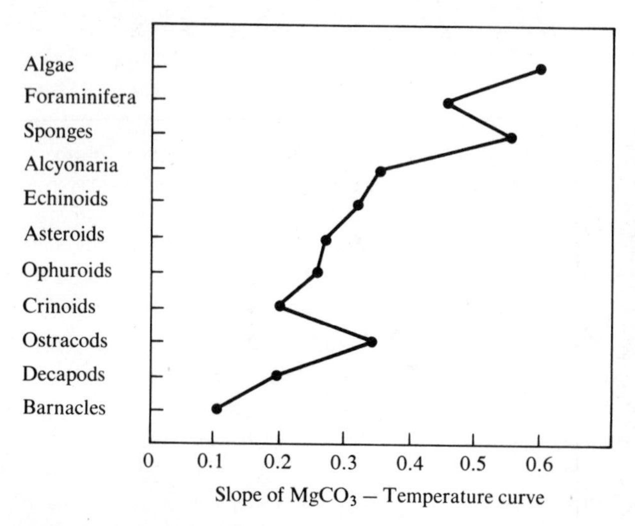

Fig. 5-30 Relation between sensitivity of magnesium concentration to temperature in the skeletons of various organisms plotted in order of their approximate level of organic complexity. (After Chave, 1954; from *The Journal of Geology,* the University of Chicago Press, copyright © 1954 by the University of Chicago.)

Although Sr concentrations are characteristic for different higher taxa, they do not correlate very well with taxonomic level (Fig. 5-29). Instead, a notable difference exits between Sr in aragonite skeletons of most mollusks and fishes (relatively low Sr) and those of chitons and other invertebrates and algae (relatively high Sr). Among some, but not all, closely related species there are differences in Sr content referable to interspecific physiological differences (Lerman, 1965).

In addition to these mineralogical and taxonomic effects, trace-element concentrations sometimes correlate well with environmental parameters. Usually, magnesium concentration is directly proportional to temperature, other factors being equal (Chave, 1954). Temperature effects tend to be most marked in primitive groups, with higher invertebrates displaying only small Mg variations with temperature (Fig. 5-30). Differences in temperature-Mg responses are documented between congeneric species and probably occur between different populations of the same species at times. Among mollusks, Mg may vary inversely with salinity, as in outer calcite layers of some species of *Mytilus* (Dodd, 1965). However, among some echinoids (*Dendraster*; Pilkey and Hower, 1960) and articulate brachiopods (Lowenstam, 1961), Mg is reported to vary directly with salinity.

Strontium concentration also correlates with temperature and salinity, although a wide variation in response is found. In calcite skeletons of *Crassostrea virginica*, Sr increases directly with temperature (Lerman, 1965), whereas in the echinoid *Dendraster*, also calcite, it increases inversely with temperature (Pilkey and Hower, 1960). Even more interesting is the situation in shells of both *Mytilus californianus* and *M. edulis*, for in these species Sr increases with rising temperature in the outer prismatic (calcite) layers and decreases with rising temperature in the nacreous (aragonite) layer (Dodd, 1965). The shell of the bivalve *Cardium edule* has two layers, both aragonite. Hallam and Price (1968b) found that Sr concentration is significantly different in each layer, although it decreases in both layers with rising temperature.

Other trace elements detected in invertebrate carbonate skeletons include manganese, iron, barium, sodium, copper, aluminum, and boron, but they are not yet well studied. A positive correlation between sodium and salinity has been recorded for *Crassostrea virginica* (Rucker and Valentine, 1961). The sodium concentration may reflect the entrapment of seawater in the shells and its subsequent evaporation (Dodd, 1967).

The organic matrix that pervades invertebrate carbonate skeletons is proteinaceous. The proteins seem to break down rapidly, geologically speaking, but the amino acid building blocks of which they are constructed may persist. The thermal stability of amino acids is low, so they are degraded fairly rapidly at temperatures that are common within some fossiliferous crustal rocks. Nevertheless, at temperatures of about 20°C, some amino acids may persist for 10^{10} years (Abelson, 1957). Altered organic matrices have been identified in fossil shells that are about 450 million years old (Grégoire, 1967), and amino acids are known from shells about 360 million years old (Abelson, 1957). They are preserved by virtue of their inclusion within protective carbonate skeletons.

It is well known that among individuals of the same species, analogous tissues of a given organ contain similar suites of amino acids, whereas between individuals of separate species there are often profound differences in amino acid suites, which are to some extent correlated with their degree of taxonomic relationship. The general correlation between organic composition and phylogenetic relationship has been extended to include amino acids in invertebrate skeletons (Roche, Ranson, and Evsseric-Lafon, 1951; Hare and Abelson, 1965; Ghiselin *et al.*, 1966). Shells of forms that are morphologically similar and are therefore believed to be closely allied, tend to contain similar suites of amino acids, whereas less closely related forms have distinctively different amino acid contents, quantitatively if not qualitatively.

Differences in organic shell materials among the Mollusca seem partly related to shell structure. Shells that have both nacreous and prismatic portions contain the most organic material, whereas shells with the complex structure called crossed lamellar contain the least. It happens that the shells with the nacreous and prismatic structures are primitive; therefore, high organic content may be partly a primitive feature. The primitive Amphineura (chitons) have a relatively high organic content although they possess crossed lamellar shells.

Qualitative differences and relative concentrations of amino acids found in shells of different structure and taxonomic level also suggest that matrix composition is closely related to shell structure, level of evolution, and taxonomic affinity. For example, the amino acids of primitive mollusks resemble those of the primitive brachiopod *Lingula* (Hare and Abelson, 1965). However, even closely related species are somewhat different in the amino acid composition of their shell proteins on a relative basis (Degens, Spencer, and Parker, 1967). Variation is also recorded between different shell units in the same species. The organic matrix associated with different structural units of the mussel *Mytilus californianus* seems to differ in composition from unit to unit; outer prismatic and nacreous layers, for example, yield slightly different amino acid suites (Hare, 1963). Presumably all these differences reflect differences in protein content of the original matrices.

Organic matrix compositions of *Mytilus californianus* have been shown to change with environmental temperature and salinity (Hare manuscript, 1962; see Hare, 1963). The matrix of the aragonite layer is affected more markedly than that of the calcite layer. Temperature and salinity effects are also reported in a number of bivalves and gastropods by Degens, Spencer, and Parker (1967). They note that the shell protein of each species reacts in a different way to environmental changes.

The organic matrix of mollusk shells thus displays, so far as is known, a pattern of variation similar to that of inorganic shell mineralogies and compositions; in addition, it seems to show primitive and advanced states within some large, long-ranging lineages and between major lineages.

It seems likely that most aspects of the secretion of exoskeletons and endoskeletons alike are under rather strict physiological control. That the physiology of organisms is sensitive to environmental fluctuation is obvious. It is likely, then, that the environmental correlations noted in skeletal mineralogy, structure, and trace- and minor-element concentrations are not simple physical and chemical responses to changes in such environmental factors as temperature or salinity, but are chiefly mediated by the physiological processes of the organisms. It is these processes that are responding to environmental changes. One of the more common physiological changes induced by environmental fluctuation is metabolic rate. There is evidence that some ecologically sensitive changes in trace-element compositions in some taxa are related to changes in the rate of shell secretion; shell-secretion rates are, in turn, related to temperature, and probably often to other environmental factors (Moberly, 1968).

An exception of sorts to the physiological control of trace-element concentration is the relation between the concentration of cations relative to calcium in water and in carbonate shells. This relation may be expressed as a *distribution coefficient*:

$$\text{Distribution coefficient} = \frac{(M\ \text{Ca})\ \text{skeleton}}{(M\ \text{Ca})\ \text{water}}$$

where M is the molar concentration of the element under consideration, and Ca is the molar concentration of calcium (Dodd, 1967). For Sr, the only element carefully studied in this respect, the Sr/Ca ratio in shells of fresh-water gastropods and bivalves varies directly with that in the water (Odum, 1957), so that the distribution coefficient tends to remain constant. If this is true, it generally implies that discrimination against trace elements is accomplished chiefly by regulating the distribution coefficients. For marine invertebrates, this would effectively set the general level of concentration of a trace element, since variations in M/Ca ratios in the oceans tend to be slight. It is interesting that mollusks have among the lowest distribution coefficients of invertebrates, even though they deposit their shells from extrapallial fluids rather than from body fluids.

The adaptive significance of the various mineralogical and structural states and trends of carbonate skeletons has not been demonstrated. Perhaps the bulk of evidence at present suggests that most are not adaptive in themselves as changes and differences in the skeletons, but rather reflect the evolution of the biochemical and biophysical systems of the organism. The minor- and trace-element compositional patterns in taxa with carbonate skeletons may also be chiefly side effects of evolution of the biochemistry of the organisms and not themselves adaptive characteristics. The distribution coefficient of each taxon would be fixed according to the constraints of its own biochemistry and of the chemistry of the mineral phases. A decreasing Sr content through time has been reported within certain lineages. Together with the general decrease in trace-element content observed with increasing phylogenetic level this may result from an increased biochemical effiicency attained through the anagenetic evolution of excretory and circulatory systems, lowering the distribution coefficients as metabolic efficiency increases (Odum, 1957; Lowenstam, 1963b; Dodd, 1967; Hallam and Price, 1968a).

Although most compositional change in skeletons seems to be one or two steps removed from environmental cause, their correlations are sufficiently high to give hope that the changes may be employed empirically as tools in reconstructing some parameters of biospace inhabited by fossil organisms. We shall not be on perfectly safe ground in our interpretations, however, until we understand the principles governing the effects. Obviously, many basic questions remain unanswered, and thus there are many opportunities for research.

The Isotopic Compositions of Skeletons Are Sensitive to Many Environmental Parameters

Many of the elements that form invertebrate skeletons are represented by more than one isotope. All isotopes of an element have the same number of protons in their nuclei and therefore the same nuclear charges; they also have identical chemical properties and are interchangeable in chemical reactions. However, atoms of different isotopes of an element do differ in having different numbers of neutrons in their nuclei; consequently, they have different atomic weights. This mass difference among isotopes may lead to fractionation, whereby one isotpoe is preferentially concentrated during some processes. Take, for example, the evaporation of water from the sea surface. Molecules of water vapor may contain the common oxygen isotope 0^{16} (8 protons and 8 neutrons in the nucleus) or the rare, heavy, stable isotope 0^{18} (8 protons and 10 neutrons in the nucleus). Isotope 0^{17} also exists, but it is so rare that it may be ignored for present purposes. These isotopes of oxygen are stable—that is, they are not radioisotopes subject to spontaneous decay. Each molecule of oxygen, heavy or light, has identical chemical properties, but the lighter molecules have a higher vapor pressure than the heavier ones. Therefore, evaporation of a given number of water molecules results in proportionately more light molecules (those containing 0^{16}) being evaporated than heavy molecules (those containing 0^{18}), relative to their concentration in the water. The water becomes enriched in heavier molecules, and isotopic fractionation of 0^{18} and 0^{16} has occurred between water and water vapor. At equilibrium, this relation may expressed as

$$0.992 \frac{[\mathrm{H_2o^{18}\ vapor}]}{[\mathrm{H_2o^{16}\ vapor}]} = \frac{[\mathrm{H_2o^{18}}]}{[\mathrm{H_2o^{16}}]}$$

There is also a differential between O^{18} and O^{16} in dissolved carbonate ions and water molecules. This differential is temperature dependent; the relation may be expressed as

$$\frac{[\mathrm{Co_3^{=}}^{18}]^{1/3}}{[\mathrm{Co_3^{=}}^{16}]^{1/3}} = K(t) \frac{[\mathrm{H_2o^{18}}]}{[\mathrm{H_2o^{16}}]}$$

where K is an exchange constant and t is the temperature.

The properties of the oxygen isotopes O^{16} and O^{18} in water and in carbonate at different temperatures and isotopic cocentrations were studied by a team of scientists at the University of Chicago in the late 1940's and early 1950's. This team, led by the chemist H. C. Urey, investigated the isotopic fractionation, built a sensitive instrument (a specialized mass spectrometre) to detect minute differences in oxygen isotope ratios, and studied isotopic concentrations and distributions in seawater and other waters and in skeletal carbonates (see Bowen, 1966, for a review).

In analytical work on oxygen isotopes, it is usual to express the results as the difference between the ratio O^{18}/O^{16} in the sample analyzed and a standard O^{18}/O^{16} ratio. This difference is given in per mil—parts per thousand—and is symbolized by δO^{18}. Most investigators use as a standard the one employed at the University of Chicago, the "PDBI" standard, which has an isotopic ratio near the present mean ratio of seawater. In terms of this standard, the following relation between temperature and oxygen isotope ratios in $CO_3^=$ and H_2O was developed empircally (Epstein *et al.*, 1953):

$$t°C = 16.5 - 4.3(\delta O^{18} \text{ of } CO_3^= - \delta O^{18} \text{ of } H_2O) + 0.14(\delta O^{18} \text{ of } CO_3^= - \delta O^{18} \text{ of } H_2O)^2$$

It follows that O^{18} concentrations in $CO_3^=$ increase relative to O^{16} concentrations with rising O^{18} concentrations in water, or with falling temperatures, or both. Some organisms, including many living foraminifera, mollusks, and brachiopods, have been shown to secrete their carbonate skeletons in isotopic equilibrium with seawater; there is no fractionation owing to physiological processes. Therefore we can measure the O^{18}/O^{16} ratio of these skeletons and estimate either the temperature at which the skeletons were deposited or the oxygen isotopic composition of the water in which they were secreted. However, we must independently determine one of these factors, or assume a value for it, to estimate the other.

Other groups of organisms, including algae, coelenterates, and echinoderms, do fractionate oxygen isotopes during metabolism (Lowenstam and Epstein, 1954; Keith and Weber, 1965; Weber and Raup, 1966). The amount of fractionation is not always consistent within a group, but may vary between lower taxa and even between different parts of the same skeleton. Metabolic fractionation has been shown to be the most probable cause of unusual oxygen isotope ratios in the shells of Cretaceous bivalves of the extinct genus *Inoceramus* (Tourtelot and Rye, 1969). In these forms, there are even isotopic differences between aragonite and calcite shell layers, which are suggested to be due to metabolic processes. Thus, it is dangerous to assume that fossil groups do not engage in metabolic isotope fractionation, even though their living relatives do not.

Variations in the concentration of O^{18} in waters of a variety of environments have been studied (Epstein and Mayeda, 1953; Lloyd, 1964). Although the isotope is stable, its concentration in water varies considerably from place to place, chiefly due to fractionation during evaporation and precipitation. Usually, highly saline marine or marginal waters have undergone much evaporation and are thus enriched in O^{18}. The evaporated water vapor, however, is relatively impoverished in O^{18}, and so are the fresh waters that originate as precipitates of the impoverished water vapor. Ice and snow in high latitudes are notably poor in O^{18}, and marine waters diluted by runoff from melting glaciers and snowfields therefore have reduced O^{18} concentrations. Thus, the source and history of a marine water mass, as well as its contemporary processes, help determine its range of O^{18}/O^{16} ratios.

Usually salinity and O^{18} concentration will vary directly. However, Lloyd (1964) has described a case in the Florida Everglades where prolonged evaporation of fresh water results in its marked enrichment in O^{18}. When this fresh but isotopically heavy water runs off into the sea and intermixes with marine water, a tongue of brackish, O^{18}-rich water is created. Such a runoff occurs in Florida Bay, where waters are normally richer in O^{18} than in the adjoining open ocean, even when there is no runoff, due to evaporation in the shallower bay. The shells of mollusks living in the bay therefore have higher O^{18}/O^{16} ratios than those from outside the bay; additionally, the shells of mollusks living in parts of the bay occupied by the brackish water, rich in O^{18}, tend to have even higher O^{18}/O^{16} ratios than those of other bay mollusks. These variations in shell isotopes are thus not primarily caused by a temperature effect in isotopic equilibrium, but are due to variations in isotopic concentration in the water.

A somewhat similar result was obtained by Keith and Parker (1965), who studied isotopes in mollusks from such marginal marine envrionments as Laguna Madre along the coast of the Gulf of Mexico. There, the temperature effects on O^{18}/O^{16} ratios were completely masked by variations in the isotopic water composition that were due to factors such as fresh-water dilution from rivers and high rates of evaporation in restricted bodies of water.

In view of these many difficulties, it is evident that paleotemperature investigations must be carefully designed so as to eliminate the many possible sources of error. First, it must be determined that the original isotopic composition of the fossil is likely to be preserved. Criteria include (1) lack of any evident alteration of the skeleton, such as by recrystallization; and (2) presence of cycles of temperature variation, such as seasonal cycles, indicating that the natural progression of climatic or other events is preserved. Second, the possibility of metabolic fractionation should be evaluated. Fractionation may be insignificant if (1) there is agreement of paleotemperature estimates in all parts of a skeleton, and especially if (2) there is agreement of paleotemperature estimates among a number of distinct taxa from the same locality, allowing in both cases for natural variations. Third, the original isotopic composition of the water must be closely estimated. Criteria include (1) evaluation of the biogeographic setting to suggest, for example, whether conditions were restricted or open to the ocean; (2) evaluation of the biotic composition to indicate normal or abnormal salinities; and (3) cross checking of paleosalinities by other aspects of shell structure and composition. Finally, the isotopic paleotemperature estimates that appear to be the most reliable may be compared with each other, with regional or world-wide data, and with knowledge of general climatic patterns, to check for incompatibilities. Temperatures much below 0° or much above 30° are immediately suspect, as are anomalous estimates that depart widely from regional trends and have no obvious special explanations. Clearly, caution is required in interpreting oxygen isotope ratios, for none of the criteria is definitive.

In addition to oxygen isotopes, stable carbon isotopes have been studied in recent and fossil shells. These are the common C^{12} and the rarer C^{13}. It is not necessary to employ an extremely sensitive mass spectrometer (such as is required for oxygen isotope work) to study carbon isotopes. Carbon isotope ratios are generally reported as the difference between the ratio in a sample and the ratio in a standard, often the PDBI standard; the difference between C^{13}/C^{12} in the sample and that in the standard is symbolized by δC^{13}, expressed in per mil.

In nature, variations in δC^{13} arise through fractionation of the heavy and light carbon isotopes during physical or chemical processes, such as occur when a carbonate ion is formed. The reaction

$$C^{13}O_2 + C^{12}O_3^{=} \rightleftarrows C^{12}O_2 + C^{13}O_3^{=}$$

goes to the right normally, thus favoring C^{13} in the carbonate ion.

The δC^{13} for recent marine carbonates is high, nearly reaching +6 and ranging to a little above −4. Although the equilibrium constant for C^{13} in the system CO_2—$CO_3^{=}$ is not precisely known at present, there is evidence that a δC^{13} value of between about +2 and +5 results from precipitation of carbonate in seawater that is in equilibrium with atmospheric CO_2. The variations in δC^{13} in marine carbonates probably arise when sources other than the atmosphere have contributed significantly to the local CO_2 supply. From Fig. 5-31, it is evident that CO_2 derived from the decay of land or sea plants or from organic animal detritus will be poor in C^{13}, relative to the PDBI standard, and water surrounding such material will have a relatively low $\delta C,^{13}$ especially if the supply of organic material is large. If dissolved Co_2 is considerably depleted, however, additional supplies may be released from bicarbonate ions in the seawater, which are relatively enriched in C^{13} (Degens, Guillard, *et al.*, 1968). Since the solubility of CO_2 is inversely proportional to temperature, there is less CO_2 dissolved in warm water, and δC^{13} in organisms such as plankters, which live where CO_2 may be depleted, tends to be higher in warm waters than in cold. Indeed, such variation in δC^{13} between antarctic and tropical plankton has been reported (Sackett *et al.*, 1965; Degens, Guillard, *et al.*, 1968).

Biological complications in the C^{13}/C^{12} regime are numerous. As different biochemical pathways may fractionate the isotopes differently, there is quite a range of C isotope ratios even among different biochemical products of the same organism. Degens, Behrendt *et al.*, (1968) have studied C^{13}/C^{12} in plankton samples. They found that lipids and cellulose and lignin compounds tend to be relatively enriched in C^{12}, with correspondingly low C^{13} values, whereas amino acids and hemicellulose compounds are relatively poor in C^{12} (Fig. 5-31). As they point out, the C^{12}-rich compounds happen to be the more stable ones, so that

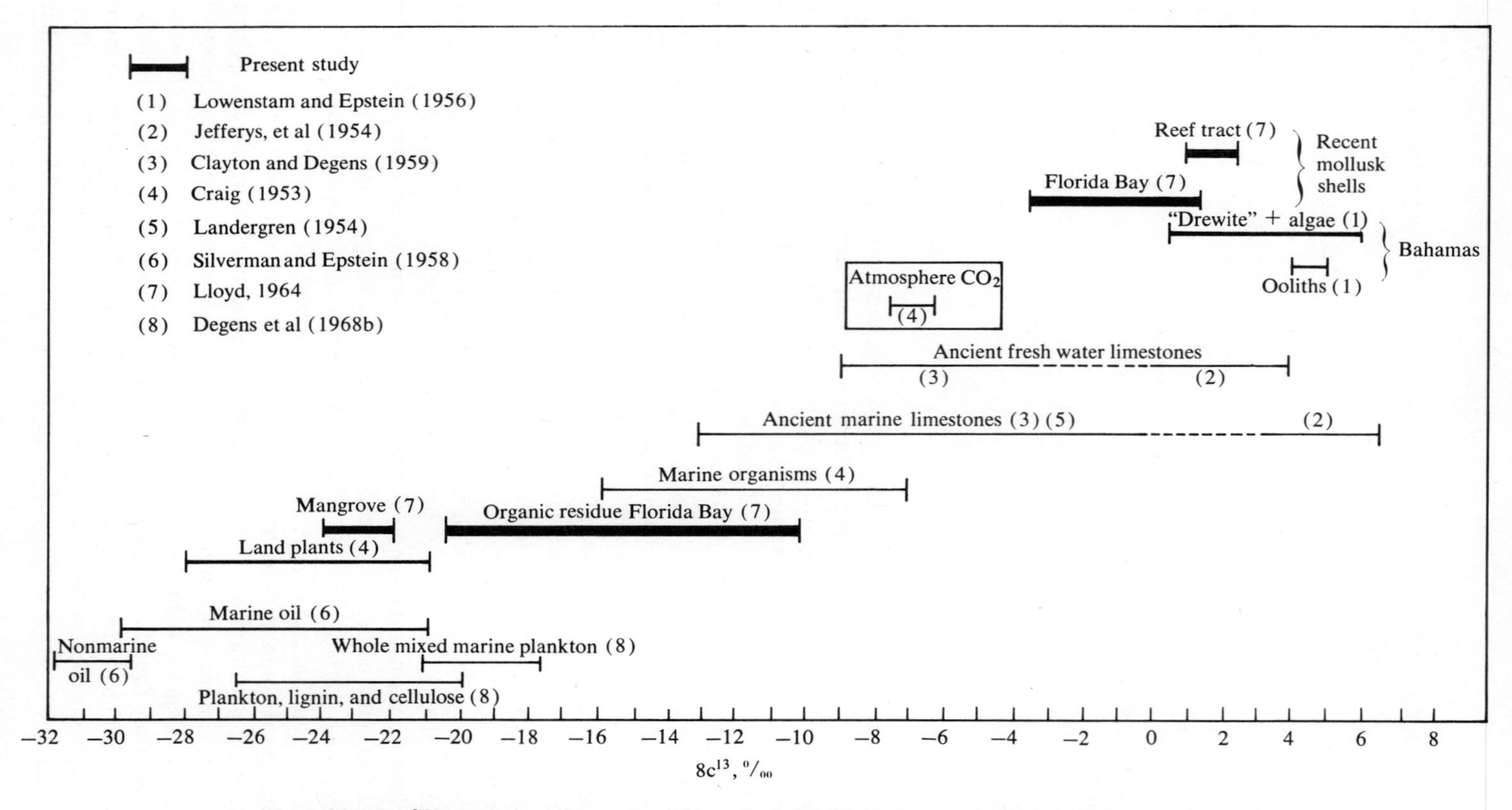

Fig. 5-31 The δC^{13} range in various natural substances. (Modified after Lloyd, 1964; from *The Journal of Geology,* the University of Chicago Press, copyright © 1964 by The University of Chicago.)

fossil organic detritus derived from these organisms would tend to be biased in favor of C^{12}, with δC^{13} perhaps as low as -23 or less (Fig. 5-31).

Lloyd (1964) was able to demonstrate that mollusk shells secreted in the waters of Florida Bay, where there is much land plant detritus, are lower in C^{13} than shells secreted in open ocean waters nearby, which are relatively free from land plant debris. Between these extremes, a gradient of seaward increase in δC^{13} was found that reflected either a seaward decrease in organic detritus, a seaward increase in water (and CO^2) exchange with the open ocean, or both. A study of C^{13}/C^{12} in mollusks in marginal marine environments along the Gulf coast by Keith and Parker (1965) records a similar seaward increase in δC^{13}. Many ancient marine limestones seem to have undergone a reduction of δC^{13}, probably as the result of C^{13} loss through exchange with C^{12} in C^{13}-poor ground water or soil, and perhaps by addition of carbonate cement low in C^{13} (Fig. 5-31).

In general, it appears that δC^{13} in shells will fluctuate with changing carbon sources. The low-C^{13} sources may be chiefly land plants or marine plankton (stable constituents). Evidently the effects of contributions from ancient limestones and other sources are secondary (Keith, Anderson, and Eichler, 1964). Figure 5-32 summarizes the relationships between environments and oxygen and carbon isotopes in mollusk shells.

We are clearly very far from being able to employ any of these physical and compositional properties of skeletons in a routine manner to interpret their

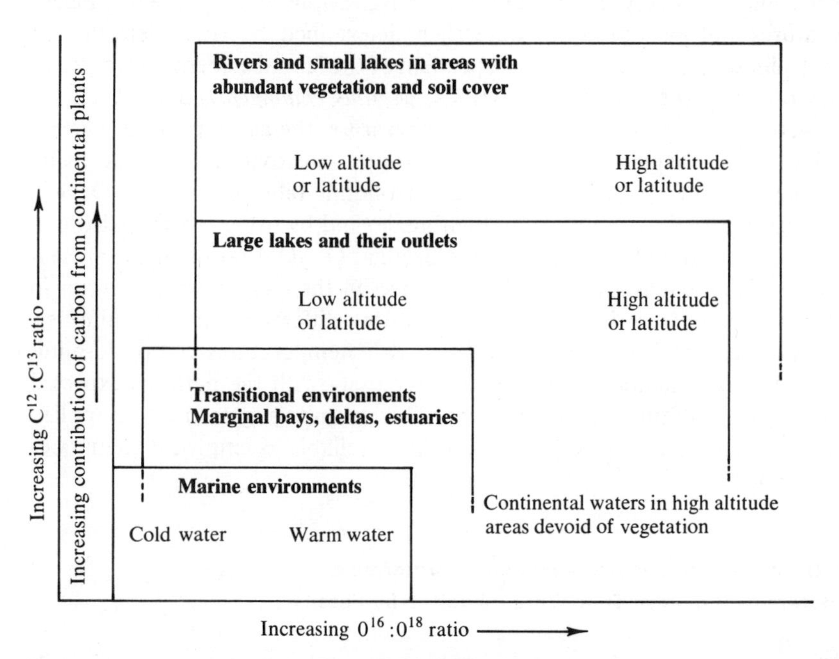

Fig. 5-32 Generalized distribution of carbon and oxygen isotopes in skeletons deposited in different environments. (After Keith, Anderson, and Eichler, 1964.)

former living conditions. Many of the compositional/environmental relations that are easy to measure are restricted to a single species and cannot be given general application. For extinct taxa, a solution to this problem would be to calibrate such relations by means of a taxonomically insensitive property, if one could be found. Isotope ratios may work for some taxa. However, exchanges of ions and of isotopes between fossils and their postdepositional environment, usually through the action of ground water, are common, and may occur without leaving evidence that alteration has occurred (Turekian and Armstrong, 1961; Turekian, 1963). Compositional features reflected in the texture or the internal structure of skeletons, such as seasonal mineralogical changes, are the most likely to be preserved (Dodd, 1964), but they still require calibration before they will yield numerical estimates of environmental parameters.

Stanton and Dodd (1970) have compared environmental interpretations based on shell biogeochemistry with those based on faunal methods. They studied a thick sequence of fossiliferous Pliocene and early Pleistocene sediments from the Kettleman Hills, San Joaquin Valley, California, where large faunal associations had been previously described and interpreted, chiefly by Woodring, Stewart, and Richards (1940). Considering the results of previous studies, as well as their own work, Stanton and Dodd concluded that the fauna, when interpreted by empirical methods, indicated little climatic change. However, a variety of salinity regimes were represented, from normal marine to brackish and fresh water, oscillating in response to repeated transgressions and regressions. Paleotemperatures and paleosalinities were then determined by trace-element and isotope techniques that are partly independent of the faunal evidence. Strontium/calcium ratios in shells of fossil mytilids, *Mytilus coalingensis* and *M.* cf. *M. edulis*, were used as a criterion of paleotemperature, the assumption being that these fossil shells displayed the same temperature sensitivity to Sr as do living Californian mytilids—a case of empirical taxonomic biogeochemistry! Oxygen isotope analyses of fossil shells were then made, and by using the Sr paleotemperatures it was possible to calculate the original O^{18}/O^{16} ratios of the ancient waters in which the fossils had lived. Differences in the oxygen isotope ratios in the ancient waters were interpreted as indicating differences in paleosalinities, since this correlation is common (Fig. 5-33). Paleotemperatures and paleosalinities inferred from chemical evidence agreed closely with the patterns expected from the empirical faunal interpretations. These results suggest that, if applied with care, biogeochemical techniques can be as reliable as empirical techniques in paleoecology.

Many Other Techniques Are Available to Determine Parameters of Biospace That Were Inhabited by Fossils

Numerous other methods of determining the paleoenvironments of fossils have been employed, but most are actually studies of paleoenvironments by nonbiological methods. In a sense, all such evidence that leads to a paleoenvi-

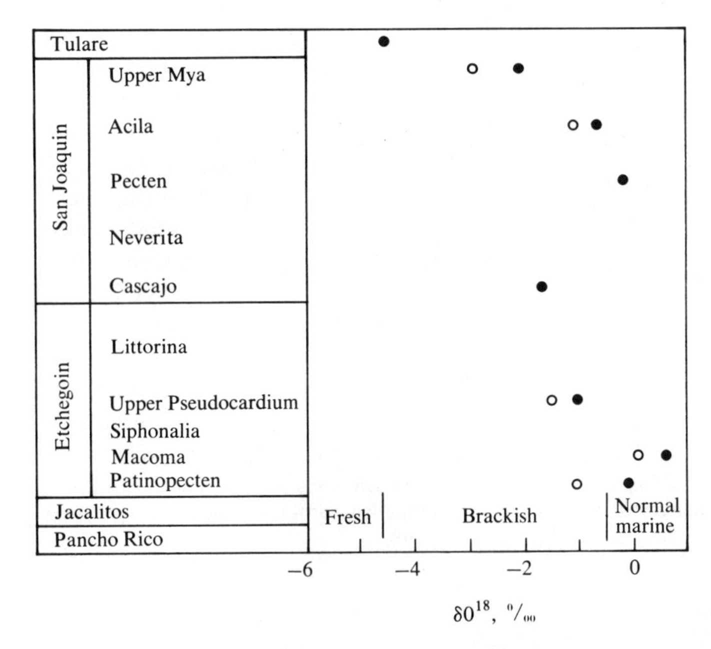

Fig. 5-33 δO^{18} values in the succession of faunal zones in Pliocene and Lower Pleistocene formations in the Kettleman Hills, California. Probable salinity ranges indicated. Solid circles are shell samples; open circles are calculated values for the waters in which the shells were secreted. (After Stanton and Dodd, 1970.)

ronmental interpretation for a stratigraphic unit bearing indigenous fossil remains is contributing to an understanding of the biospace of the fossils. Such evidence is beyond the scope of this book. In practice, of course, the field of evolutionary paleoecology benefits greatly from the results of these studies.

A few paleobiological methods of interpretation have not been specifically covered here. Many of them rely upon detailed observation of fossil orientations in the field (for example, Johnson, 1960); these are, in fact, simply very good empirical methods. Others result from interpretation of trace fossils, the trails, burrows, borings, and other marks left by organisms. They furnish direct evidence of the activities of organisms, and can often be related to feeding and to other significant ecological functions. They also permit the extension of ecological interpretations into stratigraphic units that lack skeletonized fossils but contain traces. However, the methodology of their interpretation is similar to that of skeletons. Seilacher (1964) has written a clear introduction to these fossils, and the scope of recent work is suggested by Crimes (1970).

CHAPTER SIX

ECOLOGICAL FUNCTIONS OF POPULATIONS AND THEIR EVOLUTION

Population, when unchecked, increases in a geometrical ratio. Subsistence increases only in an arithmetical ratio. A slight acquaintance with numbers will shew the immensity of the first power in comparison of the second.
—T. R. Malthus, *Population,* London, 1798.

Of all the levels of the ecological hierarchy, that of the population is probably the most difficult to deal with in the fossil record. A population may be defined simply as all the individuals of a given species living in a certain region. The extent of a population (and therefore of the region) is usually considered to depend upon its reproductive pattern; interbreeding within a population is ordinarily greater than crossbreeding between that population and other populations of the species. The limits are therefore located at partial discontinuities in gene flow. Many of the ecological factors of the population level are associated with the reproductive regime, which operates to maintain a viable population size, geographic range, density, and dispersion pattern in order to balance the various sources of mortality tending to alter or destroy these population properties. Other population factors are associated with the pattern of variation of ecological factors in space, from one part of the population's geographic range to another.

The evidence required to reconstruct many of the population parameters that depend upon reproductive regimes is of a quality not common in the fossil record. It is preferable that the evidence dealing with demography is based upon the state of a population at a given moment or over a relatively short range of time, such as a generation, whereas evidence dealing with spatial patterns of a population should preferably be representative of the state of a population throughout its living area. These kinds of evidence require samples of a fossil generation that are relatively unbiased with respect to the age of individuals and with respect to the original living area—evidence that is difficult or impossible to obtain for most fossil populations. Nevertheless, it has proven possible to investigate fossil population parameters in a number of instances, and there is no reason the believe that these pioneering efforts cannot be extended to include a somewhat representative sample of the fossil record.

The population level is especially important and merits much attention because of two interrelated features that are owing to its function as a reproductive unit. One is simply that, through reproduction and heredity, a population becomes potentially immortal due to its ability to produce new individuals that resemble their parents, whereas, on the other hand, the population is the level at which evolution occurs and at which novelty arises. In Chapter 2, we considered the evolution of populations from the standpoint of genetic change. Now we shall examine some of the ecological features of populations for their possible adaptive significance, and for the accuracy with which they may be inferred from the fossil record.

Reproduction and Mortality: Internal Population Dynamics Reflect the Age-specific Effects of Death on Birth Systems

Obviously, births must balance deaths to maintain the continuity of a population through time. The *natality rate* at which new young appear (N_y) is expressed as the number of young produced in a given time (t),

$$\frac{\Delta N_y}{t}$$

or as the number produced in a given time by a given number of individuals, such as births per year per thousand individuals,

$$\frac{\Delta N_y}{(1 \text{ yr})(1000)}$$

or births per year per average individual,

$$\frac{\Delta N_y}{(1 \text{ yr})N}$$

where $N =$ the total number of individuals in the population during t time. Natality rates expressed in the last two ways may be used to compare fairly the rates in populations of different sizes.

In considering population properties, it is useful to employ idealized models. Let us imagine a population that lives in a perfect environment for reproduction, with ample food and optimal conditions for every environmental factor. The resulting natality rate would be limited only by genetically determined physiological attributes of the population, which vary somewhat among individuals but which have a modal or average level. This *maximum natality* (N_m), although seldom realized in nature, provides a referent with which to compare the natality that does in fact occur. The difference between maximum and realized natality is a measure of the restriction imposed upon natality by the environment. So

$$\frac{\Delta N_y}{\Delta t} = \frac{\Delta N_m - a}{\Delta t}$$

where N_y is realized natality, N_m is maximum natality, and a represents the environmental restriction. Since a great many factors influence natality, such as various causes of mortality among females of reproductive age, or various causes of the inhibition of ovulation (food shortages, extreme temperatures, and so on), the restriction a may be partitioned into the individual causes, so that $a = a_1 + a_2 + a_3 + \cdots a_n$. Of course, the effects of each environmental factor will vary greatly according to the reproductive biology of the population, and often with changes in other environmental factors.

Rates of mortality (N_d) in populations can be expressed by notations similar to those for natality, as

$$\frac{N_d}{t} \quad \text{or} \quad \frac{N_d}{tN}$$

If a population lives in the most optimal of environments and is free from accidents, one can imagine the mortality rate being reduced to an absolute minimum, when nearly every individual dies of "old age." Minimum mortality therefore provides another useful conceptual referent, and actual mortality may be contrasted with it to provide some measure of the effect of environment on mortality. The age of individuals at death in most populations with minimum mortality would lie chiefly within a relatively narrow range, grouped about the average potential lifespan of the population, except for individuals that possess defective genotypes.

Few populations ever reach minimum mortality. Instead, environmental factors prove lethal to most individuals long before they reach their potential age span; mortality is distributed among different age groups within a population, and is usually concentrated within the very young groups. If we know the age distribution of mortality within a given population, we know a great deal about the dynamics of that population, and the next step is to determine the relative effects and age distributions of various causes of death.

A useful method commonly employed to organize and present mortality data is the *life table*, which has been applied to natural and laboratory populations by Pearl and collaborators (1921, 1935), Deevey (1947), and others. Table 6-1 is a life table prepared by Sellmer (1967) for *Gemma gemma*, a marine bivalve that does not have a planktonic larval stage; the young are retained within the shell of the mother until they begin an independent benthic life (that is, they are *ovoviviparous*). The table shows, for each age class listed: the number of individuals dying (per thousand of the initial population); the number (per thousand of initial population) surviving from younger ages to reach each age class; the

Table 6-1 Life Table for the Gem Clam, *Gemma gemma*. Age reckoned from time of liberation. Data based on 1954–1957 year classes at Union Beach, New Jersey. Mean length of life = 1.13 months.

x *Age (months)*	x^1 *Age as % deviation from mean length of life*	d_x *Number dying in age interval out of 1,000,000 liberated*	l_x *Number surviving at beginning of age interval out of 1,000,000 liberated*	$1000q_x$ *Mortality rate per 1000 alive at beginning of age interval*	e_x *Expectation of life, or mean lifetime remaining to those attaining age interval (months)*
0–1	−100.0	79,000	100,000	790.0	1.09
1–2	−11.5	8800	21,000	419.1	2.32
2–3	+77.0	3700	12,200	303.3	2.63
3–4	+165.5	2850	8500	335.3	2.56
4–5	+254.0	1850	5650	327.4	2.59
5–6	+342.5	1240	3800	326.3	2.61
6–7	+431.0	850	2560	332.0	2.55
7–8	+519.5	550	1710	321.6	2.70
8–9	+608.0	355	1160	306.0	2.75
9–10	+696.5	191	805	237.3	2.74
10–11	+785.0	116	614	188.9	2.43
11–12	+873.5	137	498	275.1	1.88
12–13	+962.0	188	361	520.1	1.40
13–14	+1050.5	92	173	537.6	1.38
14–15	+1140.0	43	81	530.1	1.39
15–16	+1227.5	20	38	526.3	1.40
16–17	+1316.0	9	18	500.0	1.39
17–18	+1404.5	5	9	555.6	1.28
18–19	+1493.0	2	4	500.0	1.25
19–20	+1581.5	1	2	500.0	1.00
20–21	+1679.0	1	1	1000.0	0.50

SOURCE: After Sellmer, 1967.

mortality rate (per thousand reaching the age class); and life expectancy for an individual of the age class. The initial population is considered to have been a *cohort*, that is, a class of individuals produced during a single reproductive event, such as a short spawning period. The symbols at the head of the columns are conventional. Note that if the figures in any of the columns d_x, l_x, $1000q_x$, or e_x are known, the remaining columns can be calculated readily (see Deevey, 1947). Thus, if one knows the age at death of a sample of a population, one can class the deaths by age intervals and form a column like d_x; therefore, if the sample is large enough, it may be assumed to represent an average cohort and be employed in an estimate of the life table of the population.

In Fig. 6-1, a survivorship curve representing the tabulated population is

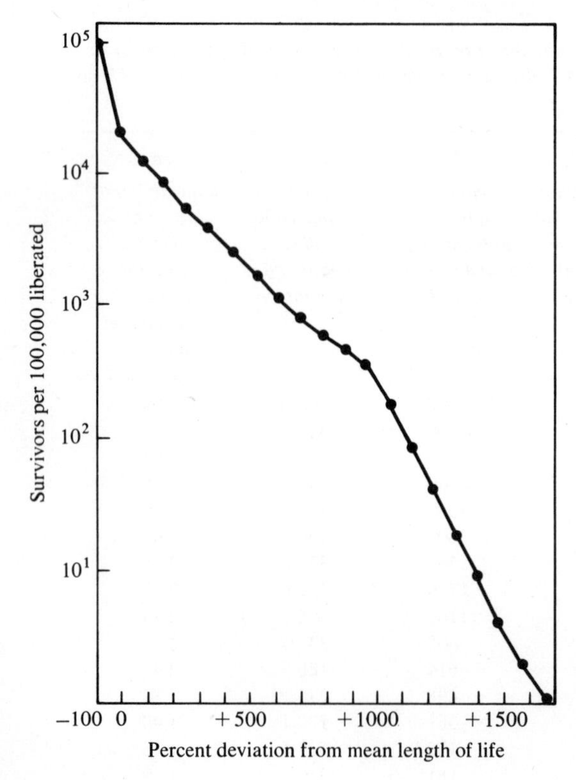

Fig. 6-1 A survivorship curve for *Gemma gemma*, plotted for an imaginary cohort numbering 100,000 individuals. See Table 6-1. (After Sellmer, 1967.)

depicted, based on the figures in column 1. The number of survivors (per thousand initial population) are plotted on the vertical scale, and on the horizontal scale is a series of age classes that are expressed, not as ages in months or years, but as percent deviation from mean length of life, which is 1.13 months for the *Gemma* population. This method of expressing ages permits a direct comparison of the age distribution of mortality in populations of organisms having markedly different longevities, such as insects that live weeks with mammals that live decades. Obviously, the steepness of the curve indicates the severity of mortality. From both the life table and the survivorship curve for *Gemma*, it is evident that mortality is high in early life (over 80 percent died within the first month), then is nearly constant at an intermediate level until nearly ten times the mean length of life, and finally is higher again among very old individuals, although still rather constant. On the average, about 40 percent of the population died each month.

Pearl and Miner (1935) have described three types of survivorship patterns, illustrated by the curves in Fig. 6-2. In populations with survivorships of type

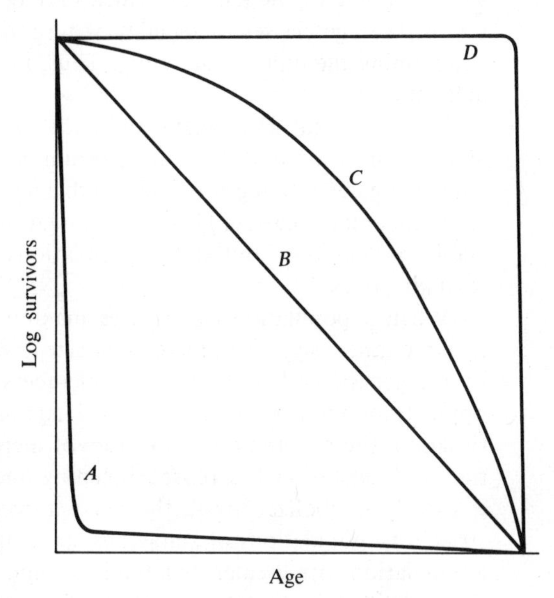

Fig. 6-2 Some hypothetical patterns of survivorship. *A*: High mortality at early ages. *B*: Constant mortality at all ages. *C*: Constantly increasing mortality with age. *D*: Little mortality until some terminal age, and then all mortality concentrated in a narrow span. (From E. S. Deevey, Jr., The probability of death. Copyright © 1950 by Scientific American, Inc. All rights reserved.)

I, an individual has little chance of surviving infancy; once past this critical stage, the expectations of longevity improve dramatically. This pattern is common among benthic invertebrates with planktonic larval stages, such as the oyster. In populations of type II, an individual has as much a chance of surviving at birth or late in life as at any other time; in fact, the curve implies that there is no upper physiological limitation to longevity. Some coelenterates, such as *Hydra*, have been considered to have curves of this type. Populations of type III tend to survive until they reach a physiological limitation of some sort, when they suffer a relatively abrupt mass mortality. The usual example of a type III curve is afforded by a population of starved fruit flies that are never fed but survive only until nourishment from their eggs is exhausted and they reach their limits of starvation tolerance. A curve of similar form would also be produced by a population at minimum mortality.

If natality is fairly constant through time, perhaps fluctuating narrowly about a mean, then the number of individuals reaching any age class will depend upon the sum of the mortality of younger age classes. If the age distribution of mortality is also fairly constant, then each age class will contain a fairly constant number of individuals. The proportionate distribution of individuals by age in a population is called the *age structure.*

The absolute number of individuals found in any age class depends partly upon the supply and thus ultimately upon natality. But although natality does not have an age distribution as far as the young are concerned, all starting life at the same age for practical purposes, natality does have an age distribution of parents. Usually fecundity is restricted more or less to a certain age range within the life of the females, which may be far shorter than the lifespan. The size of the age classes of females of reproductive age is an obvious factor in determining the number of young. Thus, interrelations exist between mortality and natality.

The biological mathematician Lotka has shown that if certain simplifying assumptions are fulfilled, an equilibrium will be established among the factors controlling age distribution, and a stable age structure will occur within a given environmental situation. Growing populations have relatively more young than stable populations, and stable populations have relatively more young than shrinking ones.

When a population experiences maximum natality and minimum mortality simultaneously, it will increase in size at the maximum possible rate allowed by the genetic constitutions of its members. If Lotka's model is correct, and populations tend to achieve a stable age structure in a given environmental situation, the maximum possible rate of increase will be constant, at least until the first generation has passed breeding age. This maximum rate of increase is usually symbolized by r_i, the *intrinsic rate of natural increase,* which is restricted to populations of stable age structure. Actually, r_i can be achieved by a population with greater than minimum mortality, if mortality is so distributed that natality is not affected. Symbolically, r_i can be represented in the expression

$$\frac{N}{t} = r_i N$$

where the environment is unlimited so far as maintaining maximum natality is concerned and the population has a stable age structure.

The *reproductive potential* of a population may be thought of as expressed by the intrinsic rate of increase r_i. The restriction imposed by the environment on the reproductive potential to produce the actual rate of increase that occurs is called the *environmental resistance* (E_r). When biotic potential and environmental resistance are equal, the population is not increasing but is stable in size, and natality (N_y) and mortality (N_m) must be equal. Symbolically,

$$\frac{\Delta N}{\Delta t} = r_i N - E_r = N_y - N_d = 0$$

when the population is stable (see Odum, 1971).

Thus, maximum natality, minimum mortality, and the intrinsic rate of natural increase are all theoretical population factors that provide a basis for measuring the environmental effects in real situations. The rate and direction of change in size, the age structure, and the age distribution of mortality are related features of a population; if any two are known, the third may be calculated.

Intrinsic rates of increase occur only under very special conditions in nature and then for short times. More usually the realized rate of increase is at a lower rate, r, as symbolized by

$$\frac{\Delta N}{\Delta t} = rN$$

From this expression it is evident that any positive rate of increase whatsoever, however modest, will result in a population that increases by a greater number of individuals in each succeeding generation. Population increase is "geometric," or follows an exponential curve, for since N is larger in each succeeding time interval, rN is also larger, so that the change in numbers (ΔN) steadily increases in magnitude. The implications of this situation were well appreciated by Malthus (1798; see quotation at beginning of chapter) and eventually led to the theory of natural selection. If, on the other hand, r is negative, even if it is very small, extinction will eventually ensue. Evidently controls of some sort regulate r, so that populations may persist without overwhelming the biosphere. The nature of these controls is one of the central problems of population biology.

Internal Population Dynamics Are Partly Regulated by, and Partly Adapted to, External Factors

No single process regulates population growth and size. On the contrary, a great number of factors operate within nearly all populations, and different suites of factors commonly operate within different populations. It is common to treat all the factors contributing to environmental resistance as comprising two major classes: those that increase their resistance to population growth as the number of individuals increases; and those that operate to control population size, despite the number of individuals present. The former are called *density-dependent* factors and the latter *density-independent* factors, and although it is clear that both classes operate frequently in nature, there has been a considerable controversy over their relative importance (for an excellent review, see Clark *et al.*, 1967).

A typical example of a density-dependent factor is food supply when the amount of food available is partly controlled by the population itself. As the population grows and consumes more and more of a food source that is in limited supply, growth is eventually checked by lack of food. If population growth is checked gradually, the population may approach an equilibrium at which r is 0. In this event, the growth of the population to equilibrium may often be represented by the expression

$$\frac{N}{t} = rN\frac{(K - N)}{K}$$

where K represents the maximum population that can be supported, sometimes called the *carrying capacity*. This is the *logistic equation* of population growth (due to Verhulst, 1838). In some cases, population growth will *not* be checked gradually but will continue at about the same rate of increase until the food is nearly gone, when famine will occur and drastically reduce the population size, at times even to extinction. In either of these cases, food is obviously a density-dependent factor. Any factor that is consumable by the population may become a density-dependent factor under certain conditions; oxygen, nutrient minerals, burrowing or attachment sites, and numerous other factors may be nearly or quite used up at certain population densities. Even density itself may be a density-dependent factor at times; individuals may be stressed by the presence of an abundance of other individuals of the same population, quite apart from any competition for resources, and consequently natality may fall. Since density-dependent factors cause competition within populations, and since natural selection will tend to favor superior competitors, these factors are of special interest from the standpoint of the evolution of population properties.

Density-independent factors include such environmental parameters as temperature, salinity, water turbidity, turbulence, and many others that can limit the population without regard to its size. Many factors can be either density dependent or density independent, according to circumstances. Oxygen, for example, is density independent if the supply is controlled primarily by the velocity of currents, but is density dependent if the supply varies critically because of its consumption by members of the population. Even food, usually density dependent, seems to be density independent in some marine situations, such as with suspension feeders on the continental shelves in intermediate latitudes. Predation may also be either dependent or independent of density, according to the prey-predator systems (Chapter 7). Populations limited by density-independent factors undergo selection for phenotypes that are tolerant of the stresses involved.

Density-independent factors may control population size through their effect on the length of time that the geometric effects of a positive rate of population increase r is permitted to operate. If a population repeatedly suffers heavy mortality owing to, say, recurrences of temperatures that are lethal to most individuals, the size of the breeding population may remain small. Therefore, even though r may be positive for much of the time, the episodes of mortality effectively regulate the population size. Population sizes, however regulated, may fluctuate constantly, and thus the concept of an equilibrium population would not be applicable on the scale of one or several generations. However, the concept may usually be justified over a length of time on the order of 100 generations (Istock, 1967).

The regulatory aspects of only a few marine population systems are well known. Case histories of the regulation of a variety of insect populations which convey a sense of the multiplicity of factors that can be involved are given by Clark *et al.* (1967). Actually, for a given population in a given situation there are often only a few key environmental factors that determine population size trends.

As an example of a marine population system, we may use *Gemma gemma* (Table 6-1; Fig. 6-1). Sellmer (1967) has provided natality and mortality data and a tentative interpretation of the mortality factors, which, as he stresses, are not proven and require further study. Figure 6-3 depicts the age-mortality distribution. *Gemma* is ovoviviparous; the high juvenile mortality rate may be due in large part to the stresses of assuming an independent life on soft bottoms after liberation, but a density-dependent factor (possibly food) associated with the crowded conditions of juveniles is suspected, as there are commonly over 100,000 and sometimes 1 million juveniles per square meter. The rather even

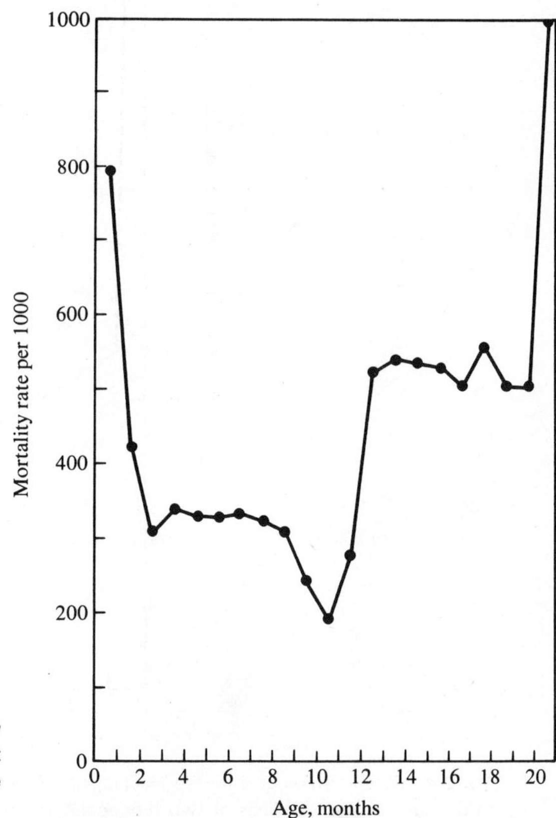

Fig. 6-3 Rate of mortality for *Gemma gemma*, based on an imaginary cohort numbering 100,000 individuals. See Table 6-1. (After Sellmer, 1967.)

mortality during much of the first year may result chiefly from predation; parasitism is known to be a relatively insignificant cause of mortality in *Gemma*. The second year is also characterized by a rather even mortality, but at a significantly higher rate than during the first. The second year mortality may also be chiefly due to predation as a proximate cause. Sellmer (1967) suggests that this rise in the mortality rate is owing to the drain on physiological resources by reproduction. The high mortality at the end of the second year presumably represents physiological breakdown due to aging.

A second example is afforded by a population of the inarticulate brachiopod *Glottidia pyramidata* studied by Paine (1963). An estimated survivorship curve is depicted in Fig. 6-4. *Glottidia* has a planktonic larval stage, and the larval

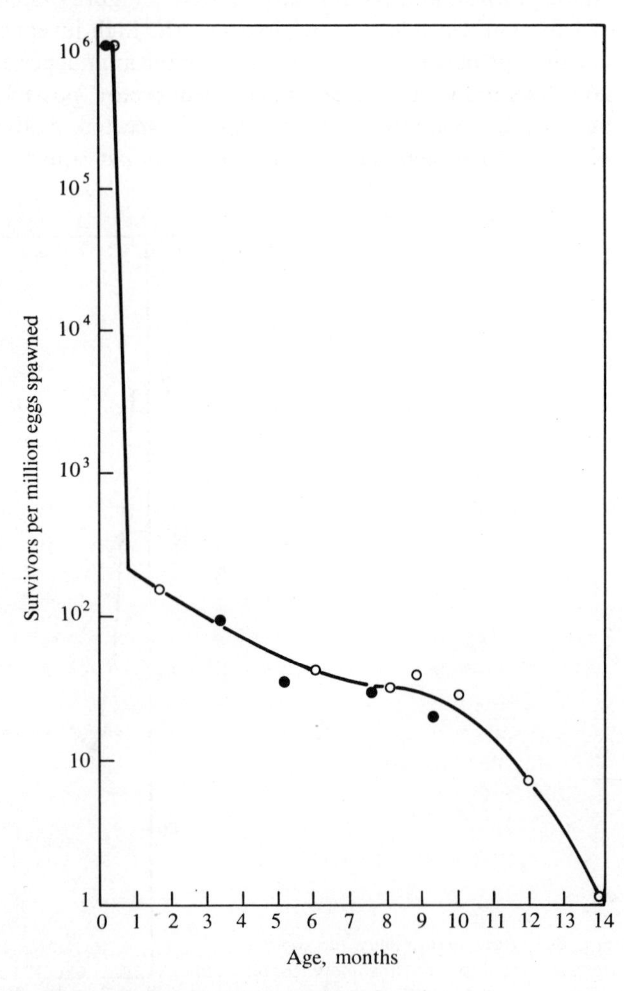

Fig. 6-4 Survivorship curve for the brachiopod *Glottidia pyramidata*, based on the age-specific densities of two subpopulations (open and closed circles) and on egg production. (After Paine, 1963.)

mortality rate far outstrips that of juvenile mortality in the ovoviviparous *Gemma*. Note that it was necessary to calculate the curve on the basis of an original cohort of 1 million to show a few individuals surviving to the end of the first year. Actually some may survive for several months more than is indicated on the curve (Paine, 1963). The great loss of planktonic larvae is doubtless owing partly to predation and partly to outwashing of larvae from shallow water to areas unfavorable for settling; *G. pyramidata* settles onto shallow sandy bottoms. Upon successful settling, life expectancy increases dramatically. The relative contributions of various postlarval mortality factors are not known, but sometimes the factors include predation by gastropods and birds; in addition, *G. pyramidata* is heavily parasitized by a trematode. The sharp increase in mortality at about nine months correlates with spawning and suggests that the physiological drain of reproduction contributes effectively to mortality.

Instead of sacrificing the adult population for the advantages of dispersal, or of dispensing with a planktonic stage altogether, we find that a third sort of reproductive regime is possible: to reproduce in some moderation but over a relatively large number of years (iteroparity). Thus, although larval mortality may be high, so high that successful recruitment into the benthic population occurs only occasionally, the population has so many chances that reproduction is normally successful. Fotheringham (1971) has described this situation in two predatory muricid gastropods from the Californian province. These gastropods are comparatively long lived, and their adult populations, although relatively small, suffer only a few percent mortality annually from predators, chiefly crabs.

In summary, a population has inherent (genetically based) properties, including potential rates of increase, that interact with any given set of environmental parameters to produce a certain pattern of mortality. This interaction governs the size of the population (which may be eliminated, be stable, or fluctuate) either by its influence on r or by density-dependent control of numbers. The environment-population interaction may involve adjustment of natality rates also due to mortality of reproductive age females, to parasitism of gonads or eggs, or to some other factor. From these interactions, a certain regime emerges, and the population age structure and survivorship are stabilized or fluctuate in a characteristic manner. Similar populations in different environmental settings may have quite different life tables. Minor or even major perturbations of a population system by "unusual" events, either favorable or unfavorable, are not uncommon. Finally, natural selection may operate so as to favor the production of individuals that are especially well adapted to a particular population regime (Nicholson, 1957). Such evolutionary changes will alter the inherent population properties, and a new population size equilibrium may be established, with corresponding adjustments of appropriate properties. Certainly there is a large element of self-regulation in population systems, both in the sense of adjustment of the system to a given environment based on inherent properties and in the sense of the evolution of new inherent properties in response to selective pressures.

Fossil Population Dynamics Are Difficult but Sometimes Possible to Reconstruct

To understand the natality and mortality of a fossil population, we must obtain the sort of data that permit the construction of a life table. Because this requires the determination of the distribution of mortality, preferably by age if accurate comparisons and contrasts with other populations are to be made, several major difficulties arise in dealing with the fossil record.

Most of the major sources of bias in fossil assemblages are associated with skeletal size. Once they are empty, small shells tend to be destroyed rapidly unless they are buried (Birkett, 1959; Hallam, 1967). Even then, bioturbation and solution by ground water and other diagenetic effects must differentially destroy the smaller and more fragile skeletons. It is clear that such size-related bias can be important. For example, the mortality distribution of *Gemma gemma* (Table 6-1) indicates that nearly eight out of ten dead shells represent animals less than one month old, and the removal or destruction of these minute shells will eliminate 80 percent of the mortality record. On the other hand, the mortality record of adults might remain essentially intact, permitting reconstruction of that part of the life cycle. The fossil record as a whole must be biased in favor of larger skeletons. Furthermore, transporting agents normally move small skeletons more easily than large ones, and thus size sorting must occur commonly when there is much postmortem transport. There is some dispute about the relative importance of size sorting, compared with size-preferential destruction without appreciable transport, in biasing population mortality records (see, for example, Fagerstrom, 1964, and Hallam, 1967), and the full extent of bias of any sort is not clear. In fact, it is not possible to be certain when a fossil population does or does not represent an unbiased sample of the mortality suffered by the ancient living population. However, if the entire association appears to be in place at its living site, and if individuals of minute as well as of large species are present in abundance, so that no size sorting is apparent in the fossil association as a whole, then some confidence might be placed in the quality of a population sample. Some fossil populations appear to be relatively unbiased with regard to size at death, at least for postlarval stages. Nevertheless, there is never a final proof that mortality parameters are preserved for a given population.

Another difficulty is that most fossil associations include representatives of numbers of generations. Indeed, a sample of a moderately thin interval from a continuously deposited sequence may embrace thousands of years. On the other hand, populations that probably represent only a single season may be preserved, as conchostracans that occur on bedding planes of Permian ponds (Tasch and Zimmerman, 1961). Mass mortalities can cause the local accumulation of entire populations, killed more or less simultaneously and therefore preserving the age structures of the time.

The way in which a sample is collected will obviously influence the type of population record obtained, and both may depend upon the nature of the fossil deposit and upon the kind of exposure. Ideally, a population sample should be taken from as homogeneous and narrow a stratigraphic interval as possible as a first requirement, and then from as small a lateral distance as possible. Since the collection must be of a certain size in order to include an adequate representation of specimens, it may have to include a rather thick interval over an extensive area, in some cases. At any rate, it is rarely possible to sample contemporaneous individuals that belonged to an ancient population. Rather, the samples usually represent a mixture of generations from which the demographic structure of a hypothetical cohort that is representative in some ways of an average during the time interval collected may be calculated. This can have important advantages, since the fluctuations of short-term inequilibria will be eliminated and the long-term mortality distribution may appear. On the other hand, important population changes may be overlooked and the average may be totally unrepresentative of any actual historical situation.

Finally, an additional difficulty is simply determining the age at death of the organisms represented by fossils. The most direct and accurate method is to study the growth-increment patterns of skeletal structures. By this means, the ages of the organisms at death may be determined to within weeks or even days, the season of death and possibly that of birth can be determined, and such important additional information as the spawning regime may sometimes be inferred (see Chapter 5). By studying incremental growth of entire population samples in thin section, and by collecting from the most favorable localities where size bias seems to be least, it should be possible to obtain rather accurate data on fossil populations. Yet at this writing there are no published studies of this quality.

In skeletons containing growth increments, the annual variations in skeletal growth are commonly on a scale sufficient to permit their recognition on the skeletal surface. This is true of many bivalves, for example, where "growth checks" or "growth rings" may be well developed; the check or ring usually indicates winter growth. Brachiopods, corals, and echinoids also display growth lines. If growth rings are assumed to be seasonal, the age of the shell can be easily determined without extensive preparation, and the season of death estimated. Unfortunately, growth rings are also caused by factors other than seasonal growth-rate changes. A severe storm, unusual turbidity, or the attack of a predator can cause a temporary cessation or diminution of growth, and may be recorded in the shell as a ring; these episodic rings cannot always be distinguished from seasonal growth rings by external morphological criteria, although it is an easy matter to distinguish them in thin section (Pannella and MacClintock, 1968) when growth increments are preserved. In the absence of growth-increment studies, however, growth rings provide an estimate of age that may be used, with caution, in demographic calculations (Craig and Hallam, 1963).

Size is an obvious feature of fossil skeletons that is easily measurable and correlates in some way with age (Fig. 5-1). Size-frequency distributions may be used as estimators of age-frequency distributions, and to the extent that they are reliable they can serve as a basis for the reconstruction of demographic parameters. Most studies of age frequency in fossil populations have been based at least partly on size-frequency distributions, but, as we shall see, the results are not altogether encouraging. Nevertheless, for the many nonincremental skeletons or for incremental skeletons that do not preserve the growth-increment records, the only criterion of age may be size, which can be helpful under certain circumstances.

Not all members of a population that are of the same age are of the same size. The variation in size, a result of both genetic and environmental factors, is usually normally distributed at any age. The variability of the age-size relationship changes from age to age and population to population. If size is used as if it is directly correlated with age, then smaller individuals of an actual age class are tallied in size classes that are interpreted as younger age classes, and larger individuals are tallied in size classes that are interpreted as older age classes. If all actual age classes contain the same numbers of individuals and have the same sort of age-size variation, then the number of individuals of similar ages that would be lost from a given size class because they were too large or too small would be exactly replaced by the larger individuals of younger ages and the smaller individuals of older ages that fall into the given size class. Only the smallest and largest size classes would not contain the same numbers of individuals as the age classes they are intended to represent. However, when there are progressively fewer individuals of successively older ages, then the number of smaller individuals of older age classes that are counted in a given size class is inadequate to replace the larger individuals lost from the age class that is intended to be represented by the given size class. Therefore, if size-frequency curves of populations that decrease steadily with age are interpreted as age-frequency curves, they are regularly biased in favor of younger age groups. Also, slight differences in growth rates among individuals will produce increasingly greater differences in size between individuals of the same age. Owing to this effect, the variability of size with age tends to increase in older age classes, so that the size classes representing them lose progressively more individuals with increasing age that are larger or smaller than the class limits. Changes in the mortality trend that alter the rate of population decrease with age complicate these biases, for the contribution of size classes to age classes becomes irregular. These biases can only be corrected if the trend of the mortality curve is known.

A further complication arises from the individual growth patterns of the organisms. If skeletal size increase is constant, then size classes that are all of the same size interval will represent equal age intervals (in their biased way). If growth slows with age, however, then the average size difference between older age classes becomes progressively smaller and the degree of size overlap between them becomes exaggerated. In this situation, tallies in equal size classes bear less and less relation to the actual age distribution of mortality with increasing size. All these biases are difficult to correct without some knowledge of the growth rate and the age distribution of mortality.

One property possessed by numbers of marine invertebrate populations actually helps in determining the size-age relationships—periodicity of reproduction. Many populations have a relatively short reproductive season once (or more) a year; thus, during each reproductive season, a distinctive size class is formed (Fig. 6-5). Both *Gemma* and *Glottidia* have an annual reproductive

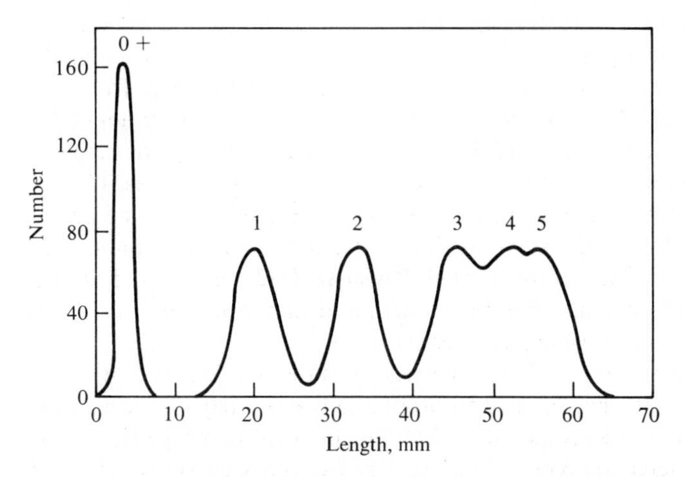

Fig. 6-5 A size-frequency distribution of a hypothetical population of *Mytilus edulis* that was recruited from five year classes, derived from data by Savage. (After Craig and Hallam, 1963.)

pattern of this sort. The size-frequency distribution of skeletons representing each season's age class can be readily determined for the younger age groups; rising size-age variability and decreasing growth rates soon cause the year classes to begin to overlap in size, but there is little difficulty in distinguishing the size-frequency distribution of year classes until the overlap becomes very large. However, some invertebrates live for decades, some possibly for a century (Chia, 1970, attributed to Pearse), and year classes could not be distinguished throughout most of their age ranges.

Craig and Hallam (1963) studied the size classes of skeletons resulting from discrete seasonal reproduction of the mussel *Mytilus edulis*. The size-frequency distribution of a *Mytilus* population from the Conway Estuary in Great Britain is illustrated in Fig. 6-5. This curve is artificial in that it assumes a constant natality for a short period each year with no mortality; however, it is based upon actual monthly measurements of 500 specimens by Savage (1956), and it illustrates some important features of seasonally reproduced populations. In Fig. 6-5, the high narrow peak near 4 mm represents the youngest age class, which was recruited into the population the previous month; the one- and two-year-old populations are distinctive also, although size variation is noticeably greater in these older age classes, and the spreading out of individuals into more size classes lowers the height of the modal peak. The third-, fourth-, and fifth-year age classes overlap progressively in size range, and their peaks are progressively closer together, although they are still recognizable.

Now, if mortality among the *Mytilus* were chiefly restricted to the rather severe and stormy winters, but constant among the age groups, then the population of Fig. 6-5 would generate a size distribution of dead shells that would closely resemble the peaked curve in the figure, but with peaks displaced toward the right and decreasing in height with size. By contrast, if there were no mortality until the fifth year, then the population would generate a unimodal size-frequency curve centered around 57 mm. In fact, the sorts of size-frequency peaks generated by *Mytilus* beds along storm beaches in Scotland are unimodal, as Fig. 6-6 depicts. The storm beach deposits contained numerous small to minute shells of other marine organisms, thrown up by storm waves. If small mussels had died in abundance, their shells would have been expected on the beach along with the other small skeletons from the marine community. Craig and Hallam (1963) therefore inferred that small dead specimens of *Mytilus* were lacking because they had not been killed, and that mortality was largely restricted to mussels which were a few years old.

As a large number of population properties affect the size-frequency distribution of skeletal remains, they interact to produce a wide variety of size distributions. Nevertheless, it seems possible that the interacting properties might produce some characteristic types of skeletal size-frequency curves which could be used to class the demographic properties of the original populations. Investigations of typical population models have been carried out by Olson (1957) and more recently by Craig and Oertel (1966), who used computers to investigate a wide range of population property combinations. By holding all factors but one constant in various models, they examined the effects of changes in individual factors in different population types. From this work, a number of generalities have emerged concerning the production of size-frequency distributions of dead skeletons.

If the growth rate is constant, then a mortality rate that decreases with age leads to a positively skewed size curve similar to that of *Gemma*, whereas a mortality rate that increases with age leads to a flatter curve, which, if the increase is great enough, can become negatively skewed. If, by contrast, the mortality rate is constant but the growth rate is decreasing, a negatively skewed distribution results, since the age classes resemble each other more and more in size with increasing age; if the growth rate were to increase (which is not known to occur), then the size distribution would be positively skewed. Finally, if the mortality and growth rates are both constant, the size distribution of the dead population is a "mirror image" of the size distribution of the living population. Size peaks in the living population cause peaks in the dead one, which are spaced

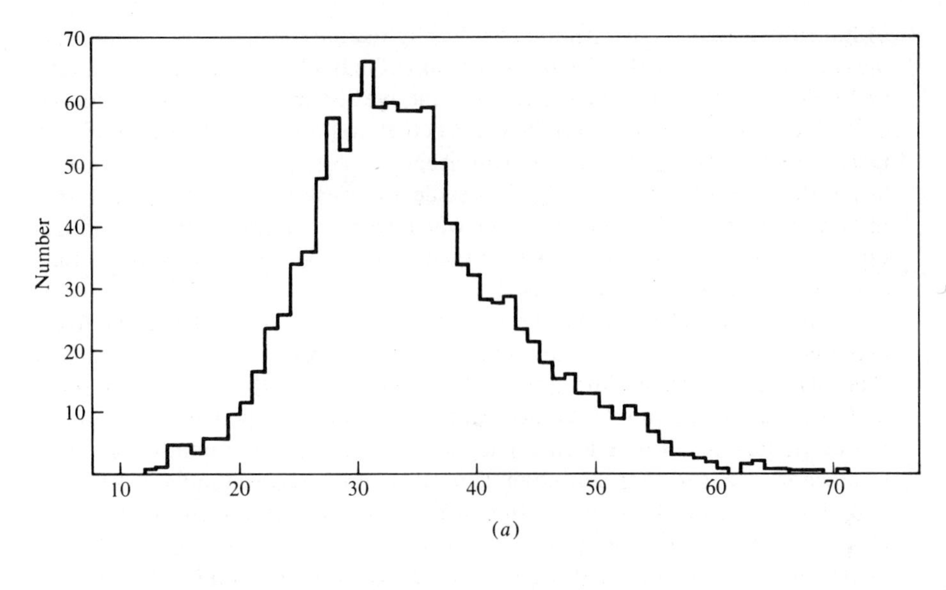

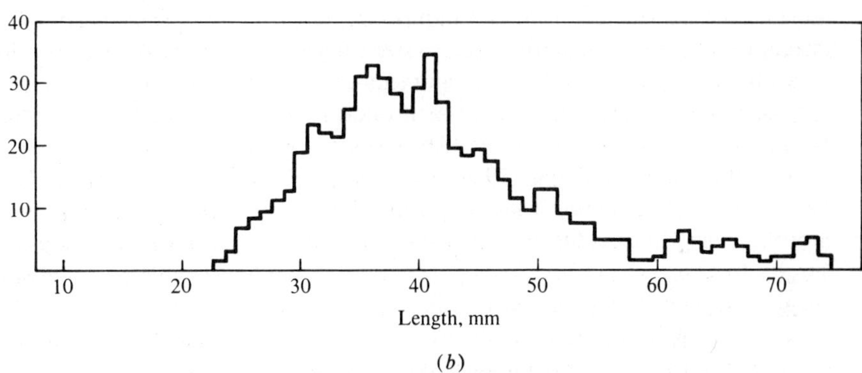

Fig. 6-6 Size-frequency data on valves of *Mytilus edulis* washed onto storm beaches in Ferny Ness (*a*) and Craigelaw Bay (*b*), Scotland. (After Craig and Hallam, 1963.)

similarly but have their modes offset. Inasmuch as such peaks are commonly annual, they may be used with caution to calibrate age-size relations and to determine absolute rates of growth.

Some additional general considerations apply to the study of the internal dynamics of fossil populations, regardless of the source of the age estimates at death. First, the appearance of the youngest age group in invertebrate benthic populations frequently results not directly from natality but from larval settling; the new benthic generation is recruited after some growth has already occurred, and the minimum age is set by recruitment and not by natality.

Second, individuals may be recruited at any age, and a steady immigration of individuals that do not belong to a locally recruited cohort may occur, and may also be coupled with a steady emigration of individuals. Unless immigrating and emigrating factions have exactly the same age distributions, the age structure of the local population will be affected. Whether migration could be a significant factor in population age structure can usually be estimated from the mode of life of the population; obviously, for sessile and weakly locomotive organisms migration will be small or lacking. Even when migration is important, it is probably often controlled in some manner related to the general size controls that operate for the population involved.

Third, the assumption has been tacitly made heretofore that each fossil skeleton represents a dead individual. However, many organisms produce skeletal remains at times during their life cycle other than the terminus. Arthropods are the prime example. Molted carapaces of an arthropod age class do not reflect the number of individuals dying at that stage, but the number that live to reach it, and it is often not possible to distinguish between molted skeletons and those skeletons remaining after death. To calculate the mortality for a given stage between molts, called an *instar*, requires that the number of individuals reaching the next oldest instar, as indicated by the abundance of their remains, be subtracted from the number representing the given instar. Since instars usually occur in fairly discrete size ranges and commonly are morphologically distinctive as well, the assignment of individuals to age classes can be very accurate and the estimates of arthropod mortality can be far better than estimates based on size-frequency distributions of permanent skeletons.

Levinton and Bambach (1969) have prepared survivorship curves for a Devonian bivalve, *Arisaigia placida*, a nuculanid that appears from functional morphological interpretation and from taxonomic inference to be ecologically similar to living species of *Yoldia*, such as *Y. limatula*. They both presumably share a deposit-feeding mode of life, respiring through siphons and feeding primarily by palp probiscides. Relative ages of the fossil shells were estimated from their sizes by assuming a logarithmic growth model. A survivorship curve was calculated from the size-frequency distribution of a large sample (178 specimens); it indicates that for the ages represented, death was about as likely at any time of life—mortality was fairly steady (Fig. 6-7). This survivorship pattern is similar to that of *Y. limatula*, but differs from that of bivalve suspension feeders that live with *Yoldia;* they tend to suffer a considerably higher mortality at younger ages than at older ones. The suspension feeders are not as well adapted to contending with soft muds, which are the natural habitat of many detritus feeders. Thus, the survivorship curve inferred for the fossil population is probably real and suggests that its ecological relations were within their optimal range.

There are as yet few other published studies of the population dynamics of fossils, despite all the preliminary work on recent death associations. For marine organisms, studies have been carried out on Ordovician bivalves by

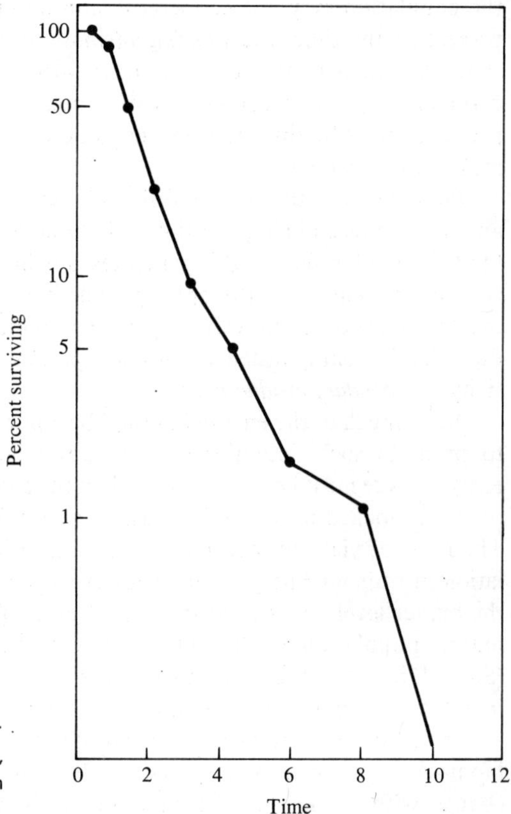

Fig. 6-7 Inferred survivorship curve for *Arisaigia placida*, a Devonian bivalve, based on 178 specimens. (After Levinton and Bambach, 1969.)

Snyder and Bretsky (1971), on Silurian ostracodes by Kurtén (1953, 1964), and on Permian fusilinids by Kaesler and Fisher (1969). Here is an interesting field for further work.

The Productivity of Fossil Populations Is Important in Theory

There is a complicated terminology associated with productivity that is confused by inconsistent usage among authors. We shall require only the few terms mentioned here for our purposes. The total amount of living matter represented by a population is called the *biomass*, which is usually measured as the total dry weight of the living population. The reproduction and growth of organisms result in the creation of new biomass, whereas the deaths of organisms remove it. A population that has high reproductive and growth rates is producing biomass at a high rate. However, if the mortality rate is also high, the population may not be increasing in numbers, and the amount of biomass present at any time (the *standing biomass* or *standing crop*) may be nearly constant, although in this case biomass turnover is high. Since the production, maintenance, and consumption of biomass involves energy exchanges, these processes may be discussed in energetic terms; the energy involved is usually measured in calories.

Productivity is the rate at which biomass is produced, and *production* is the amount produced in a given time. The productivity of autotrophs is of a special kind, for the biomass that it generates is being synthesized from inorganic materials in large part. Heterotrophic productivity, on the other hand, depends upon the consumption of previously produced biomass. Therefore, autotrophic productivity is distinguished as *primary productivity*, and heterotrophic productivity as *secondary productivity*.

Now, much of the energy ingested by a population is employed in respiration to promote such activities as locomotion and reproduction. Therefore, the energy intake must be much higher than the energy *yield*, that is, than the amount of energy locked up in the biomass of a population and available to predators. The ratio of yield to ingested energy is a measure of the efficiency of a population in utilizing energy from a lower trophic level and making it available to the higher levels; it is termed the *ecological efficiency*. Ecological efficiencies in natural populations are found to range chiefly between about 5 and 15 percent (Slobodkin, 1960, 1962). In other words, between 85 and 95 percent of the energy ingested by populations is employed in respiration or lost in other ways.

The relations between the number of individuals, their sizes, and the standing biomass of a population are obvious. It is also obvious that these relations will change with changes in the survivorship and even in the age structure of a population, for the age structure is related to the size-frequency distribution of individuals. However, the relations between productivity, size, and standing biomass are a bit more complicated. Partly this results from smaller organisms tending to metabolize at a higher rate than larger ones; therefore, for two populations of organisms of different sizes having the same biomasses, the population with the smaller individuals will usually require a higher amount of primary production to sustain it, since its ecological efficiency is relatively low. It also results partly because in two populations of equal size, standing biomass, and ecological efficiency, but with unequal turnover, the one with the higher turnover is the more productive.

It is exceedingly difficult and probably will always remain impracticable to estimate the parameters of productivity for fossil populations in absolute terms. However, the energetic aspects of populations are probably the most important aspects of all those that play a role in the ecological regulation and evolutionary adaptation of populations. This being the case, paleontologists are forced to deal at least semiquantitatively with problems of population energetics. Paleontological observations that might suggest high productivity would include high density of fossil remains, although unless the region from which they were assembled and the length of time that they represent are fairly well known, this could not be confirmed. Perhaps ancient productivity will always have to be judged from a quasitheoretical basis through models of ecological and evolutionary processes in which energetics play major roles. These models are based upon community ecology and will be considered in Chapter 7.

Population Dispersion and Density Reflect Both Internal and External Factors

Other properties related to the internal dynamics of a population are *dispersion*, the way in which the living individuals are spaced relative to one another, and *density*, the number of individuals in a given area or volume. The three main sorts of dispersion patterns are *random*, *regular*, and *clumped* (Fig. 6-8).

In perfectly random dispersion, no part of the area of dispersion is more favorable than any other, and the presence of an individual at any place has no effect on the presence of any other individual, so the distances between

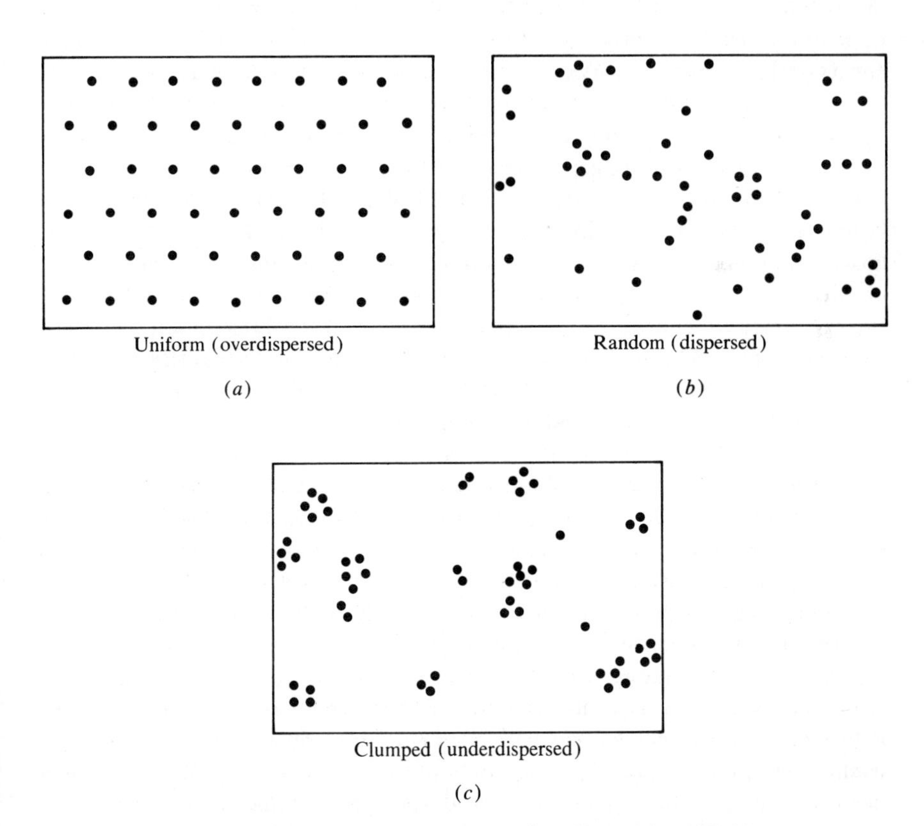

Fig. 6-8 Three patterns of dispersion. (After Odum, 1971.)

individuals are strictly "matters of chance." In clumped dispersion patterns, the chances of occurrence of an individual are higher in the proximity of another individual. Since clumped individuals are less well dispersed or less spaced out than in a random distribution, they are said to be *underdispersed.* In populations that are perfectly regularly distributed, each individual is found at a constant distance from its nearest neighbor(s). Since populations that tend toward regularity are more evenly distributed than if they were random, they are sometimes termed *overdispersed.*

In general, all populations are clumped; none is distributed evenly, and none randomly, over the entire earth. Nevertheless, within certain restricted regions of the earth, such as the region occupied by a given community, some populations may be random or overdispersed. But even within such restricted regions, the greater number of populations is usually clumped. Factors leading to the clumping or *aggregation* of individuals of the same population are numerous, but a few are especially important.

The overriding causes of aggregation are simply the heterogeneity of the environment and the ecological specialization of even the most generalized organisms, which exclude them from certain habitats. Within areas of suitable habitat, populations are frequently aggregated by localized food sources, as when scavengers gather around carcasses or predators around prey populations.

Commonly, aggregation in suitable benthic habitats is achieved primarily by larval selection of settling sites. In other cases, larvae are aggregated by means other than their own choice. Current sorting of larvae is probably common and can easily lead to the eventual concentration of larvae of the same population owing to their similar hydrographic properties (Baggerman, 1953). Furthermore, many marine organisms that produce young lacking a planktonic stage provide some sort of brood protection ranging from simple attachment of egg clusters to the substrate, to ovoviviparity. Thus, the young tend to be aggregated in sibling swarms when they first begin independent life outside the egg. If such broods are not dispersed by currents or through their own locomotion, they will give rise to aggregated adult populations.

Even though aggregation may thus arise as a sort of side effect of the life cycle, it is so widespread that it must entail no great disadvantage on balance; indeed, it seems likely that there are positive advantages leading selection to favor aggregation in many circumstances. This could be brought about by modifications of life cycles so as to produce clumping. One obvious advantage to aggregation in sexual organisms is that it facilitates reproduction.

Among marine invertebrates, additional advantages of aggregation are especially evident among sessile suspension feeders. For sessile organisms, suspension feeding is especially practical when currents from which organisms may remove the suspended particles constantly introduce food. For any given feeding system, there must be an upper limit to the velocity of water from which particles can be usefully extracted. Indeed, it is expected that there is an optimal range of water velocity, fast enough to provide adequate food supplies under normal conditions, and yet slow enough to permit extraction of particles by the suspension feeder (and also to permit the suspension feeder to maintain its position in or on the substrate). Suspension feeding is thus promoted by a certain amount of lateral water motion and turbulence, which tends to keep particles in suspension in some eddies where velocities are higher than average, and to permit the easy removal of particles from eddies where velocities are low. Clumping behavior that may be an adaptation for food gathering in relatively strong currents is displayed by some living unstalked crinoids in the Caribbean (Meyer, 1969, and personal communication). Two or more individuals of some species have been found clustered together with their tentacular crowns in close proximity. Suspended particles have been seen to move relatively slowly and irregularly within the irregular tentacular baskets formed by the crinoids, where water velocities are lower and turbulence greater than in the ambient current. Therefore, it has been tentatively suggested that a purpose of this behavior is to lower the effective current velocity in order to promote suspension feeding.

Current velocities and turbulence can also be enhanced by aggregations to improve feeding (Barnes and Powell, 1950; Knight-Jones and Moyse, 1961). Populations of the barnacle *Balanus crenatus*, when crowded, form characteristic growth patterns that produce a hummocky surface [Fig. 6-9(*a*)]. This surface causes turbulent motion in the water flowing across it. Presumably, the large barnacles near the centers of the hummocks have had some initial growth advantage and thereafter were more favorably placed to secure food than their neighbors, thus continuing their advantage. The barnacles in depressions between hummocks cannot secure as much food, but nevertheless contribute to the population by just being part of the topographic pattern of the population surface and furnishing lateral support to individuals in the surrounding hummocks. Even barnacles in the depressions may benefit to some extent from the water turbulence. A similar example is furnished by the serpulid worm *Filograna implexa* [Fig. 6-9(*b*)], aggregated populations of which produce a surface of relief that (1) causes water turbulence and (2) provides more room for feeding organs, thus accommodating more individuals than would a flat surface. The barnacle aggregation presumably arises from larval selection, whereas that of the serpulid worm arises from reproduction by budding.

The advantages of aggregation are carried further in populations wherein individuals remain connected by tissues and thus form true *colonies*. A major advantage of the colonial way of life is that interindividual coordination may be established by common nerve paths, by metabolites either transported internally or excreted into the water, or by other means. The resulting cooperation can lead, for example, to common orientation of tentacular or ciliary activity to cause consistent feeding currents over the entire colony, which moves a large body of water in a single direction and thus draws water with its suspended food from relatively large distances away to bathe the colony. Such an effect has been demonstrated in the Ectoprocta (Mackie, 1963). Any such more or

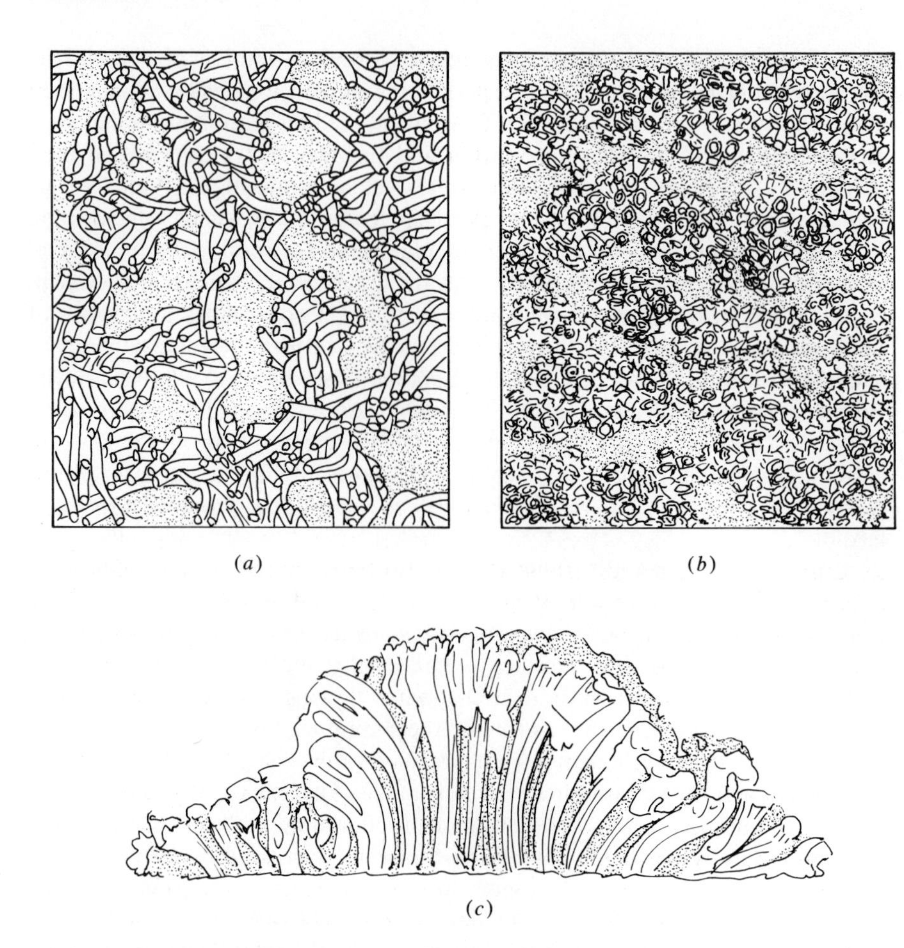

Fig. 6-9 Hummocky surfaces created by populations of *Filograna implexa* (*a*) and *Balanus crenatus* (*b*). (*c*) Cross section of a *Balanus* hummock. Some individuals in the depressions are not shown. (After Knight-Jones and Moyse, 1961.)

less unidirectional feeding current also provides an ideal medium for the removal of metabolic wastes. Colonial tunicates, for example, combine their excurrent flows to remove wastes to relatively far distances (Mackie, 1963). These advantages might conceivably accrue to noncolonial aggregations at times, although there seem to be no documented examples. Coordination that is possible within colonies has some special advantages over that possible in noncolonial aggregations. For example, warning of the presence of a predator that attacks one colonial individual can be transmitted rapidly to the remainder of the colony, as it is in some corals (Horridge, 1957).

The colonial habit has probably been evolved from aggregations in response to selection in favor of these sorts of advantages. Within many colonies, selection has continued to operate to produce differentiation among individuals that become specialized for functions such as protection, feeding, or reproduction—functions for which the colonial system is particularly advantageous. In colonies in which this trend is carried far, it becomes impossible to speak meaningfully of the ecology of the individual, and necessary to discuss the ecology of the entire colony. The colony has ceased to be only a collection of individual organisms and has become a "superorganism." An extreme example is the siphonophore colony, as exemplified by the Portuguese man-of-war, *Physalia* (Fig. 6-10). In these colonies, each individual, or "person," fills the role of an organ or organ system of a higher metazoan. Thus, the organ grade of construction is attained by these tissue-grade animals by evolving into colonial "superorganisms" (Mackie, 1963).

Fig. 6-10 *Physalia*, the Portuguese man-of-war, a colonial coelenterate of the order Siphonophora. (1) A medusoid person, the pneumatophore or float. (2) Polypoid persons, both gastrozooids, or feeding persons, and small dactylozooids, or feeling persons. (3) Large dactylozooids. Other medusoid persons include the swimming bell, or nectophore, below the float, and gonophores, which bear gonads. (After Hyman, 1940.)

There are certainly numerous other causes of aggregation, which clearly span the range from (1) simply the common restriction of genetically similar conspecific individuals by limiting factors, through (2) the similar responses of conspecific individuals to optimal conditions, to (3) the positive advantages of a clumped as opposed to a more dispersed state, advantages which may eventually lead to obligatory aggregation in colonies and to the development of "superorganisms."

Aggregation is not always advantageous, however. We have already seen that many limiting factors are density dependent, and aggregating certainly raises the population density. Probably, regular or overdispersed living patterns are usually adaptations to mitigate the deleterious effects of high densities that exceed the carrying capacity of the environment (K). Mortality arising from high densities can be reduced by spacing out the individuals at some distance so as to minimize their competition. Therefore, it is not uncommon for individuals within populations that are markedly clumped to display markedly regular distributions within the clumps. In colonial organisms, the spacing may be regulated by budding and growth patterns, so that individuals occur at optimal intervals along colony branches or around colonial masses; such individuals are usually overdispersed. Intraspecific competition is thus strictly regulated in colonies despite their aggregative nature (Knight-Jones and Moyse, 1961). Regularity may also arise from larval settling behavior; larvae that tend to settle gregariously frequently attach only at some distance from their nearest neighbors. Some larvae search the immediate vicinity of any potential settling site, and if they encounter another individual they travel a certain distance before searching again. Vagile organisms may space out simply by migrating away from any disturbing influences created by other individuals.

In sum, the members of any given population are underdispersed in the world at large, being restricted to certain habitat ranges. Within these habitats, they may be either random, regular, or clumped; if clumped, the distribution of individuals within the clumps may be either random, regular, or clumped. For some species, combinations of these tendencies are possible under appropriate conditions. For a hypothetical example, clumps might be internally random at low densities, clumped at intermediate densities, and regular at higher densities.

The dispersion of skeletal remains representing numbers of generations of a population lineage is usually rather different from that of the living organisms, even assuming that there is no destruction or transport. Imagine a biotope with uniform habitat conditions wherein sedimentation is low relative to generation length. Populations in this biotope that are randomly dispersed will have skeletal remains that are random also, although the density of the remains will continually increase so that eventually skeletons of this population will be found essentially everywhere within the biotope, but in random density distributions. For clumped populations, several patterns of skeletal dispersion are possible. If some factor is operative to cause succeeding generations to occupy the same sites as ancestral generations, then the aggregations will remain discrete and become steadily denser, so that the clumping will actually intensify. If, on the other hand, some factor causes succeeding generations to avoid the sites of ancestral occupancy, the aggregations will move around from generation to generation until, as a limiting case, the skeletal remains coalesce into a sheet; the clumping will decline during this process. Finally, if the location of a new aggregation is random with respect to ancestral living sites, then the effect is intermediate between the previous cases. Each of these situations could be caused by appropriate larval selection of settling sites. Populations that are regularly distributed will tend to generate skeletal patterns that remain regular if recruits are repelled by factors associated with ancestral skeletons, or become clumped on a regular pattern if recruits are attracted to factors associated with ancestral skeletons, or become more and more random if larvae are uninifluenced by factors associated with ancestral skeletons.

In a real biotope wherein the environment is somewhat heterogeneous, living dispersion patterns vary partly with the distribution of limiting and optimal factors, and the skeletal dispersion patterns vary partly with the changes in the distribution of such factors in time. Furthermore, in a real biotope, skeletal destruction and transportation usually occur to some extent, multiplying the possible skeletal patterns. Finally, much additional disruption of the original living dispersion pattern occurs owing to diagenetic or metamorphic destruction of skeletons and to erosion. The average fossil record of living population dispersion must be very poor. Nevertheless, we have seen that the fossil record occasionally presents us with superlative biotic data, and sometimes original dispersion patterns may be reconstructed. For example, this is true for some fossil reefs, where the living patterns of some populations may be inferred by referring the skeletal patterns to the environmental framework that can be reconstructed from the growth pattern of the reef structure and from the associated sedimentary framework.

Variability Among Individuals Is an Important Population Property

Since populations are collections of individuals that have all drawn their gametes from the same reproductive pool, they are composed of individuals with much in common. On the other hand, no two individuals are precisely alike. The variations among individuals must produce a population that has a far greater range of form and function than does any single individual. In taxonomy, it has become a well-accepted principle that the morphology of an individual specimen cannot represent the morphology of the population from which is has come, and, indeed, sometimes an individual specimen has an aberrant or unusual morphology. In describing species, taxonomists attempt to describe the pattern of variation in morphology within the species populations, including both the range of variation and the frequency with which each variant occurs. This is not always possible, of course, since the sample of the population available to a taxonomist is commonly inadequate. Nevertheless, it is an accepted goal. The same principle applies to the functional aspects of populations. The functional attributes of a single individual are not representative of an entire population, and a description or interpretation of population functions should include an account of the patterns of functional variation within the population whenever possible.

The functional variations arise from three main sources. The first source is genotypic variation, which simply reflects the different functional responses evoked in different genotypes. The other sources are phenotypic rather than genotypic. Genes do not constitute a simple plan whereby a certain ontogenetic sequence of phenotypes is produced, but rather they form a sort of operations manual for ontogeny, often with contingency plans. Environmental fluctuations that affect development are very common. Temperature changes, for example, alter the rates of development, and thus may alter allometric proportions. Obviously it can be advantageous for developmental processes to be able to adjust to environmental fluctuations.

For some species, developmental adjustments to the environment are corrective; that is, there is a tendency to produce a certain modal ontogeny, and deviations from the norm that might be produced through environmental fluctuations are corrected by adjustments in development; many alternative developmental pathways are available. This is ontogenetic homeostasis. Surely it can be a great advantage to be able to counter environmental variations that might alter the ontogeny from the course which has been "proven" by selection to be relatively fit. In populations of such species, individuals tend to resemble each other closely.

However, there are circumstances in which it is advantageous for organisms to develop in ways that do not represent the phenotypic modes of the previous generation. Such phenotypic plasticity is common in plants, especially in those adapted to disturbed habitats. These plants, such as weeds, tend to produce phenotypes that are best adjusted to the particular microhabitat in which they happen to occur. Phenotypic plasticity is also displayed by invertebrates. Among corals, for example, colonies that grow on the exposed parts of reefs are massive or have very short stout branches, whereas colonies of the same species on the less turbulent reef flats have stout skeletons but longer and thinner branches, and those that grow in quiet lagoon waters have tall slender branches or are developed as flat delicate foliae. Some species with short stout branches at exposed reef sites also range into deeper water, where they commonly grow in forms with fragile slender branches. The sort of variation displayed by these corals, wherein the genotype has the ability to produce any of a number of different ontogenies, depending upon which is evoked by the environment, has been called "multiple-choice variation" (Bonner, 1965).

Another main source of functional variation is based upon the lack of precision that naturally occurs in developmental processes, involving, as they do, astronomical numbers of reactions. Even identical genotypes in identical environments would produce somewhat different phenotypic products owing to this imprecision or "developmental noise" (Waddington, 1957). Organisms must have a certain developmental tolerance; otherwise, requirements for successful development could be so high that few individuals would survive and populations could not be maintained. This third type of variation has been called "range variation" (Bonner, 1965). All these types of variation—genotypic, multiple-choice, and range variation—appear to be of adaptive value, and all are ecologically significant.

In fossil populations, it is difficult to proceed from an interpretation of the quality of an individual function to even a semiquantitative evaluation of the variation of that function in the population. It is easiest when variations in the shape or dimensions of a skeletal character can be directly related to variations in function. For example, if certain spines function to buoy an animal up on soft muddy sea bottoms, it would be possible to determine the relative buoyancy of different spine numbers and sizes and then to measure their variation in skeletons of fossil populations. In this way, the patterns of functional variation among the skeletons could be inferred and interpreted.

To pursue this hypothetical example, imagine a set of collections of spiny benthic skeletons representing allied populations from localities that are somewhat scattered in space and time (that is, horizontally and vertically within geological formations). The average length and the variance of supporting spines are determined for each population, with the results indicated in Table 6-2.

Table 6-2 Hypothetical Example of Variations in the Means and Variances of Brachiopod Spines in Space and Time

TIME	LOCALITY 1		LOCALITY 2		LOCALITY 3		LOCALITY 4	
	Mean	*Variance*	*Mean*	*Variance*	*Mean*	*Variance*	*Mean*	*Variance*
Time 2 (younger)	26	2.2	26	2.1	28	2.3	40	1.3
Time 1 (older)	20	1.5	19	1.4	21	1.6	33	0.9

The variance is a measure of the amount of variation of a character, in this case spine length, among different individuals. When the variance is high, it indicates wide variation in spine length. Consider first the spatial patterns in mean and variance indicated in the table. At both time 1 and time 2, localities 1, 2, and 3 all have very similar means and variances; the variance is relatively high. At locality 4, however, the mean is much higher but the variance is much lower.

That is, the individuals living at locality 4 have longer spines on the average than individuals at other localities, and they are all pretty much alike. A hypothesis that immediately suggests itself is that conditions at locality 4 are very rigorous and only those individuals with a particular buoyancy can survive. This might be for many reasons; less buoyant individuals might sink into the substrate and suffocate, whereas more buoyant ones might protrude from the bottom so as to be relatively easy prey for passing predators. At any rate, there appears to be high selection for a particular buoyancy at this locality. At localities 1 through 3, individuals with a wide range of buoyancies may survive. Of course, other interpretations are possible. Instead of there being genetic differences due to the operation of selection between locality 4 and the others, the morphological differences might reflect "multiple-choice variation"; both populations might have similar gene pools, but phenotypes growing on the less stable bottoms might develop more spines, and so on.

The difference between populations at time 1 and time 2 seems consistent with a general increase in spine length, and therefore buoyancy at all localities. Again, several interpretations can be suggested; a likely one is that the sediments are simply becoming less supportive and there is consequently a selective pressure for increasing buoyancy. One way to test this hypothesis would be to find an independent criterion to measure the original softness of the sediments, perhaps involving the texture of structure of the sediments, and to contrast this property with the spine data. An inverse correlation between mean spine length and sediment buoyancy would be consistent with the hypothesis. The change between times 1 and 2 certainly suggests that the differences among the populations are genetic rather than strictly phenotypic. Additional lines of evidence might be developed by investigating other characters that might be expected to evolve in concert with spine length for additional support—perhaps spine number or position on the skeleton. If the population is actually undergoing a significant functional change, it might be possible to reconstruct a complex of interrelated skeletal changes. Assembling a number of individual functions into a picture of a coadapted form-function complex is precisely the sort of process employed in interpreting productoids (Chapter 5), except that they were interpreted on the level of the individual, whereas this example operates on the population level, matching patterns of population characters with environmental patterns.

Some of the other methods of functional interpretation of individuals (Chapter 5) can be employed to estimate population variability that may be functional. Morphological features that correlate empirically with environmental patterns may be quantified, for example, and their patterns of variation described. Patterns of biogeochemical variation may also be established. However, with the more empirical methods, the functional ranges of individuals are not clearly defined, so that their patterns cannot be established, but the functional range of populations or groups of populations can be interpreted.

Adaptive Strategies Fit Populations to Patterns of Environmental Variation

Population properties clearly vary from population to population, from place to place, and from time to time. Their variations are correlated with environmental variations, and appear to be adaptations to these varied conditions. The ways in which populations have adapted to patterns of environmental variation are called *adaptive strategies* (Levins, 1968). The term "strategy" is not intended to suggest that populations plan ahead, but merely that natural selection adapts them to certain eventualities, such as seasonal environmental fluctuations.

Adaptive strategies are strictly population properties as employed here; they do not occur as adaptations for the individual level. However, the term is occasionally applied to the individual level so that usage is not uniform. For example, phrases such as "the insectivorous habit as an adaptive strategy" are not uncommon. But feeding is of advantage to the individual, and occurs in populations simply because they are composed of individuals. Reproduction, by contrast, is only of advantage to the population (it is often a disadvantage to the individual, causing its demise directly or indirectly), although it occurs in individuals, since they compose populations. Because reproduction is very definitely a population property, any adaptive features of reproduction related to environmental patterns are adaptive strategies. The semantic argument involved here is rather academic, but the difference between individual and population adaptations is important and deserves emphasis.

Population properties must commonly be coadapted and certainly seem to harmonize in the examples we have reviewed. It is to be expected that natural selection would act upon these properties to promote population fitness, bringing the reproductive regime into such a relation to the environmental regime as to optimize replacements for mortality while permitting appropriate dispersal, maximizing food sources, and minimizing losses to predation. The responses to these and similar factors result in a characteristic state or regime of age structure, productivity, dispersion, geographic range, population size and density, and similar features for any given population in any given environmental regime.

Patterns of environmental variation occur in both spatial and temporal dimensions. Where temporal fluctuations in environmental parameters are great, the organisms must have broad tolerances, or have behavioral responses that permit them to escape the consequences of inclement periods, such as dormancy or retreat to a microhabitat where they can weather the unfavorable conditions. If the variable parameters are not density-dependent factors, then they will not necessarily affect population size so long as they do not exceed the tolerances of the individuals. If the fluctuations do exceed the tolerances of a portion of the population, then they will act as a source of mortality, and to maintain a given size the population must either evolve so as to broaden the tolerance of more of its numbers or make additional reproductive efforts.

If the variable parameters are resources that form density-dependent factors, they may regulate the population size and thus call forth a reproductive strategy, through selection, to make optimal use of the resource regimes. Food (or nutrient and solar energy resources for photosynthesizers) is a basic resource that can be density dependent and that appears to be responsible for many features of adaptive strategies. When the supply fluctuates upward or is at a high level, individuals with high fecundity leave more offspring, and thus a high *r* is favored. This process has been termed *r selection* (MacArthur and Wilson, 1967). When the trophic resources available to a population are stable but are at a low level, perhaps because many are consumed by competitor populations, there tends to be a selective emphasis on the quality rather than the quantity of offspring. Since relatively few individuals of the population can be supported in a given region, it seems reasonable that the individual able to function most efficiently (that is, to perform life functions most satisfactorily on the least expenditure of energy, and therefore to have relatively low energy requirements) will have a selective advantage. Any such selective focus on superior functional accomplishment implies a trend toward a continuing refinement of a few functions that can be performed very well—in other words, a trend toward specialization.

There are probably nearly as many ways of specializing as there are species, but many of them have similar effects on niche geometry and can be considered as characteristic strategies of specialization leading to particular niche styles. One strategy for specialization is to become very well adapted to a very narrow habitat range. Thus, a burrowing suspension feeder might adapt to life in a very restricted range of sediment textures, evolving the precise burrowing systems that permit most efficient penetration, the sorting devices that permit maximum efficiency at coping with a narrow range of grain sizes, physiological optima to deal with the ambient ranges of physical parameters associated with the burrows, and so on. This specialization would tend to lower its population size relative to an unspecialized burrower, for it could occur only in patches of appropriate sediment type, whereas the unspecialized form could burrow over a great range of substrate types and so could support a relatively large population.

Another common strategy for specialization is to feed on a narrow range of prey. To do this requires an ability to search, discover, and capture the prey particularly efficiently. This may call for highly specialized coordination of locomotory, sensory, and feeding functions, narrowly adapted to the special characteristics of a particular prey. For prey that are widely dispersed, predators must possess the ability to cross a variety of habitats that might intervene between their prey organisms, and thus must tolerate a rather broad habitat range. Because a strategy such as this involves a restriction of food sources, it can be advantageous only if the specialized feeding habits are so efficient that they result in an increased probability of successful feeding. Presumably, generalized feeding involves much competition for food, whereas specialized feeding usually involves little or no effective competition. Specialized feeding strategies would therefore seem to be most advantageous in situations where prey species have specialized defenses so that generalized predators cannot easily secure them, and where specialized predatory feeding is widespread so that competition is minimized.

Still another strategy that would limit population size is to develop direct reproductive restrictions, so that only a few young are produced. It has already been noted that this is most easily understood from the standpoint of selective advantage when the young develop and grow up in aggregates, so that individuals in small broods are well fed and have a greater survival potential than the poorly developed, underfed individuals in large broods. In low-density, low-trophic resource situations, such as the deep sea, it is easy to imagine this selection by inverse brood size in operation. Certainly, more and more specialized reproductive adaptations would be expected at lower and lower population densities in order to ensure a level of fertilization adequate to meet the natality levels required by the population system.

This suggests another major selective process that may limit population sizes by reducing reproductive potentials. Reproduction requires much energy; therefore, populations that produce large numbers of young require relatively large food supplies, and the young themselves must be fed. When food is limiting, a population with a high r will put a higher percentage of its energy supply into reproduction, and therefore may have less energy available for other functions, relative to a population with a low r. Other functions that could be enhanced with additional energy sources might include those associated with competition, feeding, and protection from predation, processes that affect survival greatly. Enhancement of these functions would lead to an increasingly efficient population and, usually, an increasingly specialized one, which in many situations would have an overall survival potential superior to a generalized, high-r population. Selection that leads to relatively specialized populations that ordinarily maintain themselves near the carrying capacity (K) of the environment has been termed *K selection* (MacArthur and Wilson, 1967). There seems to be a clear advantage in producing enough, but no more than enough young to ensure population replenishment and survival (see Lack, 1954).

Many different combinations of these strategems exist in nature, and different species approach them from different directions. Clearly, the precise adaptations of form and function will be different among different species even though they are pursuing similar strategies. For example, a species that is adapting to a restricted habitat in a rocky intertidal biotope will exhibit different modifications than one that is adapting to a restricted habitat on fine, soft-sediment bottoms. Furthermore, a species of one phylum adapting to a resticted, soft-sediment substrate will commonly exhibit sorts of modifications different from those of a species of a different phylum that is adapting to the same restricted substrate, since their ground plans and structures are distinct. Nevertheless, from a functional standpoint, the latter two species (of different phyla) are undergoing parallel evolution, and all these species are developing similar niche styles—all are narrowing the dimensions of their ecospaces that are associated with habitat tolerances and requirements.

The aspect of trophic resource supply that seems to exercise a major control of niche style is resource stability. This aspect can be further subdivided into two components: (1) the amplitude of resource fluctuations; and (2) their regularity or predictability. Imagine a perfectly stable regime wherein resource levels are constant night and day, summer and winter, year in and year out. Under such a regime, population densities could be adjusted to the resources through adjustment of their internal dynamics and could then exist in a steady state. Populations could be maintained even at extremely low resource levels (although they would probably be very dispersed and very specialized), since the resource supply would never fail. Now consider an environment wherein trophic resources fluctuate periodically, such as seasonally. During times of high resource levels, relatively high population densities can be maintained, but during times of low resource levels, only much lesser densities can be supported. A number of adaptive strategies are common in this situation. First, a population density may become adjusted to the higher resource levels, with feeding reduced or eliminated during the low-resource season. The individuals must undergo a general curtailment of activity during the "starvation" period, with a consequent seasonality in reproduction and other functions and probably a heavy mortailty during the inactive period; population losses are replaced in the richer resource season. A second strategy is to develop an annual (or shorter) life cycle based on the seasons, and to pass through the lean period as a zygotic resting stage. During the rich season, the population may grow rapidly to maturity and reproduce, dying out with the onset of lean conditions but leaving eggs behind. A third strategy is to develop a complex life cycle with a sequence of stages that are each adapted to the peculiarities of the appropriate season. Regardless of the strategy pursued, the shorter the rich season the greater the premium placed on rapid growth and reproduction during the period that food is available.

To the extent that the trophic resource fluctuations are unpredictable, it becomes less and less practicable for populations to establish rigid periodic activity rhythms, and there is an increasing premium on flexibility. The probability that low trophic resource levels will occasionally last inordinately long will place a high value on robustness and survival potential under deleterious conditions, and the probability that high resource levels will occasionally be inordinately short places value on very high rates of population increase (r) and rapid growth. Clearly, when reproductive periods can be very short, the chances of some individuals surviving deleterious environmental conditions are maximized by having the largest population possible.

In most cases in the sea, low resource levels are commonly associated with high environmental stability and high resource levels with instability. Perhaps the most important single reason is that fluctuations in solar energy supply vary with latitude, from essentially stable at the equator to seasonal elsewhere, so that nights and days are half a year long near the poles. During the long winter's night at very high latitudes there is essentially no primary productivity because there is no energy source for photosynthesizing autotrophs, although nutrient levels may be high. Primary productivity is therefore essentially restricted to half a year; in fact, in many arctic regions, sea ice develops during the winter and does not break up until well after the sun has returned, further restricting the period of productivity. Nutrients accumulate in the euphotic zone during the seasonal darkness, and when light returns productivity is extremely high. One estimate of average annual zooplanktpn production in a colder sea (Labrador Sea) is three times that in a subtropic region (Sargasso Sea) (Fish, 1954), despite the seasonal restriction of productivity in the subarctic and the relatively high turnover in subtropical waters. Thus, there is commonly an inverse correlation between stability (solar seasonality in this case) and resources, and we should expect to find a strong latitudinal variation in adaptive strategies and niche styles, correlating with this latitudinal variation in seasonality of productivity. This is certainly the case at present. Species living in high latitudes tend to have broader habitat tolerances, broader food tolerances, and larger populations than those in low latitudes. Assuming that in the past the earth's axis was tilted as it is today relative to the plane of its orbit, and that latitudinal variation in seasonality was therefore present, we should confidently expect that niche styles and associated biotic features varied with latitude then as now.

Another important reason that high environmental instability correlates with high resource levels is that nutrients are continually used up in the euphotic zone and must be regenerated in situ or resupplied. Much of the resupply is provided by upwelling, which is commonly the result of seasonal disturbance of the water column by winds. Thus, even in low latitudes, coastal upwelling regions may have high energy flows and high seasonality if the winds are seasonal. An additional source of nutrients in the nearshore waters is in runoff from lands, which also tends to be seasonal. Some seasonal estuaries and lagoons, for example, are characterized by seasonally high nutrient and detritus supplies, often derived from bounding marshes, and by marked seasonal fluctuations in salinity and temperature. Lagoonal organisms tend to be tolerant of variation in physical environmental parameters and to maintain large populations, at least seasonally.

The populations in stable environments tend to be fairly stable themselves, although some certainly display biological rhythmicities that may involve fluctuations in energy intake and storage, in growth and reproduction and in many other physiological rhythms, and in population size and structure. There are obviously many ways that a population may be stable or unstable, so that there may be a variety of definitions of stability. As a datum for a definition, we shall assert that a perfectly stable population is one in which population parameters do not vary. To achieve this, recruitment, aging, and mortality must coordinate perfectly, and the distribution of individual functional ranges must be statistically identical from moment to moment. Clearly, perfect stability of this sort is highly unlikely in the real world. In fluctuating environments there is a tendency for population parameters to fluctuate in concert to take optimum advantage of each environmental change. On the other hand, organisms contain many homeostatic mechanisms, and it is at least conceivable that populations could remain relatively stable in environments that were moderately unstable, if the individual organisms regulated their environmental responses so as to maintain rather constant population parameters.

Figure 6-11 depicts some relations between environmental fluctuations and

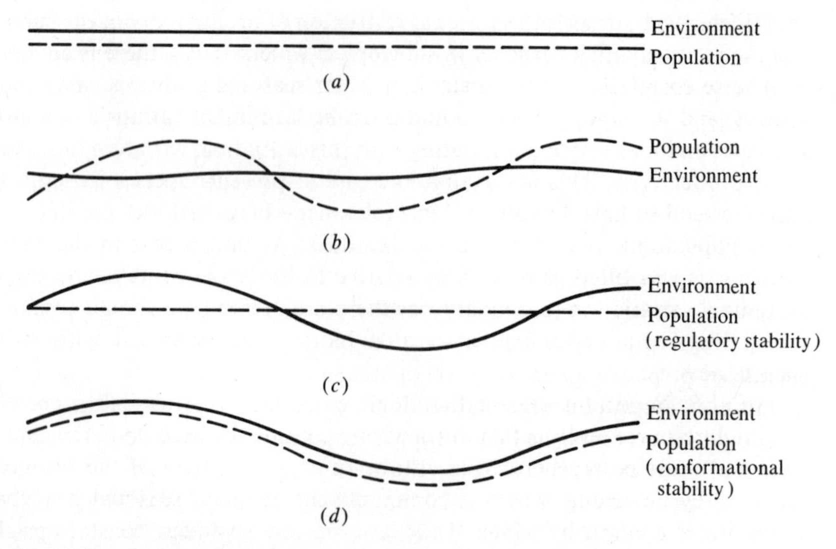

Fig. 6-11 Relations between stability of the environment and stability of populations. (*a*) Both stable. (*b*) Environment stable but populations fluctuating owing to some innate periodicity. (*c*) Environment fluctuating but populations remaining stable through homeostatic regulation. (*d*) Environment fluctuating and populations conforming thereto.

population stability. In case (*a*), both the environment and the population are stable. In case (*b*), the environment is stable but the population is fluctuating—for example, because of an innate periodic cycle of reproduction and numerous other functions that are integrated therewith. In case (*c*), the environment is fluctuating but the population remains stable, whereas in case (*d*), the environment and the population fluctuate together. Note that in case (*d*) we are dealing with conformers, and that each time the environment returns to a given state the population also returns to a given state. In a way this population is stable also, for with regard to some reference state of the environment the population does not change. Therefore we can distinguish between two types of population stability: *regulatory stability*, in which the population remains stable regardless of whether the environment fluctuates; and *conformational stability*, in which populations fluctuate in response to environmental fluctuations but retain a constant relation to each environmental state. Superimposed upon either of these types may be innate rhythms, which cause the population to fluctuate somewhat independently of the environment. Further complications arise because populations require time to achieve an equilibrium condition, so that in a rapidly fluctuating environment conformers may never "catch up"; therefore, their relations to a reference environmental state that may be approached rapidly from different directions at different times may be in disequilibrium. Furthermore, a population may adjust by regulating one function and conforming with another.

Spatial variations in the environment also pose adaptive problems that may be solved by different strategies. In a large region that is spatially homogeneous, a population that finds the habitat within its limits of tolerance may spread widely and have dispersion and density patterns that are regulated according to internal population optima and the temporal regime. But in a spatially heterogeneous environment with considerable habitat diversity, a population with a narrow habitat tolerance may be restricted to only a few sites where the appropriate habitat occurs. Problems of reproduction, dispersal, and recruitment may be met by special adaptations. A population with very broad habitat tolerances, on the other hand, might be able to range throughout strongly heterogeneous regions. This could be accomplished by equipping each individual with a very generalized or plastic phenotype, so that each could occupy a wide variety of habitats even though he might suffer higher mortalities than a specialist would; or it could be accomplished by having a large number of different but more specialized phenotypes available to the population and reproducing them so abundantly as to fill the environment. The genetic strategies that might underly these and other phenotype patterns have been treated theoretically by Levins (1968). The foregoing responses to a spatially heterogeneous environment are in an order that requires increasingly greater energy to be spent in reproduction. Which strategy is pursued by most populations may depend upon the resource regime; small populations require more stable regimes, whereas large populations require higher resource levels.

It Is Commonly Possible to Describe the Relative Difference in Niche Sizes and Shapes Between Populations

The environmental interactions of a population that we have conceptualized as the population ecospace are usually called the population niche. The prospective niche includes all the ecological functions possible at any time or for any specific time period; the realized niche includes the ecological functions that actually occurred during the specified time (Fig. 3-3). The limits of the prospective niche are limits of those individuals that are most tolerant for the respective environmental factors. The limits can be visualized as a sort of skin enclosing the population ecospace within the biospace lattice; the skin will be called the *ecospace boundary*. The prospective ecospace boundary places an absolute limit upon the environmental distribution of a population; obviously, a population may only live in those parts of the world where conditions included within its prospective ecospace are realized.

The ecospaces of two different populations have somewhat different shapes and sizes. If these populations are conspecific and occupy similar dimensions of biospace, but one is on balance rather more tolerant of some factor—such as temperature or salinity—than the other, then the hypervolume displaced by the prospective ecospace of the more tolerant population will be the larger, and it can be said to have the larger prospective niche; it is the less specialized of the two. If two populations are not closely related taxonomically (or, in many cases, even if they are), they may occupy rather different environmental dimensions or have considerably different ranges of tolerance along different dimensions. Their respective ecospaces have different "shapes" within the biospace lattice. It can be difficult to decide which population has the larger niche under such circumstances, for there is no clear way of equating a given temperature range with a given salinity range. An additional problem in judging niche sizes is that some populations have a greater number of environmental interactions than others, which is especially evident for biological interactions. Some species feed in more than one way, perhaps as both predators and scavengers, and even as deposit feeders. Other species utilize a single feeding mode and may even prey upon only a single species. Does an organism that is involved in a large number of ecological interactions have a larger niche than one involved in fewer interactions but having a greater range for those few?

To standardize the concept of niche size, we shall employ the following conventions. All biospace parameters are considered as calibrated between the minimum and maximum values that could ever be tolerated or utilized by organisms. A given proportion of the total range can be assumed to represent an equivalent distance along different dimensions. The niche size of a population is computed by adding all the proportions of all the dimensions that are occupied; the higher the total, the larger the niche.

Actually, we do not know the ranges of tolerance of even a single population for all environmental parameters, and for the overwhelming majority of living marine invertebrate populations we do not know any of the limiting values at all, insofar as possessing definite experimental evidence is concerned. Therefore, we cannot precisely determine their niche sizes. Nevertheless, it is possible to compare the occurrences of populations with the distribution of environmental parameters and with each other, and to determine for many parameters which populations have the broader limits. With care, the relative sizes of different realized population niches may then be estimated. Clues as to the sizes of prospective niches of populations may be gleaned from observation of the ecological and geographic distribution of the species to which they belong. If the species lives in a wide variety of habitats associated with different life assemblages over a broad range of climatic conditions, it certainly indicates a large prospective species niche and suggests that the prospective niches of various populations of the species may be large as well. The same reasoning may be applied to fossil species.

However, populations of species with large niches do not necessarily have large niches themselves; this will depend upon the mode of niche structure (Fig. 3-4), the strategy of the population system based in part upon the adaptive breadth required by the environment.

Populations Are Polythetic, and They Belong to Species That Are Polythetic

At certain times, local populations may have some characteristic functions setting them apart from other conspecific local populations in other regions. Commonly, these functional differences are correlated with morphological differences. Usually the functions are not restricted only to members of a given population, although this is occasionally the case. However, even when a function is universal within and restricted to a local population at a given time, we can be sure that such has not always been the case. It is essentially impossible for all individuals of a population to possess unique characters that are necessary and sufficient for membership in the population throughout its existence.

Consider the evolution of a function (Fig. 6-12); it appears in one or several individuals as the result of mutation, immigration, or some novel gene combination. It may then spread if it is of positive selective value. If a given population develops a function that becomes universal [as between (*b*) and (*b*′), Fig. 6-12], there are nevertheless times during the spread of this function when members of the population do not possess it, or when members of other populations also possess it. For example, in Fig. 6-12, if population (*b*′) is defined as becoming separated from (*b*) at time 5, then no members of (*b*) possess the new character, but some members of (*b*′) do *not* possess it. If the lineages are defined as separate at time 7, then all members of (*b*′) possess it, but so do some members of (*b*). Thus, the function is not exclusive to either (*b*) or (*b*′) and cannot be employed as a definitive character in a monothetic classification. However, it may be employed as part of the definitive set of (*b*′) in a polythetic classification. The only way that a population function can be truly monothetic is if it appears in a single founder individual that reproduces, perpetuating the function in all offspring to found a unique population that never receives immigrants lacking the function. This situation can be found among clonal populations, but among

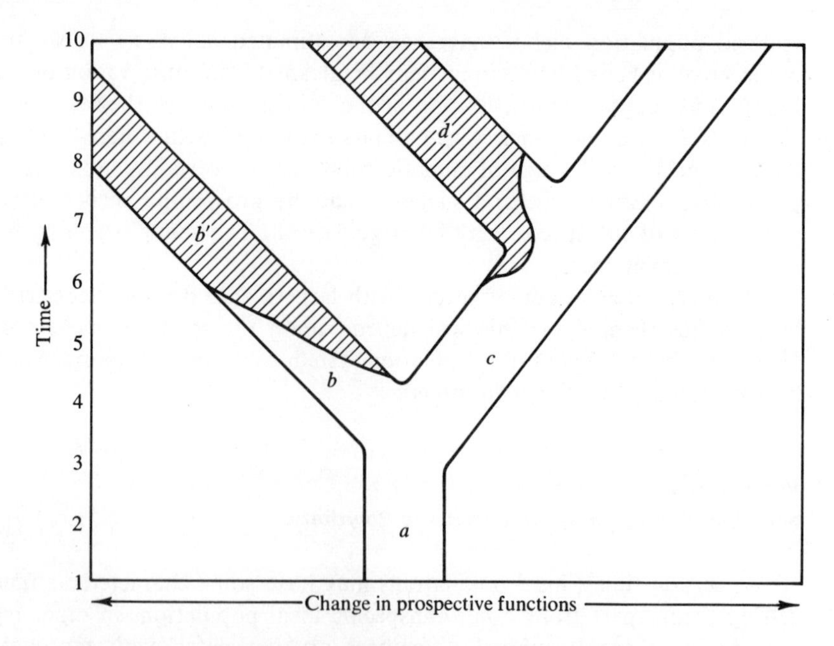

Fig. 6-12 A simple case of the origin and spread of a new function within a population, illustrating the polythetic nature of populations and of species. A species lineage *a* is depicted, which fragments into two local populations, *b* and *c*. Population *b* evolves into *b'* through the gradual introduction of a new character (and function) between times 4 and 6. Population *c* gives rise to population *d* owing to the appearance of a new character (and function) between times 6 and 9.

Mendelian populations it requires a set of circumstances that could conceivably occur but would be very rare. It probably occurs occasionally in plants, owing to the appearance of polyploids that are infertile with their parents. However, each population lineage of a species is usually distinguished by possessing different proportions and associations of functions that are found within other population lineages of the species at times, although at any given time a local population may indeed possess unique functions that occur in virtually every individual.

Like populations, species are polythetic units, and the principle illustrated in Fig. 6-12 can be applied to them as well. Entire species are actually more likely than populations to have unique functions at a given time, since the immigration of functions is not common between species involving interspecies hybridization. The property of possessing unique qualities at some particular time is found among all the units in the ecological hierarchy and lends a monothetic aspect to ecological units. This sort of *temporal monotheticism* has sometimes led to misunderstandings of the true nature of these units as being evolutionary as well as ecological.

Use of an Ecological Hierarchy Helps to Sort Out and to Avoid Some of the Difficulties in Functional Interpretation

The basic features of the processes of the evolution of populations both as phyletic units and as units in the ecological hierarchy were discussed in Chapters 2 and 3. It is useful to look on populations as evolving ecologically by means of change in the frequency of gene combinations. Three major aspects of the frequency shifts are of major interest: proportional changes brought about through increasing or decreasing the relative frequency with which given combinations are represented; qualitative changes brought about through the appearance of new gene combinations or the disappearance of old ones; and changes in diversity, in the number of combinations represented. The genic changes and recombinations accompanying ecological evolution commonly result in the appearance of new and different functional ranges among the individuals with novel genotypes; this, in turn, results in a shift in the niche of the population (and the species).

A notable factor of each step from gene to niche, however, is the lack of perfect correspondence between genetic features, the physical features or processes that they produce, and their associated functional roles. At each step in the development of ecological functions on the population level are either several alternate processes that can lead to similar results or complications that can produce results that mask, modify, or mimic the "expected" outcome of the step. Living systms have fantastic operational complexity. Interpretation of the causes underlying a population feature is clouded by the host of alternative possibilities that may have operated at each step in the production of that feature from the genic level. Feedback occurs between most levels, producing regulatory mechanisms that in the final analysis represent the machinery of adaptation—machinery that often seems as devious in proceeding from cause to effect as the wildest contraption conceived by Rube Goldberg. A certain percentage of the differences between populations in the record has not been created by selection at all, but by genetic drift or some other process. For that matter, population functions, however created, may be adaptive in themselves, but the differences in functions between populations need not be adaptive. Furthermore, evidence of only a certain percentage of the ecological functions of populations is even fossilizable under the best conditions, for some functions are simply not recorded in either skeletal morphology or patterns of fossil-paleoenvironmental correlation.

The employment of a carefully defined hierarchy of ecological levels makes possible the routine use of methodological ploys, which partially evade some of these very real difficulties in causal interpretation in evolutionary ecology. The chief ploy, from which others derive, is to consider that a process on any hierarchical level is explained if it is understood insofar as the interactions of the major subsystems of that level are concerned. In other words, we use the near-decomposable property of hierarchies to reduce the complexity of our problems. Therefore, if it is possible to explain a population pattern as arising from the adaptive advantages to individuals, we can be satisfied so long as we are concerned only with the population. If we are dealing with individuals, however, then we must be concerned with the bases of *their* adaptations, which we may explain in terms of characters. Of course, any *complete* explanation of an ecological phenomenon from an evolutionary viewpoint would require an understanding of all the levels concerned, which would ordinarily mean all the levels that exist.

Basically, we trace the evolution of ecological units by discovering the functional explanations for the units at any time, employing evidence such as we have considered thus far, and then by tracing the changes in the configuration of the units in a sequence of time horizons and interpreting them by comparable methods, or, alternately, by interpreting two temporally distinct configurations and then interpolating. Again, it is desirable to trace units on each hierarchical level independently and to fashion explanations for each level on the basis of its major subunits. If we do this for more than one level, we may then examine problems of feedback between levels.

CHAPTER SEVEN

COMMUNITY ECOLOGY AND EVOLUTION

Evolution cannot be understood except in the frame of ecosystems.
—Ramón Margalef, *Perspectives in Ecological Theory*, University of Chicago Press, Chicago, 1968.

We conclude that community characteristics are to be understood as cumulative effects of species evolution and multiple steady-state processes.
—R. H. Whittaker and G. M. Woodwell, in *Ecosystem Structure and Function*, Oregon State University Press, Corvalis, 1972.

In the ecological hierarchy, communities represent the next higher level of organization above the population, and include populations as their first-order subsystems. There have been a wide variety of definitions of the ecological community. MacFadyen (1963) has reviewed some of the more notable and historically important definitions and has summarized the main concepts involved in them. The concepts range from that of a more or less accidental association of organisms which happen to share certain tolerances to that of a highly integrated, self-regulating entity. We shall examine these concepts one by one in the light of the properties they each predict for communities. In order of the increasing amount of organization required, the main concepts are as follows:

1. A living association of populations of species.
2. A recurrent association (that is, recurrent in space or time) of populations of particular species.

3. A recurrent association of populations, internally regulated to some extent so that it acquires a certain dynamic stability.

4. A recurrent, internally regulated association of populations containing integrative features that are more than simply a result of the collective properties of the populations.

5. A recurrent, internally regulated association of integrated populations that forms a sort of superorganism.

First, communities can certainly be living associations of populations; at least, such associations are found all around us on land and in the sea, and we can call them communities if we wish. However, such a general definition makes of the community concept a mere truism.

If the association is recurrent, in that the same association of species is represented by populations in different places and/or different times, then at least we have a descriptive unit with some general significance, which can be identified by its taxonomic character. If such a community is described at one locality, however, and then another nearby locality is investigated, even one at which conditions appear to be similar and where the taxa appear to be the same at first glance, upon close investigation we shall nearly always find that the association of taxa has actually changed. Usually, a few species are represented at the new locality that were absent from the old, and a few species at the old are not present at the new. Even different samples of what appears to be the same community from the same area will often prove to be somewhat different taxonomically. Indeed, simply enlarging the sample size at the same locality will usually introduce representatives of additional species into the faunal list. This is all partly a consequence of the polythetic nature of communities (as discussed in Chapter 3). Communities are not perfectly recurrent associations of any particular set of species, as they would be if they were monothetic units. Rather, communities must be defined in terms of the presence of a certain number of the species belonging to a definitive set. Thus, we may accept the concept of a community as a "recurrent association" with the proviso that the association is flexibly defined and need only include a certain percentage of the species in some definitive set of species in order to qualify.

The next community concept on the scale of increasing organization specifies that there are processes of internal regulation within communities. Evaluation of this possibility requires first an examination of interpopulation interactions and their relation to the flow of energy within community structures.

Trophic Structure Is a Fundamental Attribute of Communities

The populations that form communities, by any definition, tend to be closely associated spatially; therefore they commonly interact. As emphasized in Chapter 3, many of the more important interpopulation interactions involve feeding, and thus energy transfer from one population to another. Since all populations must have an energy source, usually food or sunlight, the flow of energy through a community is a basic attribute, and the pattern of energy flow must be a major key to the way in which communities are organized (or disorganized, if that is the case). The pattern of feeding interactions and of the resulting energy flow is called the trophic structure.

Basic elements in the trophic structure of most marine communities are represented in the energy-flow chart of Fig. 7-1. Energy derived from solar radiation is employed by autotrophs to synthesize the complicated organic constituents of protoplasm. The autotrophs are then eaten, either while alive by browsing or grazing or suspension-feeding herbivores, or after they are dead and have become part of the organic detritus. As detritus, they are utilized by selective detritus feeders or deposit feeders, and by such decomposers as bacteria. The herbivores and detritus-feeding organisms are themselves prey to predators while alive, and can be consumed by scavengers and detritus feeders when dead. Feeding upon these first-level predators and scavengers are second-level predators. In many communities, third-, fourth-, and even fifth-level predators are present.

Some fairly general characteristics of this structure should be noted. First, the structure has two main subdivisions—the consumer "chain" and the recuperator "chain" (Gere, 1957). Part of the energy in organisms not eaten by consumers is retrieved by recuperators—the scavengers, detritus feeders, and decomposers—who make the energy available again to be reutilized in the consumer chain. In this way the diversity of the community is greatly enhanced.

Second, recall that much of a population's energy is expended in metabolism; therefore, only a fraction of the production of all the populations that occupy a given trophic level is available to consumers or recuperators at the next higher level. The metabolized fraction commonly amounts to as much as 80 to 90 percent. Probably there is less loss at the lower trophic levels, because plants do less work and expend less energy on the average than do herbivores, and herbivores probably expend less energy on the average than first-level carnivores, and so on. At any rate, the energy available decreases by large steps at each level. The rate of energy storage resulting from the manufacture of biomass by all autotrophs is called *gross primary productivity*. Inasmuch as much of the energy involved in gross primary productivity is expended by metabolism, only a fraction of it is available to the next trophic level; this fraction is called the *net primary productivity*. At each subsequent trophic level, the rates of energy stored are termed *secondary productivity* or *assimilation*. The percentage of the energy assimilated at any trophic level that is assimilated at the next higher level indicates the ecological efficiency of the energy transfer between these two levels. In other words,

$$\text{Ecological efficiency (as \%)} = \frac{\text{energy assimilated at higher level}}{\text{energy assimilated at lower level}} \times 100$$

This is analogous to the ecological efficiency of an individual population.

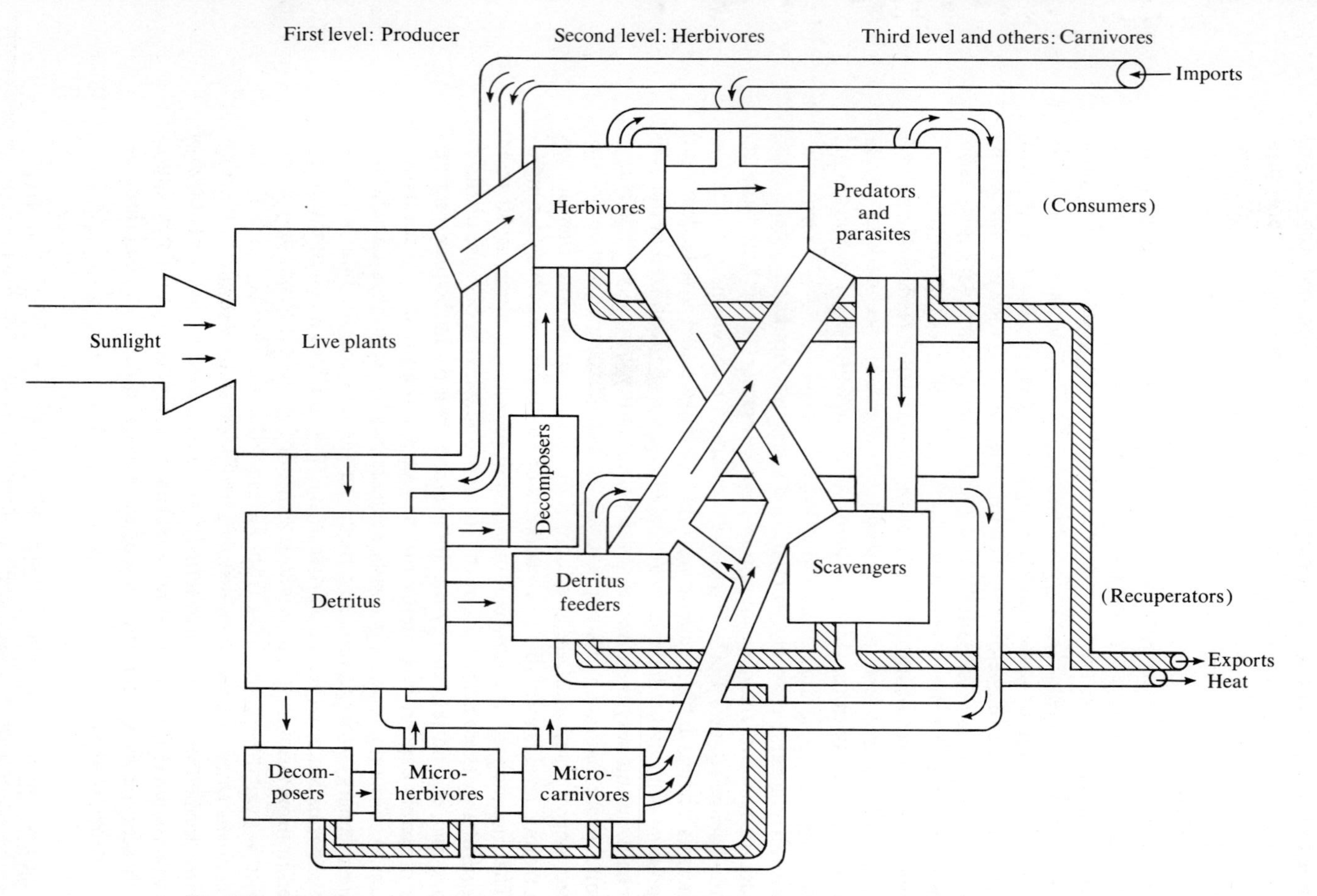

Fig. 7-1 Energy flow in ecosystems. Organisms that eat living plants or animals compose the consumer chain; those that eat dead plant or animal material compose the recuperator chain. There may be three or four levels of carnivores, each repeating the pattern of flow at the level depicted.

Ecological efficiencies in the consumer part of the trophic structure are often reported to range around 10 percent (Odum, 1971; Slobodkin, 1962). The efficiency of energy transfer between several trophic levels is sometimes called the *food-chain efficiency*. Thus, for five successive levels between which energy transfer occurs at an ecological efficiency of 10 percent, the food-chain efficiency is 0.01 percent. At this efficiency, a net primary production of 1 million tons would support a biomass of only 100 tons at the fifth trophic level.

Third, the proportion of energy that flows through the separate chains and through organisms playing any particular ecological role varies greatly from community to community, and in some communities there are no representatives of ecological roles that are dominant in others. For example, in the deep sea there are no live photosynthesizers and no browsing or grazing herbivores. Energy input is chiefly from an import flow, which takes the form of either microscopic detritus that slowly settles to the bottom or larger animal carcasses and remains. Detritus feeders and decomposers at the second trophic level and scavengers at the third recuperate this energy. In contrast, open-coast rocky-shore communities are based on plankton, minute benthic algae, and larger benthic algae and phanerogams; they contain abundant herbivores, both browsing-grazing types and suspension feeders. These, in turn, commonly support a wide diversity of predators and scavengers.

A final general character of trophic structures is that the number of levels must be considerably limited in any real community, since much of the energy available to any given level is unavailable to higher levels. Probably six levels is about the maximum number found in the sea (see Slobodkin, 1962).

Interactions Between Trophic Levels ("Vertical Interactions") Are Commonly Density Dependent

Since communities are associations of populations among which energy flow occurs, it is clear that understanding the interrelations among populations on succeeding trophic levels is basic to an understanding of community structure and function, and understanding the evolution of population interrelation is basic to understanding community evolution. Populations must survive despite or because of their interactions with other populations. In the consumer part of the trophic pyramid, prey populations must have a sufficiently large reproductive capacity to replace themselves despite losses to predators. (The general remarks in this section about predators and prey can apply equally well to herbivores and autotrophs.) They may manage this by developing a high reproductive potential (r_i), or by developing habits or structures that make them difficult for predators to find, or by developing defensive mechanisms that make them difficult to eat once found (size increases, skeletonization, and toxicity are obvious examples). Various combinations of these and other adaptations are often employed. On the other hand, predators develop ways of discovering and securing the prey that are essential to *their* population survival. Since predators and prey have both existed throughout Phanerozoic time and doubtless earlier, checks and balances obviously operate to permit coexistence.

Consider the effect of predators on population growth. The logistic equation

$$\frac{\Delta N}{\Delta t} = rN\frac{K - N}{K}$$

for a prey population N may be modified by the addition of a predator P, to beome

$$\frac{\Delta N}{\Delta t} = rN\left(\frac{K - N - \alpha P_1}{K}\right)$$

where P_1 represents the effects of predation by population P_1 on population N, and α is thus a *coefficient of predation* or, more generally, a *coefficient of consumption.*

If it is assumed that a predator population is limited only by food, a number of results are possible from the predator-prey interactions that depend partly upon the relative simplicity of the system. In the simplest experimental system, with one predator and one prey population, the predator may continue to expand its population until all the prey have been eaten; then the predator population starves. If the system contains some refuge, however, wherein at least a few of the prey may survive, then the predator may eat all but the refugee prey and starve to extinction, whereupon the remaining prey are able to repopulate the habitat. An additional system is generated when the predator population does not go to extinction, but when a few individual predators survive until the prey population again expands. In this event, the predatory population would also expand again until the prey became largely depleted. The predators would then also become largely depleted until the next re-expansion of prey; an oscillating system has developed. Still another possibility is that the prey remains abundant in the refuge and continuously repopulates the predator's habitat, so that a steady food supply is available. This situation has been described in nature for the barnacle *Balanus glandula* in the San Juan Island region, Washington (Connell, 1970). *Balanus glandula* settles thickly over the entire intertidal zone, where it is preyed upon principally by the starfish *Pisaster* and three species of the predatory gastropod *Thais*. *Balanus glandula* is more resistant to emergence and ordinarily lives higher in the intertidal zone along rocky shores than do these predators. Therefore, an adult population may survive in the highest intertidal, which serves as a refuge. The reproductive rate of these refuge dwellers is high enough not only to replace their own losses but also to supply young to low zones, young which serve chiefly as food for *Thais* (Fig. 7-2).

Further complications arise when there are alternative prey populations available to a predator. In this event, the predator may switch his prey when the density of one prey population falls to such low levels that finding prey becomes

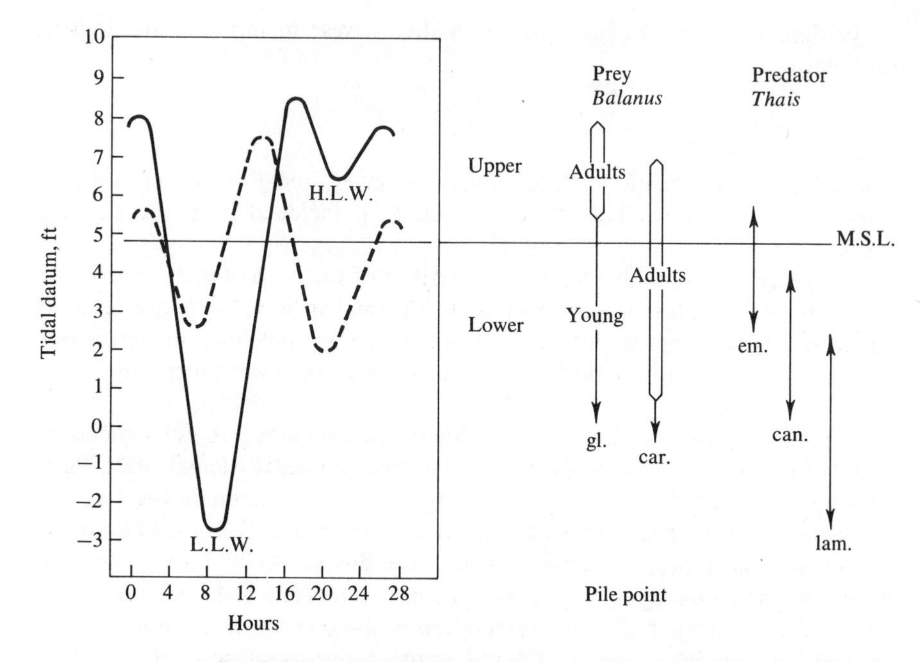

Fig. 7-2 Intertidal distributions in the San Juan Island region, Washington, of two species of *Balanus*, *B. glandula* (gl.) and *B. cariosus* (car.), and their predators of the genus *Thais*, including *T. emarginata* (em.), *T. canaliculata* (can.), and *T. lamellosa* (lam.). *Balanus glandula* settles over a wide vertical range but survives to breed only in the upper intertidal region; its young in lower zones feed the species of *Thais*. (After Connell, 1970.)

difficult. While the predator attacks the new population, the former prey population may recover; when the densities of these two prey populations have reversed, a switch may be made back to the first prey (Elton, 1927). When predation acts as a density-dependent factor, switching of this sort can act to stabilize the population size of the prey under some conditions (Murdoch, 1969). Predators sometimes become "trained" to a given prey population because they come across a dense patch of this prey and thus make several or many consecutive meals of one population. Thereafter, they may seek out other patches of individuals of that population until it is depleted, when "retraining" may occur (Murdoch, 1969). These effects must vary greatly among different prey and predator combinations, depending upon the strength of prey preference, the patchiness of the environment, and on a number of other parameters.

Finally, in many natural situations, there are several predator populations on each prey and several prey to each predator, simultaneously. The amount of each prey population consumed by any given predator sometimes depends simply upon the proportion of prey populations present, but, in other cases, a predator has marked feeding preferences and therefore exploits prey populations disproportionately to their relative abundances.

Predators commonly have predators themselves; in terms of the logistic formulation,

$$\frac{\Delta P_1}{\Delta t} = rP_1\left(\frac{K - P_1 - \beta P_2}{K}\right)$$

where P is a predator on N, P_2 is a predator on P_1, and β is the coefficient of predation of P_2 on P_1. The effect of N on P_1 is included in K, the carrying capacity.

Body size is commonly an important aspect of success and even of direction in predation; predatory organisms tend to eat smaller prey. There are certainly a number of exceptions to this, many of which relate to predators on herbivores. For predators that attack predators, there seem to be fewer exceptions to the usual size relations.

Parasitism may be defined as a trophic interaction in which the higher-level population takes its meals from the lower without immediate lethal effect. Parasites do commonly have a density-limiting effect; the P terms in the last two equations could represent parasites, in which case α and β are coefficients of parasitism. The action of parasites is not very distinct from that of predators insofar as the trophodynamics of ecosystems are concerned. In both cases, the flow of energy through the prey population is speeded up, to support either a predator population, usually of larger more active organisms, or a parasite population, usually of smaller less active organisms. The debilitating effect of many parasites raises the mortality rate of prey populations and requires a higher natality level to maintain a population. The ecological and evolutionary significance of parasites is obviously enormous. We must treat them in rather a theoretical way in paleontology, however, for we lack empirical evidence of their effects except in isolated cases.

Herbivores, predators, and parasites are all consumer organisms that eat live prey. An important fraction of trophic energy in ecosystems, however, is passed through one or more levels of the recuperators—detritus feeders, scavengers, and decomposers. These organisms have a markedly different significance from that of the consumers. Although they rely upon a supply of dead plants and animals as food, they do not themselves contribute to mortality at lower trophic levels in any direct way; thus, there can be no "coefficient of recuperation." Clearly, proportionately more energy will flow through the recuperator links in communities that have higher mortalities arising from density-independent factors (such as climate) and from competition (see following discussion).

Competitive Interactions Within Trophic Levels ("Horizontal Interactions") Are Commonly Density Dependent Also

In relationships between trophic levels, the patterns of energy transfer and flow are of chief interest; but when we examine a single trophic level, interest centers on how the trophic resources are divided among the various populations. Probably the most significant interaction between populations on the same trophic level is *competition*. Competition can be defined as the demand by two or more individuals or populations for a common resource or requirement that is actually or potentially limiting (modified after Clements and Shelford, 1939; see also Miller, 1967).

Dramatic events are occasionally ascribed to competition, from the exclusion of populations from certain communities to the extinction of entire species. However, it is extremely difficult to demonstrate that any particular population or species has become extinct due to competition. Probably the final stage of mortality in a dying population commonly has proximate causes, such as disease, that are not the fundamental cause of destruction of the viable age structure which has led to extinction. Competition may sometimes be a fundamental cause, setting the stage for extinction, as may also be such factors as predation, parasitism, or the direct effects of climatic change.

Clearly, for those who believe that density-dependent factors play little role in ecology, competition is an unimportant process and must play little role in evolution. Indeed, it is conceivable that even if density dependence is common and significant, competition itself may not be important. However, there are good reasons—theoretical, experimental, and observational—for believing that competition is an important density-dependent factor.

A theoretical model of competition in resource utilization, based on the logistic expression, was developed by Volterra, Lotka, and Gause. Recall again the simple logistic expression for population N,

$$\frac{\Delta N}{\Delta t} = rN\left(\frac{K - N}{K}\right)$$

Now, if a resource that is a component of K, the carrying capacity, is being partitioned between population N_1 of one species and population N_2 of another species, so that growth of N_1 is inhibited owing to the utilization of part of the common resource by N_2, then the growth of N_2 may by represented by

$$\frac{\Delta N_1}{\Delta t} = r_1 N_1\left(\frac{K_1 - N_1 - \gamma N_2}{K_1}\right)$$

where γ is the coefficient expressing the inhibiting effect of N_2. If N_1 is similarly affected by N_2, the similar equation

$$\frac{\Delta N_2}{\Delta t} = r_2 N_2\left(\frac{K_2 - N_2 - \delta N_1}{K_2}\right)$$

represents the growth of N_2 when δ is the coefficient expressing the inhibiting effect of N_1. In these equations, γ and δ are *coefficients of competition*. We can deduce from these expressions that the population which is most inhibited by the competition will be completely eliminated in most cases, since the resource supply will be increasingly appropriated by the other population. However, there are some cases, in particular if $\gamma < K_1/K_2$ and $\delta < K_2/K_1$, when both species may survive indefinitely.

These expressions have been tested experimentally, first by Gause (1935), who used protozoans as experimental organisms, and later by a number of investigators who mostly used insects, especially by T. Park and his colleagues. There seem to have been no experiments of this sort using marine organisms. In general, the experiments demonstrate that when two populations compete for the same food, one of them is usually eliminated as predicted. However, it is not always possible to predict which one will lose; furthermore, changing the conditions of the experiment commonly changes the outcome. At least in part this seems due to the fact that K is divisible into a wide variety of components, and presumably the environmental alterations changed K for one or both of the populations. The outcome may also be affected significantly when the environmental heterogeneity is altered, as when it is increased in a way that can be exploited by one or both of the populations. For example, Crombie (1945) experimented with two species of beetle; when populations of both were placed in a rather homogeneous flour medium, one was always eliminated, whereas when small glass tubes were mixed into the flour, both populations survived well beyond the time that elimination of one had previously occurred. The explanation seems to be that one of the species is a superior predator on eggs and pupae, which is the chief cause of mortality of the other species. Pupae are able to survive inside the glass tubes, however, and thus the population could not be entirely eliminated, in this experiment.

Competition has been defined as an interaction between populations on the same trophic level, yet this case of "competition" involves predation, a mixed trophic-level interaction. To clarify such situations, competition has been divided into two elements, *interference* and *exploitation*. Interference is applied to interactions that limit the access of one population to a resource it shares with another with which it is interfering. Thus, because there is a common resource, the predation involved between Crombie's beetles is also interference, an element in competition. Some predators compete with some of their prey, when there is a common resource, and, in some cases, parasites utilize the trophic resources of their prey rather than the body of the prey and thus could be considered competitors. Clearly, there is some gradation between competition and predation and parasitism, but it will not confuse our aims and we shall not explore it further. Exploitation, the other element of competition, is simply the utilization of a resource to which access is available. A population that cannot utilize a resource rapidly enough to maintain its size will eventually be eliminated. Both these elements are surely present in most competitions in nature (see Miller, 1967).

It has been inferred from experiments, such as the one with the beetles and glass tubes, that by increasing the environmental heterogeneity within habitats, the carrying capacity of different species populations is altered so that the condition of $\gamma < K_1/K_2$, $\delta < K_2/K_1$ can be attained for numerous pairs of competitors, all of which can then survive indefinitely, even though they are in competition.

In summary, theoretical predictions and the experimental work they prompted have both suggested that when environments are homogeneous, only one of two or more competing populations can ordinarily survive. Another way of presenting this principle is to say that when two (or more) species have identical resource requirements and similar ecological needs, only one can support a population in a given habitat. Therefore, even in highly heterogeneous environments, two species with very similar ecological requirements will be unable to coexist. This principle has been stated in a variety of ways and is usually termed the "principle of competitive exclusion," or "Gause's principle."

We shall employ a version of this principle, stated in the following terms: populations of different species in a given environment which have equal ecological responses to those components of their carrying capacities that are realized in that environment cannot coexist in equilibrium. To the extent that this is true, it means that all the populations of different species that live in association (or, for practical purposes, in the same community stand) must have different ecological requirements insofar as density-dependent factors are concerned, if they are in equilibrium. It also means that populations can be very different ecologically and still be unable to exist in association, because they happen to be ecologically identical for the factors relating to their densities.

It remains to assess the importance of competition from observations in nature. Only a few studies have been made on competition in marine organisms, although a large number of observations on changes in habitat and distribution patterns of marine organisms can be plausibly ascribed to competition. A well-studied example of marine competition is by Connell (1961), who demonstrated partial exclusion of one species of barnacle by another. *Balanus balanoides* and *Chthamalus stellatus* are intertidal barnacles whose geographic ranges overlap from the Bay of Biscay through the North Sea. They were studied in Scotland, where adult *B. balanoides* is excluded from the higher intertidal region by dessication, predation, and probably competition. At lower intertidal levels, however, it is successful. *Chthamalus stellatus* is capable of survival at all intertidal levels, except that at low levels it is crowded out by *B. balanoides*, which settles more densely and grows more rapidly and which can physically displace *C. stellatus* by wedging out or crushing it during growth. *Chthamalus stellatus* is therefore restricted to the high intertidal beyond the vertical range of *B. balanoides*, where it is successful. This is clearly an example of competition between populations of different species; in this case, the common resource is living space.

A growing number of cases are recorded involving the competitive exclusion of one species from part of its range by another; each case differs in the details of the interaction on which the exclusion is based. Miller (1967) has reviewed several cases for terrestrial and fresh-water species. In these cases, the excluded species has wider tolerances and a wider functional range, and therefore a larger niche, than the excluding species, whereas the latter has some special quality that makes it a superior competitor within its more limited functional range. For the barnacles, it is assumed that *C. stellatus* occupies a relatively large ecospace, and that, in the absence of *B. balanoides*, it can range throughout the intertidal. The vigorous settling and growth of *B. balanoides* constitute a specialization that permits success against *C. stellatus* and excludes the latter from the portion of its ecospace that overlaps spatially with *B. balanoides*.

Another line of evidence of competitive exclusion comes from gross biotic studies wherein patterns of habitat occupation of some populations change as the composition of the biota changes, although the interactions that underlie the habitat changes are not known. For example, recall that in the Baltic Sea there is a strong salinity gradient from the south where marine waters occur to the northern end where fresh-water runoff is dominant (Chapter 4). The number of species in the Baltic communities falls off regularly as salinity decreases and the tolerance of more and more marine forms is exceeded. Some of the tolerant species that range into the biotically poor brackish-water regions expand their habitats there, so that, for example, the bivalve *Macoma baltica*, together with some associated species, is confined to shallow nearshore areas outside the Baltic but ranges to depths of over 100 m in the Baltic proper. This change also involves an expansion of habitat, for the shallow-water substrates inhabited by *M. baltica* are coarser than the muds of the deeper Baltic bottoms (Segerstråle, 1957). It is possible that *M. baltica* and the associated forms are excluded from the deeper marine waters outside the Baltic by more specialized populations that do not penetrate the brackish Baltic water. It is also possible that some or all of these species are excluded by predators rather than competitors. Interpretation of these patterns is therefore equivocal, and they cannot be explained as due solely to competition without further evidence. Nevertheless, competition is a possible cause, and has been shown in some well-studied terrestrial cases to underlie patterns of just this sort.

An additional approach to the study of competition involves the investigation of those factors that permit two species with very similar ecological interactions to coexist. That is, given two coexisting species that have practically identical resource requirements, we can examine the reasons why competitive exclusion does *not* occur. In some communities, swarms of congeneric species coexist, but few of them have been studied in the marine environment. One series of studies, however, has been made of the resource requirements of numerous species of the predatory gastroped genus *Conus* (Kohn, 1959, 1966, 1968). *Conus* possesses a reservoir of poison that can be injected into prey via the radular apparatus. Off southern California, there is only a single small species (*Conus californicus*), which has a plain shell and feeds upon quite a wide variety of marine animals; Kohn (1966) records 28 prey species belonging to six classes and four phyla, chiefly Mollusca and polychaete Annelida. In Hawaii, there are nearly 30 species of *Conus*, most of which have colorful and largish shells. Six or more of the species are sometimes found associated in the same general habitat, but their feeding habits are specialized, so that each species takes different prey. For example, eight of the more abundant Hawaiian cones eat worms, but each of them specializes on a different prey species or on a few species belonging to a single prey taxon higher than species (Kohn, 1966).

The genus *Acmaea*, an archaeogastropod limpet, provides another example of a genus with several species living in close association. The species of *Acmaea* are grazers, feeding primarily upon benthic diatoms and other small plants that form films and small clumps on hard substrates. Along the Californian coast are about 17 species of *Acmaea;* some occur in close association. So far as is known, these species feed upon similar sorts of algal films, and although it is probable that the food sources of some of these species are different, this is not yet demonstrated. It is certain that the species live in separate microhabitats. Some live on certain species of marine grasses, some on certain types of algae, some on rocks. When two or more species live in close proximity, as upon the same intertidal rock, they have modal differences in station; one may live higher and another lower, or one beneath and another on the seaward face of the rock. These species are utilizing different fractions of a more or less widespread food source that is available in different habitats. Thus, in a sense, each species has its own food supply. The spatial partitioning of a food supply, as by *Acmaea*, rather than by food type, as by some *Conus*, is probably very common among marine invertebrates. Indeed, some species of *Conus* have adopted this method of resource partitioning (Kohn, 1968). In habitats that are relatively heterogeneous and support relatively small, patchily distributed prey populations, species of *Conus* tend to be generalized feeders with habitat specializations. However, in habitats that are relatively monotonous and support dense prey populations, species of *Conus* tend to feed upon special prey but range throughout the habitat.

Some intertidal species of *Thais* that utilize the barnacle *Balanus glandula* as prey have different habitats within the intertidal zone, so that their food supplies are somewhat separate (Connell, 1970). *Thais emarginata* is adapted to higher intertidal levels than other sympatric species of *Thais;* it is smaller and thinner shelled than they, and presumably can therefore subsist on a small food supply, such as is available in upper intertidal levels, where the barnacles tend to be small. The larger and thicker-shelled species of *Thais* are restricted to lower zones, where they prey upon larger barnacles. *Thais emarginata* cannot survive in these lower zones because it is easy prey for predatory shore crabs, which live at those levels and which can easily chip its thin shell. The thicker shells of the larger *Thais* provide more adequate protection. Thus, various aspects of the adaptation of these various species, chiefly related to predators, enforce a mutual exclusion that eliminates or greatly reduces competition between *T. emarginata* and other species of the same genus.

A case of niche division involving only two species has been described for amphipod crustaceans by Croker (1967). *Neohaustorius schmitzi* and *Haustorius sp.* coexist in the upper intertidal zone of sand beaches in Georgia and North Carolina. They are morphologically similar and feed upon the same general range and proportion of food items, with one notable exception: *H. sp.* is larger and has proportionately longer maxillae (which function to filter food particles) and utilizes larger food items than does *N. schmitzi*, although the size of particles accepted by these species overlaps broadly. Thus, we conclude that the resources of these two amphipods are divided by food type—in this case, size—and this is analogous to the Hawaiian cones, except that the amphipod foods are not taxonomically distinct. The phenotypic basis of the resource division between the amphipods is body size; we have stressed that body size often serves as a basis for the separation of organisms by trophic level, and this example shows that it may serve to separate them on the same level.

Since competitive exclusion operates to restrict the realized niche of populations and to effect a partitioning of biospace resources between ecologically similar species, and, furthermore, since many ecologically similar species have prospective niches that effect a division of resources in an analogous manner, it is reasonable to suspect that competition may contribute significantly to the evolution of the divisions within prospective niches. The reasoning is that when populations of two ecologically similar species become sympatric, individuals of the different populations that are most similar will be competing most sharply, whereas individuals of the different populations that are least similar will suffer the least from the interspecific competition. If the increased competition results in mutual inhibition, fewer of the individuals engaged in sharp interspecific competition will survive, and, to the extent that the differences are genetic, there will be selection in favor of interspecifically dissimilar individuals. Thus, the population norms will tend to shift away from those regions of the ecospaces of these populations that are contested between them. As a result, a genetically fixed division of functions will develop.

This is an easy process to conceive, but, like most evolutionary processes, it is difficult to demonstrate and to evaluate. When two populations evolve so as to improve resource divisions, the evolution should at least occasionally involve morphological changes. In this event, the changes should be divergent, that is, should result in a greater interspecific difference, since they reflect an ecological divergence. The divergence of characters in this manner has been termed *character displacement* (Brown and Wilson, 1956). It need not be morphological. No studies of character displacement seem to have been published for marine organisms, most alleged cases being of birds. The displacement is defined by comparing allopatric populations of different species with sympatric populations of these same species; when the sympatric populations show divergence relative to the allopatric, character displacement is suggested. However, some species display no change whatever, and others actually display morphological convergence when sympatric (Mayr, 1963). This is not unexpected, since a common environment might evoke a common adaptive tendency among related species, and they would therefore converge—in fact, a classic case of convergence.

Therefore the role of competition in evolution remains in dispute. Some investigators express doubt that it has much importance, and believe that most evolution represents adptation to density-independent factors. Others believe that niche differentiation is at least maintained by competition as an element of stabilizing selection. Still others suspect that many niche dimensions are actually determined in large part through selection for the mitigation of competition. In this view, competition plays a major role in defining the evolutionary pathway of a population. The path would run in a direction that is chiefly determined by the nature of all the other associated populations, especially those that share limiting, density-dependent resources. Therefore, the coevolution of actually or potentially competing populations would be a major process in community evolution.

Regardless of how it has arisen and how it is maintained, niche differentiation clearly exists for the vast majority of species. In terms of the ecospace model, this means that each of nearly all associated populations possesses a unique volume of realized ecospace. We shall call this unique realized ecospace simply the *exclusive ecospace;* it represents the ecological functions that differentiate the niches of populations within communities. Use of this term does not imply that niche differences have necessarily arisen as a result of competitive exclusion or are necessarily maintained thereby.

Other Interpopulation Interactions May Contribute to the Regulation of Community Structure

In the previous sections, we have discussed predation, parasitism, and competition as they affect the flow of energy resources between trophic levels (vertically) and the division of trophic resources within a single level (horizontally). Other population interactions affect this flow and its division; they have been tabulated according to whether their effects favor (+), inhibit (—), or do not affect (0) population growth in one or both populations. In Table 7-1, the possible combinations of these effects are depicted; some of the classical ecologic terms for each type of interaction are included.

The first column includes interactions that have no effect on the growth of

Table 7-1 Summary of the Chief Interpopulation Interactions

One population neutral			*One population negative (the other not neutral)*		*One population positive (the other not neutral or negative)*
0	0	0	−	−	+
0	−	+	−	+	+
Neutralism	Amensalism	Commensalism	Competition	Parasitism Predation	Mutualism Protocooperation

at least one of the populations. Neutralism implies that although the populations are associated within the same community, they in no way affect each other's density. This may mean merely that they have very distinctive niches, being organisms of much different sizes or being separated by two or more trophic levels. In this case, the neutral relation carries no special evolutionary implication. On the other hand, neutrality may sometimes be the *result* of evolution. We have already reviewed the sort of niche division that can lead to resource partitioning and neutralism. Evolution toward neutralism is probably a common adaptive strategy to escape competition for limited resources. This aspect of neutralism will be most common between populations on the same trophic level, although evolution to escape interlevel interactions could lead in theory to "vertical" neutralism.

In amensalism, one population is inhibited; in commensalism, one population is favored, by (usually) obligatory interaction with another population, which is itself unaffected. Amensal relations are a sort of one-sided competition, in which one population holds the upper hand completely; if the successful lower intertidal barnacle *B. balanoides* discussed earlier does not suffer any population density reduction from the presence of *Chthamalus*, which it regularly excludes from its habitat, then this is actually a case of amensalism. Commensalism is chiefly the result of the evolution of a specialized adaptation in which one species exploits some characteristic of another but without detrimental effects. A probable fossil example is provided by the gastropod *Platyceras*, which used a crinoid calyx as a substrate and probably derived nourishment from crinoid excretions, presumably without affecting the crinoids.

The interrelations depicted in column 2 of Table 7-1 have already been discussed. In competition, both populations have negative coefficients in their logistic equations as a result of their interactions. Predation and parasitism involve a positive coefficient for the predator or the parasite (normally this positive effect is simply included in K, the carrying capacity) and a negative coefficient for the prey.

Finally, column 3 of Table 7-1 contains the cases of populations that interact by favoring each other's growth. This is sometimes called "symbiosis," and the organisms involved are "symbionts." However, the term symbiont is also commonly used to denote the favored population in commensal and even parasitic interactions. When a mutually favorable interaction is obligatory, it may be termed "mutualism," and when facultative, "protocooperation" (see Odum, 1971). A classic example of mutualism is displayed by the minute algal zooxanthellae and the various organisms (chiefly coelenterates, but also foraminifera, bivalves, and others) that harbor them within their tissues. The algae obtain CO_2 and nutrients from their hosts, and the hosts obtain such photosynthates as sugars from the algae (Muscatine and Cernichiari, 1969). Protocooperation is exhibited by hermit crabs and the many coelenterates, bryozoans, and other sessile organisms that are commonly attached to the crab's home shell; the crab gains protection and/or camouflage, and the sessile organisms gain a mobile homesite.

The classification of population interactions in Table 7-1 is exhaustive, and includes the classic forms of population interrelation. Nevertheless, the classic ecological terms and the concepts they traditionally embrace do not fairly represent the population interrelations within large portions of ecosystems, mainly because a major flow of energy occurs in the recuperator chain, from which important amounts of energy are fed back into the consumer side. What is the relation between a detritus feeder and the plant that "contributes" itself to the detritus? Is it commensalism? This is the proper box in the table; the plant is unaffected by the detritus feeders, which certainly benefit from the plant. But the plants are dead when utilized; are they even to be considered as a "population"—are trophic interactions on the recuperator chains "interpopulation interactions"? Certainly the utilization of a detritus feeder as prey by a predator qualifies as predation, so that accounting for the energy flow within a trophic web often involves alternating between classical interpopulation interactions such as predation and such recuperator functions as scavenging.

In addition to the interactions between specific populations indicated in Table 7-1, interactions between whole groups of populations may occur; these may have a profound effect on community organization. Most of these species-group interactions have to do with a conditioning of the environment by one group of species which affects another group, but which does not really involve a competition for common resources. Such an interaction is a form of amensalism. Rhoads and Young (1970) described a possible case of group amensalism—in a benthic marine invertebrate community associated with muddy bottoms in Buzzard's Bay, Massachusetts. These muddy bottoms support a variety of infaunal deposit feeders, chiefly protobranch and tellinid bivalves, but are poor in epifaunal organisms and suspension feeders. The activities of the deposit feeders produce an unstable surface layer a few centimeters thick that is easily resuspended by even low-velocity currents. Rhoads and Young suggest that the muds, when resuspended by tidal currents, clog the filtering structures of suspension feeders, smother the larvae of any species not particularly well adapted to such soupy bottoms, and generally inhibit the occupation of the unstable habitat by epifaunal organisms. This would account for the peculiar trophic composition of the community. Because there is abundant organic material in the water column to support suspension feeders, it appears that their exclusion from the habitat is not due to competition for a common resource but to the activities of the deposit feeders in conditioning the bottom sediments; this interaction can be best described as group amensalism.

Internal Stability in Marine Communities Arises from the Possession of a Multiplicity of Alternate Limiting Interactions

We can now examine the notion that biotic communities may be internally regulated so as to achieve some sort of dynamic stability. In dealing with populations, we considered stability to be of two types: (1) conformational, describing a population that alters its size and structure with environmental changes but returns to similar configurations under similar environmental circumstances; and (2) regulatory, describing a population that tends to retain its size and structure despite environmental changes. Community stability may also be divided into these two types.

Internal stability regulation in communities is commonly attributed to the availability of a multiplicity of alternate limiting interactions to the average population. The effect of alternate limiting interactions is illustrated in Fig. 7-3. In Fig. 7-3(*a*), two prey (or autotrophic) populations are each utilized and

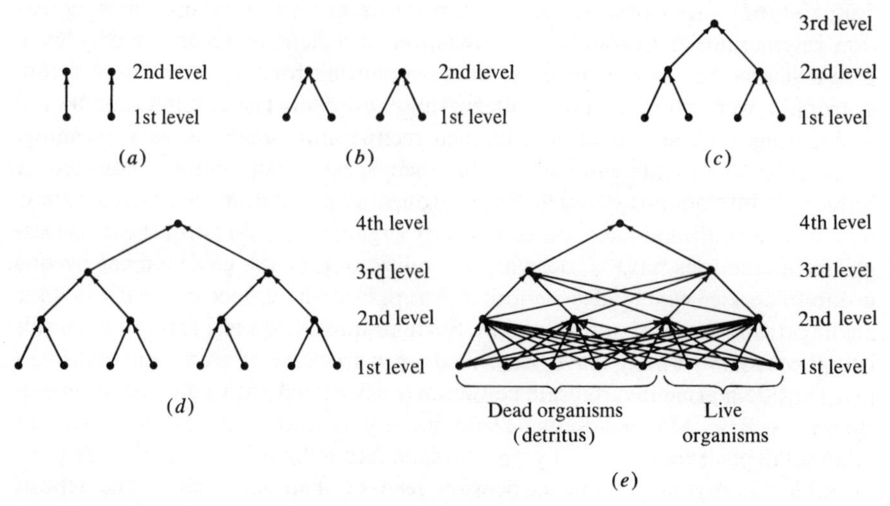

Fig. 7-3 Some hypothetical trophic structures. Direction of arrow indicates direction of energy flow.

partly limited by a separate predator (or herbivore) population. The two trophic links are independent, and if one prey population suffers either a great size reduction or increase, its special predator must or may alter its population size accordingly. Conversely, if the predator expands rapidly, the prey may be nearly or quite consumed, whereas if the predator population shrinks, the prey may expand rapidly. The system is very unstable. In Fig. 7-3(*b*), there are two prey per predator population; thus, if one prey fails, the other may be utilized. Furthermore, if the predator has a sudden population increase, the resulting increase in predation pressure is shared by the two prey, the extent of population fluctuations is damped down, and the probability of survival of the system is increased. Although each of the two predator-prey systems is somewhat more stable, they remain independent. In Fig. 7-3(*c*), a predator is added at a higher level, preying upon the other two predators. Now the fluctuations of each lower-level predator-prey system are no longer independent but can be regulated by the common top predator. Increase of a first-level prey leads to increased predation on it by its predator, which will itself increase, leading to increased predation by the top carnivore. Thus, the intermediate predators cannot undergo unchecked population increase. Increase or failure of the other populations is similarly controlled, except for some interactions involving the top predator.

In Fig. 7-3(*d*) and (*e*), progressively more stable structures are depicted. The principle involved can be summarized as follows: if a predator has numerous prey, failure of one will not doom the predator nor place undue pressure of exploitation upon any one of the remaining prey; and if a prey has numerous predators, failure of one will not engender a population explosion by the prey, since other predators would take shares of the surplus. In some cases, the predators may feed on each other when their normal prey is in short supply (Green, 1968). The principle may be extended to limiting factors other than predation. If one check on population growth or population maintenance should fail, but other alternate checks are available and come into play, the probability of stability is increased so that communities with the greatest number of alternate limiting interactions per population tend to be most stable. Indeed, MacArthur (1955) has shown that trophically linked communities containing more species are more stable than those with fewer, providing that the number of prey populations per predator population does not decline as diversity increases. This principle may be extended to limiting interactions in general, as long as they are compatible between communities. Of course in nature there are numerous complications in regulating interactions of these sorts, so that the simple models in Fig. 7-3, while displaying the principle, ignore the complexities and are vastly oversimplified.

Since such stability regulation is a mechanism that is developed internally through the evolution of multiprey and multipredator populations, which must each develop internal dynamics of its own to harmonize with the sorts of mortalities and stresses to which it is subjected, we may conclude that communities do, indeed, possess a dynamic internal stability regulation. From the foregoing discussion, it appears that the stability of a community may result from the adaptive strategies of its populations. Such strategies are a consequence of the environmental regime. Thus, community stability is adapted to a given regime but is presumably not evolved as an independent community-level entity.

Indeed, it is difficult to see how this would be possible by natural selection, unless communities were selected by some sort of survival superiority as communities; thus, there would be, in a sense, community competition (see Dunbar, 1960). Although the evidence available is far from conclusive, communities cannot at present be said to have qualities that transcend the qualities of their component populations. The natural selection of individuals and, occasionally, of populations appears to be a sufficient process to account for community organization. Therefore, it is unlikely that communities can be considered as "superorganisms" in any real sense, even though it is possible to find roughly analogous qualities which communities and organisms do share.

The Potential for the Accommodation of Species Diversity in Communities Varies Among Environments

Among the more important aspects of community organization and structure is *diversity*, the number of different sorts of components of which a community is composed. Diversity is defined and measured in a number of different ways. Commonly, the number of different species represented within a community is taken as a measure of its diversity. In other cases, the distribution of population sizes among the species is included in the diversity concept, so that a community is considered to be highly diverse if the populations of the represented species have a great variety of sizes. To distinguish between these concepts, we shall call the first, simply, *species diversity*, and the second *species equitability*. Certainly many other features of communities can be considered as contributing to diversity in the broadest sense. Diversity exists in the size of organisms, in reproductive strategies, in trophic types, in generation lengths, and in many other aspects, all of which can be important in the community structure and function.

Factors causing species diversity to be higher in some communities than in others are not certainly known, although there is a voluminous literature on this topic. Many of the ideas as to the regulation of diversity have been based upon the correlations of observed patterns of diversity variation with patterns of variation in such environmental factors as temperature. From such evidence, a model of diversity regulation can be pieced together that explains the bulk of the observations and that has some theoretical basis. However, it must be emphasized that all aspects of the model have not been verified; indeed, very few experimental data bear on this problem.

One almost certainly true aspect of the model is that special diversity in any community is partly a function of the variety of habitats present—a function, therefore, of the spatial heterogeneity of the biotope. This effect can be seen clearly when biotopes of markedly different degrees of heterogeneity are contrasted. A large-scale example is the continental shelves, which possess a relatively heterogeneous array of benthic biotopes. Patches of mud, sand, and rock bottoms are interspersed in mosaiclike fashion, and the intertidal-subtidal gradients in turbulence and light intensity provide additional variables. Furthermore, the interface between the water and the substrate forms a major ecological boundary. By contrast, the pelagic biotopes are highly homogeneous; horizontal heterogeneities are low and seem to be important only at the boundaries between water masses or currents, whereas vertical gradients in light, temperature, and associated parameters are important but do not compare with those on the shelf. It is expected, then, that more populations live on the shelves. In fact, far more do; nearly 90 percent of the marine invertebrate species live on the shelves, whereas probably less than 5 percent live in the pelagic zone. An example on a smaller scale is provided by the contrast of a rocky shore with a sandy beach. Rocky-shore habitats range from extremely turbulent to protected, from light to dark, and frequently, with a range of substrates, from rock to sand. As a result, the rocky-shore biotopes harbor biotas that are several times more diverse than the biotas of the more homogeneous sandy beaches. However, even between biotopes of similar spatial heterogeneity, species diversities vary considerably (for example, see Sanders, 1968), perhaps by an order of magnitude when, say, arctic and tropical environments are contrasted.

The biospace model provides a useful way of looking at this problem. Regions that are heterogeneous have larger biospace volumes than those that are more homogeneous, and therefore can accommodate more species. However, for any given biospace volume, greater or lesser numbers of species populations seem capable of being accommodated in different regions of biospace, raising the question as to why more population niches can be packed into one biospace than into another of equal size. It seems reasonable that the potential for diversity within a community is greatest when the populations have divided the available resources into the smallest possible allotment. If the Gaussian principle is generally correct, and each population normally requires some exclusive ecospace, some unique share of resources in order to coexist with other populations at equilibrium, then the resources should be *partitioned*—the resource subdivisions should be exclusive, at least in part. In highly diverse ecosystems, then, biospace should be narrowly subdivided among populations along the dimensions that represent potentially limiting resources. If this line of reasoning is valid, then to discover the regulators of diversity we must examine the factors that permit or inhibit the subdivision of such resources.

Some of the leading hypotheses of diversity regulation have grown out of attempts to explain the fact that biotas range from low diversity in high latitudes to high diversity in low latitudes in both polar hemispheres (well documented by Fischer, 1960). Many of the hypotheses have been summarized by Pianka (1966); it is worth considering several of the more widely supported ones. One hypothesis is that the diversity pattern reflects, at least in part, the length of time that various environments have been in existence or have been available to organisms. High latitudes, then, are considered to represent relatively young environments, so that evolution has not had time to develop a diversified fauna with appropriate adaptations. Yet the temperate environments on the margins of the tropics have probably been present throughout the Phanerozoic, but are not as diverse as the inner tropics, habitat for habitat. It would seem that, if the time hypothesis is to be employed to explain diversity patterns in high latitudes, it should hold for intermediate and low latitudes as well; there is no reason to expect a latitudinal gradient in age of environment, even if very high latitudes are newly refrigerated.

Other hypotheses seek to discover some latitudinal environmental gradient to match the diversity gradient. Temperature level has long been a favorite hypothesis of diversity control, since it is cold in high latitudes and warm in low latitudes, but the mechanism of temperature regulation has usually not been carefully considered. One suggestion has been that evolution proceeds more rapidly at higher temperatures, leading to more species; however, the underlying causes of this rapid evolution have sometime been referred to a higher and more stable productivity in warmer regions rather than to the temperature itself (Stehli, Douglas, and Newell, 1969). The discovery that biotopes in the deep sea, where it is certainly very cold, support species associations that are about as diverse as similar biotopes on the tropical shelves and more diverse than similar biotopes on the shelves in intermediate and high latitude (Hessler and Sanders, 1967), demonstrates that diversity within communities is not totally temperature dependent, nor can it always be dependent upon a high trophic energy level.

Temperature fluctuations have also been suggested as a control of diversity, for the tropic shelves have a lower temperature range than shelves in midlatitudes. The diverse deep-sea environment also has a stable temperature regime, which fits the pattern. However, the correlation of fluctuating temperatures with low diversities does not hold at all well in other regions. The most highly fluctuating marine thermal regimes are in the shallow waters of midlatitudes, especially along western ocean margins, as off southern Russia and North Korea in the Sea of Japan and along the northeastern United States and southeastern Canada; the same pattern exists in the southern hemisphere, but is not as extreme, since continentality is lower. At high latitudes, temperatures become much more stable than at these midlatitudes. for water temperature does not fall below freezing, but summer temperatures decline to poleward, thus narrowing the seasonal range. Thus, diversity should decline from the equator into midlatitudes and rise into high latitudes. It does nothing of the kind; it falls from equator to poles and fails to match the predicted pattern.

Another environmental factor that varies with latitude, and therefore correlates with the latitudinal diversity gradient, is seasonality, which is low equatorially and increases poleward, and is therefore inversely related to diversity. In the deep sea, seasonality is probably essentially absent, and there, again, diversity is high. One suggested explanation for this correlation is that environments which fluctuate widely (and somewhat unpredictably) are difficult to adapt to, only a relatively few species succeed, and therefore diversity is low. Stable environments, on the other hand, require only narrow tolerances and are relatively benign; therefore, many species can adapt and diversity is high.

A major difficulty with this explanation is that natural selection has only to invent an adaptive mode once; after that, diversification may proceed by cladogenesis to develop a rich (if taxonomically compact) flora or fauna. Furthermore, if a strategic adaptation can be made by one lineage, it can usually (although not always) be made by others as well; similar adaptive strategies are commonly followed by lineages with radically different body plans. A difficulty shared by all these explanations of diversity regulation involving temperature and seasonality is that they fail to identify factors which act to limit the resource consumption of populations, either through ecological controls or through the evolution of adaptations having this effect, and therefore which permit the partitioning of resources in environments where diversity is high and inhibit it where diversity is low. Such factors may be termed *diversity-dependent* factors. They operate to control the number of species in a community at a certain level, so that they limit or favor populations, depending upon how many different sorts are already present. In general, resources may probably be partitioned only so many times to support so many different species populations in any given environmental regime, because of the adaptive strategy requirements for both reproductive and vegetative success that have been discussed previously. Since these requirements involve population sizes and reproductive potentials, density-dependent factors become implicated in diversity regulation.

Environmental factors that act to exclude or favor populations without regard for whether species diversity is high or low may be termed *diversity-independent* factors. These factors usually cannot be partitioned, but are pervasive, such as temperature or salinity. For example, in environments wherein temperature fluctuates greatly, individuals of many species may be excluded because they are unable to cope physiologically with the wide temperature extremes. Thus, diversity in such environments may be low only because a relatively few populations are adapted to them. Similarly, salinity or oxygen fluctuations or abnormalities lead to the exclusion of many species, and therefore to low-diversity associations. Still, if appropriately tolerant species are available, there is no special reason associated with temperature, salinity, or oxygen levels or fluctuations that precludes high diversity within environments where they prevail. In theory, at least, if the entire planetary ocean were fluctuating markedly in temperature or some other similar parameter, then evolution could eventually produce a highly diverse biota to inhabit it, a biota entirely well adapted to the "normal" fluctuations. Of course, there is clearly a spectrum of environments that gradually become intolerable to all metazoa and, in fact, to nearly all living things, so that in essentially anaerobic conditions or in boiling water, higher organisms as we know them could never live, and evolution would have to take an entirely different tack in order to richly populate them. At any rate, diversity-independent controls clearly operate today, and when they cause diversity to drop it is because the diversity-independent conditions are unusual and few species are adapted to contend with them. Diversity-independent factors may be biological as well as physical; the immigration of a voracious general predator is an example. The quantitative effects of diversity-independent regulation are obvious from this discussion. Qualitative effects arise simply from the unequal distribution of adaptations for any given factor among different taxa. The members of some higher taxa are better adapted to tolerate certain unusual environments, and therefore are favored whenever these environments occur.

Diversity-dependent regulation has a fundamentally different significance from that of diversity-independent regulation; in the former case, evolutionary processes can do little to enrich the biota when it reaches the regulatory maximum, in theory. This point should be stressed: when diversity-independent factors change to cause a diversity decrease, evolutionary processes may act to return diversity to its former level, even though there is no further change or amelioration of conditions; but this is not so for diversity-dependent factors.

The most common resources that can be exhausted by populations within a community, and that must be shared out and can thus operate as diversity-dependent factors, are trophic and spatial resources—food and habitat—as discussed in Chapter 6. We have already seen that diversity is frequently higher in environments with the greater number of habitats. Perhaps the variations in diversity among environments with similar habitat properties are due to diversity regulation by trophic resource regimes. It is possible to form a model of diversity regulation based on this assumption that predicts the observed patterns of diversity very closely. The model is not yet verified in all particulars, however.

The model assumes, as suggested in Chapter 6, that adaptive strategies are commonly a function of population energetics, and that niche sizes may commonly relate to the stability of trophic resources. Populations in stable environments may persist at low densities and abundances, and tend to have small niches and to be stable. They are restricted to a narrow habitat range, freeing other habitats for occupation by other species, or to a narrow selection of prey, freeing other prey for utilization by other species, or to both. Since communities must have properties that arise from the species of which they are composed, communities in stable regions should be relatively stable themselves and may be highly diverse. Furthermore, the stabilization of energy flow through the community structure that results when a community is composed principally of stable populations permits the development of relatively numerous trophic levels, resulting in even greater species diversity [Fig. 7-4, box(*e*)].

In environments where trophic resources fluctuate, an entirely different community structure is found, according to this model. The average population there must maintain a large size, has a high *r*, requires a relatively large share of food to survive, and has a relatively broad habitat range or can utilize a wide range of food, or both. Diversity must remain low because of the high drain on resources by each population. At each higher trophic level, the fluctuation in available biomass (and therefore in trophic resources available to the next higher level) is progressively greater, owing to a lag effect in utilization of the prey populations of preceding levels. Therefore, relatively fewer trophic levels are developed in fluctuating communities, which lowers their diversity also relative to communities in stable environments [Fig. 7-4, box(*a*)]. Intermediate resource stabilities should lead to communities of species that have intermediate qualities, perhaps that are polymorphic for *r* and K adaptations (Roughgarden, 1971).

In the ocean, nonfluctuating environments are most commonly found in certain well-defined circumstances (Chapter 4). The most stable region of all is probably the deep sea. In shallow water, solar radiation is relatively constant throughout the year in low latitudes, and therefore photosynthesis can be relatively constant there, providing a stable energy base for the trophic pyramid.

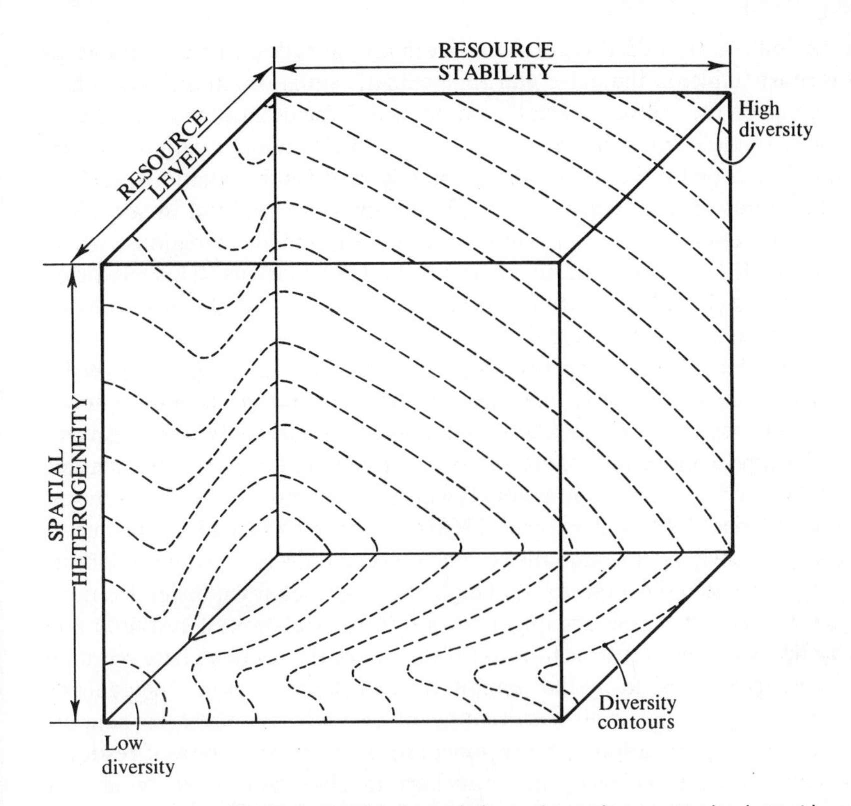

Fig. 7-4 Possible interrelations of trophic resource level, trophic resource stability, and spatial heterogeneity in controlling species diversity. In this model the lowest diversities are associated with the lowest spatial heterogeneities, resource stabilities, and spatial heterogeneities, while the highest diversities are associated with the highest spatial heterogeneities and resource stabilities but with intermediate resource levels.

The stable supply of energy from solar radiation need only be coupled with a stable supply of nutrients to support a stable regime of productivity. Stability of the nutrient supply in a given region is commonly a function of the relative degrees of continental versus maritime climates that prevail. Highly continental climates are those with much temporal heterogeneity (Sanders, 1969); there tend to be considerable differences between seasons, with seasonal winds and storms causing upwelling or mixing of water columns at times, whereas, during alternate seasons, there may be little wind and stable water columns. In this event, nutrient supplies may be quite high during upwelling and mixing but quite low and limiting during stable conditions. Continental climates can have the same sort of effect on trophic resource regimes that high latitudes do, that is, they can create seasonal fluctuations in primary productivity.

If the model of trophic resource regulation of diversity is correct, we should expect to find the most diverse communities in low latitudes, where solar energy input is most stable, in the more maritime climatic situations at any given latitude, such as on the shelves of small continents in large oceans, and, in general, in the most stable habitats in any given situation. Communities with low diversity should be found in high latitudes, in highly continental situations, such as along the shores of large continents bordering small oceans, and in such fluctuating environments as coastal lagoons, estuaries, and delta regions, where seasonal runoff causes significant environmental fluctuations. This predicted pattern agrees well with the pattern observed at present on a global scale (Fig. 8-8), lending support to the model.

Another aspect of the trophic resource regime that may affect species diversity is the level of resource supply. The effects of different resource levels are not at all clear and may vary under different circumstances. Certainly, high resource levels can support more individuals than low levels, but it is not predictable whether these more numerous individuals will belong to more or to fewer species. Connell and Orias (1964) and Tappan (1968) have assumed that adding resources to an ecosystem will lead eventually to higher species diversity, while Valentine (1971a) has suggested the opposite, that high resource levels favor lower diversity. A simplified version of the argument for a direct correlation of resource and diversity levels can be stated as follows: If the size of the resource base required for a species is established by the stability of the resource regime, then raising the resource level will permit partitioning of the new resources to support additional species populations. The argument for inverse correlation of resource and diversity levels rests partly upon analogy to observations on living ecosystems, and a simplified version can be stated as follows: Increasing resources do not ordinarily support additional species, but rather are used to increase the sizes of those populations capable of utilizing the added resources, that have the highest reproductive potentials. These enlarged populations overwhelm some of the more narrowly adapted ones, reducing diversity. Something of this sort happens when living communities are contaminated by high nutrient inputs. However, this may be a diversity-independent effect which natural selection can overcome in time.

Perhaps the long-run effects of resource levels on diversity depend upon the state of the ecosystems affected. If an ecosystem with a certain spatial heterogeneity contains much unoccupied biospace at a given resource level, then raising the level should eventually lead to enhanced diversity. However, if a given region of biospace is saturated with populations and resource levels are raised so as to create surpluses, it seems possible that the more opportunistic species would indeed increase their population sizes and create disturbing effects that would lead to lowered diversity at equilibrium. Such a relationship would go far toward explaining some otherwise obscure situations. Coral reefs, for example, have both high productivity (Sargent and Austin, 1949; Odum and Odum, 1955; Kohn and Helfrich, 1957) and high spatial heterogeneity, and are of *high* diversity relative to non-reef community complexes in otherwise similar environmental regimes. Upwelling waters, on the other hand, are of high productivity but low spatial heterogeneity and are of *lower* diversity than comparable non-upwelling waters (of lower productivity; Ryther, 1969).

Direct Ecological Interactions May Raise as Well as Lower Diversity

The limitations of resource requirements by populations, which can lead to higher diversity by freeing resources for other species, are presumably, by and large, evolutionary features that result from the adaptive strategies of populations. Competition for resources can be severe, and when environments are stable, a strategy of concentrating on a certain place to live or on capturing a certain prey, so as to establish a superior capability of utilizing specialized resources, would seem to be a reasonable outcome of natural selection. However, it is also possible that ecological processes, such as competition, predation, or diversity-independent disturbances, can limit the range of habitats or prey which a population can exploit within a community. The barnacles studied by Connell are an example of how a diversity increase can result from competition. A specialized barnacle excludes a generalized one from part of its habitat range; thus, the habitats become partitioned, and two species coexist within the habitat range of the more generalized.

An example of predation that raises diversity is recorded among some intertidal invertebrates from the outer margin of the subtropics, in the northern Gulf of California where the environment is far from stable (Fig. 7-5). Most of the carnivores have a number of different prey, and one very general feeder that is large—the starfish *Heliaster*—serves as top carnivore. Paine (1966), who recorded these relations, calls the trophically interrelated populations in Fig. 7-5 a *subweb.* A subweb is a group of populations within a community that is tied together by feeding relations, dominated by a top carnivore (or in some cases by more than one), and is somewhat independent of other subwebs found within the community. This *Heliaster*-dominated subweb demonstrates a simple fact; when species diversity is high, any trophic generalists present may feed upon large numbers of different species and thus create complex webs.

Paine has studied subwebs at other localities. One of these was experimentally manipulated by removal of the top carnivore, and diversity subsequently declined. Paine interpreted this as due to the exclusion of some populations by competitively superior ones with higher reproductive or at least higher recruitment potential in the community, which therefore took over space formerly occupied by the excluded populations. So long as the trophic dominant was present, the competitive superiority of the excluding populations was masked, for they were preyed upon by the dominant to the extent that they were prevented from reaching population sizes at which they exercised competitive exclusion. Thus, top carnivores may permit communities to contain more species popula-

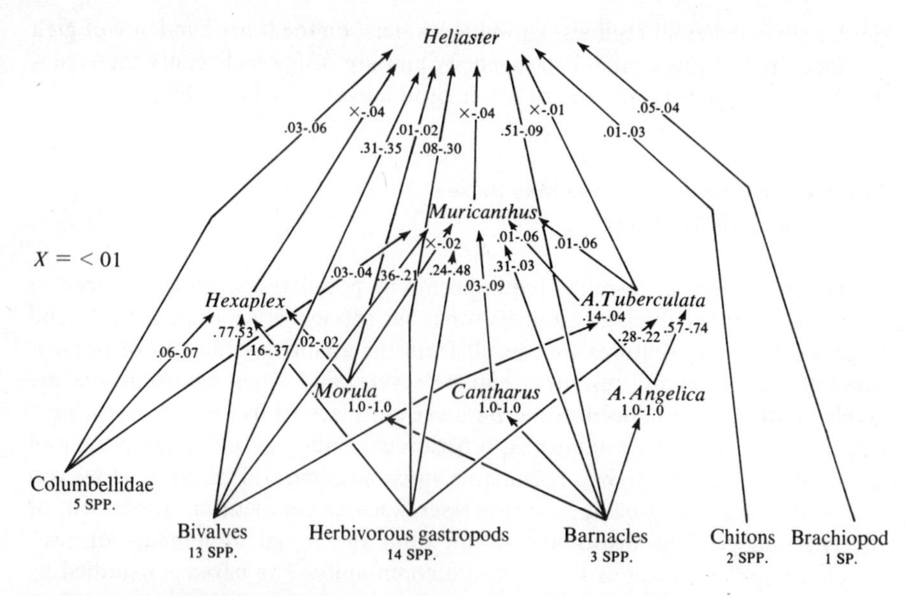

Fig. 7-5 Feeding web (or subweb) dominated by the starfish *Heliaster* in the northern Gulf of California. Numbers indicate calories. (After Paine, 1966; from *The American Naturalist,* the University of Chicago Press, copyright © 1966 by the University of Chicago.)

tions by permitting some resources, which would otherwise be utilized by only a few species, to be shared among several. Paine (1969) has called such top carnivores, on which a certain community state depends, *keystone species*.

A marine example of physical perturbations that permit diversity higher than a stable situation has been reported by Dayton (1971), who studied a rocky intertidal shore along the Washington coast. In this community, space is most limiting. Monopolization of space by the more successful competitors, especially *Mytilus californianus* and also *Balanus cariosus*, is prevented partly by heavy predation by the starfish *Pisaster*, which thus plays the role of a keystone species. In addition, the continuous damage to the mussel and barnacle beds by drift logs and by waves which attack the margins of gaps left by log damage, also helps significantly in preventing a takeover of intertidal space by these dominants. Instead, other species are permitted to inhabit this space, and thus diversity is enhanced.

Therefore, processes that reduce species diversity under certain circumstances—competition, predation, and environmental disturbances, all of which can exclude species from a community—can all act to raise species diversity under other circumstances. In effect, these ecological processes are causing species to forego some resources just as if they were too specialized to utilize them. In these communities, direct ecological interactions (as opposed to evolutionary adaptations) are causing the release of resources for utilization by additional species. Presumably, the resource regime will ultimately control the extent to which the ecological effects can operate, however, for if a population is reduced below the size required for reproductive replacement by these ecological effects, it will become locally extinct.

The Quality and Proportions of Feeding Types Within Ecosystems Vary with Environmental Factors

A significant aspect of ecological structure in communities is the relative diversity of feeding types among trophic levels. This is usually a reflection of three main sets of environmental features: (1) the physical nature of the association of habitats within a biotope; (2) the qualitative nature of the trophic resources; and (3) the trophic resource regime.

The physical nature of the biotope greatly affects the type of food available to a community, and therefore is tied in with the quality of trophic resources. For example, biotopes on rocky substrates have relatively few habitats for burrowers; those that do occur there are chiefly suspension-feeding rock-borers. Infaunal detritus feeders are therefore essentially absent. Detritus from living macroscopic plants is common, however, and is strewn over the rocky surfaces or collected in depressions, where it is available to epifaunal foragers. The live macroscopic plants also support browsing populations. Biotopes on soft bottoms sometimes lack larger living plants and larger detritus, but may receive abundant pelagic detritus; they then support relatively few browsers and epifaunal foragers, but ordinarily have large numbers of infaunal detritus feeders.

Table 7-2 shows the proportions of feeding types found in fossil gastropod associations collected from distinctive depositional environments in the Pleistocene of Los Angeles Basin. These proportions are representative of scores of similar assemblages from analogous habitats in California and along Pacific Baja California, Mexico. An estimate of the total proportions of gastropod feeding types present during the Pleistocene is also given, based upon two lines of evidence: the average for the present gastropod fauna of the same region, and the feeding-type proportions found among the entire recorded gastropod fauna of the Pleistocene of the region. Clearly, the available feeding types were not evenly or randomly distributed but varied in proportion from environment to environment. Browsers are proportionately more diverse in rocky-shore assemblages and less diverse on soft substrates. Scavengers and predators are proportionately least diverse on rocky shores, more diverse on coarse sediment substrates, and most diverse on fine sediment substrates. Suspension-feeding gastropods (a relatively rare adaptive type) are most diverse in shallow turbulent waters and least diverse in quiet and/or deep waters. Now, these fossil assemblages are known to contain intermixtures of species that have lived in different habitats and have been brought together after death (Valentine, 1961; Valentine and Mallory, 1965). Nevertheless, they display differences in the proportions of feeding types that clearly relate to original differences in the trophic structure of the principle life associations from which they have been assembled. Notice

Table 7-2 Gastropod Feeding Types in the Californian Pleistocene (in Percent)

Feeding types	*Exposed Rocky Shore: U. Pleistocene, Palos Verdes Hills*	*Exposed Rocky Shore: M. (?) Pleistocene, Sta. Barbara Islands*	*Quiet shallow bays, silts: U. Pleistocene, Potrero Canyon, Sta. Monica Mts.*	*Offshore sands: L. Pleistocene, San Pedro Sand, San Pedro Bluffs*	*Quiet deeper water, silts: L. Pleistocene, Timms Pt. Silt, San Pedro*	*Entire fauna: Estimated total Californian Pleistocene*
Browsers and grazers	21–25	15	5	9	10	7
Detritus feeders	35	34	18	20	18	23
Suspension feeders	7	13	6	9	0	4
Scavengers	0	3	12	11	25	5
Parasites	7	5	20	17	0	21
Predators	26–30	30	38	34	47	40
Total herbivores	63–67	62	29	38	28	34
Total carnivores	33–37	38	71	62	72	66

that the herbivore/carnivore ratio alone is usually sufficient to distinguish gastropod associations from fine, coarse, and rocky substrates (Table 7-2).

The physical nature of the biotope and the quality of trophic resources are thus obviously interrelated and affect the trophic structure of the community. The third main environmental feature that probably contributes to feeding-type proportionality is the resource regime. In environments with great resource instability, there are occasionally long periods when primary production is low or absent, and therefore assimilation by the populations that support higher trophic levels is low or absent as well. The organisms must then adapt to occasional long periods of famine, or seek to reduce the fluctuation by securing sources of food that are more stable than the living populations. Such sources are, of course, chiefly detrital.

For example, a supply of phytoplankton may be present in the water column at high latitudes for only a few months or, in regions of sea ice, for even less time. Suspension feeders can therefore acquire energy only during a short part of the year. However, unutilized plankters fall to the sea floor, and some are incorporated into the sediments, together with organic debris from other sources. It is now available to detritus feeders or decomposers for a much longer period of time. The fluctuations in photosynthesis thus severely affect organisms that exploit living phytoplankters, but have less effect on organisms that exploit sedimentary detritus or detrital decomposers. For predators, the fluctuations in prey are great, but scavenging is an excellent strategy in unstable regions where mortality is heavy. The proportions of feeding types in the recuperator portions of trophic pyramids are usually very high in fluctuating environments (Odum and De la Cruz, 1963; Arnaud, 1970; Valentine, unpublished data).

In stable environments, the constancy of supply of primary producer populations permits them to be exploited directly by numerous suspension feeders and grazers. Indeed, a good adaptive strategy in stable environments might be to exploit the plankton before potential competitors do. This would lead to specialization for filtering or trapping plankters high in the water column. Similarly, carnivores need not resort to scavenging dead prey but may become highly specialized on some living populations, since a steady supply would be assured. The consumer portion of the trophic pyramid should be more diverse proportionately in stable regions.

A fourth aspect of the environment that affects diversity differentially within the trophic structure is the severity of density-independent cases of mortality. In an environment wherein fluctuations are so severe that population sizes are partially or completely controlled by mass mortalities, there are fewer resources available for browsers, predators, and parasites and larger resources available for recuperators. Normally, the environmental fluctuations that lead to variation in productivity also lead to mortalities of this sort, so that these aspects are highly correlated. Finally, even those density-dependent causes of mortality that do not involve consuming live prey will lead to an increase in the proportion of energy in the recuperator chains. An example is provided by the competing barnacle populations studied by Connell; one species (*C. stellatus*) was crowded out by another (*B. balanoides*) during growth, and this density-dependent mortality provided energy for recuperators. These latter causes of energy increase in the recuperator chain should not necessarily imply an increase in diversity, however, unless they increase the stability of resources there.

Qualitative effects of diversity-dependent regulation are not well studied. They seem to arise from the unequal facility with which different higher taxa produce species with certain adaptive strategies. That is, some taxa contain chiefly detritus feeders or generalized carnivores, whereas others are composed chiefly of trophic specialists; some contain chiefly habitat generalists, and still others habitat specialists. Therefore, a distinct correlation exists between some higher taxa and certain environments, such as protobranchs (deposit-feeding bivalves) with high latitudes and the deep sea. Perhaps these qualitative differences are frequently due to differences in ground plans and general physiological mode.

Species Equitability Is Important in Ecological Theory but May Be Lost in the Fossil Record

Living communities include some species that are abundant and some that are rare. The distributions of abundances among species has long tantalized ecologists, some of whom have sought for general mathematical descriptions for them. Because it is difficult to estimate the former abundance of species in the fossil record, we shall only touch briefly on this large literature, although it does merit attention because of its theoretical importance.

When all the species of some higher taxon are censused in a given region, often a great many individuals of a few species dominate the biota, and few individuals of many species. Primarily on the basis of collections of Malayan butterflies and English moths caught in light-traps, Fisher, Corbett, and Williams (1943) suggested that the size-frequency distribution of species is described by a logarithmic series. It is possible to calculate the expected number of individuals for each species in a series on the basis of a logarithmic model. When this is done, observed abundances are commonly close to the expected abundances; discrepancies could possibly be due to sampling problems. However, Preston (1948), on the basis of bird census data, believed a better description of species frequencies to be a lognormal distribution, where the logarithms of the species frequencies form a normal curve. Since very rare species are the most difficult to sample, Preston assumed that the observed lognormal curve would be truncated on the rare side. Many sets of observation have fit Preston's model tolerably well, including some light-trap moth data. Both logarithmic and lognormal models are discussed with numerous examples by Williams (1964).

MacArthur (1957) considered the effects of three ecosystem partitioning models on species frequency distributions. In these models, population sizes are analogized to niche sizes within an available biospace. In the first model, populations' size frequencies are made proportional to the lengths of fragments of a stick broken at random points. This is analogous to a biospace that is completely partitioned into contiguous, nonoverlapping niches of random sizes. In the second model, the stick is divided into overlapping segments between randomly determined pairs of points, the segment lengths being proportional to population sizes. This is analogous to a biospace filled by overlapping niches. In the third model, units of populations' abundance are distributed randomly into a set of containers, so that the corresponding population sizes are analogous to niches that are noncontiguous and nonoverlapping, leaving some biospace unoccupied between the containers. When expected curves calculated from these models are compared with observed curves, only the first "broken stick" model approximates any of the observed species' frequency distributions. Commonly, there are more very abundant and fewer very rare species observed than estimated, but species frequencies of some taxa have fit the expected curves very closely. When a close fit is obtained, the species abundances should be random, and this situation may be amenable to a biological interpretation (see King, 1964). Macdonald's observations (1969) of molluscan species abundances in tidal creeks and salt marshes along the Pacific Coast of North America were tested against calculated curves for the logarithmic, lognormal, and broken-stick models. Only the logarithmic model yielded a tolerable fit, presumably indicating that the species abundances were *not* random. Whether this indicates some regulation, and whether any of these distributions actually describe a situation toward which ecological regulation ever actually tends, are uncertain.

Nevertheless, there are different numbers of abundant, common, and rare species in different communities, and it does seem from consideration of their gross patterns that the population size-frequency curves vary with trends in spatial and temporal environmental heterogeneity. It is commonly suggested that populations in high latitudes contain few species with chiefly large populations; temperate latitudes contain more species, a few of which have large, some intermediate, and many small populations; and tropical latitudes contain many species with chiefly small populations. If this is generally correct, then there is a correlation between species diversity and species equitability that may be related to ecological regulatory processes.

Since there are usually some abundant, flexible species even in tropical communities and some rare specialized ones in high-latitude communities, the question is raised as to how these exceptional species are able to avoid the effects of the regulatory processes, chiefly based on adaptive strategies for resource regimes, that have been postulated. There is no definitive answer, but the following considerations may be pertinent. Diversity within communities may be regarded as an equilibrium between the rate of appearance and extinction of populations (Preston, 1962; MacArthur and Wilson, 1963). Even in generally fluctuating environments, some fraction of the resources may be present at all times. This stable fraction may support species that further minimize the instability by occupying the more stable microhabitats, and such species could be relatively specialized and have small populations (Fig. 7-6, upper curve). The probability of survival of these species may be large enough, relative to the rate at which they become extinct, that a certain equilibrium level is maintained. Proceeding toward more stable environments, the proportion of resources that is stable increases, but even in the most stable environments there is a fraction that fluctuates. Therefore, in stable environments there can be a small number of generalized species adapted to exploit fluctuating resources, since the specialists cannot (Fig. 7-6, lower curve). The generalized forms become "opportunistic" species that appear at some localities in great numbers under favorable

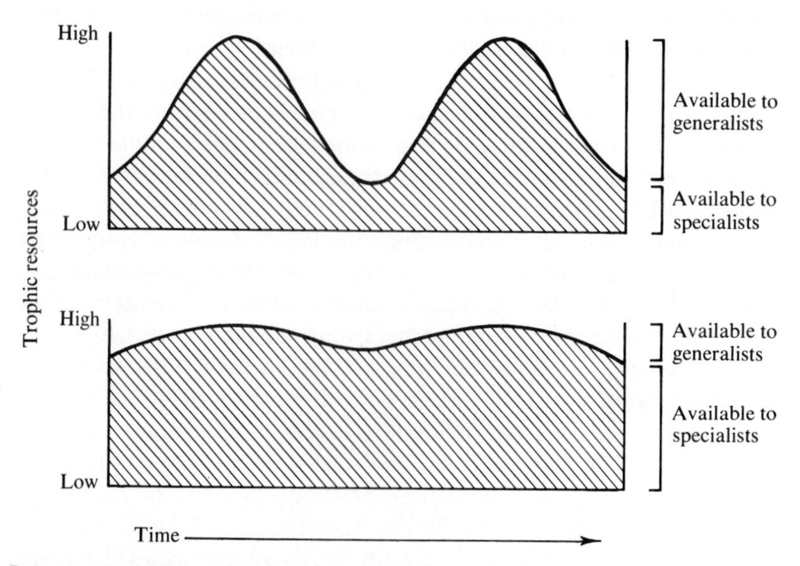

Fig. 7-6 Resources available to generalists and specialists in highly fluctuating (upper curve) and highly stable (lower curve) environments. A few specialists may occur in fluctuating environments, so long as some of the resource supply is stable, whereas the small fraction of the resources that are fluctuating in stable environments is not available to specialists but can be utilized by opportunistic generalists.

circumstances and then disappear, perhaps owing to depletion of surplus resources and ensuing competition with the established specialists. Levinton (1970) has considered the impact of such opportunists on the fossil record; they have not participated in the normal energetics of the communities they invade, yet their remains can be extremely abundant in the preserved fossil record of the communities.

Accurate estimates of the original living sizes of populations within fossil communities can be obtained only under very special conditions of preservation. Ordinarily, the mixing of individuals that lived at different times, together with differential destruction or removal of skeletons, would preclude such estimates. However, the relative abundances of populations may be preserved much more commonly. Warme (1969, 1971) has shown that the relative order of abundance of populations in living communities is very closely approximated in the death assemblages at Mugu Lagoon, California. However, Johnson (1965) has found a considerable disparity in rank of living and dead bivalve populations in Tomales Bay. Criteria to indicate when a fossil assemblage is likely to preserve the living rank orders of its populations are not yet worked out, although they would obviously include criteria suggesting little or no postmortem disturbance.

It Is Best to Measure Species Diversity and Equitability Separately as Well as Together

The measure of simple species diversity is a count of the species present, although sampling conditions make the measurement of diversity a difficult problem. It is hardly meaningful to compare diversity of samples from several communities to those from a single community, or of large samples to small samples, for example. The measure of species equitability is not such an obvious quantity, but its measurement is equally beset with sampling problems (MacFadyen, 1963; Williams, 1964). Two indices that are sensitive to species equitability seem especially useful in marine ecology and paleoecology. One is the Simpson index,

$$I - \sum \frac{n_i(n_i - 1)}{N(N - 1)}$$

In this index, I is the probability that two individuals sampled at random from an assemblage of species would belong to the same species, N is the total number of individuals in the assemblage, and n_i is the number of individuals belonging to the ith species. If all the individuals belong to the same species, then the probability is 1, but if all belong to different species, it is 0; the lower the index I, the higher the diversity. Clearly, if species with large populations appear, then the diversity is lowered by this measure.

Another diversity index sensitive to equitability is derived from the Shannon information function, and can be represented as

$$D = \sum_{i=1}^{s} p_i \log_2 p_i$$

where p_i is the probability that an individual in the assemblage under study belongs to species i, and s is the number of species observed (the simple species diversity). Equitability may be represented as e^D/s. This expression has been used by Margalef (1968) in experimental and theoretical diversity studies and by Sanders (1968) in studies of modern benthic marine diversity, whereas this or a related expression has been applied to diversity interpretation of fossil mammals by Van Valen (1964) and fossil marine invertebrates by Beerbower and Jordan (1969).

Each of these indices has properties that may aid or hinder the interpretation. The Simpson index is relatively insensitive to the presence of rare species; thus, if sample size is increased so that more rare species appear, the index doesn't change much. The index therefore seems to be relatively free from effects of sample size but, in fact, contains a sample-size-related bias of the population size-frequency distribution. The information function is rather independent of sample size except in small samples, and is not greatly affected by rare species either. Sanders (1968) reviews these and some other diversity measures that involve population sizes, many of which are derived from predictions of diversity models such as those of Preston and of MacArthur. For most paleontological studies, an information function seems preferable. In addition to the Shannon formula, there is one by Brouillon; the properties of the information functions are treated by Pielou (1969).

Differences between species diversity and equitability components of biotas have been illustrated by Buzas and Gibson (1969), who investigated diversity in benthic foraminifera of the present western North Atlantic. Their results are summarized in Fig. 7-7. Simple species diversity, equitability, and diversity as measured by the Shannon function, which is sensitive to both these properties, are plotted against depth. The Shannon function shows a peak near 40 m and another near 150 m. The simple species numbers show no peak near 40 m but do peak near 150 m; equitability shows a high peak near 40 m but no peak near 150 m. Clearly, the 40-m peak represents a region where species populations become unusually similar, whereas the 150-m peak represents a region where there are many different species. Examining all three measures is much more informative than examining any one alone.

Since relative population sizes as reflected in the fossil record do not always reflect the original size-frequency distributions found in the living communities, equitability studies may be of limited utility in investigating the theoretical bases of paleoecological processes, although additional research is clearly justified. Species equitability, nevertheless, has a high potential as a valuable tool in empirical studies leading to paleoenvironmental interpretations (Beerbower and Jordan, 1969).

Ancient Community Structures Should Correlate with Their Paleoenvironmental Patterns

We now summarize the patterns of structure and function in communities. They appear to depend primarily upon environmental patterns in space and time. In environments wherein diversity-dependent factors fluctuate greatly, diversity must be low, and the populations will ordinarily be generalists with a

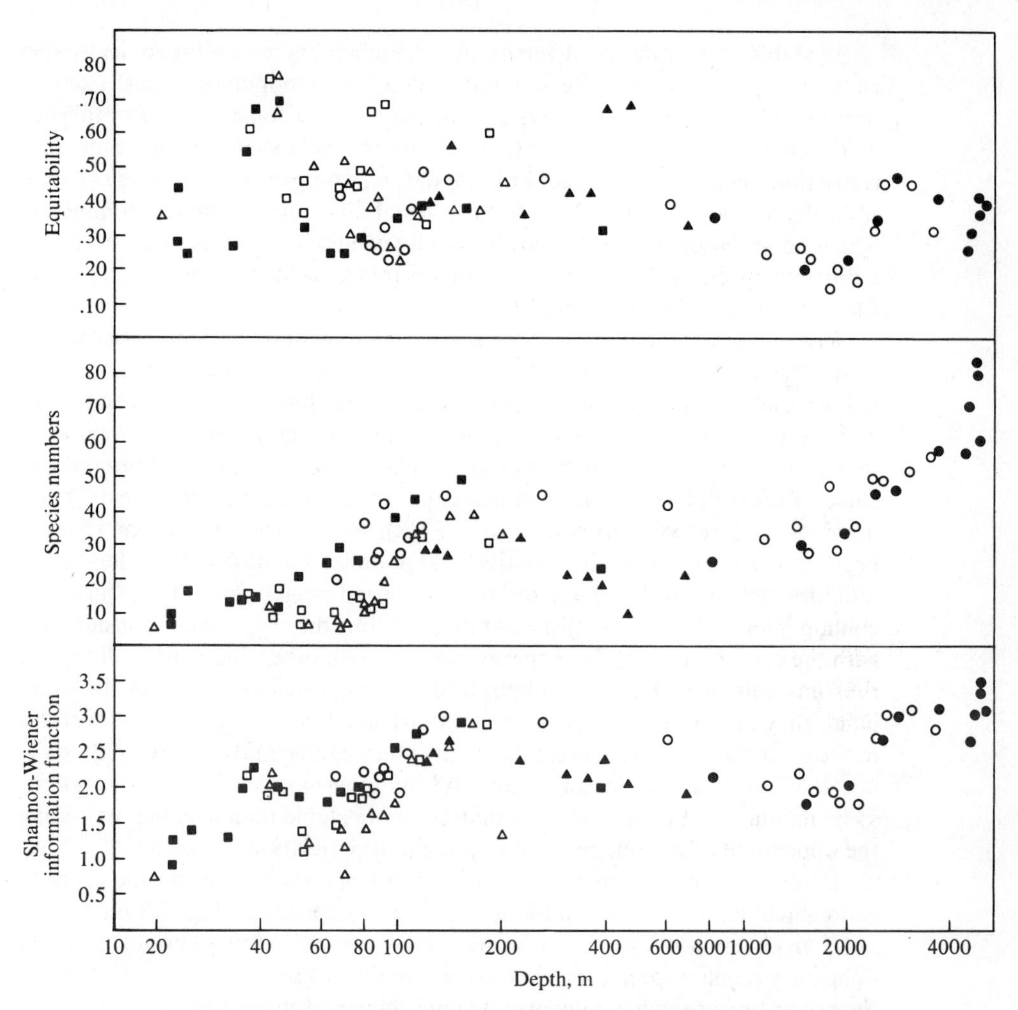

Fig. 7-7 Aspects of species diversity with depth among foraminifera off the eastern United States; the different symbols denote different sampling regions. (From Buzas and Gibson, 1969; copyright 1969 by the American Association for the Advancement of Science.)

high proportion either of detritus feeders and scavengers or of generalized feeders that can utilize organic debris. The ultimate in instability would be an environment in which every condition possible in biospace appeared in irregular but rapid temporal sequence. In theory, diversity would be limited to a single consumer species, but that species would have to be a supergeneralist, a universal species adapted to survive under all possible environmental states. As such, it would compete with and therefore exclude any other universal species, and no specialist could survive the appearance of so many conditions for which it was unsuited.

In stable environments, diversity may be much higher, and trophically specialized populations may be common, along with populations restricted to narrow habitat ranges. These restrictions seem to be commonly accomplished by the evolution of specializations, although they can also result from ecological restrictions enforced by biological (predators, competitors) or physical (episodic disturbances) agents. The ultimate in stability would be an environment in which no variation ever occurred, in which state each individual may represent a different species—there is not even an advantage in Mendelian heredity under such conditions. Diversity could be at a maximum.

Real environments are far from the ultimate in stability or instability, and contain some ecological factors that are more stable than others, and some regions that are more stable than others; they are thus mosaics of conditions, and the composition and structure of communities may vary from region to region, and even within communities from place to place. Figure 7-8 summarizes some of these differences diagrammatically. In fluctuating environments, communities have trophic structures as in 7-8(*a*), with a few populations that eat nearly everything, and with a relatively large utilization of detritus. These communities will tend to be rather similar over broad ranges of habitats, since they contain many habitat generalists, but through time they will change considerably with the ebb and flow of their energy bases or with other fluctuations. They are thus unstable in that their configuration is constantly changing. On the other hand, they are flexible and recover from inclement periods, and thus are stable in the sense that they have great durability; they can be said to be regulationally unstable but conformationally stable. Within their mosaic of conditions will be some habitats and energy sources that are more stable than average, providing the opportunity for some relatively specialized populations to occur.

In relatively stable real environments, those species having trophic specializations will tend to form trophic webs, such as depicted in Fig. 7-8(*b*). Energy flows up chains of prey and predators. To the extent that these populations are obligatory trophic specialists, they are susceptible to perturbations which affect their prey, and are rather vulnerable to environmental change. Some of the populations will be trophically restricted by ecological factors, however, and may be able to switch their feeding habits if conditions change. In these same communities are many species that are habitat specialists, or that are ordinarily restricted to certain habitats by biotic interactions. These forms may be able to eat a large fraction of the different species that pass through their habitats. Their trophic structure is depicted in Fig. 7-8(*c*). The trophic flow is canalized among different habitats but is weblike within the habitats. The trophic structure in the entire community is some combination of the diagrams (*b*) and (*c*), with energy paths tending to be canalized within habitats to an extent, and for predators that migrate between habitats, to be canalized within specialized trophic chains. Again, in real environments, there are resources that fluctuate and that can therefore be exploited by flexible opportunistic populations which will cooccur with the more specialized ones. In general, these communities will be

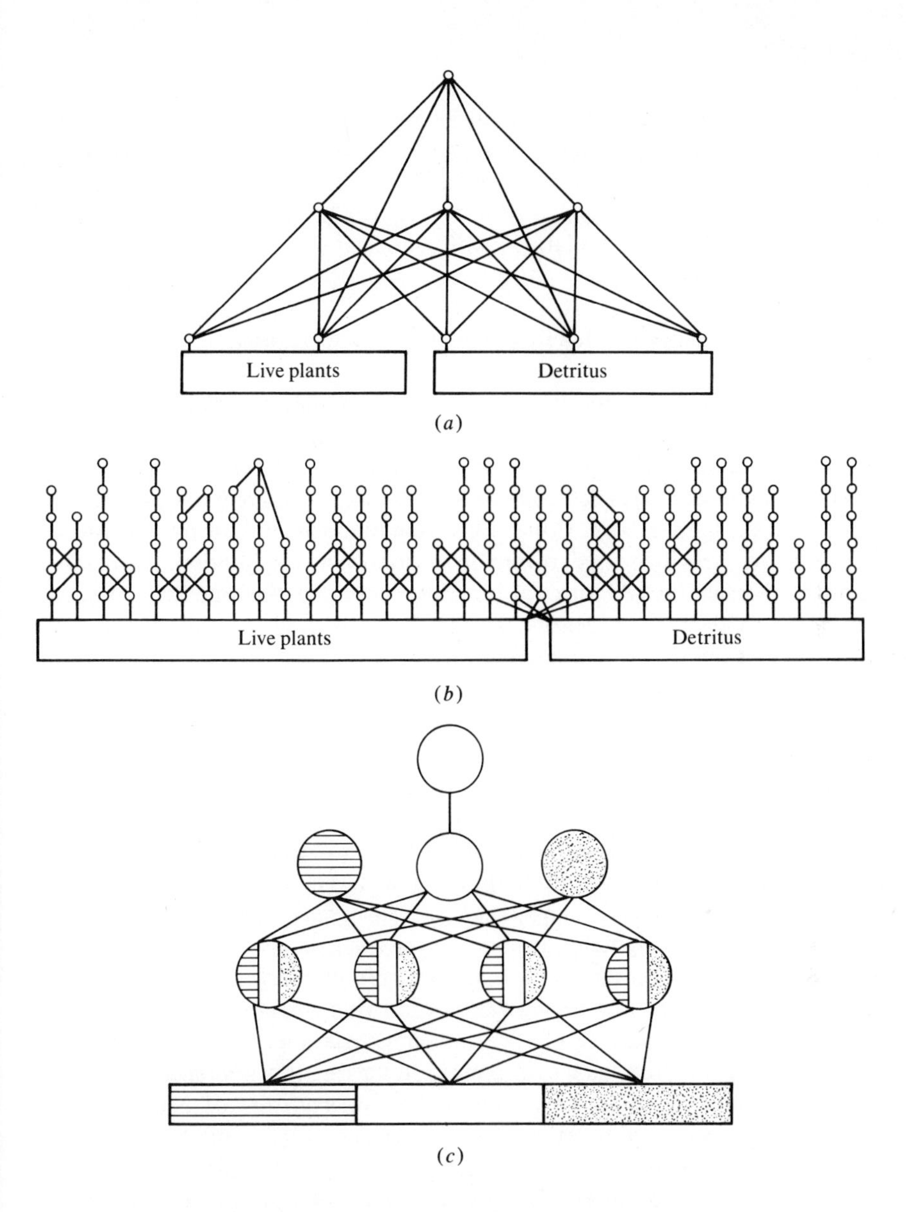

Fig. 7-8 Theoretical model of feeding webs in stable and unstable environments. (*a*) In an unstable environment the energy base fluctuates; diversity is low and species feed on a wide variety of prey with a high proportion of recuperators. (*b*) In a stable environment one fraction of populations is specialized as to prey. Among this fraction, energy flows up trophic levels along semi-isolated chains of specialized prey and predators; diversity is high. (*c*) In a stable environment another fraction of populations is specialized as to habitat. Each pattern represents a separate sub-biotope within the community biotope. Energy flows through these separate habitats along chains of habitat specialists. Communities in stable environments have a structure that is a mixture of cases *b* and *c*, and contain some generalized species as well.

relatively stable through time but will be spatially patchy, since they include many habitat specialists. If this picture of community structure is generally correct, then it should be possible to identify the spatial and temporal patterns of ancient environments from the diversity, trophic quality, dispersion, and other parameters of the populations within ancient communities.

Numerical Methods Help to Identify Ancient Life Associations

On the basis of the properties of living communities we have discussed, we may examine the problems and procedures of working with communities in the fossil record. Clearly, the fossil assemblages are even at best only incomplete remnants of what can be called communities, owing to differential contributions of species to the fossil record. Collections from numbers of different generations commonly are heavily biased owing to postmortem (taphonomic) processes, and often represent intermixtures of remains from more than one living association. Our task, then, is to identify the remains that lived together, reconstruct to some extent the sort of community structure that was present, and infer its ecological and evolutionary significance.

It is possible to employ the features of living community composition and organization in order to develop an operational definition and identification of fossil groups which very probably lived in association. These operationally defined groups may then be used as the basis of interpretations of what ancient communities were like. Of course, there is a broad spectrum of fossil assemblages, ranging from those containing specimens that appear to be largely in living positions, to heaps of skeletons that clearly have been washed together from sites that were originally dispersed over some large but unknown area, and to scattered skeletal debris that is essentially uninterpretable ecologically. When associated species can be shown to be in living position or at least to have lived very near their collecting sites, then their association is *prima facie* evidence that they belonged to the same community. In making this inference, we simply draw upon the least demanding community property, that of being a living association of species populations. Thus, Ziegler (1965) and his associates (Ziegler, Cocks, and Bambach, 1968) have recognized five distinctive benthic marine communities of Lower Silurian age. For some of these communities, the species are recognized as living associates because they are locally preserved in association in life positions.

When fossil specimens cannot be definitely interpreted as having been associated in life, it may nevertheless be possible to show that the species to which they belong have most probably lived together, from distributional evidence. This is most easily done by using another community property, namely, that component populations of communities tend to form recurrent associations. When species are found together in the fossil record at locality after locality, one suspects that they lived together (Johnson, 1962); indeed, if they did not live together, a special explanation is required for consistent cooccurrences. For example, the species associations recognized as communities by Ziegler are now known to be widely distributed in Norway, the British Isles, and eastern North America (Ziegler, Cocks, and Bambach, 1968). However, inasmuch as populations that belong to the same community are not necessarily found living together everywhere, and because the fossil record is incomplete in any event, it is difficult to specify the degree of consistency of coassociation expected of community member species.

One method of identifying recurrent associations is simply to list all the species represented by populations at each locality under study and then to group by inspection those species that appear to occur together commonly, as has been done for Upper Ordovician invertebrates of the central Appalachians by Bretsky (1970) and for Pleistocene mollusks of California by Valentine (1961). If the ancient communities are well represented in very distinctive assemblages, this procedure will obviously identify community-related groups of species. If the ancient communities are intermixed, however, or are chiefly represented by few remains, or contain many species in common, or have any of several other confusing attributes, their identification by simple inspection becomes difficult or impossible.

A number of numerical techniques have been applied to the problem of identification of recurrent associations, and these hold much promise. They employ high-speed electronic computers and they are especially useful for large bodies of data for which inspection becomes impossibly laborious. Most of them are conceptually simple but involve large numbers of calculations. A description of the computational methods involved in these techniques would comprise a large book in itself. It is only intended to show here in a general, often intuitive, way how the techniques work and what sort of results may or may not be expected. The references cited in this section contain more elaborate descriptions of the techniques, examples of their use, and further references. The usual approach is to select some measure of consistency of association and to apply it to each pair of species in the collections under study. The species are then "clustered" or placed in groups, the members of which form highly consistent associations so that the groups tend to be recurrent. The operations are usually specified in advance and are to this extent objective; they are repeatable.

One method relies upon calculating the average frequency with which each pair of species is expected to be represented in the same samples if their distributional patterns were entirely random and independent. The frequency at which the species are actually associated is obtained from the collections, and compared with the expectation. The significance of any difference between the expected and observed frequencies can then be tested by using these data to calculate a chi-square value. The chi-square statistic, which is explained in any biological or geological statistics text, is sensitive to the deviation of the observed from the expected, and a probability can be associated with any given chi-square value. For the case in hand, this is the probability that the observed frequency of occurrence is greater or less than would result from random sampling of independent species distributions. A significantly positive chi-square value suggests that the species occur together more frequently than expectable in an independent case, and this will be termed *positive bonding*. A significant negative value suggests they occur together less frequently than expectable in an independent case, and this will be termed *negative bonding*. If the probability that they are randomly and independently associated is very low, then there may well have been a biological reason—they may have belonged to the same community if positively bonded or to distinctly different communities if negatively bonded. This chi-square test of species cooccurrence was suggested by Cole (1949).

A disadvantage of chi-square is that species that appear in most samples will have reduced bonding, since even for independent occurrences they would commonly be found together and thus could not depart much from expectations, whereas species that appear in few samples but do cooccur will display exaggerated bonding for the opposite reason. Such species must be identified from the original data and treated separately. Other difficulties and peculiarities of this approach are discussed by Cole (1949), Fager (1957, 1963), and Johnson (1962).

Once pairs of species are bonded, they may be clustered into groups of species within which mutual bonding has a very high average, so that all members of a group tend to be found together. A number of different clustering rules have been suggested. Fager (1957) has suggested a clustering based on complete bonding within groups at some preselected level of significance, and his method has been most commonly used with chi-square bonds. A slightly modified version follows, the rules to be applied in order:

1. The evidence of affinity (that is, the level of bonding) is significant at an arbitrarily selected level for all species within a cluster.

2. The cluster includes the greatest possible number of species.

3. If several clusters with the same number of members are possible, those are selected which will give the greatest number of clusters without members in common.

4. If two or more clusters with the same number of species and with members in common are possible, the one which occurs as a unit in the greater number of collections is chosen.

Note that the clusters resulting from this procedure are mutually exclusive compositionally; that is, no species can belong to more than one cluster. A similar set of grouping rules, with some added features, has been employed for plants by Sørenson (1948).

Johnson (1962) has determined clusters of bonded species collected from Middle and Upper Pennsylvanian rocks of Illinois in this manner. He investigated the occurrences of 63 species from 152 collections of fossils from the marine portions of 11 cyclothems; each collection was from a single stratigraphic unit at a single outcrop. The species that were positively bonded at the 0.005 level of significance by the chi-square test were clustered by Fager's method, resulting in 19 groups that ranged in size from 13 to 1 species. Although the groups were mutually exclusive, species in any group could share bonds with members of other groups, and the number of such intergroup bonds suggested the degree of similarity between groups. In Fig. 7-9(*a*), the six largest groups are shown as boxes containing numbers, each number representing a species. The lines connecting the boxes represent intergroup bonding, and the figures by the connecting lines are the percentages of possible intergroup bonds that are

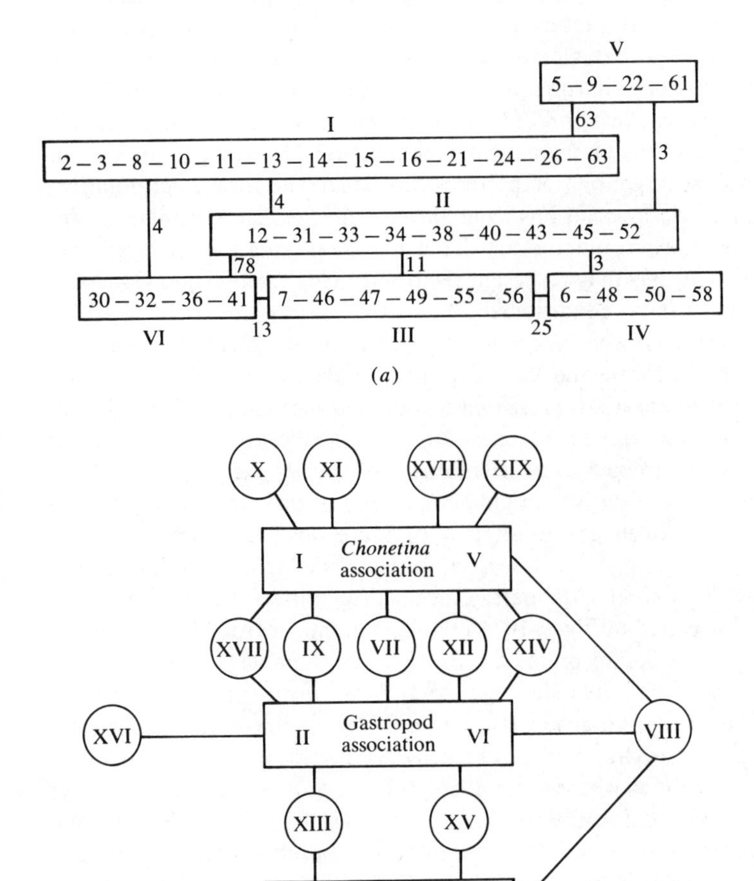

Fig. 7-9 (*a*) Major groups of fossil species whose co-occurrence was evaluated at the 0.005 level of significance by the chi-square method. The fossils are from the Pennsylvanian of western Illinois. The numbers by the lines connecting these groups indicate the number of bonds of association shared between groups. (*b*) Summary of the probable living associations represented by the major groups in (*a*), with minor group associations also indicated. (After Johnson, 1962; from *The Journal of Geology*, the University of Chicago Press, copyright © 1962 by the University of Chicago.)

realized. Clearly, groups I and V are closely related, as are groups II and VI. The remaining groups III and IV share a higher percentage of bonds together than with other groups and are quite distinct from groups I and V, with which they share no bonds at all. Reasoning from these relations, from the negative bonding between some species (several species pairs from groups I and III and from I and IV are negatively bonded), and from the stratigraphy and lithology of the rocks that yielded the collections studied, Johnson concluded that three main biologic associations were present [Fig. 7-9(*b*)]: (1) the *Chonetina* association, interpreted as part of a life assemblage from a firm substrate offshore; (2) the *Orbiculoidea* association, interpreted as from soft substrates in vegetated nearshore or shallow-water areas; and (3) the *Gastropod* association, interpreted as possibly nearshore also but from firmer substrates. The smaller groups are associated with these larger ones in patterns consistent with these interpretations (Johnson, 1962). These associations recur throughout the Pennsylvanian section studied and were clearly contemporary. They may well represent the remains of communities, in that they may represent recurrent living assemblages associated with characteristic habitat conditions.

A similar study has been made of 101 species from 225 Pleistocene fossil assemblages from California and Baja California (Valentine and Mallory, 1965). Most of these species are still represented by populations living along the Pacific coast, and the present habitats of many are known. Pleistocene communities have clearly been intermixed into assemblages of species that did not all live together. Therefore, the relatively high 0.001 level of significance was chosen as the bonding level. Sixteen groups of two or more positively bonded species were formed following Fager's rules. Each group tended to contain species that are recorded living in similar habitats. The bonding pattern between the eight largest groups is depicted in Fig. 7-10. Judging from the recorded living habitats of the species, group I is interpreted as chiefly a life assemblage from tidal flats or shallow waters at protected sites; group II from turbulent rocky shores or shallow rocky bottoms; and group III from offshore on fine-grained substrates in moderate depths. The stratigraphic and paleogeographic settings and associated species at the localities wherein members of these groups are found together support these environmental interpretations. Within each of these environmental settings today, several different life associations are found that could be termed communities. For example, shallow quiet waters contain marsh, tidal channel, tidal flat, and subtidal associations, all represented in group I.

One major anomaly appeared in the groups; two species that are recorded living on exposed sandy shores appeared in the quiet-water group I. They may represent populations that were living on the exposed sides of sand bars and were washed over the bars by large waves. Such washover shells from exposed beaches are found in quiet coastal lagoons today (Warme, 1969).

Group IV is composed exclusively of species that occur in large numbers of samples and are found regularly at localities characterized by any of several of the other groups. They are also recorded as living in a wide variety of shallow-

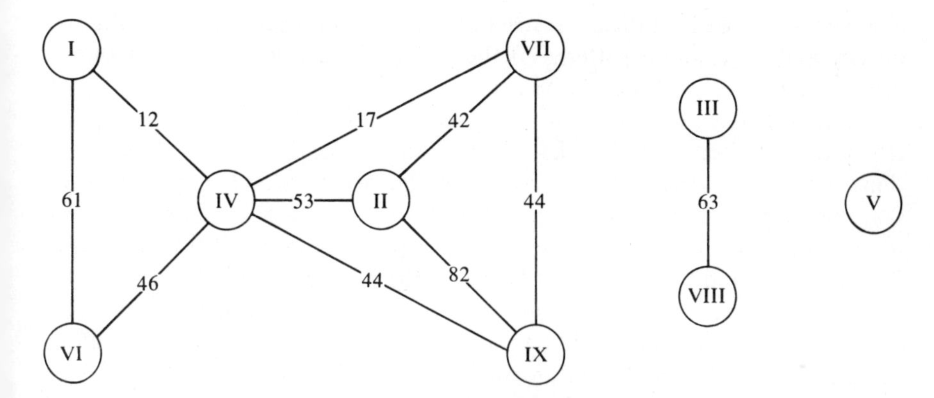

Fig. 7-10 Bonding between groups of Pleistocene mollusks from California that were grouped by Fager's method; associations were evaluated at the 0.001 level. Only bonding representing more than 10 percent of the possible bonds is shown. Arabic numbers indicate the percentage of possible bonds that are realized. (After Valentine and Mallory, 1965; from *The Journal of Geology*, the University of Chicago Press, copyright © 1965 by the University of Chicago.)

water environments. This group may be interpreted as a biological association, but it is not part of any specific community, nor of a group of closely associated communities. Instead, it is evidently a group of species that are rather ubiquitous in shallow water and thus cooccur with high frequency.

Therefore, although the evidence suggests that these analyses tend to identify living associations, the groups are often merely nuclei of the associations present at the fossil localities, since the groups must be mutually exclusive in species composition. Once assigned to a group, a species is ineligible for inclusion in other groups, and so the smaller groups are deprived of the more ubiquitous species. Furthermore, the most ubiquitous species tend to group by themselves. Inasmuch as the ubiquitous species may commonly play very significant roles in their communities, this feature presents a serious difficulty. It may be overcome in part by permitting species from larger groups to join smaller ones, following the original group formation if they would be eligible, except for the exclusiveness and the group size rules.

In summary, this technique evidently produces groups that lived in similar biotopes or groups of biotopes but may have been mixtures of populations from separate communities. The groups occasionally contain species that have lived in rather distinctive environments if the environments are consistently associated in nature and if mixing processes are fairly regular. The techniques may be most useful for establishing the nuclei of ecologically significant associations. When communities are not badly intermingled, the associations may represent the nuclei of individual communities, but when much intermixing has occurred, they may represent the nuclei of mixtures of communities that lived in proximity. In either event, the associations may represent ubiquitous forms with some common but broad ecological restrictions. By careful examination of the frequency and distribution patterns of the groups from field evidence, it should usually be possible to sort out these possibilities.

Many other bonding and clustering techniques have been proposed, and a few have been used in studies that can lead to the identification of life associations. The most widely used measures of similarity have been binary coefficients that use simple presence and absence data for the species and that are sensitive to the proportionate cooccurrences of species pairs. Simpson (1960), Sokall and Sneath (1963), and Cheatham and Hazel (1969) have reviewed the properties of a number of these coefficients; a few are presented in Table 7-3. The coefficients differ in their effects and should be chosen for appropriateness to the questions being asked and to the data. For example, Jaccard's coefficient (Table 7-3) is conservative in estimating similarities. If one species occurs in 100 samples and a second occurs in only 10, Jaccard's coefficient of similarity between these species cannot be greater than 0.1, that is,

$$\text{Jaccard's coefficient} = \frac{C}{N_1 + N_2 - C} = \frac{10}{100 + 10 - 10} = \frac{10}{100} = 0.1$$

because the species cannot cooccur at more than 10 localities. Yet it is conceivable that both species lived in the same community, and that if more fossils had been collected their cooccurrences might rise to, say, 50. Then the coefficient would have risen to 0.5, and, with exhaustive collecting, the cooccurrences might even have become identical. Simpson's coefficient, on the other hand, is a very liberal one and would consider the two original species patterns to be identical since the rarer species is always found with the more common one,

$$\text{Simpson's coefficient} = \frac{C}{N_1} = \frac{10}{10} = 1.0$$

Fager's coefficient is rather conservative but not as much as Jaccard's. The coefficients just discussed do not consider negative cooccurrences as contributing to similarity, but the simple matching coefficient includes negative matches (Table 7-3) and weights them as heavily as positive matches. To continue our example, if we had taken 150 total samples to find species 1 in 100 samples and species 2 in 10 of the 100, then

Simple matching coefficient

$$= \frac{C + A}{N_1 + N_2 - C + A} = \frac{10 + 50}{100 + 10 - 10 + 50} = \frac{60}{150} = 0.4$$

Situations in which each of these three coefficients might be especially useful are easily imagined. For example, if the collections are from rich assemblages from a variety of communities, intermixed or not, Jaccard's coefficient would be best for measuring similarity between species. However, if some of the assemblages

Table 7-3 Characteristics of Coefficients of Similarity and Difference Based on Binary (Presence-Absence) Data. Biassociational symbols are used; see text for numerical taxonomic equivalents. The expressions in column 1 result when two samples have the same number of positive features (equal diversity); those in column 2 when they share no positive features; those in column 3 when they become identical in positive features. Column 4 offers a comparison of the coefficients when one sample has twice as many positive features (presences) as the other and the number of features in common (mutual presences) is one-half the number in the less positive (less diverse) sample.

Coefficient	1	2	3	4
	As $N_1 \longrightarrow N_2$	*As* $C \longrightarrow 0$	*As* $C \longrightarrow N_t$	$\mathit{If}\ \frac{N_1}{N_2} = \frac{1}{2}\ \mathit{and}\ \frac{C}{N_1} = \frac{1}{2}$
1. Jaccard $\frac{C}{N_1 + N_2 - C}$	$\longrightarrow \frac{C}{2E_1 + C}$	$\longrightarrow 0$	$\longrightarrow 1$	$\frac{1}{5}$
2. Simple Matchings $\frac{C + A}{N_1 + N_2 - C + A}$	$\longrightarrow \frac{C + A}{2E_1 + C + A}$	$\longrightarrow \frac{A}{E_1 + E_2 + A}$	$\longrightarrow 1$	$\frac{C + A}{5C + A}\left[\text{if } A = C, \text{ then } = \frac{1}{3}\right]$
3. Dice $\frac{2C}{N_1 + N_2}$	$\longrightarrow \frac{C}{N_1}$ [= Simpson (8)]	$\longrightarrow 0$	$\longrightarrow 1$	$\frac{1}{3}$
4. 1st Kulczynski $\frac{C}{N_1 + N_2 - 2C}$	$\longrightarrow \frac{C}{2E_1}$	$\longrightarrow 0$	$\longrightarrow \infty$	$\frac{1}{4}$
5. 2nd Kulczynski $\frac{C(N_1 + N_2)}{2(N_1 N_2)}$	$\longrightarrow \frac{C}{N_1}$ [= Simpson (8)]	$\longrightarrow 0$	$\longrightarrow 1$	$\frac{3}{8}$
6. Otsuka $\frac{C}{\sqrt{N_1 N_2}}$	$\longrightarrow \frac{C}{N_1}$ [= Simpson (8)]	$\longrightarrow 0$	$\longrightarrow 1$	$\frac{1}{\sqrt{8}}$
7. Correlation Ratio $\frac{C^2}{N_1 N_2}$	$\longrightarrow \frac{C^2}{(N_1)^2}$ [= (Simpson (8))^2]	$\longrightarrow 0$	$\longrightarrow 1$	$\frac{1}{8}$
8. Simpson $\frac{C}{N_1}$	$= \frac{C}{N_1}$	$\longrightarrow 0$	$\longrightarrow 1$	$\frac{1}{2}$
9. Braun-Blanquet $\frac{C}{N_2}$	$\longrightarrow \frac{C}{N_1}$ [= Simpson (8)]	$\longrightarrow 0$	$\longrightarrow 1$	$\frac{1}{4}$
10. Fager $\frac{C}{\sqrt{N_1 N_2}} - \frac{1}{2\sqrt{N_2}}$	$\longrightarrow \frac{C}{N_1} - \frac{1}{2\sqrt{N_1}}$ $\left[= \text{Simpson (8)} - \frac{1}{2N_1}\right]$	$\longrightarrow -\frac{1}{2\sqrt{N_2}}$	$\longrightarrow 1 - \frac{1}{2\sqrt{N_2}}$	$\frac{1}{\sqrt{8}} - \frac{1}{4\sqrt{C}}$

SOURCE: After Cheatham and Hazel, 1969.

KEY: A, absent in both units compared.
C, present in both.
E_1, present in first but not second.
E_2, present in second but not first.
N_1, total present in first.
N_2, total present in second.
(When the first unit contains the fewer taxa, if units are unequal.)

are very poorly preserved so that the original associations are decimated—perhaps only calcitic skeletons are preserved at some localities, but are joined by abundant aragonitic skeletons where dissolution has not occurred—then Simpson's coefficient would indicate the original community affinities best. Clearly, there is a certain amount of subjectivity involved in the choice of a coefficient. As for negative matches, they probably do not really indicate similarities between species occurrences, for all species share nonoccurrences in a great many situations without being particularly similar themselves, and the incompleteness of the fossil record and the inadequacy of most sampling techniques further reduce the significance of negative data.

A type of coefficient that may be employed in bioassociational studies is the rank correlation coefficient, which can measure the similarity in the relative order of abundance of populations between samples. The Kendall rank correlation coefficient is a simple and effective one, for example, which is presented in most books on nonparametric statistics. Clustering of ranking coefficients would tend to group samples that contain similar species proportions (ranks, actually), and this would be a more sensitive technique than clustering by binary coefficients. It should be most useful for examining variations within a large body of samples from a rather homogeneous fossil biota.

Some coefficients can evaluate the abundances of species in different collections as well as their ranks and presence-absence patterns. The significance of abundance data for fossil populations is questionable, and the topic requires much further exploration. Many of the binary coefficients may be modified so that the contribution of each species to the coefficient is based on its proportionate representation in the samples (Greig-Smith, 1964). Other coefficients that measure abundance as well as presence are available; a promising one measures Euclidean distance in a hyperspace model (Sokall and Sneath, 1963, p. 147), but it has not been tested in bioassociational studies.

By clustering species in this way, we are measuring their similarity of occurrence among localities. Such clustering of variables in terms of cases is called the *R* mode of analysis. It is also possible to cluster localities on the basis of their similarity of species content; such clustering of cases in terms of variables is called the *Q* mode. In bioassociational studies, *R*-mode analyses produce clusters of recurrent species (biofacies) that may represent living associations or communities, whereas *Q*-mode analyses produce clusters of localities that may represent spatially recurrent habitat conditions, or biotopes. Both types of analysis are useful in paleoecology. If a set of samples are clustered in both modes, then it should be possible to evaluate the communities in terms of the biotopes where they are found and to evaluate biotopes in terms of the communities that characterize them (Kaesler, 1966; Valentine and Peddicord, 1967; Hazel, 1970). Although negative matches are inappropriate in the *R* mode, they may contribute to *Q*-mode analyses in some cases, for the absence of a species from two samples may be a measure of their similarity when a delimited range of biotopes is under study (Kaesler, 1966); however, the incompleteness of the fossil record presents a problem in this mode also.

The clustering techniques most commonly used are "pair-group" methods, and they follow a routine procedure. In the first step, the items being clustered (species, in the case of *R*-mode bioassociational studies) are either paired or not paired. Pair formation occurs only between those species that share the highest coefficients of similarity. Thus, of four species *A*, *B*, *C*, and *D* out of a set of species under study, if *A* and *B* display the highest similarity coefficients with which either species is associated, then they will pair, forming a "cluster" of two. If *C* shares its highest coefficient with *A*, it cannot pair, because *A* is already clustered, so *C* remains unpaired during the first clustering procedure. If *D* shares its highest coefficient with *C*, it cannot pair either.

In the second step, each pair is treated as a single item and pairing of all items is repeated; any item may pair or not pair. A species or cluster pairs with a cluster because it shares an average similarity with that cluster that exceeds its similarity to any other species or cluster. In this step, clusters may grow to include three or even four species, although many individual species may remain unpaired. Additional steps proceed in the same manner and clusters gradually grow by pair formation with species or with other clusters. Eventually, all items join in a single cluster.

Figure 7-11 represents the results of clustering of a set of data by two different pair-group methods. The degree of similarity of the items as they cluster is indicated on the scale across the top. When an unweighted pair-group method (UPGM) is employed, the average similarity between the units that governs the clustering is calculated from the original coefficients between all species, so that each member of each cluster has equal weight in all clustering steps. In another method in common use, the weighted pair-group method (WPGM), the average similarity between units is calculated from the coefficients of the previous clustering step. Therefore, the units being joined have equal weight, regardless of how many members they may contain. When the units are unequal in size, the species in the larger unit are each given less weight than the species in the smaller unit, for each of the many species contributes less toward the average similarity coefficient than does each of the fewer species. Therefore, late-joining items that are small tend to have more weight.

Each clustering method gives different results, and each has properties that are useful in special circumstances. For example, if a large number of samples is available and they appear to be divided rather equally among the various sorts of assemblages under study, and if the assemblages all seem to be well-preserved, then clustering by a complete bonding method, such as Fager's rules, would tend to identify the recurrent groups that most closely approximate parts of original communities. However, if many of the assemblages are poorly preserved so that only a few species remain, the pair-group clustering methods are best, for they tend to permit species that happen to be preserved infrequently to

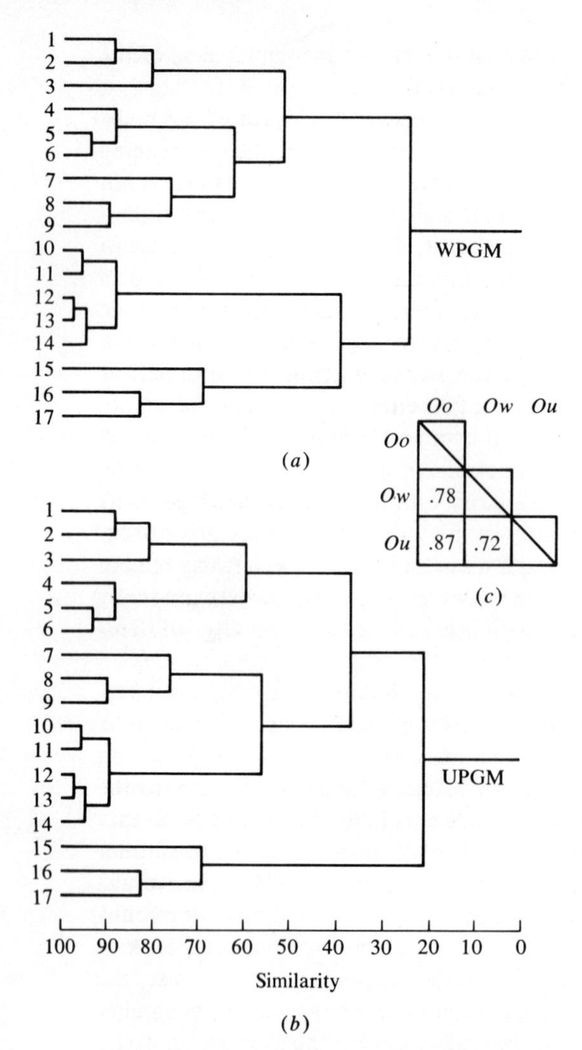

Fig. 7-11 Differences in clustering, based on different clustering techniques using the same data. In (*a*), clustering is by the weighted pair-group method, and in (*b*) it is by the unweighted pair-group method. The inset (*c*) presents correlation coefficients between the original coefficients on which the clustering is based and the averaged coefficients that were formed during the clustering. *O* represents the original coefficients; *W*, those arising from the weighted pair-group method; and *U*, those arising from the unweighted pair-group method, which has the highest correlation coefficient with *O* and therefore is most faithful to the original coefficient matrix. (After Hazel, 1970.)

cluster eventually with species from the same original assemblages that happen to be preserved more frequently. If the infrequently occurring species are poorly sampled for some special reason that biases their chances of collection, then the WPGM would be preferable to the UPGM, for it weights the late-clustering (poorly sampled) species. Methods other than these are available also, and some of them may be particularly well suited to special problems (see Sokall and Sneath, 1963).

A somewhat different approach to the identification of *R*-mode groups of fossils that may represent former living associations is provided by *ordination*, of which there are a number of variants. The most common method begins with a matrix of similarity coefficients and selects the two species which are least similar to be used as endpoints on an axis. The axis is then calibrated in terms of the range of dissimilarity (the inverse of similarity) between these species. The other species in the matrix are located along this axis at points lying at distances from the endpoints that are proportional to their dissimilarity to the endpoint species. Now, all the species may be thought of as being ordered along the axis in a fashion related to their responses to the "factors" responsible for the dissimilarity between the endpoint species. These factors may or may not be ecological. Next, two species that lie near the center of the ordination axis and that display high dissimilarity with each other are chosen as endpoints for another axis which is calibrated by their dissimilarity and erected at a high angle to the first. Now the species ordered along the first axis are spread out in the direction of the second, coming to occupy an area as they are ordered opposite the points on the second axis that represent their proportional dissimilarity to its endpoint species. Their position in the area is now controlled by two "factors." The procedure should be repeated until most of the dissimilarity in the data is accounted for; usually, this requires more than three axes, producing an ordination hyperspace, but it does not often require very many axes. Within the ordination space, species are positioned according to their relations to the "factors" that govern all the axes; those with similar relations will be grouped closely together. The distance between any two species is a measure of the dissimilarity of their relations to the axial "factors." Interpretation of what the factors are is usually a matter of reasoning from morphological, taxonomic, and stratigraphic data and pattern matching.

Ordination has been used in Q-mode analyses on living and fossil associations by Shaffer and Wilke (1965), who use a quantitative form of Jaccard's coefficient, and on fossil associations by Park (1968). In the Q mode, it positions localities, and if several distinctive environments are present the localities should group distinctly, or if environmental gradients are present the localities should be aligned along their trends in ordination space. Park (1968) discusses other varieties of ordination procedure.

One other numerical technique employed in bioassociational studies holds special promise; this is the group of multivariate methods called *factor* analysis. They are computationally complex and cannot be reviewed here in any detail. The coefficient usually employed in the similarity matrix that forms the basis of this technique is the product-moment correlation coefficient (r), a parametric coefficient which technically requires data with properties that are not present in most bioassociational data, but which appears to work anyway in many cases. This coefficient is discussed in most texts on statistical methods. Factor-analytic bioassociational studies on living material have achieved results comparable with those of other methods, including cluster analysis (Howarth and Murray, 1969). The "principle components" method of factor analysis is commonly employed. Briefly, coefficients are calculated between all pairs of variables in the data set. These coefficients may be visualized as pairs of data vectors, with the angle between them representing the magnitude of their coefficients. The angle is chosen so that the coefficient is equal to its cosine, and the correlation matrix becomes essentially a table of these cosines. In general, a multidimensional space is required to plot more than three of these vectors, so the coefficient matrix is represented vectorially in hyperspace.

Next, the vectors are projected onto "factor axes," which are orthagonal vectors positioned in the hyperspace by rules that place the first factor axis close to the mean of the data vectors. Because of this positioning, the first factor axis accounts for the greatest amount of variance possible, and subsequent axes account for progressively less. The numbers of the axes may be limited by rules related to the amount of variance accounted for. After the axes are inserted, they are manipulated so as to maximize the projections of the data vectors. In a sense, this manipulation resembles ordination, with the maximized projections providing endpoints. The factors represented by the axes are, like those in ordination, not necessarily ecological factors, although ecological factors may be involved in many of them. For any given variable—a species in an *R*-mode bioassociational study—the analysis provides a measure of the projection of its vector onto each factor axis; this measure is termed a *loading*. Species that share a common pattern of loading on the various factors may be considered as probable living associates. It is possible to use items other than species in the data set (grain size, sediment sorting, various measures of fossil condition or orientation, and so forth) and thus to have some sort of paleoenvironmental clues as to the significance of the factors from physical evidence included in the analysis. Such an approach has been employed successfully by Fox (1968). Also possible are *Q*-mode analyses. For further discussions of this technique, see Harbaugh and Demirmen (1964), Imbrie (1964), Howarth and Murray (1969), and references therein. Manson and Imbrie (1964) give a discussion of factor-analytic computation and a computer program that is especially appropriate to fossil analyses.

Temporal Community Changes That Do Not Necessarily Involve Genetic Changes May Be Conformational or Successional

Communities change in composition, structure, and function with the passage of time. These changes have been considered from a variety of viewpoints by investigators with diverse interests; however, there is no single integrated field of study that is traditionally concerned with the total range of community changes. Small-scale changes, such as diurnal and nocturnal variations, monthly variations based on tidal cycles, or seasonal variations, tend mostly to be conformational and are often studied by physiological ecologists or autecologists. It is well known that the entire aspect of animal associations change within many communities between night and day. Diurnal butterflies are replaced by nocturnal moths, for example. Some reef fishes that can be seen swimming about in lagoons and along reef margins during the day range into the ocean at night to feed. By contrast, reef cay birds, such as the mutton-bird of the Great Barrier Reef, a sort of petrel, fly afar during the day to feed but return at night to nest. Many invertebrates are nocturnal, so that shallow-water habitats have quite a different active biota at night than in the daytime. Among the oceanic pelagic communities, there is a strong tendency for diurnal vertical migration, bringing many populations to shallow depths at night, whereas during the daytime they retreat to dimmer, deeper water.

The seasonal changes in temperate and high latitudes are too obvious on land to require examples; in the sea, the most striking change in shallow zones is the reduction and virtual disappearance of many large seaweeds that grow in dense "forests" during the summer. Some of the larger kelps of temperate waters grow stipes well over 50 ft long, and yet are annuals, disappearing during the winter months. More or less regular seasonal changes in the marine plankton are well documented (see Raymont, 1963). Ordinarily, the daily to seasonal conformational changes are not regarded as forming the basis of permanent community changes.

However, changes in the physical environment that are not periodic certainly do occur continuously, and these changes also affect the composition of communities. Trends in weather or climate favor or exclude certain species and alter the population proportions and diversities within communities in the affected regions. Topographic and geographic changes in any region alter the habitats present and thereby alter the composition of the biota. Disregarding for the present any changes in selective pressures and consequent genetic changes that may flow from such environmental changes, even the strictly ecological effects are reflected in changes in community composition, structure, and function. Because these particular changes are not periodic, the accompanying changes in the communities must be regarded as potentially permanent—the community is not expected to return to its former state.

These changes in the community ecosystem are conformational, and tend to bring the community into equilibrium with the new environment. However, they result from the differential success of populations with different adaptive tactics and strategies, together with changes in population parameters within the populations remaining in the environment. They do not grow out of processes that favor new community structures for themselves; that is, there is no evidence of selection for one community structure in preference to another. Rather, the community structure grows out of processes that favor certain population properties, and the community naturally acquires structures and functions that arise from the properties of the populations of which it is composed. This is not to say that populations do not evolve some of their properties as adaptations to other populations with which they interact in a community context; some clearly do, and therefore the communities they inhabit are affected by interpopulation coadaptations.

In addition to those community changes related to noncyclic fluctuations in the physical environment, a variety of potentially permanent community changes have been studied that depend upon the activity of the organisms themselves, and that may result in qualitative and quantitative changes in community structure and function. In many cases, the sequence of such changes is rather predictable, and the process is known as *succession*, which is regarded by some workers as a basic concept of community ecology. Succession is usually described as an orderly process of community change that proceeds from the colonization of a newly opened region (*primary succession*) or a disturbed region (*secondary succession*) by pioneer organisms through a series of intermediate stages (*seral stages*), during which many of the original pioneer populations are replaced by different species, to some final *climax stage*, which is composed of populations that are in equilibrium with local conditions and will persist indefinitely. Classically, each stage in the total succession or *sere* prepares the way for the next stage by conditioning the environment. A sere has commonly been analogized to a life cycle, from birth (colonization) through development (various seral stages) to maturity (climax). This implies that the successional changes are unidirectional and proceed to some predetermined "goal." A similar philosophy once played an important part in evolutionary thinking. Called *orthogenesis*, it held that the lifespan of lineages proceeded from a youthful vigorous stage through a stable maturity and a degenerate, often bizarre old age to extinction. This idea, which appeared in more subtle forms as well, has been thoroughly discredited, and much evidence suggests that the orthogenetic aspects of succession theory are also fallacious.

Succession in marine environments has not been much studied; most work has been concentrated on the repopulation of artificially cleared rocky substrates. In the fossil record, there are only a few community sequences, such as in reefs, which have been evaluated in terms of succession. However the processes of succession have been widely invoked as a model to help explain the evolution of adaptive strategies and the evolution of community ecosystems (Margalef, 1968; Odum, 1969; Johnson, 1970; Tappan, 1971). Therefore, succession merits close attention. The term has been applied both to conformational community changes that are owing to environmental processes outside the system as well as to the changes that result from internal ecosystem processes. In this discussion, we shall use the term *allogenic succession* for the former and *autogenic succession* for the latter (see Tansley, 1935). Of course, environmental changes that are not owing to biological activity often accompany the autogenic changes, so that both types of succession may go hand in hand, and may tend to either reinforce or work against each other.

The classic work on autogenic succession is by Shelford and his colleagues, who studied gradual development of forests on regressive sand dunes that have accompanied the post-Pleistocene shrinkage of Lake Michigan. Pioneer plant communities invaded the newly exposed sands and were gradually replaced by a succession of associations leading to a climax forest; paralleling this plant succession was a succession of animals adapted to the various seral stages (see

Shelford, 1913). Clear summaries of the principal features of this example and of other work on succession can be found in Allee *et al.* (1949) and Odum (1971).

Two other well-known successional trends should be mentioned. The first is autogenic succession in grasslands. Plowed fields that are abandoned are first colonized by annual weedy plants, which are followed by grasses that are at first short-lived types and then perennials, reaching a climax grass community in 20 to 40 years. The second is the succession in ponds or small lakes, which commonly have clear, nutrient-poor water when first emponded (the *oligotrophic* condition, a "youthful" stage) and gradually become charged with nutrients and organic materials which accumulate on the bottom and in the water column (the *eutrophic* condition, a "mature" stage). The pond or lake basin commonly fills with sediments as well. The inflow of nutrients from the lake's drainage region causes an allogenic succession.

In the case of the dunes and of the plowed or disturbed grasslands, the succession begins with the appearance of pioneering populations of adaptively flexible, opportunistic species, such as weeds. Often these forms have high reproductive potentials and are phenotypically plastic; they are adapted to contend with the relatively harsh conditions of a new environment, conditions that are not at all modified or ameliorated by other biological activity (Stebbins, 1950). These early stages are of low diversity. The pioneering forms interact with the environment, perhaps stabilizing the substrate and providing organic detritus, and permit the next seral stage to develop. The successive immigrant populations tend to be more specialized and less flexible, and diversity tends to rise. The climax community probably includes the most specialized species available that are adapted to the ambient climatic and edaphic conditions. At climax, the community should be most diverse and most efficient, all other things being equal. Actually, in some cases the presumed climax association is less diverse than associations in preceding stages. Some forests have lower diversities when the dominants have spread throughout the community biotope; patches of unforested or moderately forested ground which support a variety of species that do not live in the climax forest gradually disappear. The achievement of climax throughout the biotope lowers the spatial heterogeneity and therefore the overall diversity in the region.

Allogenic pond and lake succession is normally just the opposite process. It begins with relatively stable conditions and proceeds to relatively unstable conditions; therefore, it begins with a potentially high-diversity situation and proceeds to a low-diversity situation. At the pioneering stage, the occupants are not required to be as flexible as at later stages, and the low nutrient supply favors populations with low rather than high reproductive potentials, other things being equal. As succession proceeds, the shoaling ponds become less stable and inflowing nutrients form the basis of high energy flow. Adaptation to the changing environment requires increasing flexibility and breadth of tolerance and tends to favor populations with high reproductive potentials.

Clearly, succession involves the assembling of a suite of contemporary species that are adapted to the ambient conditions of a given environment and that form a working ecosystem. The constraints on any given species association are related to ecological controls on species diversity; evidence suggests that the controls can be explained by the principles of energetics already reviewed. If the ecosystem then changes significantly, whether owing to the activities of its populations or not, the assemblage will change as the environment becomes receptive to species with different adaptive styles, providing such species are available. That is, succession is a relatively short-term process and does not ordinarily involve the origin of new species. On the contrary, succession does not involve genetic changes in populations more or less by definition; of course, genetic changes may happen to occur during succession, but they should not be associated with the successional process. Since communities are defined as a polythetic collection of species populations, those that are undergoing a marked change in composition can be viewed as a seral or chronoclinal succession of communities in a given area. The replaced communities can be reconstituted if the physical circumstances associated with those previous communities were to be repeated. This is an important point, for it would not be true if succession required genetic changes.

Some Fossil Sequences May Represent Autogenic Succession

When contemporary communities have been identified or approximated as closely as is permitted by the nature of the fossil evidence, they may be traced through time in many cases. In this way, the details of change in community composition and structure may be followed and, when correlated with paleoenvironmental interpretations, may add a fourth dimension to the model of community evolution. Presumably, they should display ecologically related changes in the proportionate representation of different populations, in species composition, and in species diversity.

Few communities have been traced in detail on the species level as yet, although numerous sequences of fossils that exhibit these sorts of changes have been recorded as biostratigraphic data, and many of these must reflect community evolution. Most such sequences require restudy with a view to establishing a knowledge of the accompanying sequence of paleoenvironmental states, in order to permit ecological inferences. Other studies have described detailed biostratigraphic changes in environmental terms, but for taxonomic groups above the species level. For example, Imbrie (1955) has described quantitative changes in the representation of supraspecific invertebrate taxa within the Permian Florena Shale in Kansas, and from lithological and geochemical evidence has tentatively interpreted the changes as salinity controlled.

A paleontological example of marine community change has been reconstructed and interpreted on a supraspecific level by Lowenstam (1957 and references therein) and cast into the framework of succession by Nicol (1962). During the Niagaran Stage of the Silurian, reef communities appeared in the shallow waters of the North American continental platform in what is now the Great Lakes region. According to Lowenstam's interpretations, some of these reefs began as local patches of organisms that grew upon the skeletal remains of their predecessors, preferring the hard substrate and perhaps the slight topographic prominences that gradually appeared and that may have created slight local turbulence. The chief pioneering species which form the substrates that became reef foundations seem to have been tabulates and bryozoa (Fig. 7-12), and associated with these forms were crinoids and brachiopods. The tabulates probably provided the structural framework, and the crinoids, brachiopods, and other organisms supplied skeletal debris to fill out the frame and often to overspread the flanks, whereas bryozoa and other organisms may have helped to stabilize the sediments in the growing buildup. Eventually, the rising prominence reached into shallow zones, where it was felt by wave surge. At about this stage, the character of the biota changed considerably; some species disappeared and many others appeared; the tabulates and bryozoa especially diversified, together with the crinoids (Fig. 7-12). The rise in local diversity appears to have been accomplished by recruiting species that were well adapted to semirough water conditions from shallow inshore zones. In other words, the changes involve the reconstitution of the life association of the buildups from stocks of species that were already extant. These changes form an autogenic succession, for they were brought about by environmental modifications (chiefly topographic in this case) caused by strictly biological activity.

During the semirough water stage, upward growth continued, and brought on a truly rough-water stage characterized by abundant tabulate framework-builders and by a highly diverse crinoid fauna. Frame-building stromotoporoids and sediment-producing brachiopods, mollusks, and trilobites also reached peaks of diversity then. Since upward growth of the buildup was finally terminated at some low tidal height, this must be considered as the climax community; subsequent growth must have been lateral, out over the skeletal debris on the flanks. At the climax stage, the community is an undoubted reef; it is self-sustaining, fixing carbonate in skeletons at a rate sufficient to maintain a physical structure despite wave attack and the bioerosive activities of "destructor" organisms (Lowenstam, 1957). The diversity rises accompanying the attainment of semirough and reef stages are probably due to the increased spatial heterogeneity achieved in the environment at these times, rather than to any overall increase in environmental stability. The creation of turbulent habitats and of microhabitats within and around the turbulent-water organisms, and the retention of quiet habitats in locally protected areas in the lee of wave-resistant structures, had the effect of greatly expanding the range of environments as the buildup proceeded into rougher and rougher water.

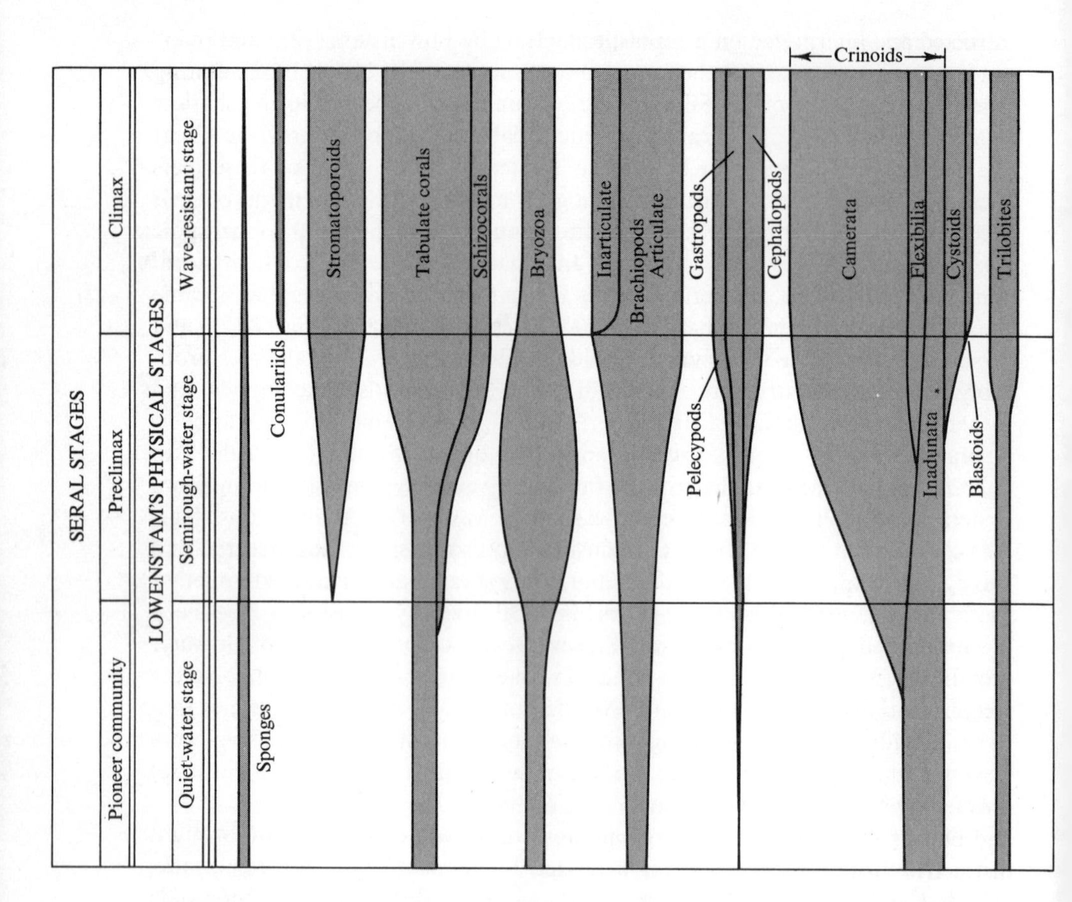

Fig. 7-12 Succession in Silurian marine communities from the Great Lakes region that have reef climaxes. (After Nicol, 1962, adapted from Lowenstam, 1957.)

The extent to which nonskeletonized organisms contributed to the resistant qualities of reef stages is not known. By analogy with modern reefs, there may well have been tough but naked forms, such as colonial anemones, that played important roles. Algae may also have been significant. A structure called *Stromatactis* was abundant during the quiet-water seral stages and still common during the reef stage; the identity of this structure is not certain, however, and its ecological role may have been anything from frame-builder (Lowenstam, 1957) to burrower (Shinn, 1968). In short, we are far from knowing the true structure of these Silurian communities. Nevertheless, the analogy of the stages of reef community change with modern seral stages is informative.

Many of the diverse taxa of the climax community seem to be represented by highly specialized species that were endemic to reef biotas. Although their appearance on any given reef buildup may merely represent recruitment from nearby reef communities, they presumably evolved at some time in response to the selective pressures accompanying the reef conditions. Not enough is known of contemporary rough-water, nonreef biotas to determine the quantitative significance of this element among the skeletonized taxa at climax. Nevertheless, it appears that phyletic evolution accompanied the development of climax reef communities and that the sequence of Niagaran reef associations was brought about by community evolution in every sense.

Fossil Communities from Similar Environments Should Have Similar Structures

Further evidence of the reality of ecological controls of community structure has been provided by detailed studies of biotic associations in two sets of communities that lived in similar environments but about 70 million years apart (Walker and Laporte, 1970). One set is from the Black River Group (Ordovician) and the other from the Manlius Formation (Devonian), both in New York.

The fossil associations occur in marine carbonates, which display strong evidence of similar depositional conditions at these two times. Criteria used to establish the environmental patterns during deposition are largely independent of ecological evidence, depending upon a variety of physical and biogenic features that have been found to be associated with certain environmental conditions by study or present-day marine sediments. These include the presence of birdseye fabric, mud cracks, intraclasts, laminations, lumpy beds, scour and fill, and other features. Within each of the two stratigraphic units are a number of distinctive lithofacies and biofacies. Four distinctive biotic associations are common to both units; they are found with characteristic associations of lithologic features that are interpreted to represent (1) supratidal and/or high intertidal, (2) midintertidal, (3) low intertidal, and (4) lowest intertidal and/or subtidal environments. They are schematically reconstructed in Fig. 7-13.

The supratidal and/or high intertidal communities are quite similar in both units, being composed of algal mats, epifaunal leperditiid ostracodes (of different genera), and small suspension-feeding burrows of unknown affinities. Both midintertidal communities include the traces and remains of a greater variety of organisms, including small deposit feeders and larger epifaunal forms. However, the common skeletal macrofossil in the Black River community is the trilobite *Bathyurus*, whereas in the Manlius it is *Tentaculites*. In both low intertidal communities, diversity is greater still, with brachiopods, ramose ectoprocts, large deposit-feeding burrowers, and other forms appearing. In each of these communities, many species appear to have close ecological analogues in the other community. For example, the strophomenid and spiriferid brachiopods are represented in the Black River by *Strophmena* and *Zygospira*, and in the Manlius

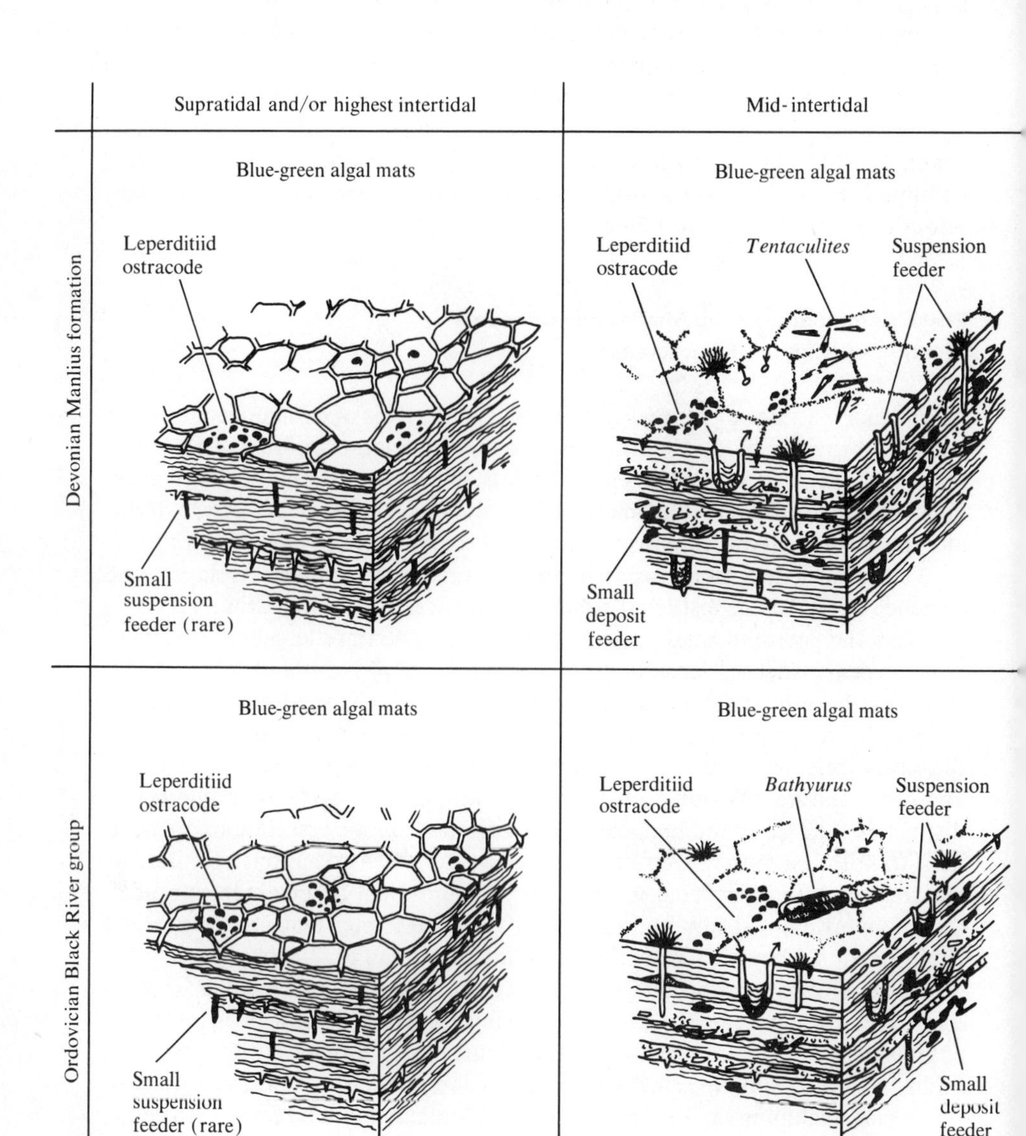
Supratidal and/or highest intertidal
Mid-intertidal
Devonian Manlius formation
Blue-green algal mats
Leperditiid ostracode
Small suspension feeder (rare)
Blue-green algal mats
Leperditiid ostracode
Tentaculites
Suspension feeder
Small deposit feeder
Ordovician Black River group
Blue-green algal mats
Leperditiid ostracode
Small suspension feeder (rare)
Blue-green algal mats
Leperditiid ostracode
Bathyurus
Suspension feeder
Small deposit feeder

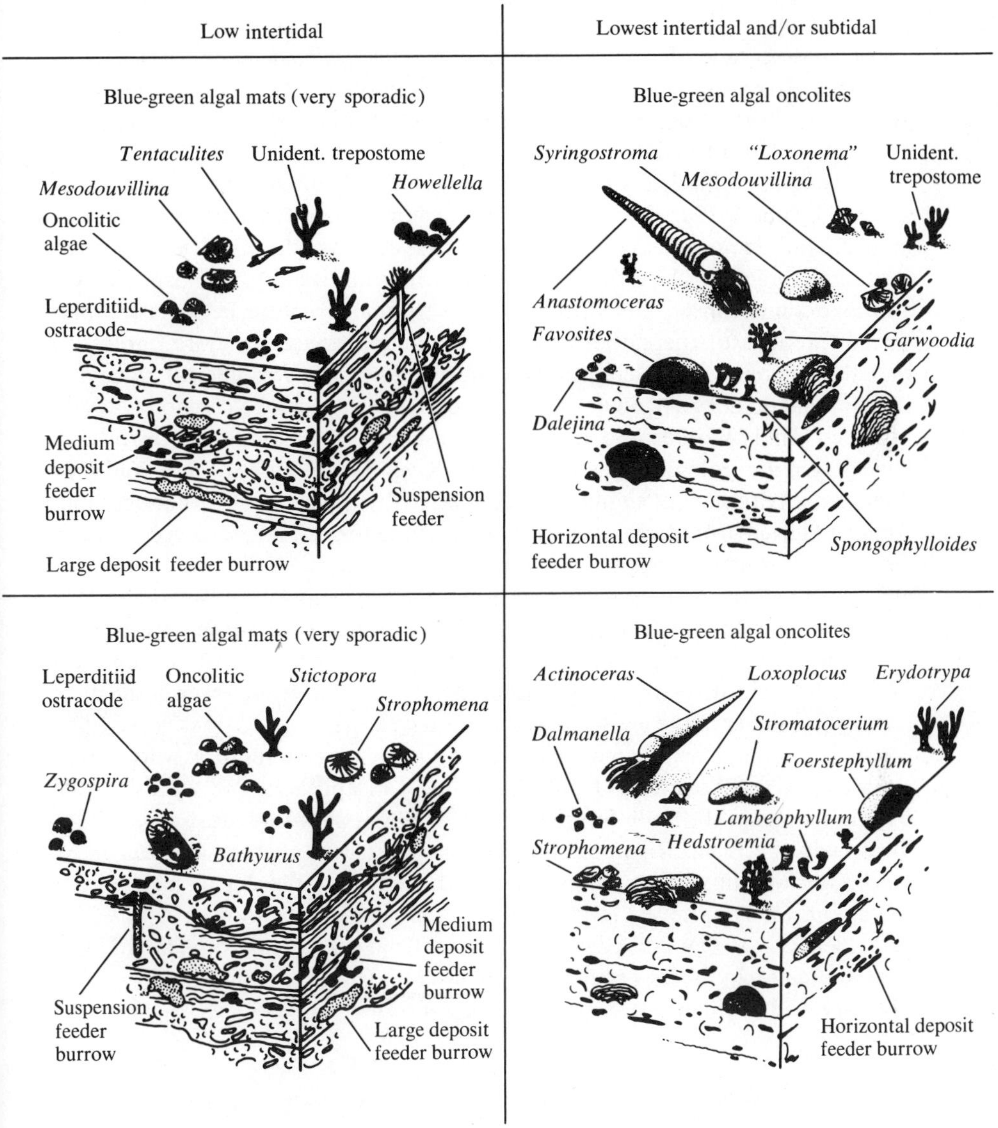

Fig. 7-13 Comparison of sets of communities shown to have inhabited analogous environments: four from the Ordovician and four from the Devonian. Their similarity in diversity and in structure, insofar as it can now be interpreted, is striking. (After Walker and Laporte, 1970.)

by *Mesodouvillina* and *Howellella*. In the fourth community type, from lowest intertidal and/or subtidal environments, there are still more taxa, again with numerous pairs of taxa that appear to represent ecological analogues, such as the tabulates *Foerstephyllum* and *Favosites*, the nautiloids *Actinoceras* and *Anastomoceras*, and the dalmanellid brachiopods *Dalmanella* and *Dalejina* in the Black River and Manlius, respectively. Even trace fossils have close analogues in each formation. However, a few taxa do not have clear analogues, such as the Black River tabulate *Tetradium*.

The degree of similarity between these two sets of communities, separated as they are by about 70 million years, is truly striking. There are no species in common and indeed no genera of skeletal fossils are considered equivalent by current taxonomic practice. Nevertheless, the similarities in diversity, in morphological and taxonomic aspect, and in the inferred trophic structure between the preserved fractions of the earlier and later communities are far more impressive than the differences. It is tempting, therefore, to try to find ecological analogues for taxa that do not have obvious close taxonomic equivalents in analogous communities. Thus, Walker and Laporte suggest that the trilobite *Bathyurus* may have played an ecological role in the Black River communities of mid- and low-intertidal environments similar to that played by *Tentaculites* in communities from similar environments in the Manlius. These forms may have been deposit feeders and/or scavengers. Of course, it is realized that the ecological function of either of these forms may have been performed by unpreserved, perhaps soft-bodied organisms in the analogous community, or it may not even have existed, the trophic material utilized in one community passing instead to decomposers in the other. The diversity trends within each formation, from lower in the most exposed to higher in the least exposed, accord well with expectations based upon control by general environmental variability or by trophic resource variability.

Many of the American Ordovician lineages disappeared and were replaced by immigrants in the early Silurian (Boucot, 1968); probably most or all of the lineages in the Black River communities were replaced then or at other times before the Manlius communities lived, although the replacement mechanisms are unknown. Despite the large turnover involving immigration, a close structural identity was maintained in communities from similar environments. Therefore, these communities appear to have been under strict regulation insofar as diversity and ecological composition are concerned.

Many Temporal Community Changes Do Involve Genetic Changes

There is evidence that the principles governing the assembling of species of certain types to form a community that is adjusted to an ambient environment apply both to the "ecological assembling" of species that are already present in the biosphere and to the "evolutionary assembling" of species through the development of environmentally appropriate strategies by selection within evolving lineages. Of course, the processes are different in these two cases.

Little is known of the sources of species populations in most communities, although in a few cases some highly suggestive evidence is available. For example, the high-diversity communities of some old fresh-water lakes are composed chiefly of swarms of closely allied species; it appears that a novel environment appeared and that immigrants were nearly or quite unavailable so that the biota was diversified through radiation of endemic lineages. These lake biotas merit special attention as possible analogues to some marine situations. A good example of species swarms is afforded by the fishes of the great lakes of Africa, especially Lakes Tanganyika and Nyasa (Fryer and Iles, 1969). The fish family Cichlidae is common in these lakes, and some genera of this family have undergone spectacular diversification into great "flocks" of species that coinhabit the same lakes; up to 200 different species of cichlids are present in some flocks. In general, the species of the flocks are highly specialized in feeding and other habitat requirements, are restricted to a single lake, and have small and stable populations. The lake environments supporting these flocks are relatively stable and have relatively low nutrient supplies, as the flocks inhabit oligotrophic lakes. By contrast, some other genera of cichlids contain relatively few species, of which only a very few coexist in the same general habitat. These species (such as of the genus *Tilapia*) are broadly tolerant and relatively unspecialized, vigorous, highly plastic phenotypically, and tend to occur over wide geographical regions in a variety of habitats, which include unstable lakes, unstable portions of large lakes that are chiefly stable, and such unusual environments as hot saline lakes. In environments where species of *Tilapia* are most usually found, eutrophic conditions prevail. Thus, the adaptive strategies of the populations are markedly different in the stable and unstable environments (Chapter 6), which is reflected in species diversity.

These lakes are old; Lake Tanganyika may date from the Pliocene and Lake Nyasa from the mid-Pleistocene (Brooks, 1950; Hutchinson, 1957b). Their large sizes, great depths, climatic settings, and other factors have served to preserve an oligotrophic condition which the activities of organisms have been ineffective in altering significantly toward the more unstable and eutrophic conditions that are normal as lake succession proceeds. These lakes support endemic species flocks of other taxa, such as gastropods, as well as the fishes. Other large old lakes also contain species flocks. Lake Baikal, the oldest (Paleocene or older) and deepest lake extant, contains a largely endemic element (especially in the more stable habitats) that has undergone diversification into species swarms, most spectacularly among the gammarid amphipods, but also among the gastropods (Brooks, 1950; Kozhov, 1963).

Presumably the stability of these lakes permits the high levels of diversity which we find today, and the general unavailability of appropriately adapted immigrants has led to the evolution of these flocks of closely allied species. A major evolutionary problem lies in determining the isolating mechanisms that permitted the rise of infertility between splitting lineages within the lakes. Most species of the flocks are endemic to only one lake and thus are presumed to have evolved in situ, rather than having been introduced as immigrants. The answer evidently is in part that these lakes are large enough and varied enough in habitat that populations are frequently isolated owing to the patchiness of their habitat. Another factor may be the necessity for populations in the regions that have the least trophic resource supplies to be very sparsely distributed. The low-resource regions would presumably be the deeper waters in Lake Baikal, and those parts of Tanganyika and Nyasa (which are deoxygenated at relatively shallow depths) that receive least runoff. To survive in these low-density situations, the populations would have to develop reproductive strategies that operate well when populations are dispersed and that entail only small energy drains. Thus, these populations might develop into series of very small populations isolated by slight environmental variations or simply by the random appearance of distributional discontinuities that could not be broached easily. These populations might eventually give rise to numbers of species. For fuller discussions, see Brooks (1950), Kozhov (1963), and Fryer and Iles (1969).

In the ocean, the most stable environment of all is in the deep sea. The deep-sea benthos probably includes only 5 percent or so of the marine invertebrate species at present, although it is too poorly known to be certain. The relatively poor species representation seems due to the low spatial heterogeneity of the deep-sea environment, which is rather monotonous. On the other hand, the high temporal stability, which is presumably coupled with low resource supplies, provides a high-diversity situation according to the model of diversity regulation presented earlier and summarized in Fig. 7-5. Indeed, the deep-sea fauna does contain a high diversity of species when it is compared with similar habitats in shallow waters; the deep-sea mud or ooze community is comparable in diversity to shallow tropical communities associated with muddy substrates (Hessler and Sanders, 1967; Sanders and Hessler, 1969).

The nature of deep-sea communities is not at all well known. The organisms tend to be small and the populations highly dispersed, so that density and biomass are very low. This probably reflects a low trophic resource supply. Well over half of the fauna is comprised of detritus feeders. Many species range widely horizontally, some being essentially circumoceanic, but they tend to occur within relatively narrow depth ranges, so that a change in depth of only about 800 m results in a faunal change as great as occurs over 16,500 km horizontally. Thus, if there are distinctive communities in the deep sea, they may be arrayed in distinctive depth zones. However, sharp faunal discontinuities have not been proven as yet. Sanders and Hessler (1969) point out that a faunal change between 2000 and 3000 m in depth, recognized by Vinogradova (1962) and others as representing the change from bathyal to abyssal depth zones, seems, in fact, to be a gradient similar to that present in other depths. The species composition of the benthos changes more or less gradually from the outer continental shelf edge, where there is an undoubted faunal discontinuity, to the bottom of the abyssal zone, so far as is known. Furthermore, the distinctiveness of the hadal zone of the deep-sea trenches has been disputed (Menzies and George, 1967). Thus, the deep sea has probably achieved its present high diversity mainly through endemic radiation, although immigration is not precluded and, indeed, probably has occurred regularly, since higher taxa that have evolved episodically throughout the Phanerozoic (including such relatively late arrivals as Neogastropods, from the Cretaceous) are regularly represented in the deep-sea biota, if their adaptations are appropriate to the energetics of deep-sea communities.

For shallow-water regions of high diversity, such as the inner tropics, a major source of the rich biotas is certainly endemic speciation, but there is also probably a steady accumulation of immigrant lineages, perhaps forming a more important element than in the deep sea. High-diversity regions that undergo occasional fluctuations would be expected to receive occasional immigrants from low-diversity regions.

It appears from these examples that evolutionary responses to resource regimes are similar to ecological responses, and a general picture of community change emerges from these considerations. As environments change, the communities affected must change, owing to the necessity for their populations to adapt, not only to changes in habitats but also to the changes in temporal and spatial regimes that involve their adaptive strategies. If, in some region, the regime changes so that the permissible species diversity is higher than actual diversity, additional species will be assembled to raise the diversity to about this new level, whereas if the permitted diversity level declines, extinctions must ensue. Since highly diverse regimes usually require different modal strategies of adaptation from poorly diverse ones, some of the original lineages will usually not develop strategies appropriate to the new situation and will be excluded whether diversity is rising or falling. Therefore, to attain the permitted diversity level, it is necessary that extinct lineages be replaced as well as that numbers of new lineages be added. Indeed, extinction must be expected as a normal process during the course of time, and replacements may be admitted to communities more or less frequently.

When an area suffers serious local extinction due to some natural catastrophe, such as a forest fire, flood, or volcanic activity, it may be repopulated with species to something approaching its equilibrium value within decades (see MacArthur and Wilson, 1967). The extinct local populations are merely replaced by immigrants from populations in nearby areas, commonly of the same species; autogenic succession will usually be involved to some extent, depending upon circumstances. In these cases, the environment has not permanently changed much, but has suffered a sharp local fluctuation and has returned to something like its previous state. When there are long-term changes that are more or less permanent, however, affecting broad regions that include all the populations of many species, the principle of an equilibrium diversity may still apply but there may well not be enough species available within immigration range to replace extinct lineages or to increase total diversity, as permitted. Phyletic evolution may now adapt the old lineages to the new conditions and produce entirely new species with appropriate adaptations.

Species found in communities following a period of change in an environmental regime may be derived from several sources. Some species may be essentially unchanged, having been endemic and preadapted to the new regime. Other species are more or less new but represent ancestral endemic lineages that successfully adapted to the new regime through genetically based functional modifications. Still others will be immigrant species; of these, some may have been preadapted, simply to immigrate unchanged, whereas some may be descended from immigrant stocks that adapted to the new regime during or following immigration. Still other species may result from endemic splitting or radiation of any of the foregoing lineage types. Presumably, the precise proportions of species derived from these several sources will depend upon the mixture or conditions that are special to each event, such as the extent of preadaptation in the local biota and in the biota of nearby regions, the ease of immigration, and the size and spatial heterogeneity of the region, which would influence the establishment of genetic isolation and thus the ease of endemic speciation.

Many More Data Are Needed on Fossil Communities

In the cases of both the Niagaran reef community study and the comparative intertidal carbonate community study, significant ecological questions were raised and possible explanations indicated, but without expansion of the scope of the investigations, definitive answers could not be reached. In the case of the Niagaran reefs, one is led to wish for data on the species composition of the biota and on ecological aspects of the origin of these species, to learn the sources from which the community components were assembled and the role of phyletic evolution in the development of the reef communities as such. For the New York intertidal carbonate studies, information on the details of species turnover in the period between the times studied would bear heavily on understanding modes of community change. Detailed information on the patterns of replacement by Silurian immigrants would be especially useful, and data on the previous community relations and phyletic history of the immigrants would be valuable in interpreting the compositional changes. Numerous other questions occur also in each case, but additional field and laboratory studies are necessary. It seems that great amounts of data are required to answer many of the questions that concern community evolution.

The main reason for this requirement at the community level is that we have reached a higher level of organization and that the scope of research is correspondingly broadened. To explain the ecology of populations frequently requires only internal data or comparative data from populations of allied taxa in associated communities, but to explain the ecology and evolution of communities

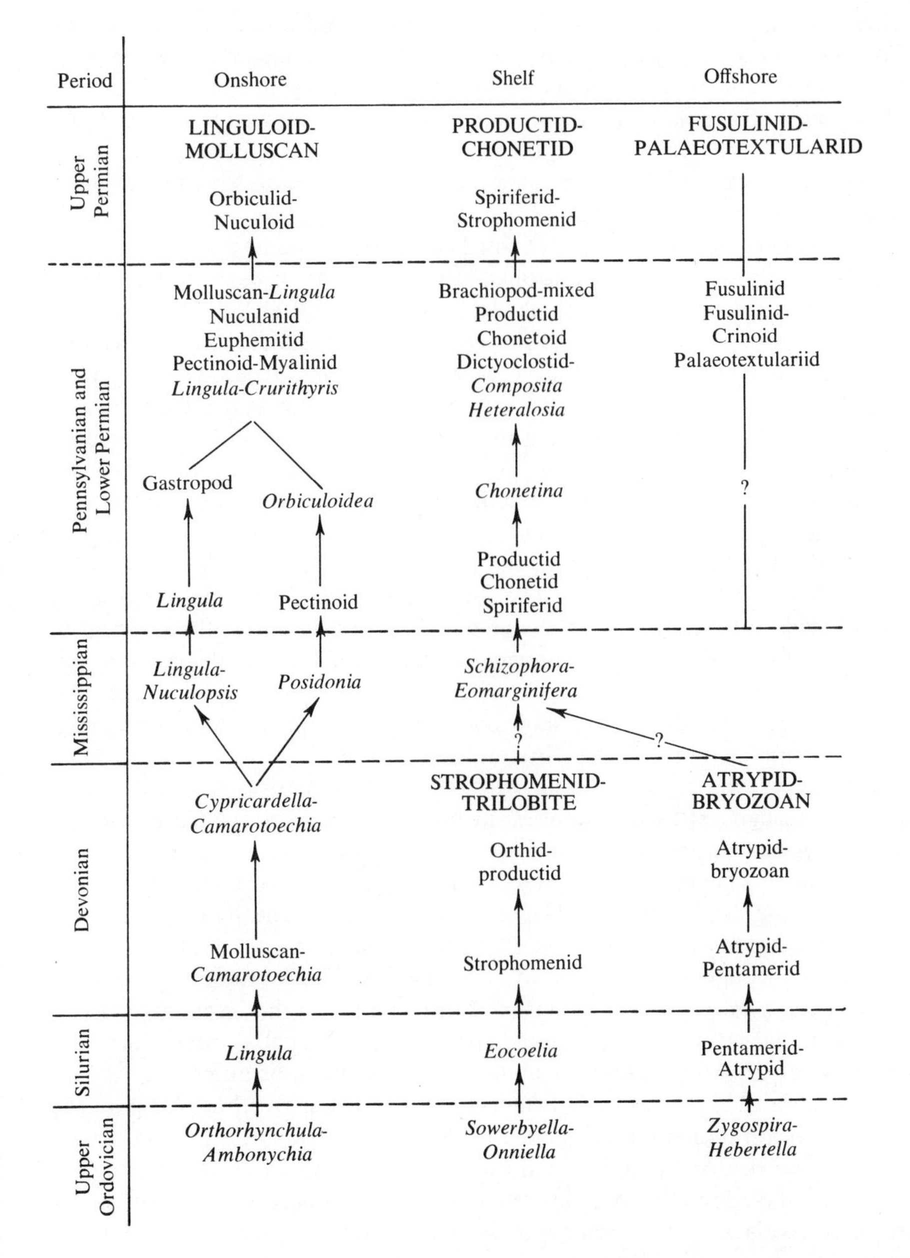

Fig. 7-14 Proposed sequence of Paleozoic marine invertebrate communities in clastic sediments. The communities and environments are both generalized. For references to community descriptions, see the original article. (Reproduced from *Palaeogeography, Palaeoclimatology, Palaeoecology,* 6: 45–59. P. W. Bretsky, Jr., 1969, Evolution of Paleozoic benthic marine invertebrate communities.)

requires comparative data from different provinces, and commonly these are on different continents. Additionally, the components of communities—species populations—are distinctly different and taxonomically diverse, whereas the components of populations—the individual members—may be treated as a statistical population. Furthermore, the incompleteness of the fossil record makes it necessary to gather much more data than would otherwise be required to test an inference.

Fortunately, excellent studies of fossil communities are now appearing regularly, although we are far from having a detailed history of benthic invertebrate associations. Enough data are available, however, that Bretsky (1969) has been able to propose a broad outline of the succession of Paleozoic community types in clastic sediments (Fig. 7-14). Five major associational types are recognized, which, together with an interpretation of their environmental settings, are as follows: (1) linguloid-molluscan, from onshore silts and sands; (2) strophomenid-trilobite, from offshore silts and muds; (3) atrypid-bryozoan, from onshore and offshore muds; (4) productid-chonetid, from offshore and possibly onshore muds and silty muds; and (5) fusilinid-paleotextulariid, from offshore lime muds. The fusilinids, atrypid, and pentameroid faunas, although recorded chiefly from offshore, evidently live onshore in suitable situations (possibly quiet water). Bretsky notes that the linguloid-molluscan associations appear to be the most conservative, even though they are in the least stable environment, and suggests that their adaptation to environmental fluctuations permitted their survival during periods when offshore communities, adapted to more stable conditions, were disrupted by environmental changes. The associational types recognized in this broadly based study embrace numerous associations that could be recognized as communities in the sense they are defined herein—as polythetic associations of particular species populations.

Clearly, fossil community studies can bear upon the most fundamental problems of ecology and evolution. They must be prosecuted locally in great detail, yet their scope must eventually be enlarged to encompass the faunas of many continents through long periods of time. Taxonomically, these studies must embrace the whole spectrum of preserved organisms, and the taxonomy must be very good, indeed, for much depends upon the recognition of species identities or dissimilarities between regions, upon valid estimates of the numbers of species in various faunas, and in understanding rates of change within lineages. The degree of erudition required of studies of this detail and scope is beyond any individual investigator, but individuals working in common cause can piece together a detailed picture of community evolution. A major problem in paleoecology has been the lack of common scientific goals among its practitioners, but the growing cohesion of the field now suggests that a common community of interest is developing which will lead to those traditions necessary for research continuity.

CHAPTER EIGHT

THE PROVINCIAL LEVEL

Everyone knows that the same animals and plants are not found everywhere . . . but that they are distributed so as to be gathered together in distinct zoological and botanical provinces, of greater or less extent, according to their degree of limitation by physical conditions, whether features of the earth's outline or climate.

—From E. Forbes and R. Godwin-Austen, *The Natural History of the European Seas,* Van Voorst, London, 1859.

Provinces may be defined as regions in which communities maintain characteristic taxonomic compositions. In other words, even in analogous environments some of the taxa are different in different provinces. Whereas in considering community composition, the chief concern was with the segregation of different species populations into different biotopes, on the provincial level the chief concern is with the segregation of entire species into different regions. A much-quoted "rule" that has been used to identify provincial regions is that at least one-half of the species living therein must be endemic (that is, native). However, there is no special reason to employ any particular arbitrary level of endemism, and it is in fact theoretically possible that a province could possess no endemic species at all and yet have distinctive communities (Fig. 8–1).

The species composition of communities inhabiting analogous biotopes in separate provinces will ordinarily be different, then, but need not be entirely

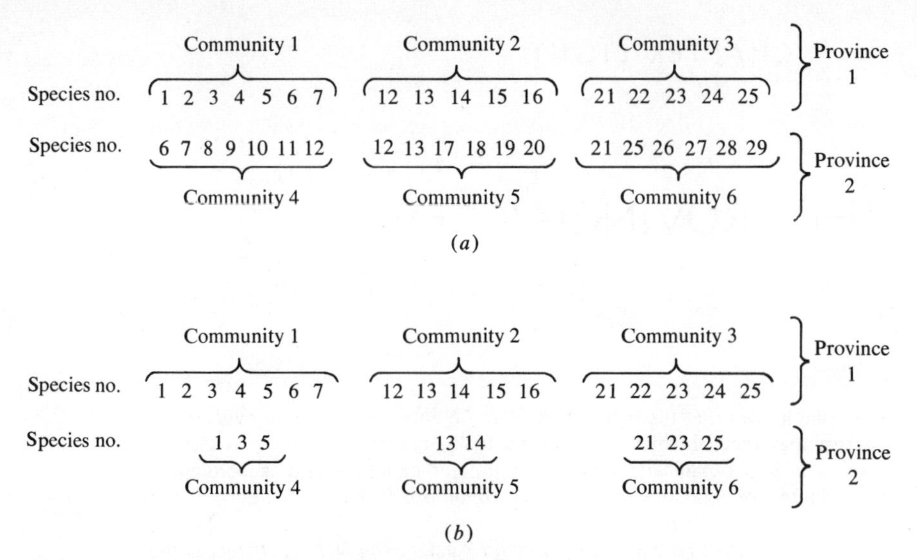

Fig. 8-1 Two patterns of provinciality. (*a*) The normal pattern, wherein communities in one province (communities 1, 2, and 3 in province 1) contain chiefly different species from communities in another province (communities 4, 5, and 6 in province 2), although some species range across the provincial boundary; the associations are nevertheless distinctive in each province. (*b*) Theoretically possible pattern, wherein province 2 has no endemic species, but province 1 has quite a number of them; therefore, province 2 has distinctive associations by virtue of lacking these endemics.

unique. Since communities are polythetic units, it is necessary that only enough of the species differ so as to form distinctive characterizing sets of species in each province. Thus, communities in analogous biotopes but in separate provinces will ordinarily be separate communities, for they will have separate characterizing sets even though they may share many species. However, it is not necessary that all communities in different regions be distinct in order for the regions to be considered provincially distinct. Provinces, like communities, must be regarded as polythetic units, so that it is only necessary that a certain number of the communities differ between provinces.

There are two major ways to regard provinces. One is as an association of taxa that must be explained in terms of the history and dispersal of their lineages and the limiting factors associated with their provincial occurrences; this is the usual content of the field of *biogeography*. The other way is to consider the functional aspects of the provincial system and the factors that regulate them, which we shall call *provincial ecology*. Clearly, these fields are closely related. It is convenient to examine some of the principles of biogeography first.

Biogeographic Patterns Have Ecological and Historical Bases

Biogeography is the study of the distribution of organisms across the face of the earth; it attempts to discover why organisms are found where they are, and why they are not found where they are not. A reasonably full answer to these questions for any particular stock would involve tracing the evolutionary events leading to the development of that stock, as well as tracing the concomitant environmental fluctuations and considering all the pertinent environmental-organism interactions. The goals of biogeography thus overlap broadly with those of ecology and paleoecology, historical geology, and with many aspects of evolutionary biology, although there is a special biogeographic approach and point of view that is related to the history of that field.

Many of the major outlines of present biotic distributions were known by the middle of the eighteenth century, and had been discussed in a more or less scientific manner as early as 1766 by Buffon. The writings of Darwin and Wallace first gave the subject a theoretical basis, however, and Wallace's volumes, *The Geographical Distribution of Animals* in 1876 together with his *Island Life* in 1880, provided a definitive statement of what was known of terrestrial animal geography and a classic formulation of principles. The main emphasis of Wallace and numerous other biogeographers has been the description of present distribution patterns with the assessment of biotic similarities among different regions. The patterns and regional resemblances may then be explained in terms of past dispersal of organisms along certain paths and the prevention of migration along others. Information on paleogeography and on distributional patterns of the past is very important in such interpretations. This approach has been called *geographical biogeography*, and the basic data are the past and present geographic ranges of animals. Good references in this field are by Darlington (1957) for land animals and Ekman (1953) for marine animals. The classic zoogeographical writings of Wallace and Darwin remain highly informative and significant.

The populations of most groups of animals are not dispersed evenly or randomly throughout their ranges, but are concentrated in those parts of the range where the environment is most favorable and which they happen to have colonized. Most species are actually absent from vast areas within their range boundaries, chiefly where appropriate habitats are not available. The factors that control these local distribution patterns have been discussed in preceding chapters. The study of locally limiting factors in order to explain the local distribution pattern is *ecological biogeography*. A good introduction to this field is given by Hesse, Allee, and Schmidt (1951) for animals.

The division of biogeography into ecological and geographical subfields is, of course, artificial from the point of view of nature, but the divisions are real enough academically as developments of different research traditions, although many workers have combined them. One subject which involves both fields equally, almost by definition, is the study of the ecological factors that serve to delimit the total geographic range of taxa. These factors are termed *barriers* by geographic biogeographers, and are sometimes classed as "physical" (oceanic barriers for land organisms, mountain ranges for low-altitude organisms, continental barriers for marine organisms, and so on) or "ecological" (such as climate); but all barriers seem more logically treated as ecological, subdivided into a number of kinds which include the "physical" kinds.

Species Ranges Are Controlled by Patterns of Limiting Factors

If we draw lines on a map closely circumscribing the areas inhabited by a species, we shall have outlined the species range. The factors that limit the spread of the species beyond its range are not necessarily or even usually different from those that restrict the species to certain habitats within its range; it is the geography of the limiting factors that makes the difference. For an example, take the present distribution of the marine snail *Tegula brunnea*, which lives along the western United States from southern California to Oregon.

Tegula is herbivorous, chiefly browsing on larger living "fleshy" algae and probably supplementing its diet with detrital plant material as well. It belongs to a rather primitive group of gastropods, the Trochacea, which are partly characterized by gills supported by a membrane that crosses the mantle cavity from the roof to a point near the floor on the left side, subdividing the cavity and forming a pocket against its left wall and roof. Half of the gill structures lie within the pocket and half within the rest of the mantle cavity. Water currents are pumped from left to right through the mantle cavity by ciliary action, and incoming water passes over the gills, aereating them. The membrane that creates the pocket provides support for gills which are relatively larger than those in allied forms that lack a complete, pocket-producing membrane. However, the pocket is so easily fouled by detritus and sediment suspended in the inhalent current that members of this group are hardly ever found on fine-grained bottoms or in other places where waters are frequently turbid.

Tegula brunnea is obviously prevented from extending its range inland by its inability to respire in air or to maintain osmotic equilibrium in fresh water, among other things—in other words, by the failure of the marine medium. *Tegula brunnea* can penetrate into deeper waters offshore only so far as firm substrates and appropriate algae are available, and is thus chiefly limited to inner sublittoral parts of the shelf. Within the inner sublittoral zone, however, large areas devoid of algae or with soft bottoms or both are unsuitable for colonization by *T. brunnea*. In the southern part of the region inhabited by *T. brunnea*, bodies of warm water appear alongshore, especially in embayments and to the north of headlands over shallow shelves, and these areas are avoided by *T. brunnea*. Further to the south, all waters are eventually above the limiting temperature regime of this species, and its range is terminated. The same temperature factors thus significantly affect the local distribution pattern of the species and form the southern range limit. To the north, temperature is again probably the chief factor forming the range endpoint. Distributional details are not available for this species, but for many forms local distribution patterns near their northern range endpoints reflect water temperature patterns, as the species become restricted to warmer sites.

Thus, the area inhabited by *T. brunnea* is a narrow, irregular band stretching along the California coast, bounded shoreward by habitat failure that is a function of the failure of the medium, bounded offshore by habitat failure that is geographically systematized by gradients of energy distribution and light penetration in deepening water, and probably bounded to the north and south by temperature barriers.

In addition to avoiding unsuitable portions of the environmental mosaic lying within their ranges, some species have *disjunct* ranges, occupying two or more distinct regions separated by some sort of barrier. For example, some species have "antitropical" distributions, occurring on each side of the tropics (Hubbs, 1952); presumably their populations are now separated by a temperature barrier. Many disjunct populations are *relicts.* A relict has been defined as a population (or taxon) that is isolated from its main center of distribution because it has been left behind when its ancestral population, which was widespread, has retreated as conditions changed (Ekman, 1953). A number of examples of marine invertebrate glacial relicts in European seas are known, remnants of species that were more or less widespread in what is now the Boreal zone during glacial times, and that have remained at isolated localities there even though their main populations are now in the arctic region (Ekman, 1953). Such disjunct ranges may be possible because of the recurrence of habitat factors characteristic of the species' normal range in a distant region, because the disjunct populations have distinctive gene pools, or because of some combination of these. The total local distribution pattern of a species can be viewed as the result of an interplay between the pattern of variation of the species and that of the environment. Of course, ranges are dynamic, and many of them fluctuate seasonally or periodically with other environmental or biological changes. Range borders may be treated as extreme positions, or as mean positions during average conditions of a given environmental regime; or, in the case of species whose ranges regularly fluctuate widely or who migrate long distances, the range must be specially treated, for example, as having two (or more) phases.

Many species have close ecological analogues with which they share a common border. In numerous cases, it is very likely the outcome of competition between two species with similar requirements and overlapping tolerances that determines the precise location of their common species border. We have already seen that populations of some species (usually the more highly specialized form) may displace or exclude other species from habitats which they could otherwise inhabit. Commonly, the geography of the contested habitats is such that they form a species range border and not merely a local population border. Although such competitors are often closely allied taxonomically, they certainly do not have to be so; all that is necessary is that they have a common limiting resource, and that one of them succeeds in utilizing the resource to the exclusion of the other in habitats which have such a geographic pattern that a distributional barrier occurs.

Range Size and Pattern Depend Chiefly upon the Vagility of Species and upon Geography and History of Limiting Factors

Taxa with limited ranges are called *stenotopic*, and taxa with wide ranges, *eurytopic*. Some taxa are so wide ranging that they occur in essentially every part of the world where their habitat requirements are met; these are termed *cosmopolitan*. *Vagility* is the ability of a form to disperse. Other factors aside, a relatively vagile form will attain a wider range than a less vagile form, or at least attain it sooner. Eurytopic forms may be of two kinds, important for our purposes to distinguish: those forms that are rather adaptable, and therefore achieve a wide range; and those that, although narrowly adapted, nevertheless achieve a wide range due to the broad geographic distribution of the environments to which they are adapted. Taxa of the latter sort can be called *ecologically eurytopic*. Similarly, taxa that are narrowly restricted by an extremely heterogeneous environment, even though broadly adapted, can be called *ecologically stenotopic*. A number of detailed dispersal models and examples are given by MacArthur and Wilson (1967).

Some species, such as those that originate by splitting, and owe their inception as distinct taxa partly to the conventions of definition, may possess wide ranges from their inception. However, many species with relatively small founder populations inherit narrow ranges that they may expand, and eventually the expanding populations will establish borders controlled by limiting factors which will act as barriers to further dispersal. The species will now have a range that includes all suitable habitats more or less contiguous to its area of origin. For species of high vagility, contiguity is not so necessary, and they may range into suitable habitats that are separated from the native range. For a given degree and type of vagility, the successful migration of a species to a separated habitat will depend upon the distance, the nature of the inhospitable environment to be crossed, the nature of the medium of transport, and the size of the habitat to be invaded. Now, if all these factors are identical for several suitable habitat areas that are separated from a species' native range, except that they are at different distances, the habitats nearer to the native ranges will tend to be invaded first. The longer the time available for colonization of any of the separated habitats, the greater will be the chances of success. Furthermore, environments fluctuate through time, so in any model approximating reality we must allow for changes in the nature of the barrier, in the size of the suitable habitat areas, or in other factors, such that the chances of colonization fluctuate in time; some periods are especially propitious for migration.

In other words, migration success is probabilistic, and so is the species' *total range* size, although the *local range* size is more deterministic. It is useful to consider this difference as definitive for these two types of range. Obviously, a given habitat separation so wide for a species of low vagility that migration has a low probability may be so narrow for a highly vagile species that migration is practically obligatory.

Pioneer populations of immigrant species will not ordinarily occupy all the suitable area in the newly invaded region, but will soon expand to fill the potential local range. When a new region is invaded, the chances of further migration to other suitable distant areas are often enhanced, and a species may spread in a stepping-stone fashion, from one area to another. Spread of a shallow-water species along the shores of an island archipelago or among scattered islands is a good example of this mode of range expansion. Another mode that may have especially important effects involves the broaching of a barrier separating an unusually tolerant or ubiquitous species from a vast region, such as the shelf of an entire continental coastline, from which it has formerly been barred; the resulting range increase can be spectacular.

Simpson (1940) has roughly classed dispersal routes with regard to their effectiveness in permitting organisms to pass. His terms were designed for terrestrial situations but are applicable to the marine environment; they are *corridors*, *filters*, and *sweepstakes routes*. Corridors provide a high probability of dispersal of most animals from one region to another. Thus, the Arctic shelf represents a corridor for the circum-Arctic dispersal of marine benthos. Filters provide a high probability of migration to some organisms but a low probability to others. A stretch of deep water between islands or shoals may be a filter to benthic invertebrates, being crossed most easily by those species with the longest pelagic larval lives. Sweepstakes routes provide only a low probability of migration to nearly all species, although given enough time some will cross. Many of the shelves on different continents are separated by sweepstakes routes across their intervening oceans, as between west Africa and eastern South America. If a region is isolated from all others except by sweepstakes routes, it will receive only a very odd, biased sample of the biota of other regions. Newly created islands with no biota of their own commonly support ecologically unbalanced communities, called *waif* biotas; their assembling has been in large part a matter of chance and not of selection under the processes that normally regulate community composition and structure.

As species will tend to spread from their area of origin to fill their native local ranges and then to invade suitable distant regions, where they will spread to fill the local ranges, there can be a relation between the age of a species—that is, the time since its appearance as a biological entity—and the area of its total range. In fact, a famous hypothesis by Willis (1922), the "age and area" hypothesis, holds in part that groups spread more or less regularly and steadily from their point of origin and that the area of their total range at any one time is a measure of their age. In this event, short-ranged species would all be newly evolved forms. However, many short-ranged species have been extant for long periods; some are members of ancient lineages that are relicts living in refugia after being eliminated from much of their ranges. There is even a hypothesis, due to Rosa (1931), suggesting that groups originate over broad areas and then become more and more restricted; in this view, part of a theory called "hologenesis," range area tends to be inversely related to age. As Simpson (1940) has pointed out, each of these two opposing views contains elements of truth. In combination, they describe a common pattern, which may be generalized as follows:

1. When a distinctive gene pool arises that fits its population to a certain habitat and also provides protection from contemporary biological pressures, the population will tend to expand throughout the local range to fill the habitat.

2. As time passes, the species expands into distant habitats by overcoming barriers to dispersal. The expansion need not be symmetrical—among the marine benthos it will hardly ever be—but will correspond on the average to the distribution pattern of dispersal probabilities associated with the barriers between suitable environments and the realized species range at any time. Many species will never enter perfectly suitable areas because of migration barriers, and some will not enter suitable areas well within the range of their potential dispersal owing to "chance." For example, suppose 100 closely contemporaneous species have an average duration of 5 million years, and that they have nearly the same powers of dispersal. There are regions suitable to each one of the species to which dispersal of that species has a probability of $p = 0.99$ in 5 million years. That is, the chances in favor of colonization are 99 to 1. Despite this highly favorable chance, one of the species will probably fail to disperse to a suitable area. See Simpson (1952) for an extended discussion of dispersal probabilities.

3. As more time passes, environments change; the original gene pool (or its derivatives) no longer provides adequate fitness for its populations, some of which are eliminated by climatic fluctuations, some by the rise or invasion of superior competitors. The total range is thus decreased and local ranges are compressed. Frequently, populations will survive in exceedingly disjunct regions, separated by barriers that are essentially impenetrable by the species in question without the presence of intermediate stepping-stone populations that have become extnct.

4. Since nearly all species seem to become extinct eventually (that is, few gene pools seem adequate to maintain a stock "forever" but must be revised to meet changing conditions), their ranges are eventually zero.

To summarize, we state that most populations are involved in range expansions and contractions not necessarily associated with species origins or extinctions. For example, the temperatures of much of the world ocean have undergone fluctuations during the series of Quaternary glacial and interglacial ages. When cool waters expanded, northern stenotherms were permitted to expand their ranges; no doubt many potential competitors had retreated before the falling temperatures. When warm waters expanded, many tropical species extended their ranges northward, but northern stenotherms had to retreat. These climatic changes resulted in significant changes in range sizes of many species and sometimes led to the invasion or abandonment of large areas—especially east-west trending shelves—and to the develpoment of disjunct patterns typical of species with shrinking ranges. Yet the expanding ranges do not necessarily indicate a youthful species, nor are the contracting ranges necessarily preludes to extinction.

In this example, the species that attained expanded ranges became ecologically eurytopic. There is no reason to believe that similar range fluctuations are not brought about by changes in gene pools and in environmental factors limiting range borders, so that species may become alternately more and less eurytopic through time owing to continual revision of their tolerances in interplay with a changing environment. Perhaps we can summarize this point by noting that range sizes and patterns may change in any direction and may change repeatedly, the changes depending upon the duration of a species' existence and the relative rapidity of appropriate environmental changes that happen to be associated historically with the species. There does not seem to be any biogeographic "cycle."

One important consequence of range fluctuations is that populations living on the periphery of the ranges are liable to evolve. Different genes may be favored in peripheral and central populations, especially during times of range expansion when the species is invading new territory, thus giving rise to genetically distinctive populations. During times of range restriction, peripheral populations may become isolated and follow different evolutionary pathways from the central population, leading eventually to speciation.

The Major Topographical Barriers in Modern Seas Are Related to Tectonic Processes

The biogeographic barriers of provincial magnitude usually involve major environmental discontinuities, at which the changes in a great variety of environmental factors become limiting for many species. Thus, a number of species on either side of the barrier are prevented from penetrating it. Of the barriers based chiefly on major changes in habitat, those between the land and the sea, and those between continental shelves or shallow island flanks and the deep sea, are most important for marine organisms. Since these barriers are clearly related to the topographic pattern of the earth's surface, which is largely under tectonic control, it is important to examine the current understanding of the tectonic processes that generate the patterns.

Figure 8–2 illustrates some of the major tectonic features of the world's oceans. According to the general theory of global tectonics, the outer part of the solid planet, the lithosphere, is divided into a few large rigid plates that rest on a plastic asthenosphere. The plates are bounded at their margins either by ridges (where crust is created by material rising from the earth's interior),

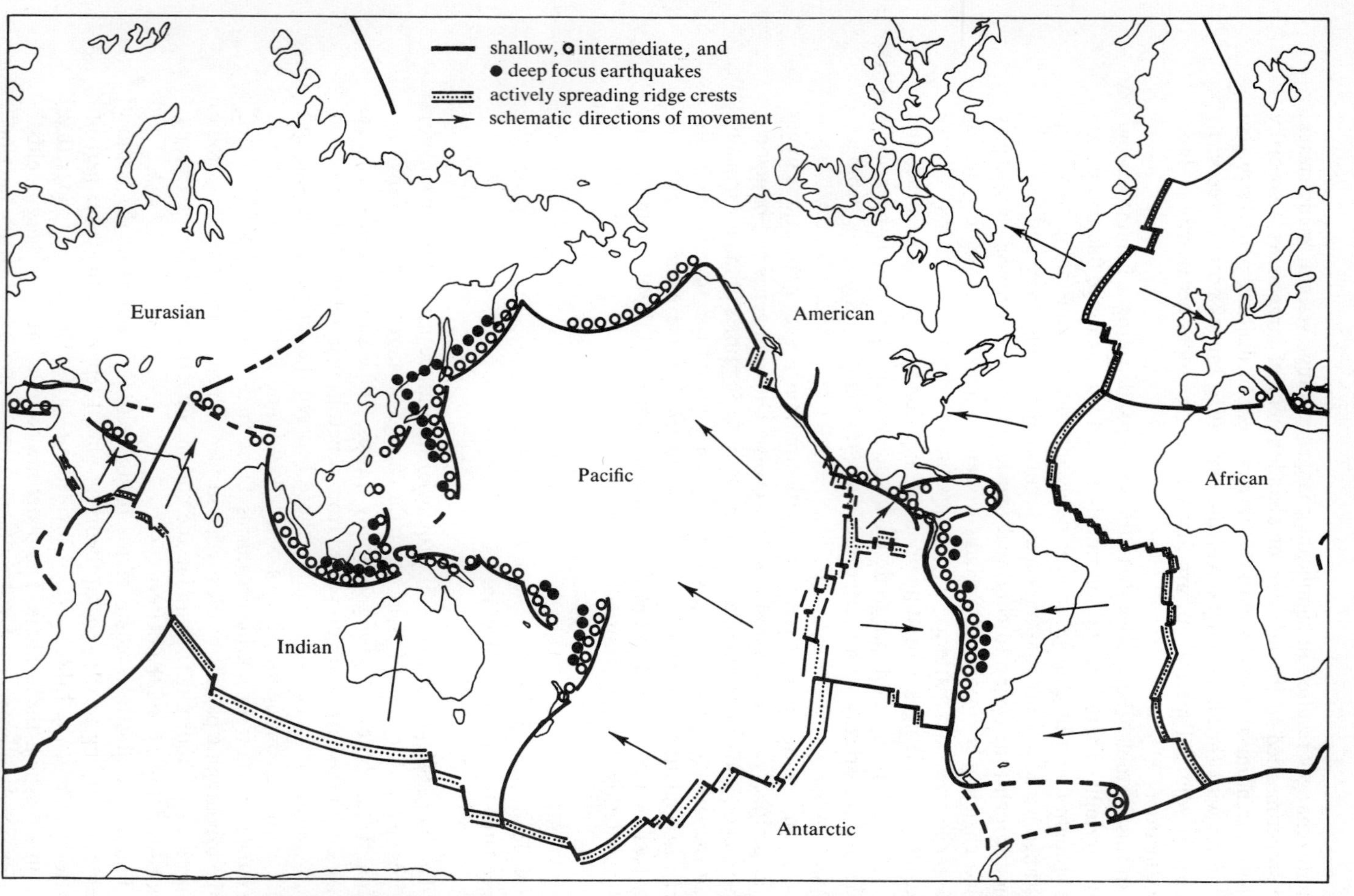

Fig. 8-2 The major lithospheric plates of the earth. Spreading rates at ridges vary from 1 cm per year (near Iceland) to 6 cm per year (equatorial Pacific). (After Vine, 1969, based partly on Le Pichon, 1968.)

by deep-sea trenches (where crust is consumed by sinking into the interior), or by fracture zones (which accommodate the differential movements between lithospheric plates or plate segments). Oceanic crust is generated chiefly at the ocean ridges and spreads laterally away from the generating zones on either side (Fig. 8–3). Most large plates bear one or more continents, which are relatively light and are passive riders on the thicker and denser plates. The plates eventually descend into deep-sea trenches and are consumed; the zones of sink-

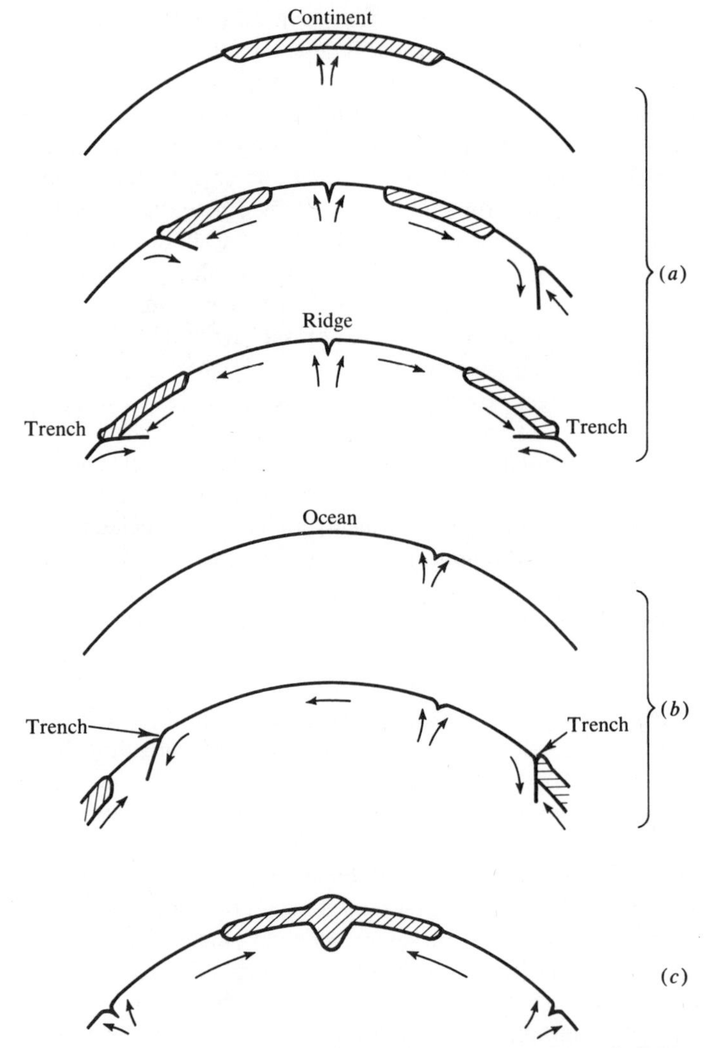

Fig. 8-3 Simplified diagram to illustrate continental fragmentation, assembly, and possible continental relations to ridges and trenches. (*a*) Rifting beneath a continent; (*b*) rifting beneath an ocean; (*c*) continental collision. (After Vine, 1969.)

ing plates are termed *subduction zones*. However, the lighter continents are not consumed, but when they collide with a subduction zone, the continental margin is deformed and the subduction zone is modified. Vine and Hess (1970) give a short account of the evidence that has led to this theory; see also Morgan (1968), Le Pichon (1968), and Isaaks, Oliver, and Sykes (1968).

Today, ocean ridges rise well above the general abyssal sea floor, but do not usually reach shallow waters, although isolated island peaks occur occasionally, such as the Azores on the mid-Atlantic ridge. In no case, however, are extensive chains of islands associated with ridges at present, although there is one large island on a ridge—Iceland—that probably aids significantly in the dispersal of shallow-water marine biota. In general, the appearance and spreading of new oceanic crust from ridge crests create broad topographic slopes inclined gently away from the crests and forming deep ocean floor. The ridges and their spreading flanks are cut by numerous transverse fracture zones (Fig. 8–2). These zones are transform faults, created by differential spreading of oceanic crust from different segments of the ridges and accommodating the differential movement. Volcanoes are generally not present along the fracture zones today. As a result, the ocean ridges, together with their flanking slopes and associated fracture zones, are usually parts of abyssal barriers that block the dispersal of most shallow-water organisms.

Subduction zones, like ridges, are primary tectonic features. If continents lie close to them but opposite their consuming margins, such as where South America lies close to the Peru-Chile Trench, mountain systems, such as the Andes, are formed by vulcanism related to crustal consumption, producing chiefly andesites and rhyolites. When no continents are closely associated with the plate margin formed by subduction zones, however, volcanic island arcs, also chiefly of andesitic and rhyolitic rocks, are formed opposite the consuming margin of the subduction zone. An extensive subduction zone is marked by the east-west chain of trenches, mountains, and island arcs lying between Burma and the New Hebrides, bordering continental plates to the northwest (Burma and India) and in their south-central part (Australia–New Guinea). Subduction associated with this chain chiefly accommodates northward spreading from the east-west trending Indian-Antarctic ridge (Fig. 8–2). Because this zone forms the northeastern margin of the Indian lithospheric plate, it must be fairly continuous when active. Thus, the associated island arcs must provide a fairly continuous and extensive shallow-water area. Subduction zones, then, are the biogeographical opposites of ridges and fracture zones in that their associated islands form excellent biotic dispersal routes rather than barriers for shallow-water marine organisms. Continents themselves provide clear pathways to dispersal of shallow-water biota along their shelves, which although narrow in places are essentially continuous; at the same time, their shorelines form terrestrial barriers to marine organisms. Dispersal between continents should be chiefly a function of their geographic proximity, or of their connection by subduction zones.

One additional oceanic feature deserves mention here; this is the midplate volcano, which usually occurs in a swarm or line and provides shallow-water habitats that are associated neither with continents nor with island arcs. These volcanoes are chiefly of basaltic and tholeitic rocks and must be generated above thermal highs in the lithosphere or asthenosphere, the origins of which are unknown; they have not yet been integrated into the theory of plate tectonics. The volcanoes themselves must move with the crust in which they lie (see Wilson, 1963, and Menard, 1969), although if the hot spot is in the asthenosphere it may generate a whole succession of volcanoes as the lithospheric plate passes over it. Thus, an island chain may be formed that is *not* associated with a subduction zone; the Tuamotu and Hawaiian Islands appear to have such origins. If a sublithospheric hot spot were to occur beneath a ridge, it might cause a local topographic high (such as Iceland), and, if persistent, it might generate a chain of volcanoes migrating off the ridge in each direction. However, the volcanoes would presumably move down the ridge slopes and into deeper water, so that even if they continued to grow somewhat during their ridge flank existence, they might well not reach shallow depths. We can only assume that island chains of this sort probably do not commonly extend up rises to broach barriers at ocean ridges, since none do today. However, the presence of numerous midplate volcanic systems in the northwestern Pacific, a region of older crust, suggests that they may have been generated more commonly at times in the remote past than in the more recent past, so that their potentials as transridge dispersal routes cannot be dismissed completely.

The ridges generate abyssal ocean floor that acts as a barrier to shallow-water biotas, the island arcs provide dispersal routes, and the continents form barriers, whereas their marginal shelves form dispersal routes. We would therefore expect to find distinctive provincial biotas on opposite sides of oceans containing ridges and on opposite sides of continents, and similar biotas along continuous stretches of either island arcs or continental shelves. Biotas also should extend from chains of island arcs to continental shelves or between either of these and midplate volcanoes, whenever they happen to be in proximity. Figure 8–4 depicts these relations. Clearly this picture is very incomplete, for an additional major component of provinciality is associated with the latitudinal thermal gradient.

The Major Thermal Barriers in Modern Seas Are Related to Climatic Determinants

In Chapter 4, we have seen that temperature adaptation is of fundamental importance to most invertebrates because temperature controls their metabolic activity rates; moreover, so many physical parameters in the sea are temperature dependent that oceanic temperature changes involve changes in whole arrays of environmental factors. It happens that the ocean today has a high thermal gradient, from near 30°C at the equator to below 0°C in high latitudes;

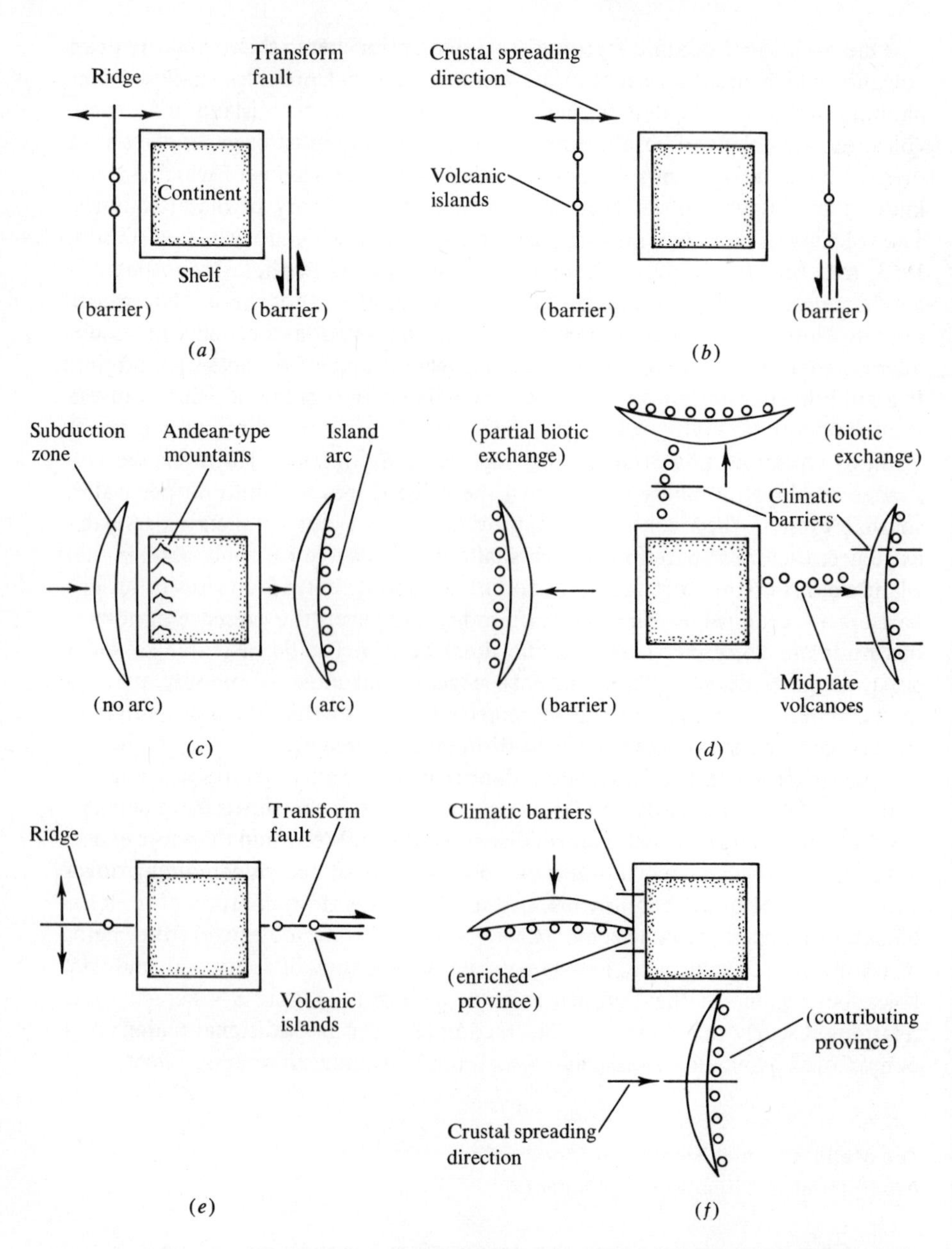

Fig. 8-4 Some biogeographic possibilities of plate-tectonic elements. (*a*) When ridges are near and parallel to continents, they form barriers to shallow-water organisms, since they are creating deep-sea floor. When transform faults are near and parallel to continents, they also form barriers when one side at least represents deep-sea floor. (*b*) The same conditions occur if these elements are distant from the continents, assuming no other elements intervene. (*c*) Subduction zones dipping beneath continents form barriers, for they have deep-

thus, few invertebrate species can tolerate this entire temperature range. Instead, most species tolerate only a narrow range that limits their latitudinal distributions, as in the case of *Tegula brunnea*. Of course, the latitudinal range end points of many species are determined locally by factors other than temperature, so that these species can expand their latitudinal ranges if their other limits are expanded. Nevertheless, experiments have indicated that many species range near the limits of their temperature tolerances.

If the thermal regime graded gradually or evenly from high- to low-latitude conditions along a north-south continental shelf, then we might expect that the compositional changes in the shelf biota would also be gradual and regular within any given biotope, and each community would form a sort of cline, with no sharp breaks in compositions that could lead to the discrimination of distinctive biotic assemblages. If on the other hand, the latitudinal change in thermal regime is uneven, with long stretches of shelf displaying very similar conditions being bounded by areas where temperature changes rapidly, then we would expect a large number of species to find the sharp changes limiting. That is, the discontinuities in the gradient would concentrate the limiting factors of numerous species within a small area, causing a sharp latitudinal change in the composition of the biota.

This is precisely the situation along continental shelves today. In some places, the change in thermal regimes is very marked—for example, where two thermally distinctive water masses converge, as between tropical and temperate waters in the eastern margins of oceans—and there the biotic changes are equally sudden. As mentioned in Chapter 3, changes from tropical to temperate waters may involve a change in over 80 percent of the species. Where such changes occur, eight out of ten species are different in the average temperate community than in the analogous tropical one. Clearly, these temperate and tropical associations form different communities, which change at the provincial boundary. Provincial boundaries may be very sharp or, more often, comprise zones of tens of miles or even hundreds of miles in width, within which the distinctive biotas interfinger or intermingle.

The present topographic configuration of the lithospheric plates and of the

---

sea floor on their seaward side and create Andean-type mountain systems on their landward side. If they are near to but dip away from continents, there may be some faunal exchange with the organisms living on their associated island arcs. (*d*) If subduction zones are parallel to but far from a continent, the intervening deep-sea floor will ordinarily form a barirer. However, if midplate volcanoes lie in the intervening ocean, the possibility of biotic exchange is opened. (*e*) Ridges or transform faults abutting continents at high angles will not ordinarily provide for faunal exchange. Of course, there would ordinarily be a transform associated with the ridge-continent junction, and a subduction zone and a ridge associated with the transform-continent junction. (*f*) Subduction zones at high angles to a continent provide clear pathways of dispersal; the character of the interchange would depend upon climatic barriers and other factors. (After Valentine, 1971b.)

continents is such that extensive north-south coastlines are developed, as along both sides of the Americas. The coasts support chains of latitudinally distinct provinces along each north-south shelf, the number of provinces depending to a large degree upon the steepness and extent of the latitudinal temperature gradient, and their boundaries determined by the distribution of water types over the shelf and of latitudinal changes in the hydrographic regime. One of the best-studied north-south coastlines is that of temperate-to-Arctic latitudes along western North America, which may serve as an example of latitudinal provinciality. Figure 8–5 presents the results of a numerical analysis of the recorded latitudinal ranges of all the shelled benthic mollusks known to be living along the continental shelf of western North America between the southern tip of Baja California, Mexico, and Point Barrow, Alaska (Valentine, 1966). The analysis is simple; the shelled molluscan species living within each degree of latitude are compared with the species living within every other degree of latitude, and a coefficient is calculated to express the similarity in each case (Jaccard's coefficient was employed for Fig. 8–5). The more heavily shaded squares are those between degrees of latitude with the highest species similarities, as measured by the coefficient—in other words, they are a function of the number of species in common between the latitudes. Adjoining degrees of latitude are compared with each other along the diagonal of Fig. 8–5. There are obviously regions with high species similarity, as between latitudes 38 and 46°, and regions with low similarities, as between 34 and 35°. Clearly, the fauna on either side of the 34–35° region will be rather different, owing to the faunal change that is indicated by the low similarity there. In Fig. 8–6, the coefficients of similarity along the diagonal of Fig. 8–5 are clustered, so that regions of greatest faunal monotony and those of greatest faunal change can be seen in a different perspective. Seven segments of coastline have relatively monotonous molluscan faunas, separated by regions where the faunal change is great. These correspond to provinces (or subprovinces) and their boundary regions, respectively.

The environmental basis of this pattern is clearest in the south. The faunal break at 27–28° accords with the change from Pacific equatorial water to the transitional water of the California Current in summer (Chapter 4). The next important break, at 34–35°, marks the position of isothermal clustering localized by the semipermanent gyre south of Point Conception (Chapter 4). The breaks at Monterey Bay (37–38°) and Puget Sound (47–48°) seem related to local changes in the hydrography due to nearshore topography, and not to changes in the general hydrographic regime, and they form less distinctive boundaries. The break at Monterey Bay seems to be based on the protection afforded the waters of the bay from intermixing with cool upwelling waters by the local topography, resulting in the occurrence there of many species otherwise known only to the south. The break at Puget Sound appears due to the presence of numerous glacial inlets or fjords to the north, within which summer water temperatures are quite high, giving this stretch of coast a characteristic temperature regime. The more distinctive change near the Queen Charlotte Islands

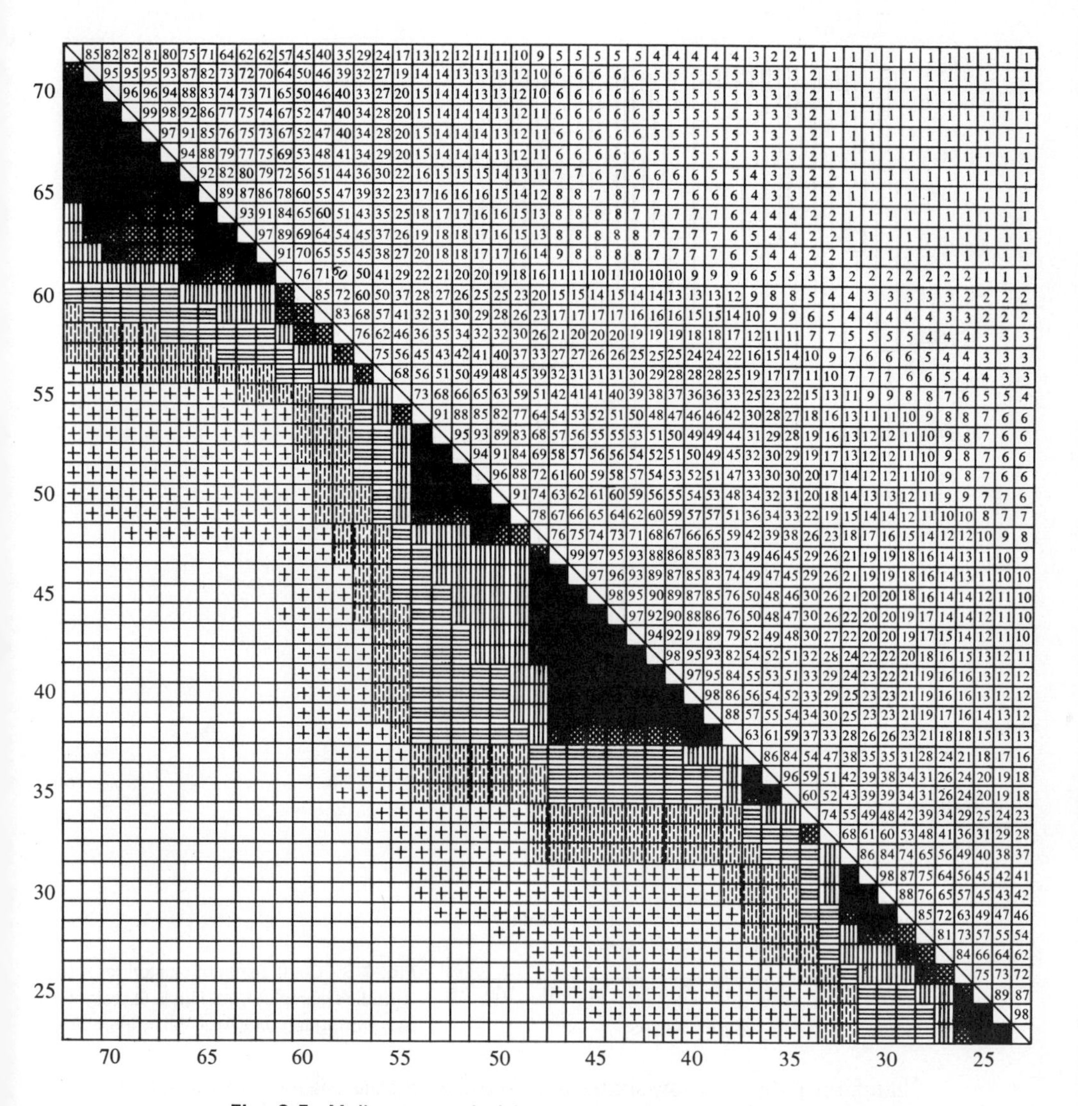

Fig. 8-5 Molluscan provincial patterns of the extratropical northeastern Pacific shelf. The shaded patterns in the matrix represent 15-point classes of Jaccard's coefficient; the squares each represent one degree of latitude; all possible latitude pairs are found in the matrix. The average species difference between adjacent provinces exceeds 50 percent. (After Valentine, 1966.)

may be owing to the change from California Current water to Alaska Current water in that region. The low similarities found along the latitudes between 54 and 60° are probably artifacts of the coding system employed in calculating the similarity coefficients, for between these latitudes are two coastal segments for each degree, one on either side of the Gulf of Alaska (Fig. 8–5). On the other

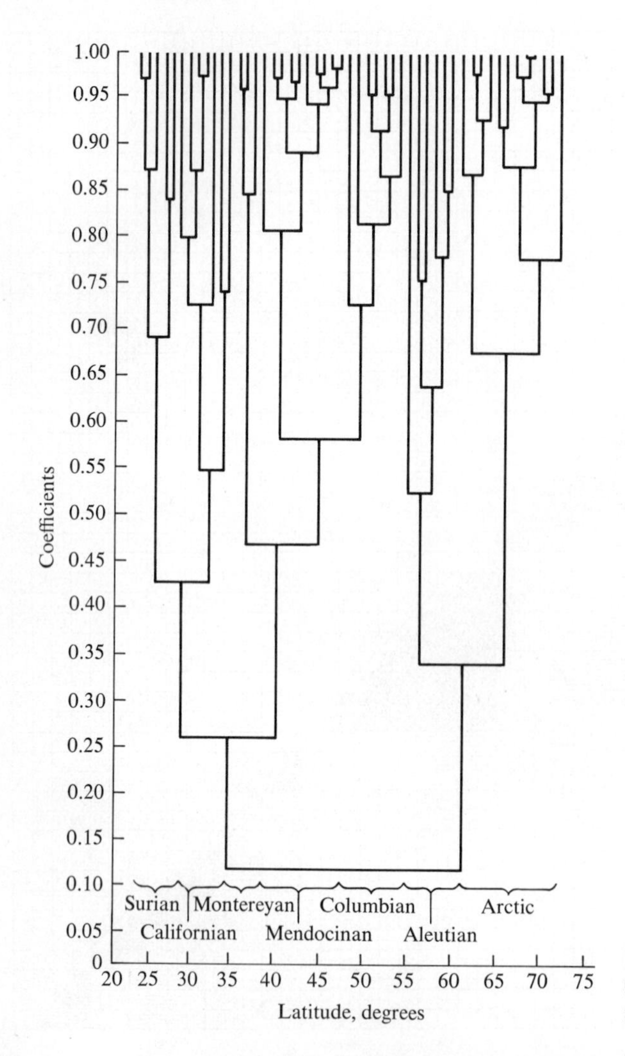

Fig. 8-6 Coefficients of clustering of the latitudinal samples from Fig. 8-5; weighted pair-group method. (After Valentine, 1966.)

hand, the patches of low similarities found along the coast of southern California and Baja California are probably real, and reflect the presence there of unstable upwelling regions alternating with more stable patterns, and of gyres and eddy currents.

Provinces, then, can be recognized on the basis of their species compositions. Provincial biotas include endemic species found nowhere else. They usually also include numerous species found elsewhere but not in association. For example, the Californian province includes numerous species that range southward into the Surian region but not northward into the Montereyan area, together with species that range northward into the Montereyan or beyond but not southward into the Surian. These two distributional elements are provincially associated only in the Californian. Clearly, different provinces frequently share many species in common, and the latitudinally contiguous provinces are separated by thermal barriers that are only filters. Even for shelf provinces that are separated by oceanic barriers, such as those on opposite shores of the tropical Altantic, dispersal of planktonic larvae is fairly common and numbers of species live on both sides (Scheltema, 1971). Nevertheless, the species associations are unique in each province.

Thus, the faunal pattern fits the hydrographic pattern rather closely. Whenever the hydrographic regime changes, or is modified by local features, such as large or deep bays, numbers of species find the change limiting. Where the regime is relatively monotonous along wide stretches of shelf, the fauna is relatively monotonous also. These principles appear to hold throughout the world, at least wherever provinces have been studied with care. The provinces of western Africa and Europe, the Altantic coast of North America, and Australia, for example, are relatively well known insofar as their general climatic-hydrographic regimes and their boundaries are concerned. In all cases, the provincial boundaries are marked by marine climatic changes, usually localized by topographic irregularities. The geographical regions blocked out in shallow waters by the patterns of continents, ridges, island arcs, and so on, are, then, further subdivided latitudinally into provinces blocked out by the climatic regime.

Figure 8-7 depicts the major marine shallow-water provinces at present; a few general features require mention. First, the provinces on the equator should average twice as broad latitudinally as more poleward provinces, simply because the temperature gradient is repeated to both north and south. Second, the Arctic province has a broad latitudinal extent, probably owing to the fact that water temperatures are near freezing in this whole region in winter, and thus the regime is similar and the thermal range is low. Third, some coastlines support more provinces per latitudinal segment than others. A good example is eastern North America, with four extratropical provinces south of the Arctic Ocean, and western Europe, with five. Fourth, east-west provinces may be of vast extent; the Indo-Pacific province stretches from east Africa to the Tuamotu Archipelago and onto some rather isolated outlying islands, being prevented from reaching tropical west America by the east Pacific Barrier (Ekman, 1953). This barrier is composed of the San Andreas and related faults in the north, which are interpreted as transform faults bounding the Pacific Plate, and by the East Pacific Rise (an ocean ridge) in the south.

Knowledge of the living biota is as yet too spotty to permit quantification of the patterns of species resemblance among all the world's marine provinces. For extratropical western North America, however, enough data are available to make interprovincial comparisons. In Table 8–1, the adjacent shelf provinces and subprovinces in this region are compared for bivalves and gastropods, using Jaccard's coefficient of similarity (Table 7–2). This approach is useful in the fossil record. Williams (1969) has employed Simpson's coefficient (Table 7–3) to evaluate similarities among Ordovician brachiopod faunas, and J. G.

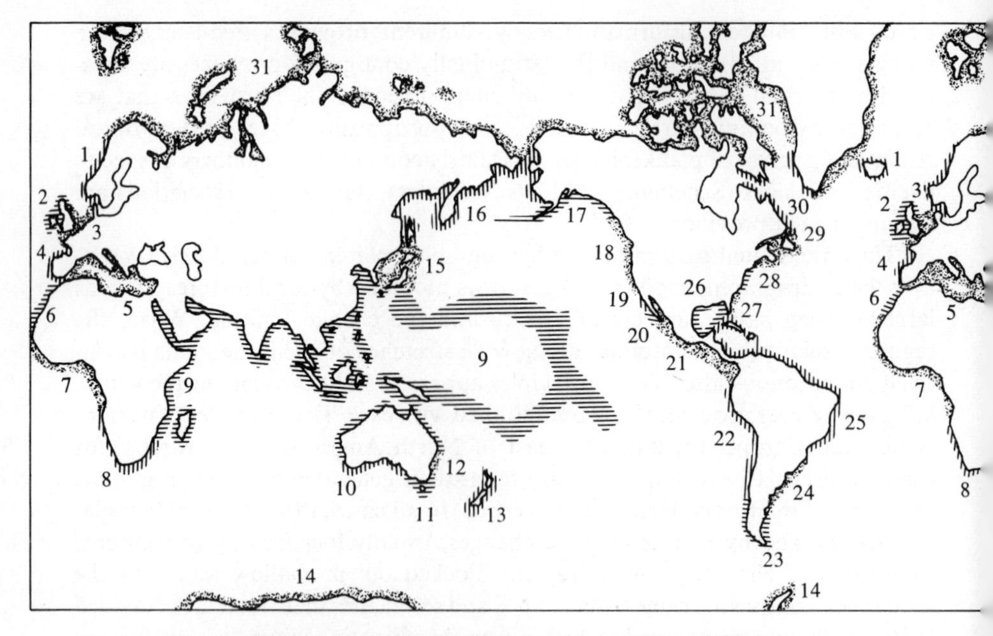

Fig. 8-7 Marine molluscan provinces of the world's shelves. (1) Norwegian. (2) Caledonian. (3) Celtic. (4) Lusitanian. (5) Mediterranean. (6) Mauritanian. (7) Guinean. (8) South African. (9) Indo-Pacific. (10) South Australian. (11) Maugean. (12) Peronian. (13) Zealandian. (14) Antarctic. (15) Japonic. (16) Bering. (17) Aleutian. (18) Oregonian. (19) Californian. (20) Surian. (21) Panamanian. (22) Peruvian. (23) Magellanic. (24) Patagonian. (25) Caribbean. (26) Gulf. (27) Carolinian. (28) Virginian. (29) Nova Scotian. (30) Labradorian. (31) Arctic.

Table 8-1 Similarity of Living Bivalve and Gastropod Species Between Adjacent Pairs of Molluscan Provinces and Subprovinces in the Extratropical West American Shelf, Measured by Jaccard's Coefficient

Provincial or subprovincial pair	*Jaccard's coefficient*
Aleutian–Arctic	33
Oregonian–Aleutian	31
Columbian[a]–Aleutian	45
Mendocian[a]–Columbian[a]	54
Montereyan[a]–Mendocinan[a]	58
Californian–Montereyan[a]	46
Californian–Oregonian	38
Surian–Californian	38

SOURCE: After Valentine, 1966.
[a] Subprovince.

Table 8-2 Similarity of Devonian Brachiopod Genera Between the Appalachian and Great Basin Regions (Column 1) and the Appalachian and Some Canadian Regions (Column 2) During a Series of Time Intervals, Measured by the Provinciality Index

TIME INTERVAL	REGIONS COMPARED	
	1 *Appalachian–Great Basin*	2 *Appalachian–Western and Arctic Canada*
Frasnian		2.50
Taghanic		1.83
Eifelian–Middle Givetian	0.28	0.29
Emsian	0.46	0.18
Middle and Late Siegenian	2.06	0.12
Gedinnian–Early Siegenian	0.58	0.47

SOURCE: After J. G. Johnson, 1971.

Johnson (1971) has devised a special coefficient, termed the Provincial Index (PI), to study similarities among fossil provinces. The PI is expressed as

$$\mathrm{PI} = \frac{C}{2E_1}$$

where C is the number of taxa in common between two regions and E_1 is the number of taxa restricted to only the region with fewer restricted taxa. Using the number of restricted taxa for the region with the fewer of these makes the coefficient conservative, for it helps to reduce the effects of unequal knowledge of the regions being compared. In Table 8–2, Devonian brachiopod genera from successive time intervals in the Appalachian Province are compared with those from two other regions by the PI. The barriers between the Appalachians and the Great Basin were most easily crossed during the Middle and Late Siegenian, a time when the similarity between the Canadian localities and the Appalachians was very low. In general, total diversity in these regions would be lowest when the faunas are most cosmopolitan. The use of similarity measures between ancient provinces can thus indicate shifting patterns of endemism and migration in time, forming a basis for interpretations of topographic and climatic barriers and of diversity trends.

Spatial and Temporal Environmental Heterogeneity Affects Provincial Structure Significantly

Provincial ecology is the study of provincial structure and function, which is regulated and mediated by the provincial environmental regime. Just as population functions arise from the advantages accrued to individuals from certain functions that are then favored by selection, and just as community functions arise from these population functions, so the functions of a province may be considered to arise from the functions of its component communities. It is possible to consider a province as a sort of super-extensive community, which inhabits the whole spectrum of biotopes within the provincial boundaries and has a characteristic taxonomic composition and ecological structure. Two different provinces may have closely analogous communities; each may have communities inhabiting rocky shores, sandy beaches, muddy bay bottoms, eel grass stands, subtidal sands, and other common biotopes. These analogous communities may resemble each other closely, far more closely than, say, a rocky-shore community from one province resembles a muddy-bottom community from another. Nevertheless, there will ordinarily be similar changes in the structure of each pair (or of the majority of pairs) of analogous communities when we proceed from one province to the other (for example, all communities in one province may contain more species than analogous communities in another province).

The provincial structure depends heavily upon the spatial heterogeneity of the provincial environment. Clearly, a province developed along a sandy coastal plain on a rather flat shelf will be relatively homogeneous, with perhaps coarse onshore sand grading offshore into finer substrates. On the other hand, a province that is topographically varied owing to being in a tectonically active region may contain a wide range of patchily distributed sediment and rock substrates along a varied coastline that may range from rock cliffs to deep embayments; thus, a great variety of biotopes is present. The former example could be said to represent a rather small biospace, and would ordinarily be inhabited by a biota that changes only gradually in composition and structure from shallow to deep subtidal zones. It is possible that in some parts of this environment the gradient could be interrupted by a discontinuity caused by biological activity. For example, group amensalism of burrowing deposit feeders might modify the substrate conditions in the deeper, finer-grained sediments so as to cause a relatively sharp environmental change and a correspondingly sharp biotic change. Nevertheless, this monotonous province would support relatively few different communities and would therefore have a relatively low provincial species diversity. The latter province with the varied biotopes would support a great number of communities, not only because of the varied biotopes but probably because of varied hydrographic conditions as well, which would be caused by the irregular topography; provincial species diversity would be relatively high. Diversity within analogous communities might nevertheless be about the same in both these provinces.

In general, there are intraprovincial diversity trends that are controlled by the geography of the diversity-regulating factors we have already discussed. Usually, high intertidal communities will be less diverse than low intertidal ones, and low intertidal ones less diverse than subtidal ones, and shallow subtidal ones less diverse than deep subtidal ones, and so on, for any given biotope type. However, proceeding from shallow rocky shore to subtidal sand flats, diversity would ordinarily rise in descending the rocky intertidal into more stable lower zones, then fall upon reaching the more homogeneous subtidal sands, and rise again upon proceeding to deeper sandy bottoms with more stable environments. Increasing diversity with depth appears to be the normal pattern in fossil communities, both in very shallow water (including the intertidal zone, as interpreted by Walker and Laporte, 1970) and from shallow water to moderate depths (Bretsky, 1970; Sutton, Bowen and McAlester, 1970; Stevens, 1971). Presumably, diversity increases further in proceeding to deep water, but the fossil record of deep-sea environments may be particularly poor because of the small size, thin skeletons, and low density of the deep-sea benthos, coupled with a generally unfavorable environment for the preservation of calcareous material. The deep-sea record, even when preserved, is really not comparable with the shelf-sea record.

Since marine provinces usually consist of regions with relatively uniform hydrographic regimes (although they may be unstable regimes) and climates, there is probably a characteristic production for each, based upon the nutrient supply from upwelling or runoff and on the solar radiation. Similarly, the character of the fluctuation of trophic resources, and thus of productivity, would tend to be tied into the hydrographic and climatic regimes, and thus to be characteristic for any given provincial region. Inasmuch as these factors are probably major determinants of the diversity of species populations within communities, there should be a characteristic diversity, community for community. It is evident, however, that certain special circumstances could provide the basis of local community structures which differ from the general provincial style—regional flooding of a few biotopes from large seasonal estuaries, for example.

Thus, provinces ordinarily exhibit differences in climate, hydrographic regime, productivity, temporal heterogeneity, and spatial heterogeneity, all major factors in determining the ecological structure. Provinces lying in similar climatic zones along coastlines that happen to be physically similar may commonly resemble each other relatively closely, even though separated by barriers to biotic dispersal. It is between two such provinces that migration, once accomplished, is most likely to be successful. The warm temperate provinces of eastern and western North America or cool temperate provinces of western Europe and Australia are similar provincial pairs at present.

Figure 8–8 depicts the species diversity patterns of marine bivalve faunas from different regions of the world. These data were assembled by a number of different workers at different times. Many of the diversity figures represent provincial diversities—the number of species of Bivalvia known to be living in a biogeographic province. Other figures do not represent entire provinces but rather some special region within a province that is large enough, and that includes a range of biotopes great enough, to be considered as somewhat representative of provincial diversity levels. Thus, the figures are not strictly comparable at all, and we must be cautious in their interpretation. Furthermore, in

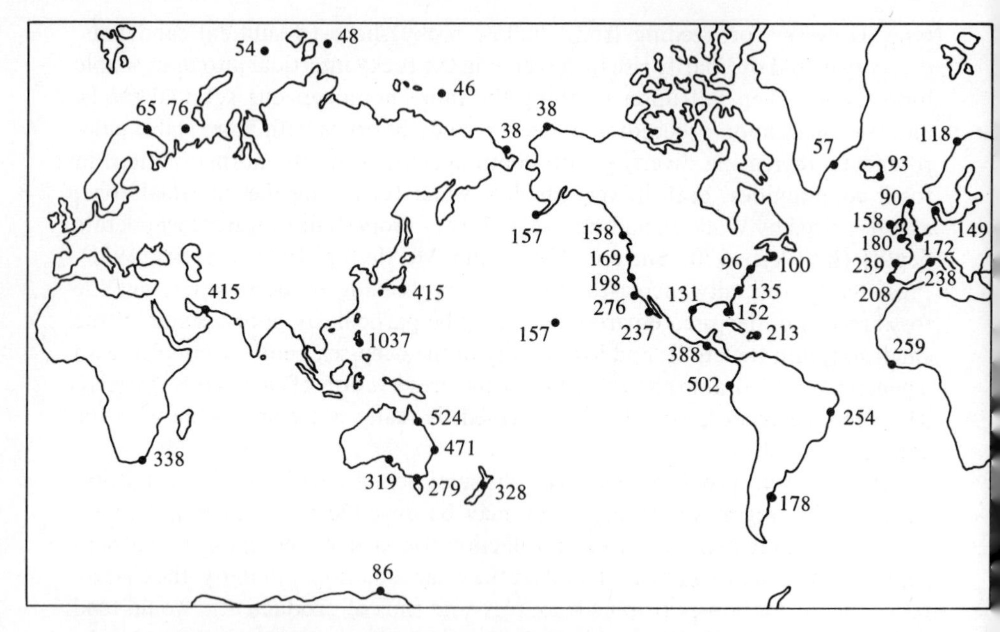

Fig. 8-8 Pattern of species diversity of marine Bivalvia on the shelves. Numbers represent total shelf bivalve fauna for provinces or for regions of subprovincial extent. The numbers thus have been compiled from lists and monographs of varying degrees of completeness, and represent regions of varying size. The significance of the numbers is therefore variable. However, the trends in diversity that these bivalves display are similar to the trends displayed by other major taxa, whenever data are available, so they are probably representative of trends within the marine shelf ecosystems. (After Stehli, MacAlister, and Helsley, 1967, with additional data from Valentine, 1971b.)

considering interprovincial diversity, it is dangerous to employ the diversity pattern of only a single taxon, such as the class Bivalvia, because special adaptive features perhaps related to the morphological ground plan of the class might cause it to be unrepresentative of the entire biota at all localities. For example, bivalves appear to form a higher proportion of the provincial biota along the southeastern United States, where the substrates are chiefly sands and muds, than along the southwestern United States, where the substrates include much rock and gravel and support a larger epibenthic fauna. The ratio of gastropods to bivalves, for example, increases from the temperate eastern U.S. shelf to the temperate western U.S. shelf, and in the reef belts of the central and western Pacific it is much higher. However, the trends displayed by the bivalve data are supported by diversity trends from other large taxa whenever data are available, and they may be used, with caution, as the basis of the following discussion. Figure 8–9 displays the diversity patterns of reef-forming coral genera in the Indo-Pacific; it will supplement Fig. 8–8 in the following discussion.

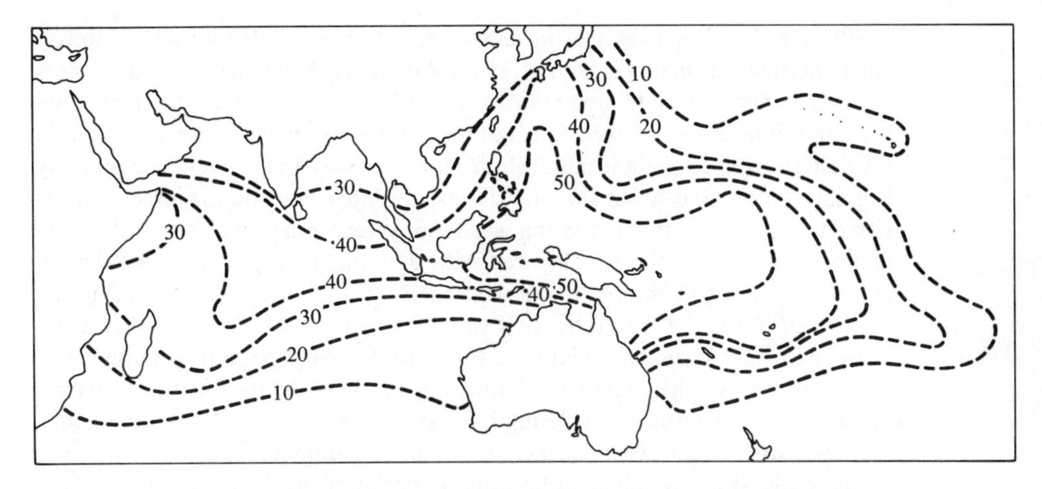

Fig. 8-9 Generic diversity of hermatypic corals in the Indo-Pacific, indicated by lines that connect regions with equal numbers of genera. (After Wells, 1954.)

In the Indo-Pacific, the highest diversity is centered on the shelves of small continents and island shoals from Borneo eastward to the Gilbert Islands in low latitudes, with a northward high-diversity extension toward Japan. The northward extension is due to the Kuroshio Current, which transports warm tropical water and invertebrate larvae to relatively high latitudes. The Indo-Pacific benthic diversity is the highest in the world. The reasons can be inferred from the principles discussed previously. First, this is a low-latitude province, so that solar radiation is relatively nonseasonal. Second, the islands and small continents surrounded by a large ocean favor a highly maritime climate with a relatively stable water column. Third, the coral reef complexes provide a spatially heterogeneous environment. Fourth, deep-sea regions between the many disjunct shallow-water areas may act as filters and favor the isolation of some populations.

Although the Indo-Pacific biota is most diverse along the more or less continuous island arcs of subduction zones and associated shelf areas, it does extend onto midplate volcanic islands (as the Gilberts, Marquesas, and Tuamotus) and onto the shelves bounding large landmasses (India, east Africa) whenever there is a clear pathway of dispersal thereto. The diversity along the large continental masses and on somewhat isolated island groups is lower, however, and intraprovincial diversity also falls in higher latitudes.

The second highest diversity known in a marine province is along the west American tropical shelf, where large continents face a large ocean. The least tropical diversities are to be found along the Atlantic coasts, where large continents face a small ocean. This longitudinal diversity variation appears to correlate well with the decrease in maritime conditions found on shelves as landmasses increase and oceans decrease in area. The principle appears to apply in all latitudes. For example, in temperate latitudes, shelf diversity is lowest where marine seasonality is highest. In the Atlantic, the highest seasonality occurs along the northeastern United States and southeastern Canada, with more equable conditions along western Europe; diversity is lower in the western than in the eastern Altantic in these middle latitudes (Fig. 8–8). In the Pacific, seasonality is greatest along the Asian shelf bordering the Japan Sea; unfortunately, diversity data are not available for this shelf, so we can only predict that diversity will prove to be low there, compared with the same latitudes along the American shelf. Although the provinces with the lowest diversities are found in the highest latitudes, the Antarctic shelf supports a more diverse biota when compared with the Arctic shelf, latitude for latitude. Presumably this is owing to factors related to their relative continentality; the Antarctic shelf lies along a relatively small continent bordering large oceanic expanses, whereas the Arctic shelf lies in a small ocean bordered by major continents. Figure 8–8 shows these trends. If gastropods had been used instead of bivalves, the trends would be even more extreme; bivalve diversity patterns are relatively conservative. Some data are available for ectoprocts, which show the longitudinal variations well but have a lower latitudinal diversity gradient than bivalves (Fig. 8–10; Schopf, 1970).

Unfortunately, there are not sufficient data on enough taxa or on such important aspects of provincial environments as productivity regimes to attempt a more quantitative treatment of these diversity patterns at present. In most cases, it is not certainly known how the diversity differences between provinces are accommodated within the ecological structure. Three of the more likely possibilities are (1) that there are the same number of communities in high- as in low-diversity provinces, but that the former contain more species populations; (2) that there are more communities in the more highly diverse provinces, and consequently more species populations; or (3) that there are the same number of communities containing the same number of species populations in high- and low-diversity provinces, but that the species in low-diversity provinces range into more communities (Fig. 8–11). Probably all three sorts of effects may be found.

Case 1 in Fig. 8–11 represents the most often discussed aspect of latitudinal diversity gradients. Communities in high latitudes are simply not as diverse as those in comparable biotopes in low latitudes. Dredge hauls from comparable mud bottoms taken by Sanders and colleagues (Sanders, 1968), for example, demonstrate that diversity increases in lower latitudes in this shelf biotope.

The second case in Fig. 8–11 is not well documented, but such data as exist suggest that it occurs and contributes to diversity accommodation. For example, Macdonald (1969) has studied the mollusks of salt marshes lying on either side of the Californian-Oregonian provincial boundary along Pacific North America. Although these salt marsh and associated tidal creek communities are of low diversity, they reflect the general diversity trends of both provinces. Of 30 molluscan species represented by living material in Macdonald's quantitative sam-

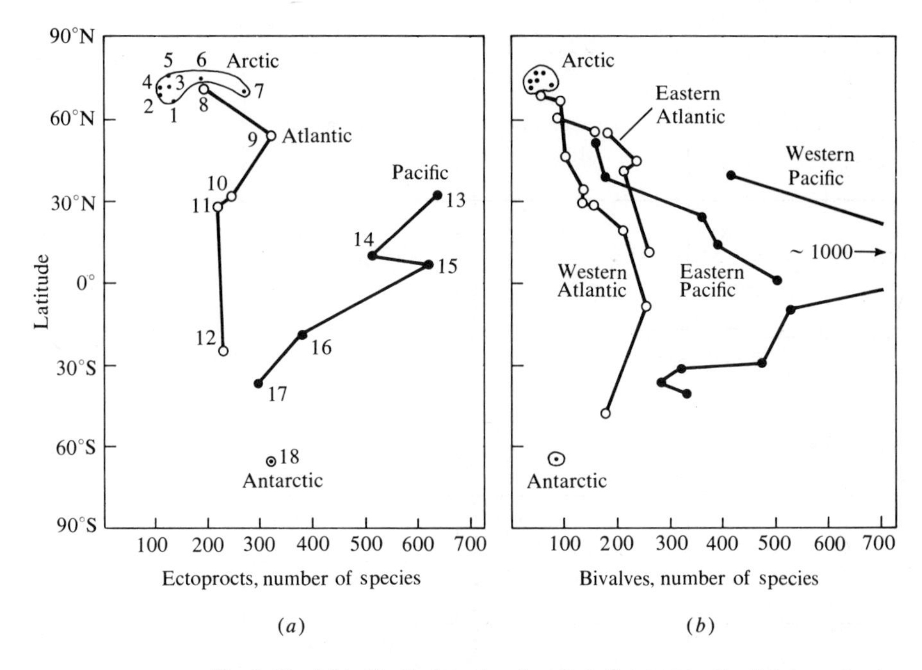

Fig. 8-10 (*a*) Latitudinal species diversity in Ectoprocta. Small dots are Arctic or Antarctic species; open circles are Atlantic species; and closed circles are Pacific species. Numbers refer to the following regions: (1) White Sea; (2) Barents Sea; (3) Point Barrow, Alaska; (4) East Siberian Sea; (5) Laptev Sea; (6) Kara Sea; (7) Chukchi Sea; (8) Greenland; (9) Western Europe; (10) Southeastern U. S.; (11) Gulf of Mexico; (12) Brazil; (13)–(16), Chinean, Mexican, Malayan, and Papuan "faunal provinces"; (17) Victoria, Australia; (18) Antarctic. (*b*) Latitudinal species diversity in Bivalvia. Symbols as in (*a*); localities as plotted by Stehli, MacAlister, and Helsley (1967) with Hawaii and Indian Ocean stations omitted. (After Schopf, 1970.)

ples, only 10 cross the provincial boundary; 7 are restricted to the northern province; and 13 to the southern. Thus, the boundary involves a compositional change and a rise in diversity to the south that corresponds to case 1, Fig. 8–11. Furthermore, there is a change in the marsh zonation across the provincial boundary; to the north, a single association of plants occupies both high and low marsh, although compositional gradients were apparent from tidal zone to tidal zone; to the south, two distinctive marsh plant associations are present, one on either side of mean high high water (MHHW). Thus, a single northern floral community may be said to have been partitioned into two to the south, as in case 2, Fig. 8–11. A comparable partitioning of the molluscan community is not found; however, only a single molluscan species is common in zones above MHHW both north and south of the boundary. In some other regions, when biotopes have been compared across provincial boundaries (as the *Thalassia*-associated communities of the southeastern United States and Carib-

Case 1

Province 1		Province 2	
Community 1 5	Community 2 10	Community 1 10	Community 2 15
Community 3 15	Community 4 20	Community 3 20	Community 4 25
50 species		70 species	

Case 2

Province 3		Province 4	
Community 1 5	Community 2 10	Community 1 5	Community 2 10
		Community 3 15	Community 4 20
Community 3 15	Community 4 20	Community 5 5	Community 6 10
		Community 7 15	Community 8 20
50 species		100 species	

Case 3

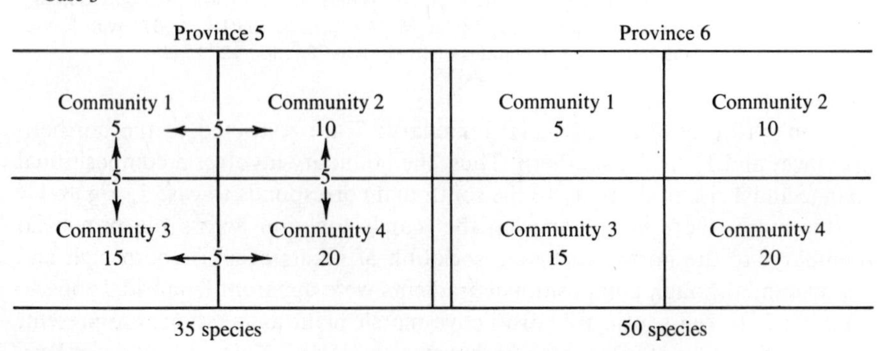

Fig. 8-11 Hypothetical models of accommodation of increasing species diversity between provinces. Case 1: More species live within communities in the more diverse province, but number of communities is the same. Case 2: The same number of species live in communities in the more diverse province, but there are more communities. Case 3: The same number of species live in communities in the more diverse province and the same number of communities are present, but fewer species are shared between communities.

bean; J. Jackson, personal communication), there are more associations in a given biotope in the tropics. This seems quite reasonable on the basis of the theory of adaptive strategies presented earlier, for the narrowing tolerances of species in stable environments could cause some environmental discontinuity, one that might have been completely disregarded by the more broadly adapted species of unstable environments, to become limiting to more and more populations in increasingly stable environments. Eventually the biotic change associated with the discontinuity would require the recognition of two separate communities.

The third case has not been carefully documented either. However, the broader tolerances of many species in low-diversity environments is a matter of observation, and it might be expected that this would result in the appearance of an average species in more communities in low-diversity than in high-diversity environments. The restriction of species to particular communities is an aspect of *fidelity*, which is a measure of the consistency and restriction of a species to a particular environmental situation. The fidelity of species for some given recurrent association might correlate directly with diversity and with environmental stability. It is, however, possible that the greater environmental range expected of communities in unstable situations might keep species fidelity to a community at a fairly high level. It might be well worth studying the average fidelity of species within different recent provinces, for fidelity could conceivably provide a clue as to the original diversity of ancient provinces.

Fossil Provinces Are Best Defined by Their Communities

The recognition of biotic provinces in the fossil record may be based upon either biogeography or provincial ecology. A biogeographic provincial definition depends upon the appearance of a unique association of taxa in a given region. The taxa are commonly species but may be higher taxa if these appear to be appropriately restricted or associated, and it is common for paleontologists to employ genera in reconstructing provincial regions. From what is understood of the principles governing the distributions of organisms, the biogeographic approach to provincial definition is entirely legitimate, but owing to the nature of the geological record it must be applied with caution.

A provincial definition based on provincial ecology considers the association of communities in a given region. Most communities will be unique to a province, although some may range into neighboring provinces. Since communities are composed of species populations the definition by provincial ecology may seem to be little different from the biogeographic one, but, in fact, it requires much more information to apply, since the habitat association patterns of the species must be known. This increase in information provides more definitive criteria.

Two examples illustrate the differences well. Cambrian biostratigraphy rests chiefly upon the successions of trilobites, which are employed to establish local faunal sequences and are used for interregional correlation. Since the 1890's, it has been known that many trilobites display marked regional restrictions in their distributions. Each region contains a sequence of endemic genera that are common and widespread intraregionally, and an early explanation was that the regions were provinces and were separated by barriers to dispersal. However, the map patterns of these regions are very peculiar when plotted on a map of the continents as they are now positioned (Fig. 8–12). One well-defined sequence

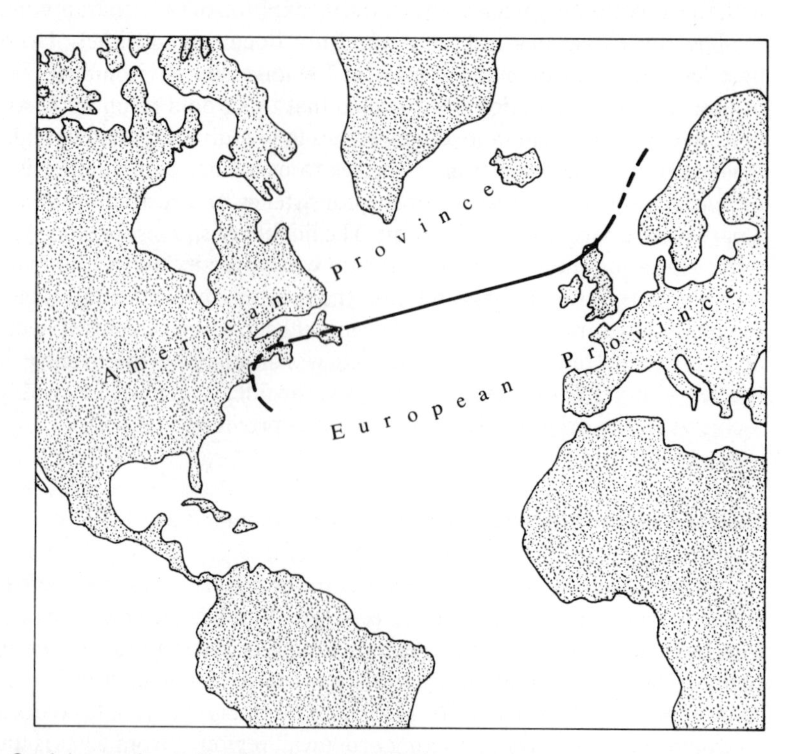

Fig. 8-12 Cambrian provinces plotted on the present continental geography.

of trilobites occurs throughout Europe, being especially well developed in Wales and Scandinavia, whereas another sequence is found chiefly in North America. However, an American trilobite sequence is known from northwestern Scotland, whereas European-sequence trilobites have been found along a narrow strip of eastern North America, from southeastern Newfoundland (northwestern Newfoundland has a typical North American sequence!) to near Boston. A barrier separating the present locations of the two sequences would have to run up the east American coast and cross the middle of Newfoundland, the North Atlantic Ocean, and Britain south of the Scottish localities, which seems highly unlikely.

A solution to this dilemma was suggested by Lochman-Balk and Wilson

(1958), who classified the trilobite sequences of the Northern Hemisphere Cambrian into three "biofacies realms," each characterized by a different trilobite assemblage: (1) a cratonic realm of variable shallow-shelf environments; (2) an extracratonic-intermediate realm, developed largely in "miogeosynclinal" settings and which contains shelf genera intermixed with 20 to 25 percent nonshelf genera; (3) an extracratonic-euxinic realm, thought to be developed largely in "eugeosynclinal" settings and to contain chiefly nonshelf genera that lived in stable, monotonous environments. The euxinic fauna is composed of the European sequence trilobites. In effect, Lochman-Balk and Wilson suggested that the differences between American and European sequences were on the community level, and not on the provincial level at all. In this interpretation, the abrupt faunal change in proceeding from northern to southeastern Newfoundland indicated the passage of the borders between two distinctive Cambrian biotopes or biotope complexes, and the species restricted to each area simply lived in separate communities or community complexes.

Continued work has weakened this suggestion, however. With the description of additional Cambrian trilobite faunas, it has been shown that the European sequence of trilobite genera is found in cratonic settings in Europe; therefore, their distribution patterns cannot be entirely controlled by the sorts of habitat requirements that were postulated. Furthermore, they are not found in "eugeosynclinal" sequences in western North America. However, the regional relations of trilobite sequences as described by Lochman-Balk and Wilson have led to a new interpretation of the ecological biogeography of Cambrian trilobites (Palmer, 1969). Using the Late Cambrian situation as typical, there were then three roughly concentric belts of sediments irregularly developed within the North American shelf seas: (1) an interior belt of sand and silts, commonly glauconitic, that lay inshore; (2) an intermediate belt, where clean limestones and dolomites accumulated—they are commonly cross bedded and oolitic and with algal stramotlites, representing a complex of shallow calcareous banks; and (3) an outer belt of silty sediments, now indicated by dark, pyritic or organic-rich siltstones and limestones. Each of these belts of sediments, which now form lithofacies complexes, seems to have been characterized by different species associations representing different proportions of trilobite genera, so that these associations now form biofacies. The substrates do not seem to have been the main limiting factors in the distribution of the trilobites, however, for the biofacies and lithofacies correlate only approximately, suggesting that each is responding to one or more different, somewhat related factors. The genera of the outer belt correspond closely to the fauna of the extracratonic-intermediate realm of Lochman-Balk and Wilson. The North American Late Cambrian fauna appears to have included at least three communities or community complexes, then; perhaps when more data are available, the details of their associations may be worked out.

The trilobite sequences of the "Atlantic" province that are found along eastern North America (Fig. 8–12) do not fit into this community level analysis of the Cambrian trilobites, but actually appear to belong to a province separate from the American sequence, as suggested so long ago. Perhaps during Cambrian time the sediments in which they are contained belonged to a European landmass and have become incorporated into the North American continental margin during a continental collision resulting from plate tectonic movements (Wilson, 1966; Dewey, 1969). If this is so, then the Scottish sequence containing North American trilobites may have been part of a North American landmass during the Cambrian that has subsequently been emplaced against the European margin during this same collision. This suggestion implies that the continents were apart during the Cambrian to permit provincialism to develop and were subsequently joined by a collision and then refragmented in such a way that some rocks of each precollision continent remained behind on the other continent. We shall discuss this sort of hypothesis in Chapter 10.

It is reasonable to conclude from this example that, first, the recognition of markedly different taxa living contemporaneously in different regions is grounds for postulating a provincial boundary on this biogeographic criterion; second, a community level analysis is required to test such an idea and to place the interpretation of the provinciality on an ecological basis; and third, when the provincial identity of faunal differences is established ecologically it can be used as a tool in paleogeographic reconstructions, leading, in turn, to the formation of new hypotheses and to a better understanding of the former biosphere structure.

A second example of ancient provinces concerns the nature of compositionally distinctive faunas that occur in separate regions during the Jurassic. These were first noticed in the nineteenth century; Arkell (1956) has reviewed the Jurassic geology of the world and refers extensively to early work. Arkell recognized three distinctive regions that he termed "faunal realms" using a biogeographic definition, the Boreal, Tethyan, and Pacific realms. The Boreal and Tethyan are distinguished by numerous biogeographic characteristics, but the Pacific is less well characterized and may not be distinctive; distributional analyses on the species level are badly needed. We shall confine our attention to the Boreal and Tethyan biotas. They differ both in quality, with a few endemic Boreal and many endemic Tethyan taxa, and in quantity, with Tethyan diversity much higher than Boreal. Table 8–3 gives some idea of the diversity differences in species or genera of several higher taxa for different stages of the European Jurassic. In addition to the provincial endemics, the two regions share many taxa that can be called consmopolitan, and at times the Boreal is characterized not by an endemic element, but by an impoverished fauna of cosmopolitan taxa, whereas the contemporaneous Tethyan includes many endemic taxa as well.

These faunal regions have usually been termed provinces and their differences ascribed to climatic causes, the low diversity, northern Boreal fauna being considered to indicate cooler temperatures than the rich southern Tethyan fauna. Hallam (1969) has presented a critique of this interpretation by studying pro-

Table 8-3 Recorded Diversity Gradients Between Tethyan and Boreal Regions at Times in the Jurassic

Species of Upper Oxfordian reef corals
239 in Southern Jura
53 in Lower Saxony
13 in Southern England
7 in Yorkshire
Species of Kimmeridgian-Tithonian reef corals
113 in Portugal
143 in Württemberg
17 in Lower Saxony
1 in Normandy
Genera of Kimmeridgian ammonites
22 in Ardèche
21 in Württemberg
9 in Southern England
8 in Central European U.S.S.R.
5 in Greenland
Genera of Tithonian-Volgian gastropods
46 in Southern Jura
40 in Württemberg
17 in Southern England
13 in Central European U.S.S.R.
8 in Greenland

SOURCE: From Hallam, 1969, quoted from Ziegler, 1964.

vincial ecology. He described three "lithofacies associations" common in the European Jurassic, each of which contains a characteristic biota: (1) a terrigenous clastic facies association, containing a moderately diverse biota dominated by bivalves; (2) an intermediate facies association of chiefly calcareous sediments and shales, containing a more diverse biota again dominated by bivalves; and (3) a calcareous facies association of reefoid limestones in shallow settings and fine-grained light-colored limestones in deeper settings, containing richly diversified reef assemblages in shallow settings and highly diverse planktonic assemblages in deeper settings. The geographic pattern of these lithofacies corresponds in general to the Boreal-Tethyan patterns, with the Boreal faunas in terrigenous and intermediate facies and the Tethyan in calcareous facies (Fig. 8–13). Even when the stratigraphic column is examined in detail, the correlation remains very high, for when the lithofacies boundaries have shifted in time the provincial boundaries have also shifted. Inasmuch as the lithologic change to terrigenous material is a reflection of the influence of rivers as sources of the sediment, Hallam suggests that the associated fresh water influx lowered the Boreal salinities sufficiently to account for most of the regional faunal differences, and

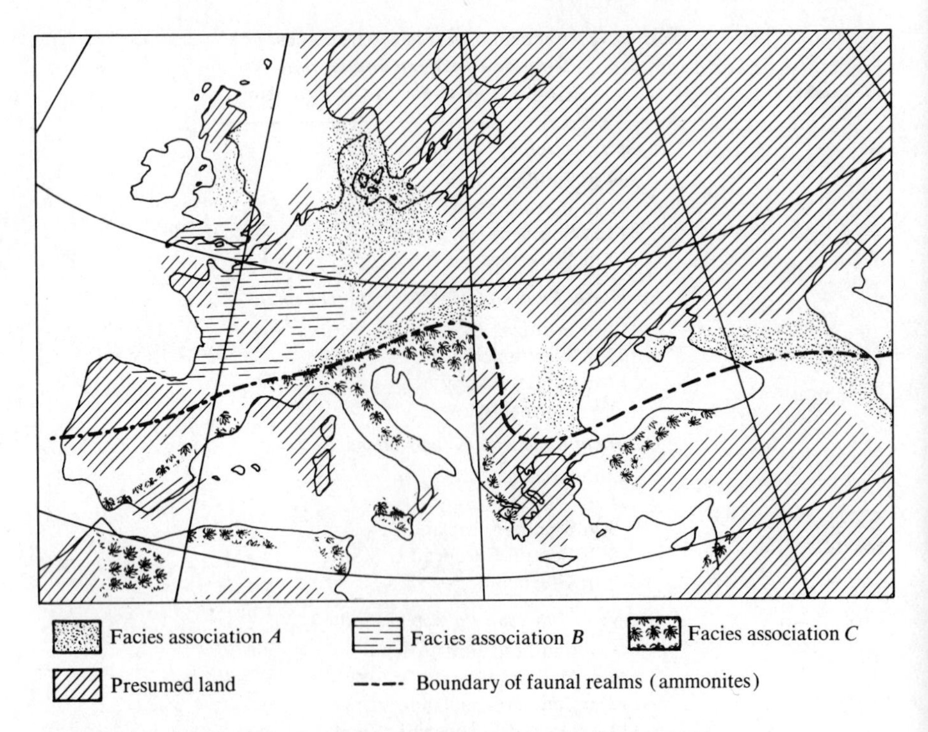

Fig. 8-13 An example of the correspondence of Jurassic faunal regions with lithofacies, for the Pliensbachian of Europe. Facies *A* is terrigenous, facies *B* is intermediate, and facies *C* is carbonate. (After Hallam, 1969; by permission of the Palaeontological Association.)

that waters of the Boreal region were of low salinities (perhaps 30 per mil or so) and not especially of low temperatures, so that the regional boundaries represent salinity barriers. This would account for some puzzling features of the Boreal province, such as the sharpness of the diversity change from Tethyan to Boreal faunas, which is greater than that between present provinces, but is easily understood as a change from one community to another. Furthermore, there is not known to be a distinctive Austral fauna developed in the southern hemisphere to complement the Boreal, as might be expected if significant latitudinal temperature gradients were present. Also, the boundary between the Tethyan and Boreal provinces is irregular and sometimes deeply interpenetrating, suggesting something other than temperature control.

However, other facts argue against interpretation of the Boreal biota as simply a low-salinity Tethyan complex (Hallam, 1969). For example, there are low salinity associations known from within the Tethyan region, and these contain Tethyan rather than Boreal components. The correlation of Tethyan faunas with calcareous facies may only reflect the ease of carbonate precipitation in warmer waters. It seems likely that the Boreal and Tethyan faunas are distinct on a provincial level. The possibility that salinity differences could lead to provinciality is very real, however, and is emphasized by Hallam's suggestions. In this instance, it remains possible that temperature control was paramount. Nevertheless, it is clear from Hallam's work that by descending to the level of the community, much light is thrown on the nature of the provinces that they comprise. Further work, perhaps actually to identify the individual communities with the community complexes associated with the lithofacies, may provide more definite evidence as to the ecological controls operating to produce these provinces.

Although in both these examples the provinces postulated on biogeographic grounds have stood up under ecological analyses, or at least have not been disproved thereby, biogeographic evidence is sometimes refuted by data on the community level. For example, the fossils from the Silurian reef rocks of Illinois are distinctively different from fossils from interreef rocks. This fact led to an early interpretation that the reef faunas represented a northern fauna and the interreef faunas a southern fauna. This opinion was overturned when community level studies were made (Lowenstam, 1957).

Provincial Systems Are Not Highly Integrated

Provinces can thus be envisioned as contemporaneous associations of communities inhabiting a region with characteristic patterns of climate, productivity, and habitats. Most communities change their compositions significantly across provincial boundaries, which represent localities where environmental features that limit the ranges of many species are concentrated. The boundaries form the margins of ecological barriers, which impede the dispersal of much of the provincial biota. Therefore, each province supports a unique taxonomic association.

Populations share strong internal genetic bonds, and display intraspecific competition among individuals. Within communities, there are frequent interactions among many of their component populations. Within provinces, however, interactions among communities should be weak, although there may be competition between species inhabiting separate communities at common community boundaries. Communities do share many ubiquitous species; therefore, there is some intercommunity gene flow, and many communities share a common pool of trophic resources. Indeed, sharing of a common pattern of productivity and climate produces among the members of communities in the same province some common features of adaptive strategy. Owing to this, analogous community pairs in separate provinces will differ structurally because of a "provincial effect."

The evolution of a province occurs through changes in the communities that form its primary subsystems. As noted in Chapter 3, there are three important classes of provincial change, one being the changes in the quality of the communities due to processes that change internal community structure. In addition, there are changes that arise from changing relations among communities, as when their proportionate representation in the province is altered, or when one or more communities are newly assembled or disappear.

Changes in community proportions must chiefly be related to changes in the proportionate representation of biotopes. On a large scale, such changes are most frequently related to changes in land-sea relations, in continental topography, or in climate, especially as they affect the quality and quantity of sediment supplied to sea-bottom habitats. Transgressions over broad continental platforms of gentle relief commonly resulted in the extension of muddy-bottom communities and of communities on bioclastic substrates over vast regions, whereas regressions greatly reduced their representation. Topographic evolution may affect shoreline configurations, create uplands or plains to alter the sediment supply and therefore the substrate type on marine shelves, and create offshore prominences or basins, in each case providing broader scope for the development of some communities at the expense of others. Climatic changes may alter sediment supply and substrate type through changing runoff patterns, and may promote or inhibit the accumulation of bioclastics through changing temperature, in either case altering the proportions of communities associated with the affected biotopes. There are, of course, many other ecological effects of these sorts of changes besides those that alter community proportions and therefore provincial structures.

Indeed, these sorts of environmental changes may give rise to changes in the number of communities, by creating the physical setting for new biotopes or removing the setting for previous ones. Thus, the appearance of offshore topographic highs resulting from deformation of bottom sediments might give rise to an association of species that were not previously living together, creating a new community. It is easy to imagine other "new" communities appearing in response to the appearance of "new" biotopes. As the environmental heterogeneity of a province increases on the scale of the biotope, community diversity will increase, and as heterogeneity decreases, community diversity will also. In theory, at least, community diversity may also change as the adaptive strategies of populations change. If the change is general and is toward habitat specialization, biotopes that formerly supported one community may be partitioned to support two or more associations of specialists, whereas if the change is toward habitat generalization, biotope boundaries that have been limiting to some species may be broached by numerous lineages, transforming two or more narrow biotopes into a broader single one.

CHAPTER NINE

THE BIOSPHERE LEVEL

Throughout the phylogenetic evolution species, genera, families . . . have appeared, changed and disappeared. By the events in inanimate nature mountains and the deeps of the sea, ocean currents and climatic zones have appeared, changed and disappeared and as a result of interactions of infinite complexity between animate and inanimate nature the present biogeographical conditions have emerged in the course of the ages.

—Sven Ekman, *Zoogeography of the Sea,* Sidgwick and Jackson, London, 1953.

The planetary biota, comprising as it does all the organisms living at any time, forms the highest level of the ecological hierarchy. The functional aspect of this biota, the biosphere, forms an ecosystem that involves the total solar energy budget of life plus minor supplementary sources of primary productivity and the assimilation and cycling of these energy resources to support the planetary biomass. At the biosphere level it is not possible to make comparative studies of contemporary systems, for we have only one world available to us, at least for the present. It is nevertheless possible to make comparative studies of different states of our biosphere in time.

There are two useful approaches to such studies. One approach is to study the distribution patterns of taxa in time. The other is to study the changes among the primary subsystems of the biosphere, namely the provincial systems. It is convenient to begin from this last viewpoint.

The total planetary environment has been conceptualized in Chapter 3 as biospace, which changes its shape and size within the environmental hyperspace lattice as the environment fluctuates in time. Inhabited biospace is conceptualized as the planetary ecospace, which similarly fluctuates in size and shape through time. It is not possible to imagine the complexities of ecospace-biospace interrelationships for the entire world biota, even in general terms. However, by regarding provincial systems as primary subsystems of the biosphere and by defining biosphere structure and function in terms of the provincial systems, the task of considering biosphere properties is greatly simplified. We are therefore going to be concerned at this level with the quality, proportion, and diversity of the provinces and of their changing patterns through time. Inasmuch as the qualities of provinces were discussed in Chapter 8, we shall concentrate here on the latter aspects of marine biosphere structure on the continental shelves.

The chief factors that regulated provincial proportions and numbers are those limiting factors which serve as distributional barriers and therefore as the determinants of provincial boundaries. Thus, for latitudinal boundaries, the chief factor is the temperature regime, and provincial number and proportion in a north-south direction depends chiefly upon the steepness and regularity of the latitudinal temperature gradient. For longitudinal boundaries, the chief factor is habitat failure, such as the interruption of the shelf realm by deep-sea or terrestrial environments, and provincial number and proportion in an east-west direction depends chiefly upon the number of times shelves alternate with deep-sea or land barriers. As we have seen, the present-day patterns of provinciality correlate very well indeed with the climate regime and tectonic framework of the present oceans, and the present pattern of species diversity within provinces correlates well with the patterns of environmental stability trends and gradients.

The Biological Structure of the Biosphere Changes with Continental Geography

It seems reasonable that past patterns of marine provinciality and diversity should correlate with the past tectonic and climatic framework of the oceans. But we have seen that provinciality was markedly different in the past—for example, that during the Middle and Late Jurassic there were only two well-defined faunal realms. It is possible that endemism within these realms was pronounced enough locally to create distinctive provinces, although this is not yet demonstrated. During the Early Jurassic, there seems to have been essentially no provinciality at all. The ammonites, and even fully benthic groups such as the Bivalvia, were cosmopolitan (Arkell, 1956; Hallam, 1965; Imlay, 1965). Today, we have over 30 marine shelf provinces. The Early Jurassic situation clearly calls for a radically different biosphere than the present one, a biosphere without significant barriers to marine shelf organisms. What kind of a world can be imagined that lacks such barriers? Continents now separated by deep-sea barriers must have been in close proximity, or connected by dispersal corridors, and sharp climatic zonation must have been absent.

Both climatic and topographic barriers can be modified through changes in the geography of tectonic features. From the theory of plate tectonics it can be inferred that major lithospheric plate margin features—such as ridges, subduction zones with island arcs, and transform faults—have all had vastly different relative positions in the past, and, in fact, that entirely different lithospheric plates were formerly present. The continents that ride upon these plates must have changed their relative positions greatly. These changes require corresponding differences in the sizes of continents and oceans, in the location and quality of dispersal routes and barriers, and in climatic regimes. All such changes must cause fundamental changes in the ecological state of the biosphere.

Of all the tectonic features, the continents are the easiest to locate in the past. One of the strongest lines of evidence for past continental positions comes from paleomagnetic data. The principle is simple, although the technical problems have been great (Irving, 1964). The stable remnant magnetism measured in rocks should indicate the bearing and inclination of the magnetic pole at some time in the past. For igneous rocks, this may be the time that they cooled past the Curie point and their magnetism became stable. For sediments, the magnetism is chiefly imparted during diagenesis, hopefully shortly after deposition. If the rock sample is dated and oriented with respect to the continent upon which it rests, then the bearing of the ancient pole indicates the former orientation of the continent, whereas inclination indicates the latitude of the continent with respect to the magnetic pole. If it is further assumed that the magnetic pole was closely associated with the pole of rotation, the continent may be positioned latitudinally on the earth and rotated to agree with its indicated orientation, although its former longitudinal position cannot be estimated. Paleomagnetic data are difficult to measure and interpret, and results are not always consistent, so that refinements and improvements are continually being made. Nevertheless, the agreement of paleomagnetic data is sufficiently good that magnetic polar directions and distances from localities on the continents were clearly different in the past than they are today. Either the poles, or the continents, or both have moved. When the continents are arbitrarily held stable, then the pole positions appear to wander across the surface with time. However, the polar wandering tracks are different for different continents. If the poles are arbitrarily held stable, then the continents move across the surface, in different directions. As plate tectonics provides a mechanism for continental drift, this solution is the more appealing.

Further evidence on past continental positions comes from geometric arguments. Many of the continents have complementary outlines, as if they represented fragments of a once-larger continent that has split up. Because it is probable that their outlines have been somewhat modified since any past splitting, the refitting of continental outlines cannot expect to be perfect. Fitting has been done by eye and by computer to find the best match under given assumptions (Bullard, Everett, and Smith, 1965; Smith and Hallam, 1970). The fits may then be tested by comparing the geology of those portions of the continents that are brought into proximity by the reconstruction. Structural trends that are at least as old as the time of continental proximity should cross from one continent to the other, and rock sequences should be similar or should have patterns that make sense in the reconstruction. Indeed, such geological tests can identify one of several alternate geometric fits as the more likely. The times of splitting of these larger former continental masses to produce smaller continents can be estimated from the age when the bearings and latitudinal distances of the poles from the continents became different, from the time of severance of structural and stratigraphic contiguity between continents, from the maximum age of the sea floor between them, from times of other widespread tectonic and of igneous events, and from faunal evidence (see Hallam, 1971, for an example of dating a continental breakup).

With intracontinental rifting and sea-floor spreading to provide the mechanism for changing sizes, shapes, and numbers of continents, it is possible to imagine a variety of continental configurations and to evaluate their environmental effects and paleobiological consequences. Perhaps the best approach is to imagine the effects of a series of simple configurations (Fig. 9-1, Table 9-1) and then to attempt to apply the results to the kinds of complex configurations that may correspond more closely to actual historical situations. In considering the simple models, we shall neglect islands.

The simplest configuration of all is one in which all continents are coalesced into a Pangaea, which is circular in plan and is centered on a pole [Fig. 9-1(*a*)]. In such a situation there would be only a single shelf province, because the entire coastline is at the same latitude and is contiguous. Continentality would be high and diversity within the province would be low. The shelf water would probably be rather warm, however, for it lies below 30° in latitude and the opposite pole is open ocean and presumably temperate. Another simple configuration would be a ring continent that runs east–west [Fig. 9-1(*b*)]; this ring isolates a polar sea at fairly high latitudes, where net radiation loss exceeds incoming radiation today. This sea would be cold, highly seasonal, and of low diversity. The equatorward coastline would be warm, much less seasonal, and of high diversity; even so, each coast would support only a single province, so that altogether there would not be many shelf species. A Pangaea centered on the equator [Fig. 9-1(*c*)] would extend to fairly high latitudes (about 65°30′) and should support at least 4 provinces, even though poleward waters would probably not be very cold; total diversity should be low, however, owing to the high continentality.

If there were 2 continents some distance apart longitudinally and offset latitudinally so that the south coast of one and the north coast of the other lay at 30° latitude [Fig. 9-1(*d*)], then the low-latitude continent would probably support 2 provinces, an equatorial and a midlatitude one, whereas the poleward continent would probably support 2 or even 4, depending upon the steep-

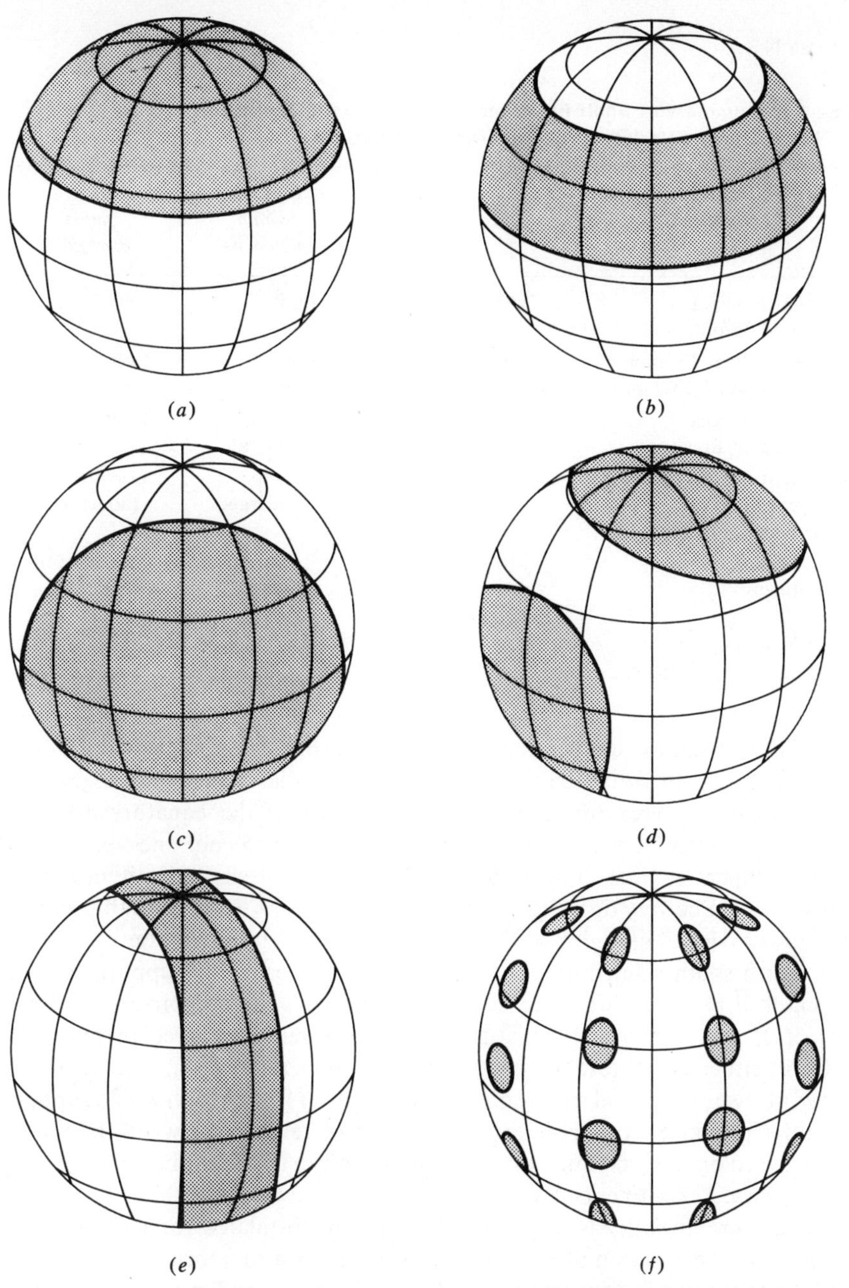

Fig. 9-1 A few simplified models of possible continental configurations. The area of the continents (shaded) is equivalent in all figures to the area of the present continents. (*a*) A circular Pangaea centered on a pole. (*b*) An east-west ring continent centered on 27°30′. (*c*) A circular Pangaea centered on the equator. (*d*) Two continents tangent to 30°, one lying to the north of this latitude and the other to the south. (*e*) A north-south ring continent. (*f*) Small continents well scattered; there are 32 in this model. See Table 9-1 for explanation of associated diversities.

Table 9-1 Summary of Shelf Provinciality and Diversity Expected Under a Variety of Continental Configurations

Configuration	*Provinciality*	*Diversity within provinces*	*Total marine diversity*
Circular Pangaea centered on pole	Very low (1)	Low	Very low
East-West Ring Continent centered on 27°30′ North	Very low (2)	Low to high	Low
Circular Pangaea centered on Equator	Low (4)	Very low to moderate	Low
Two continents tangent to 30°, one North, one South	Low (4–6)	Low to moderate	Low to moderate
North-South Ring Continent	Moderate (8–16)	Low to high	Moderate to high
32 continents well scattered	High (32)	Low to very high	High

ness of the latitudinal temperature gradient. The southernmost province of the poleward continent would be protected from cool polar waters and might well be subtropical, whereas the northernmost province of the equatorward continent is protected from equatorial waters but exposed to cool ones and might be cool temperate; thus, even though these coasts lie at similar latitudes, their climates would be very different. Continentality is still rather high in this model, and intraprovincial and global diversity would be low to moderate.

A north-south ring continent [Fig. 9-1(*e*)] would probably support at least 4 equatorial (1 on each side of each ocean) and 4 poleward provinces, for a total of 8, and if intermediate-latitude provinces were developed there would be 16; in either event, total diversity would be moderately high. Finally, if the continents were small and widely dispersed [Fig. 9-1(*f*)], then each would support a distinct province, with poleward continents of relatively low diversity and equatorward ones of very high diversity; global diversity would be very high. In the figure are 32 continents, 8 of which are in very low latitudes.

Today, provinciality is about as high as that postulated for the model in Fig. 9-1(*f*). The isolation of polar latitudes from warm equatorial waters (in the north by topographic barriers around the Arctic Ocean, and in the south by the west wind drift around Antarctica) has created a high latitudinal gradient, and the north-south trend of ridges and continental coastlines generally separates chains of latitudinal provinces along isolated coasts. However, marine shelf species diversity would probably be higher than today in a world such as Fig. 9-1(*f*) shows, because of the multitude of separate provinces in low latitudes under relatively maritime conditions, so that many endemic tropical biotas of high diversity would be present. Of course, if the continents were connected by islands, dispersal between them would reduce the endemism greatly and depress global diversity correspondingly. Although none of the real continental configurations appears to simulate any of these models, it is worth keeping the models in mind when we review the history of biosphere structure in this chapter and in Chapter 10 as well.

Changes in Biosphere Structure During the Mesozoic and Cenozoic Indicate Increasing Provinciality

Reasoning chiefly from paleomagnetic, paleogeologic, and topographic evidence, a reconstruction of the approximate continental positions for the beginning and end of Jurassic time is presented in Fig. 9-2. In Permo-Triassic times, all the continents appear to have been assembled into one supercontinent, "Pangaea," which by early Jurassic time had fragmented somewhat. The Atlantic Ocean was either closed or was represented by a narrow, blind seaway. Three oceanic regions were defined: (1) the Pacific Ocean, which stretched from the western shores of the Americas around to the eastern shores of Asia and Australia-Antarctica; (2) the Tethyan Ocean, which in early Jurassic time lay between southern Asia and northern India-Antarctica-Australia, opening out into the Pacific to the east and extending westward north of Africa; and (3) the Boreal Gulf, a high-latitude sea opening between northern Europe and North America. At times, the Pacific and Tethyan Oceans together extended for over 18,000 miles around the equator, forming a very wide stretch of oceanic conditions. During and after later Jurassic times, India evidently drifted northward across the Tethyan Ocean, complicating the geography and hydrography there.

Inasmuch as most of the continents are in close juxtaposition, there are few purely topographic barriers. The fossil evidence of low provinciality thus fits well with Jurassic geography insofar as topography is concerned. The fossil evidence indicates that there were no significant marine climatic barriers either, except at times in the Boreal Gulf. Jurassic plant fossils suggest a rather uniform terrestrial climate also. The lack of temperature barriers may be ascribed to the compactness of the continental masses and the extent of the Pacific Ocean, which stretched from high latitudes north of the continents right across the pole and southward to within about 10° of the south pole.

Reconstructions of ancient ocean current patterns must be provisional; the continents are not yet precisely located, and current patterns have not yet been calculated from first principles for any given geography. Nevertheless, we can employ general hydrographic principles and analogy with the patterns of today (Chapter 4) to reconstruct generalized patterns of ancient oceanic surface circulation that seem to be of biogeographic interest. Such a reconstruction for the early Jurassic is sketched in Fig. 9-3. A southerly drift along the eastern coasts of the southern continents is part of a great Tethyan-Pacific gyre and flows from the

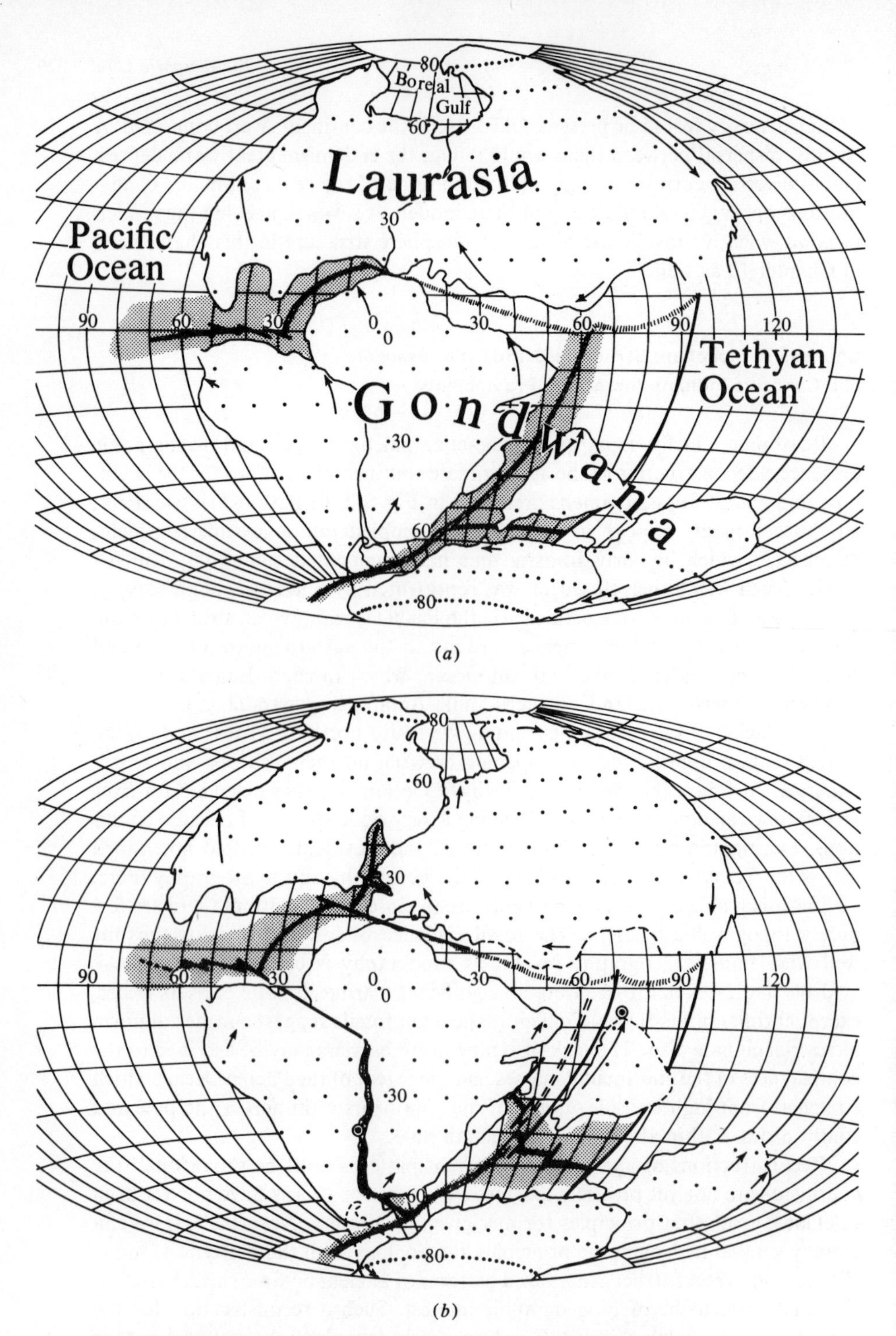

Fig. 9-2 Approximate continental positions near the beginning (*a*) and close (*b*) of Jurassic time. (Modified after Dietz and Holden, 1970.)

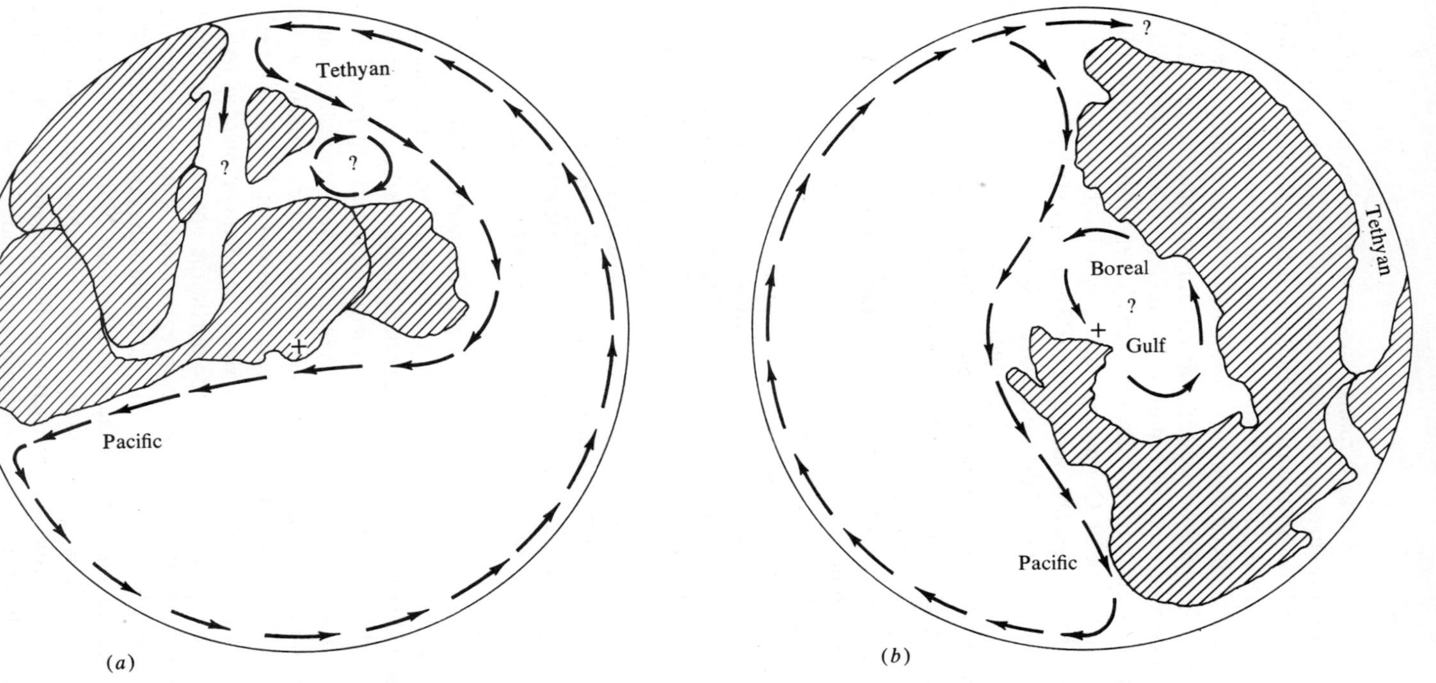

Fig. 9-3 Inferred major current patterns of the oceans during early Jurassic time. (*a*) South (Gondwana) polar view. Narrow oceanic gulfs divide India and other continents, and intervene between Africa and Antarctica; however, the chief current should run south and east around the main Gondwana landmasses, bringing warm waters to high latitudes. (*b*) North (Boreal) polar view. Note that the region of Russia east of the Verkhoyansk Mountains is depicted as part of North America. The Boreal Gulf is relatively isolated from the major Pacific gyre, which may well have partially broken down into an eastern and a western gyre.

tropics, and thus is interpreted as forming a warm western boundary current. Much of the water in this current must have rounded the Antarctic Cape of Gondwana and then drifted northward as an eastern boundary current along South America. As there was not yet an open ocean connection across southern Gondwana, there was no circumpolar west wind drift, such as encircles Antarctica today (compare Fig. 4-9). At present, this circum-Antarctic current lies in latitudes where outgoing exceeds incoming radiation, and it effectively isolates the Antarctic shores from currents flowing from lower latitudes. The high-latitude shores of early Jurassic Gondwana, however, could be washed directly by the warm current from low latitudes, and shelf waters could have been relatively warm. The surface water off the west coast of South America should have been cooler due to its longer stay in high latitudes, and it was probably diluted by upwelling, which should have been significant in this region. Therefore, southeastern Pacific shelf water should have been cooler, latitude for latitude, than southwestern Pacific shelf water.

In the north, an entirely different situation existed (Fig. 9-3). The Boreal Gulf was a re-entrant cut off from direct access to warm currents flowing northward along the east Asian coast. From Hallam's (1965) arguments, it can be considered likely that rivers flowed into the Boreal Sea and that precipitation and runoff exceeded evaporation. This would create an estuarine situation, with surface waters moving northward out of the Boreal Gulf and marine inflow restricted chiefly to deeper layers. With little heat supply from lower latitudes, it would appear that the coolest waters in the Jurassic Ocean were to be found in the Boreal Gulf. The main features of expected Jurassic provinciality, then, judging solely from these marine climatic reconstructions, would be a Boreal Province of low diversity with cryophilic endemics, a Tethyan Province of high diversity with thermophilic endemics, and perhaps a Pacific Province of intermediate diversity.

During the Cretaceous, the continents fragmented further and drifted toward their present positions, although India remained separate and Australia and Antarctica were still joined at the close of the period. In the south, the northward drift of Africa and then of South America opened a clear pathway for circumpolar circulation. The date of this important event is not yet established, but it permitted the appearance of a westerly drift encircling the polar continent, which had the effect of partially blocking the poleward transport of heat in southerly currents (Fig. 4-9). Lowering of temperature along the polar continent as well as upon it must have ensued. Meanwhile, in the north, the opening of the North Atlantic moved Eurasia and North America progressively apart into opposing positions, enclosing the Boreal Gulf, which evolved into the Arctic Ocean. The same movement opened a seaway to the south between North America and Europe, but the relatively narrow communication there has been more than balanced during the Cenozoic by a restriction of flow between the Pacific and Arctic Oceans. Thus, the progressive isolation of a north polar sea has probably reduced the heat flow into high northern latitudes, even as it was reduced into high southern latitudes for different reasons. Cooling poles must have resulted in increased latitudinal temperature gradients.

Shelf provinciality has increased progressively since the Jurassic. Tectonic barriers appear in the form of progressively widening deep-sea regions flanking the ocean ridges at which the continents split apart; in place of one east and one west coast along a nearly continuous landmass, a series of east and west coasts appeared, alternating with oceans. Longitudinal endemism appeared during the Jurassic (Hallam, 1971) and became quite pronounced by Late Cretaceous time (Sohl, 1961). Concomitant polar cooling and a rising latitudinal temperature gradient narrowed the latitudinal ranges of stenotherms and gave rise to increasing latitudinal provinciality, restricting provincial biotas to narrower shelf segments cut off by thermal barriers from circumcontinental dispersion routes. By late Cretaceous time, there were two or three shelf provinces along the more extensive north-south seaways and continental shelves in the nothern hemisphere (Sohl, 1971); today there are six or seven along most north-south-trending coasts. Taken together, these trends can account for a considerable multiplication of provincial biotas during the late Mesozoic and Cenozoic. Naturally, this raised the number of communities and species in the biosphere significantly.

Certainly, factors other than the ocean current patterns contribute to the development of barriers and latitudinal thermal gradients. The number and pattern of high mountain ranges and the degree of continental emergence, for example, must play important roles. Nevertheless, the model presented here demonstrates the sorts of processes resulting in changes in the number and proportionate development of provinces and constituting evolution of the structural aspects of the biosphere from the standpoint of its first-order subsystems.

The Taxonomic Categories May Be Regarded as Nearly Decomposable for Certain Purposes

The second approach to studying the states of the marine biosphere through time is to examine the temporal history of marine taxa. To evaluate this history, we must first consider the nature of the taxonomic hierarchy, which has many of the general properties of hierarchies that were discussed for the ecologic hierarchy. In phylogenetic taxonomy, a taxon is a set of organisms presumed to be descended from a common ancestral lineage within that taxon. The set of all taxa at any level within the hierarchy is called a *category*. Thus, the category "genus" is composed of all genera, whereas any special genus is composed of a cluster of related species, or, in some cases, of a single species.

The more inclusive categories have the better chances of being represented in the fossil record, simply because a higher category includes more individuals than a lower category, and a higher taxon includes more individuals (or, in special cases, an equal number, but never fewer) than any of its component lower taxa. Most genera, containing as they do several species on the average, stand a better chance than do species of achieving representation in the record; relatively fewer will be completely unrecorded. Yet, since it will rarely happen that the species of a genus all leave a comprehensive fossil record, fewer genera than species will be at all completely known. The same trends continue at higher and higher categories; a larger percentage of all families are represented than of all genera, but fewer of them are very completely known, and so on. Probably all skeletonized animal phyla are represented in the fossil record, and possibly all classes, but the record of such higher taxa is very far from complete. Evolutionary changes on a lower level of a family tree—such as, for example, on the species level—penetrate to higher levels in the sense that evolution within a species lineage is part of the evolutionary pattern of the higher taxa to which the species belongs. The amount of change that the species lineage undergoes is a smaller proportion of each larger taxon, however. The morphological or functional mode of a species may evolve rapidly, whereas the mode of the phylum to which it belongs may hardly be affected, and normally will undergo proportionately similar changes only over much longer periods.

This gradient of evolutionary rates of taxa, from high rates at low taxonomic levels to low rates at high levels, results in a higher turnover of lower taxa through time. For example, records of fossil Brachiopoda indicate that there have been about four genera per family of Brachiopoda all during the Phanerozoic (Fig. 9-4); as the number of families has increased or decreased, so have the number of genera, to maintain the proportion of about 1:4. However, only about 200 families are recognized, but over 1700 genera are recognized, a ratio of about 1:8½. On the average, the turnover in genera has been over twice as fast as the turnover in families.

The rapidity of evolution at lower taxonomic levels makes the standing diversity at these levels more volatile than at higher levels. It is possible for enormous numbers of species to be evolved relatively rapidly during a time when many barriers to gene flow are being created, but such speciation may result in relatively few new genera and in practically no new families. Similarly, the diversity of lower taxa may fall markedly without much affecting the numbers of higher taxa. For example, the extinction of widespread communities that are rich in endemic species, such as marine reef communities, may significantly lower the number of species living in the ocean. Yet many of the genera and most of the families represented in the extinct communities will ordinarily contain other species that survive in different habitats; the standing diversities of taxa are progressively less affected at higher taxonomic levels. We should expect that the maximum diversity of progressively lower categories would ordinarily, although not necessarily, be attained at progressively later times.

The taxonomic hierarchy, like the ecological one, is nearly decomposable. If we wish to study the evolutionary pattern of orders, we can describe it in terms of the component familial patterns, and need pay little or no attention to the evolution of genera and species as such, if we are *careful*. We merely treat the families as "black boxes," and are not interested in their internal processes.

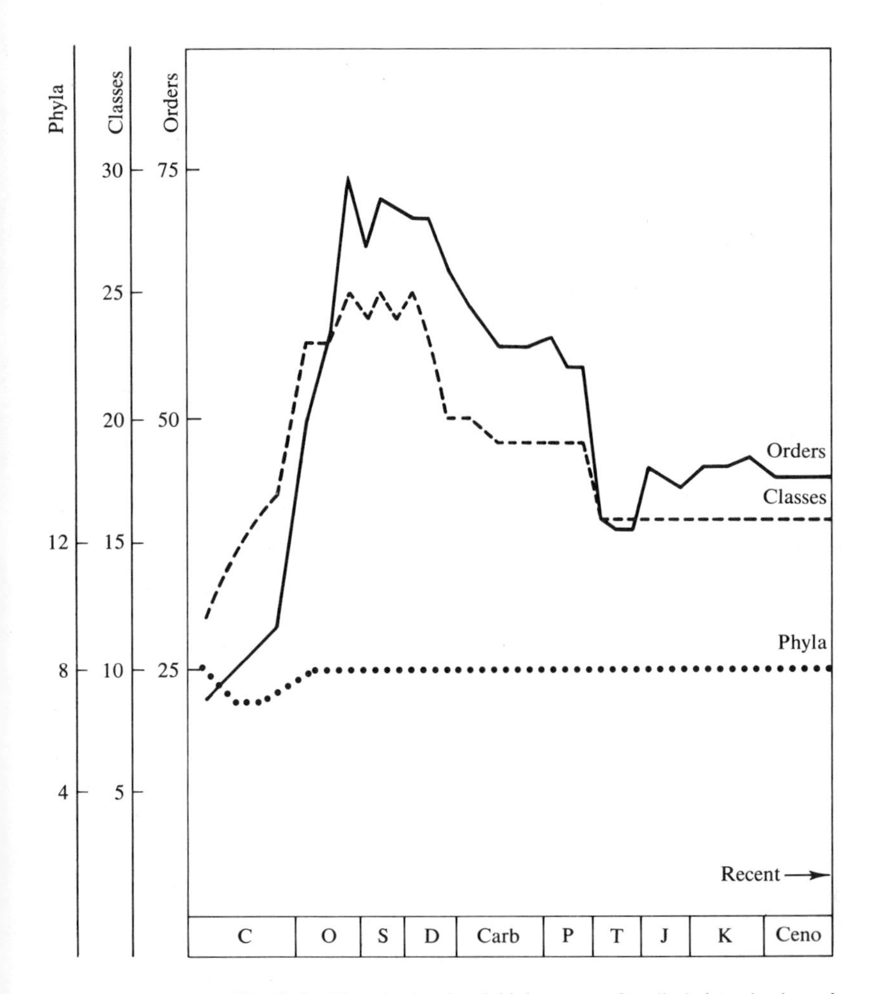

Fig. 9-4 Diversity levels of higher taxa of well-skeletonized marine shelf invertebrates during the Phanerozoic, plotted by epoch. (After Valentine, 1969; by permission of the Palaeontological Association.)

The care is excercised in correctly specifying the input and output conditions of the family. This property of near-decomposability means that if we specify the evolutionary processes that occur in speciation, and then the processes that occur in the formation of genera, we may approach the origin of families in terms of the generic patterns, and need not refer back again to the species level. If we continue up the hierarchy and arrive at the category of kingdom, we need refer only to the phylum level. We shall then have a working model of the origin of all levels.

On the other hand, the ranking of taxa in categories is not governed by any natural laws, at least for taxa above the species level, and it therefore depends entirely upon the predilections of the taxonomists who make the decisions. Categories are collections of taxa, ranked by a number of different investigators, with no assurance that everyone has had similar notions about the nature of the categories. Indeed, in reading systematic monographs, it becomes clear that different investigators have different notions. Furthermore, evolution has clearly favored some lineages with great diversity, whereas others have produced few divergences or radiations. There is no way to judge the similarity of hierarchical ranks between groups of taxa in widely separate lineages. Families of brachiopods may be quite different things from families of ammonites. These factors should be kept in mind when considering the significance of trends in the taxa described in different categories.

These factors are not so important as they might seem at first, however, because of the way the taxonomic hierarchy is handled in practice. If a taxon is revised and perhaps raised from one category to another, its subtaxa are usually elevated also, so that the relative trend of change from category to category is not ordinarily affected, except in special cases. The general trends that appear when large numbers of taxa are examined are probably real, with categories serving to measure the diversity and other properties at certain admittedly irregular intervals along the pathways of branching and diversification of lineages. It must be stated plainly that the precision of the measurement is not great. Nevertheless, it is adequate to support significant inferences if they are made with care.

The Structure of the Taxonomic Hierarchy Changed Throughout the Phanerozoic

Keeping the nature of the taxonomic hierarchy in mind, let us examine its fossil record. We shall first examine the quantitative aspects, which are so important in biosphere structure, and later the qualitative aspects. It will help to minimize the bias in the fossil record if the examination is restricted to well-skeletonized taxa. Furthermore, if it is restricted to a single ecological realm, such as the benthic shelf environment, then changes in the hierarchy may be interpreted in terms of ecological events within a relatively restricted range of habitats, and later the principles inferred to apply to this realm may be extended to the rest of the biosphere. In any case, the shelf is the best-represented major environment in the fossil record.

Figures 9-5 and 9-6 depict the numbers of taxa in the categories phylum, class, order, and family that are recorded during each epoch of the Phanerozoic, for the nine best-skeletonized benthic marine phyla; only taxa that were well-represented in the shelf benthos are included. Several trends are evident. First, the higher the category, the more stable has been its diversity; the number of classes has fluctuated more than that of phyla, orders more than classes, and families more than orders, even on a percentage basis. This would

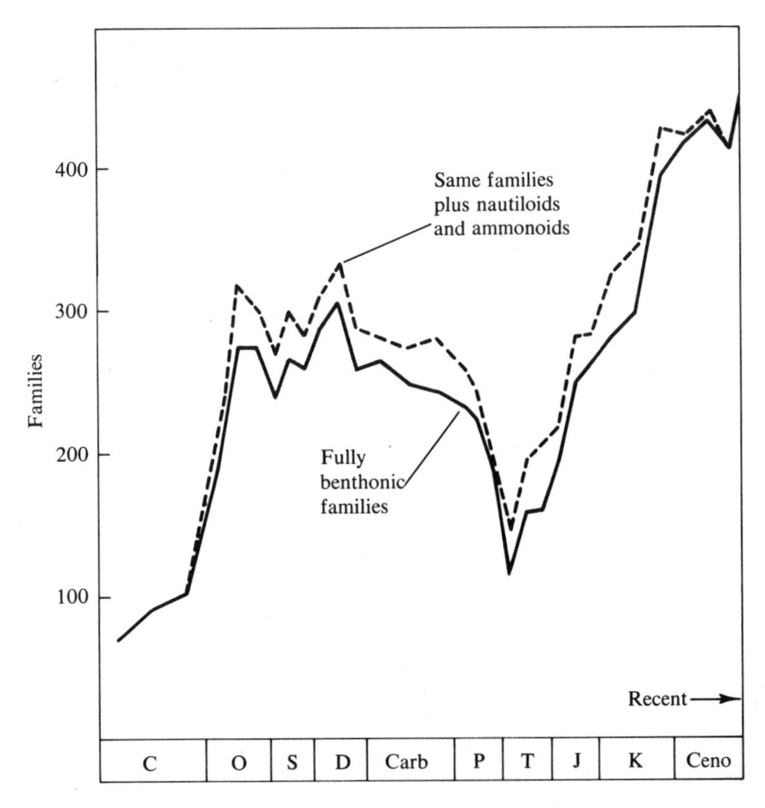

Fig. 9-5 Diversity level of families of well-skeletonized shelf invertebrates during the Phanerozoic, belonging to the phyla included in Fig. 9-4, plotted by epoch. (After Valentine, 1969; by permission of the Palaeontological Association.)

appear to be an outcome of the principle that lower hierarchical levels have higher-frequency internal dynamics than higher levels. Second, below the level of phylum, the temporal pattern of diversity of each succesively lower category is a modification of the next higher one, and the trend of modification is persistent. The phyla exhibit only a single diversity drop, in the Cambrian (owing to the disappearance of the Archaeocyatha), and a single diversity rise, in the Early Ordovician (owing to the appearance of the Ectoprocta). Classes display a great diversity rise in Cambro-Ordovician time, but suffer significant diversity declines in the Devonian and Permian. They do not display diversity rises following these losses. Orders have a similar spectacular diversity increase in Cambro-Ordovician, a Devonian decline persisting into the Early Carboniferous, and a Permian decline. They do achieve a slight post-Paleozoic diversity rise in the Jurassic. Families also have a strong Cambro-Ordovician diversity rise, a mid-Paleozoic diversity decline, a Permian decline, and then a marked post-Paleozoic rediversification that raises the number of families to new heights.

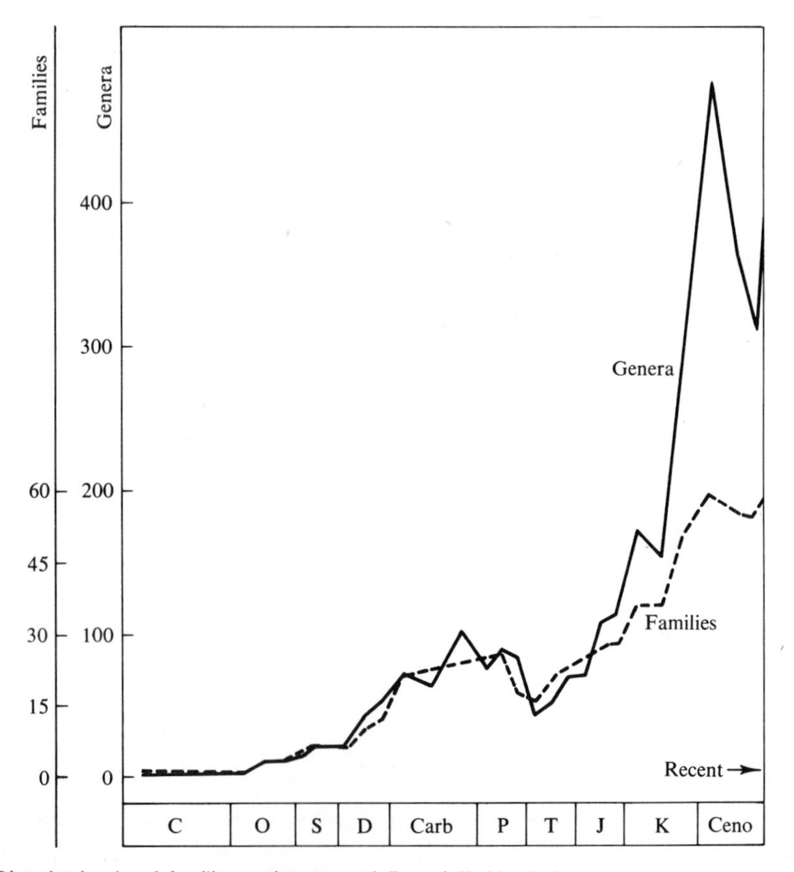

Fig. 9-6 Diversity levels of families and genera of Foraminiferida during the Phanerozoic, plotted by epoch. The poorly skeletonized Allogromiina are excluded. (After Valentine, 1969; by permission of the Palaeontological Association.)

The trends from higher to lower categories include a somewhat obscure shift in the time of diversification, becoming younger in lower categories, a shift in time of mid-Paleozoic diversity decline, also becoming younger in lower categories, and an increase in the post-Paleozoic diversity rise, becoming greater in lower categories.

Since the diversity of progressively lower categories becomes progressively less adequately represented in the fossil record, it is probably not possible to rely upon the recorded diversity of genera or species as an indication of real diversity levels within these low categories. For genera, a rough estimate of diversity trends can be based upon the records of the best-known groups. In Fig. 9-7, the recorded diversity of benthic shelf families and genera of Foraminiferida are shown, classed by epoch. This group was chosen as an example because it is relatively well known, has been monographed several times, and

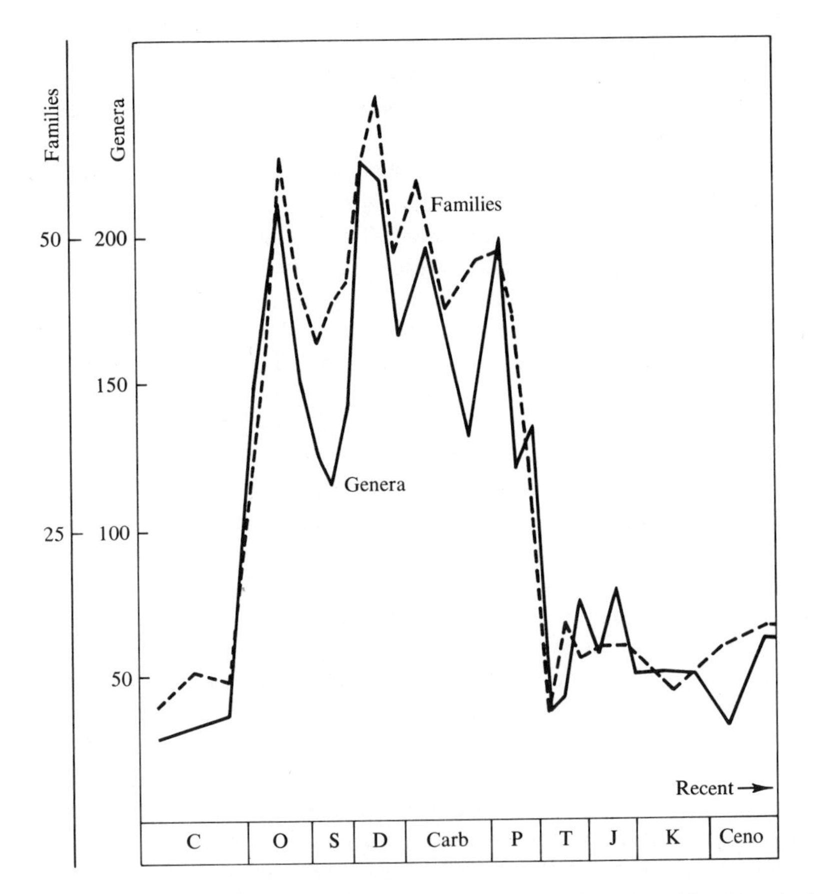

Fig. 9-7 Diversity levels of families and genera of brachiopods during the Phanerozoic. The ratio of genera to families remains near 4 during all eras. (After Valentine, 1969; by permission of the Palaeontological Association.)

because the diversity trends it displays have been evident since early work and have tended to be confirmed by modern revisions. This particular order was not very diverse at the family or generic levels during the early Paleozoic, so that it did not participate in the general Cambro-Ordovician diversification. What diversity it did achieve at these levels was reduced during the Permo-Triassic decline, and then a rediversification raised the number of families to new levels. The foraminifera contributed significantly to the Mesozoic-Cenozoic diversity rise of shelf families depicted in Fig. 9-6. Note that the numbers of genera outnumber the families by about 3 to 1 all during the Paleozoic, but that during the Jurassic the number of genera described per family increases and thereafter generic rediversification is disproportionately greater than for families. A similar trend can be seen among the other groups that contribute significantly to the post-Paleozoic rise in family diversity—the Ectoprocta,

Gastropoda, Bivalvia, and Echinoidea, for example. Therefore, there is in general a disproportionate increase in genera relative to families during the late Mesozoic and Cenozoic, and the shift in diversification to lower categories continues to the generic level.

Since the fossil record is biased, what is the chance that the generic curves are so biased as to be misleading? Perhaps the apparent diversity rise is simply a case of younger rocks being better represented and yielding more genera and more taxa of lower categories in general. This possibility cannot be discounted entirely. However, there is some evidence that the general trends are real. For example, the Brachiopoda were very important members of the Paleozoic benthos and persist today in some diversity, with 18 living families (Williams *et al.*, 1965). Unlike the foraminifera, the brachiopods did not participate much in the Mesozoic-Cenozoic diversity rise (Fig. 9-4). Nevertheless, if the bias of the fossil record is controlling the increasing genus/family ratios displayed by the dominant post-Paleozoic invertebrate groups, there would seem to be no reason that Brachiopods would not share in the bias as well. Instead, their genus/family ratios show no disproportionate generic rise, which suggests that their record is not positively biased in younger rocks, unless it is assumed that a true decline in the genus/family ratio is perfectly matched to the bias so as to mask it, which seems unlikely. Furthermore, the post-Paleozoic changes in genus/family ratios among the groups that do display increasing ratios are different from group to group. In the Paleozoic, clear trends of increasing genus/family ratios in younger rocks are not found. All in all, it seems likely that the trends are real although the absolute ratios must be subject to error.

At the species level, the incompleteness of the fossil record and of our knowledge makes it impossible to compile accurate diversity figures. However, the trends in Phanerozoic diversity that are exhibited in proceeding from higher to lower categories may be extrapolated to the species level, and the resulting pattern may then be examined for credibility. Such an extrapolation is presented in Fig. 9-8. Additional evidence to be discussed later was also used to establish this curve. At present, we can note a Cambro-Ordovician rise, a mid-Paleozoic plateau and fall, followed by a temporary peak during the Permian that is inferred because of great reef development then. This peak was reduced by Permo-Triassic extinction, and then diversity rose greatly, increasing disproportionately to genera according to the extrapolated trends, causing a rise in planetary marine species diversity of almost an order of magnitude to its present level. This general pattern of species diversity accords well with paleontological experience, and can be verified in many instances. For example, the great Permo-Triassic extinctions in species as in higher categories are surely real, for the Mesozoic-Cenozoic diversity rises that follow represent radiation from relatively few lineages, and not from a wide array of Paleozoic lineages that happened to cross the Permo-Triassic interval without leaving a fossil record in the rocks. The species curve does not reflect the many sharp localized variations in diversity

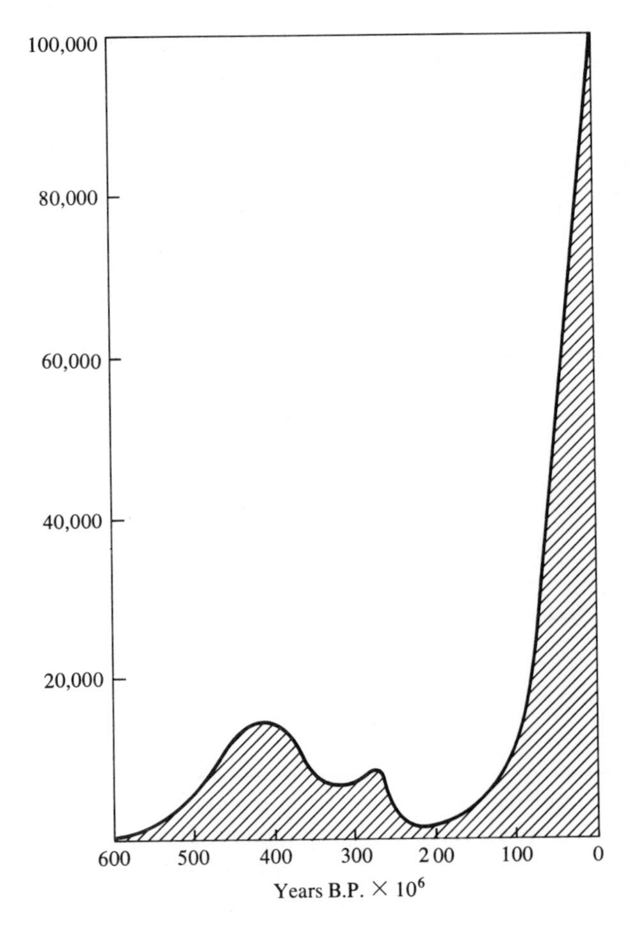

Fig. 9-8 Estimated diversity levels of species of well-skeletonized shelf invertebrates during the Phanerozoic. (Modified after Valentine, 1970.)

that must have occurred throughout Phanerozoic time, but glosses those into a smoothed sort of average curve.

Although rises in taxonomic diversity must indicate evolutionary diversifications and falls must indicate extinctions, a contemporaneous diversification and extinction may produce only a biotic turnover without affecting the diversity level. The curves in Figs. 9-4 to 9-8, then, do not necessarily indicate the intensity of biotic turnovers. Figures 9-9 and 9-10 represent the appearances and disappearances of taxa, which are assumed to represent diversifications and extinctions, in the phylum, class, order, and family categories. The higher categories exhibit greater stability, and both diversifications and extinctions occur at progressively more recent dates in progressively lower categories. In Fig. 9-9, the Cambro-Ordovician diversification and the mid-Paleozoic and Permo-Tri-

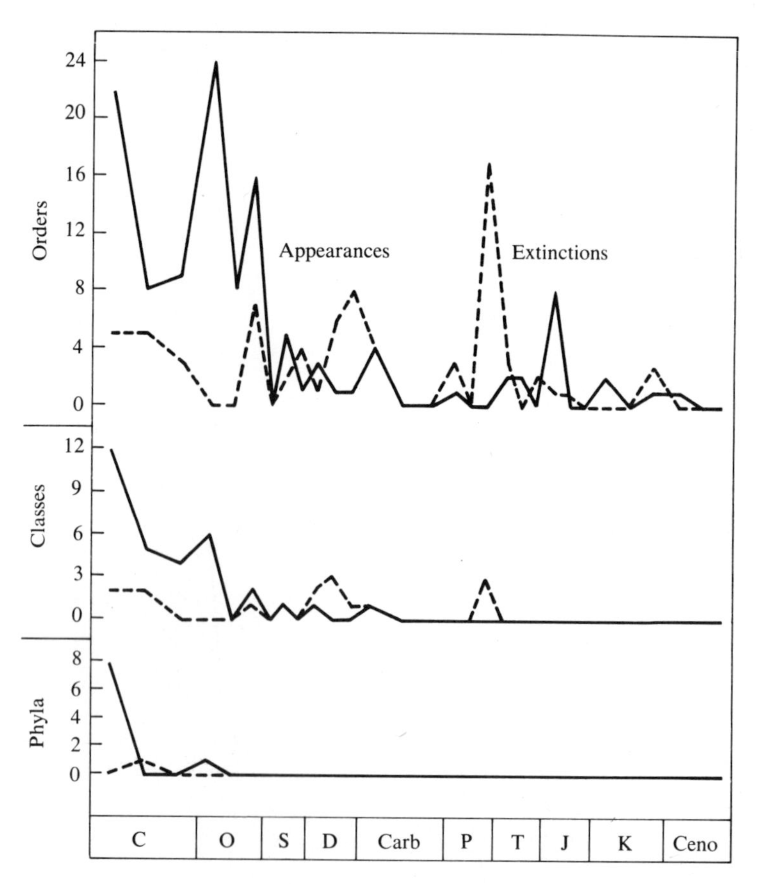

Fig. 9-9 Appearances and disappearances of phyla, classes, and orders of well-skeletonized shelf invertebrates from the fossil record during the Phanerozoic. (After Valentine, 1969; by permission of the Palaeontological Association.)

assic extinctions stand out clearly among the classes and orders. But there are other extinctions that are not reflected in diversity declines because they are accompanied by strong diversifications, especially on the ordinal level. This pattern is even clearer for families; in Fig. 9-10, two major peaks of Paleozoic extinction preceded the Permo-Triassic and one succeeded it in the late Triassic. Also, there were several additional smaller but important extinction peaks, yet their effects tend to be masked by accompanying diversifications. This is certainly not true of the Permo-Triassic extinction itself, which is unique as being unaccompanied by significant diversification at any level, so far as can be told (Rhodes, 1967; Valentine, 1969).

To interpret the changes just described in the structure of the taxonomic hierarchy, it is very useful for us to have some sort of conceptual prop to aid in devising models that serve as explanatory hypotheses. We have employed a

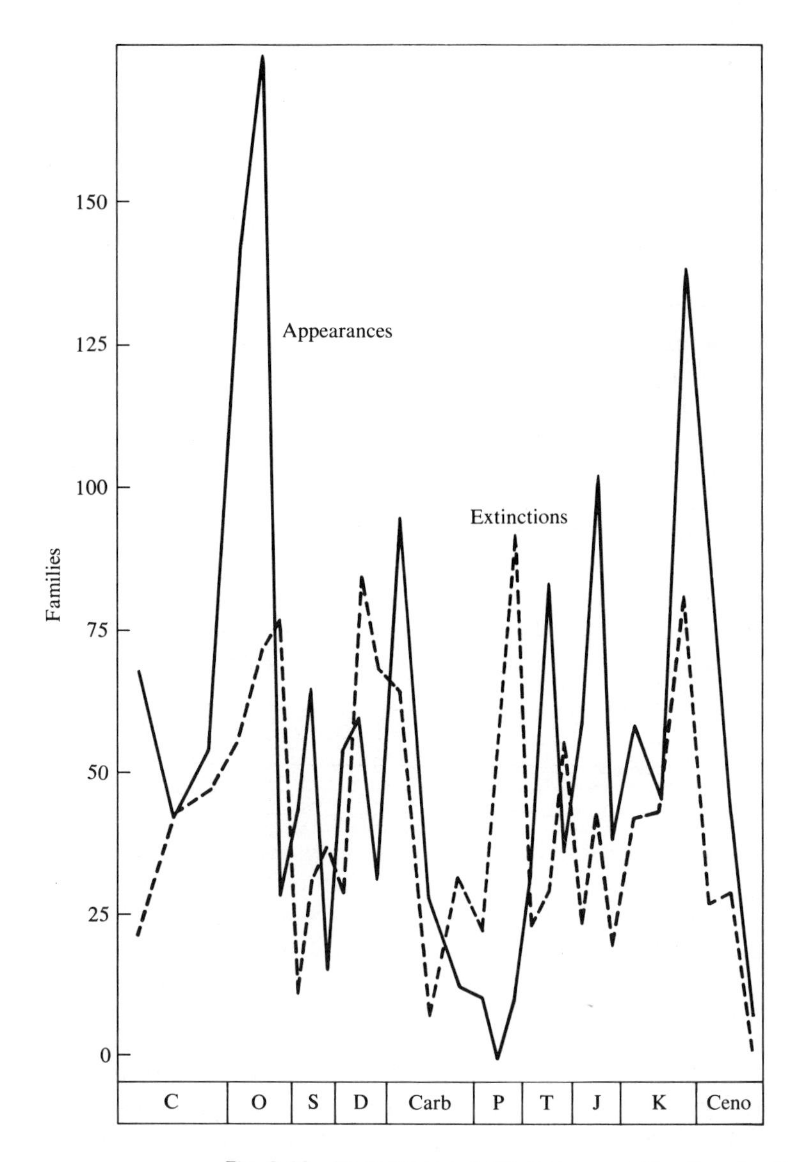

Fig. 9-10 Appearances and disappearances of families of well-skeletonized shelf invertebrates from the fossil record during the Phanerozoic. (After Valentine, 1969; by permission of the Palaeontological Association.)

multidimensional biospace-ecospace model in describing the evolutionary processes of ecological units. There is no reason that the same model will not serve for taxonomic as well as ecological units. Recall that each population has its own ecospace, prospective and realized, which lies within the environmental hyperspace lattice. The ecospaces of all populations within a species form the ecospace of that species. Similarly, the ecospaces of all the species forming a genus can be combined to constitute the ecospace of that genus. The ecospace of a family, then, is composed of the ecospace of all its members, and so on, to higher taxonomic levels. The ecospace for the entire term of existence of a phylum would be composed of all the ecospaces of every individual that has belonged to that phylum and would represent the total range of ecological functions it had ever possessed.

If we imagine the ecospace model of the hierarchy of taxa depicted in Figs. 9-4 to 9-10, it is evident that the ecospace of each category is exactly the same size and shape as the ecospace of every other category. For example, all the species depicted belong to nine phyla, and, for the taxa represented in the figures, the ecospaces of the phyla represent precisely the sum of the species ecospaces. The phylum category is divided into nine portions that may overlap and that differ greatly in size and shape, molluscan ecospace being large and archaeocyathan ecospace being small, for example. Each phylum ecospace is divided into more portions at the class level, most of which are further subdivided at the ordinal level, and so on. When studied through time, the ecospaces of the hierarchy fluctuate in shape and size, the relative proportions of different taxa within categories are altered, and the number of the subdivisions (taxa) within each category varies, especially so at lower hierarchic levels.

Mass Extinctions May Be Treated as Catastrophes or as Features of Diversity Regulation

Many examples of factors that may cause the extinction of species have appeared in previous chapters. Some are physical factors that episodically reach some unusual level lying beyond the tolerances of some species. Others are biological and involve extinction by predation, competitive exclusion, or disease. Extinctions are assumed to occur regularly, owing to the operation of such factors within the environmental regime that is "normal" for any given time. To bring about the extinction of a higher taxon, then, requires only that its species have a higher than average susceptibility to some of these factors and that when they become extinct their replacement is more often than not by species from other taxa. The gradual diminution in its species eventually leads the higher taxon to extinction. Such a gradual process appeals to the uniformitarian instinct. However, a large bulk of extinctions of all categories is concentrated in time during the waves of mass extinction that stand out in Figs. 9-9 and 9-10. Extinctions during these times are not all precisely contemporaneous, to be sure, and the disappearance of higher taxa commonly occurs in a piecemeal manner (Rhodes, 1967). Nevertheless, the occurrence of localized periods of high extinction rates is well demonstrated, and these mass extinctions require further explanation.

The disappearance of great groups of organisms from the living biosphere seems always to have excited curiosity and awe, perhaps because it raises questions as to the eventual fate of humanity. Whatever the reason, large numbers of people, many neither biologists nor paleobiologists, have proposed explanations for extinctions. In the interests of achieving an ecological discussion, we shall classify possible causes of extinctions, not according to the similarity of the causes themselves, but according to their ecological effects on the biosphere.

A general classification of possible ecological bases of mass extinction is presented in Table 9-2. Ecological effects are classed in two ways: (1) whether they are caused by diversity-independent or by diversity-dependent factors; and (2) whether the factors act to expand, shift, or shrink biospace within the environmental hyperspace lattice. A number of the more frequent or more current suggestions is indicated in the table, together with a selection of references. Rhodes (1967) and Lipps (1970) have presented good discussions of many of these suggestions. The present treatment is not meant to be exhaustive.

Table 9-2 Some Hypotheses of Causes of Major Episodes of Marine Extinction: Effects on Number and Extent of Biospace Dimensions

	Expands	*Shifts*	*Shrinks*
Diversity independent	Volcanic poisons (Pavlov, 1924) Metal poisons (Cloud, 1959) Extraterrestrial radiation (Schindewolf, 1954; Uffen, 1965; Opdyke *et al.*, 1966; Simpson, 1966; Terry and Tucker, 1968; Hatfield and Camp, 1970)	Temperature (Stokes, 1960; Hay, 1960) Salinity (Fischer, 1964; Beurlen, 1965) Climatic regime (Axelrod, 1967; Axelrod and Bailey, 1968; Bretsky and Lorenz, 1970)	Oxygen (Holser and Kaplan, 1968; Tappan, 1968; McAlester, 1970) Shelf area (Moore, 1954) Extinction of required species (Nicol, 1961)
Diversity dependent	Trophic resource level (Valentine, 1971)	Trophic resource regime (Valentine and Moores, 1970; Valentine, 1971)	Trophic resource level (Bramlette, 1965a, b; Tappan, 1968; McCammon, 1969) Habitats (Newell, 1956; Lipps, 1970) Climatic zones (Valentine, 1967, 1968a) Shelf conjunction (Valentine and Moores, 1970)

We begin with the diversity-independent factors. A number of authors have suggested that something was occasionally added to the marine biosphere that was inimical to life, such as poisonous volcanic emanations (Pavlov, 1924), metal ions (Cloud, 1959), or cosmic rays (Schindewolf, 1954). It is certain that these factors can be lethal to organisms, and it is quite conceivable, although unproven, that natural toxic substances which have poisoned ocean waters and caused local extinctions of populations have also caused the total extinction of some species that happened to be restricted to the affected region. There is no evidence of such widespread poisoning of the oceans as to lead to extinction of large proportions of the total marine biota throughout the world, however, even in selected habitats. Some times of intense volcanism do not correspond at all to times of massive extinctions, and some lineages of organisms are relatively unaffected by extinctions yet appear no less susceptible to toxins than lineages that die out; some successful and unsuccessful lineages even inhabited similar environments (Rhodes, 1967).

Extraterrestrial radiations have been invoked as a cause of extinctions. The earth's magnetic field affords protection to the earth's surface from most cosmic radiation. But at times in the past, the earth's magnetic polarity has reversed, so that the north and south poles have magnetically interchanged. During such events, the entire field may have been greatly reduced or absent for some time. In this case, cosmic ray bombardment of the surface might attain lethal proportions, causing extinctions (Uffen, 1965; Opdyke *et al.*, 1966; Simpson, 1966). The disappearance of a few planktonic species appears to correlate well with the times of some polarity reversals (Harrison and Funnell, 1964), and it is conceivable that these events are related. Other possible sources of unusually high radiation include explosions of supernovae (Schindewolf, 1954; Terry and Tucker, 1968) and movement of the earth into the region of the galactic plane, where radiation is concentrated (Hatfield and Camp, 1970). However, these proposed cosmic radiations seem inadequate to account for the massive extinctions of marine biota. For one thing, the atmosphere and particularly the water both absorb radiation, so that organisms living at some depth beneath the ocean surface would be protected (Waddington, 1967; Simpson, 1968). Suggestions that radiation would generate a load of deleterious mutations so as to produce a massive extinction seem unlikely for the same reason. Furthermore, populations are partially protected from extinction by mutational load through ordinary processes of recombination and selection.

Alterations in the quality of diversity-independent factors (Table 9-2) have been the most common sorts of explanation for extinctions. Falling temperatures that accompany climatic deteriorations have been especially popular. Probably this is due to the correlation of low polar diversities with low temperatures today, but this fact now appears to be coincidental. However, there is little doubt that temperature is the most important single factor in limiting the geographic range of species, and changes in ocean temperatures that exceed many species' tolerances could clearly lead to large-scale extinctions.

Today the polar waters are essentially as cool as possible, and further polar cooling with spreading of low temperatures into middle latitudes would compress the climatic zones. Although changes in thermal regimes would accompany such compression and might lead to some extinctions, it is likely that most species would merely alter their ranges, migrating with the changing climates. This is evidently what has happened during the Pleistocene ice ages. On the other hand, if polar temperatures were raised or tropical temperatures lowered, whole climatic zones would be eliminated which would certainly lead to extinctions, more massive when tropical regimes are cooled because more taxa live there. Although there is evidence that high latitudes were warmer in the past then at present, there is no special evidence that the equatorial regions were ever badly chilled.

Another sort of temperature change that could have caused widespread extinction is a change in the thermal regime that significantly altered the temperature ranges. As is true for all sorts of temperature changes, this change would normally be accompanied by numerous changes in other environmental factors, so that it is more realistic to speak of a change in the climatic regime as a whole. If a relatively stable regime were to be succeeded by a relatively unstable one, the tolerances of those species narrowly adapted to the stable environment would be exceeded. Inasmuch as adaptation to stability may commonly involve genetic and reproductive strategies as well as physiological ones, the requirement to develop a whole suite of harmoniously coadapted characteristics suitable to a new regime may commonly prove too difficult for many lineages. Axelrod (1967) has attributed the extinctions of some large mammals to a relatively small increase in the annual temperature range during the Pleistocene, whereas Axelrod and Bailey (1968) ascribe extinctions among the dinosaurs to an increasing seasonality in the Late Mesozoic.

An appealing feature of climatic regime changes as an explanation for massive extinctions is that they would favor the survival of lineages that were preadapted to the new regime, and thus a predictable ecological bias should appear among the survivors. There is evidence that such a bias can be found (Bretsky, 1969; Bretsky and Lorenz, 1970), for lineages that can be interpreted as the more generalized and flexible ones which live in the more unstable environments seem to survive some waves of extinction in the greatest numbers. Furthermore, events such as continental drift and continental emergence and submergence, which can cause changes in climatic regime on a world-wide scale, are known to have occurred during the course of geological time. Therefore, in climatic change we have a mechanism that can do the job and is known to occur. It therefore seems very likely that climate has played a large role in extinction.

At present, marine diversity tends to drop in waters of low salinity, and it has been suggested that widespread oceanic salinity changes account for some of the extinctions. Salinity reductions are postulated to result from glacial meltwaters of the Late Paleozoic (Beurlen, 1965) or from removal of salts in evaporite sequences (Fischer, 1964), and, of course, salinity rises would result from the reverse processes. Difficulties with the salinity hypothesis include the lack of independent evidence of any significant oceanic salinity reductions during extinctions, the presence of some scattered isotopic evidence that salinities have remained near normal (Lowenstam, 1961), the certainty that many species belonging to taxa that are stenohaline today, and which give every indication of being stenohaline in the past from their distribution patterns, were relatively unaffected by extinctions that have been attributed to salinity (Teichert, 1964), and the evidence that salinity changes associated with Cenozoic ice storage, including Pleistocene glaciations, have not led to mass extinctions. Therefore, although an important change in oceanic salinity would probably result in mass extinction, there is little evidence that such changes have occurred during Phanerozoic time.

The decrease or exhaustion of required environmental factors has been postulated as diversity-independent causes of extinction (Table 9-2). Oxygen depletion has frequently been suggested. One mechanism for this depletion is an uptake of oxygen by sulfide to sulfate conversions by bacteria (Holser and Kaplan, 1968; Tappan, 1968), but Lipps (1970) has noted that although any such depletions might be regarded as a consequence of changes which might lead to extinctions, they probably cannot be regarded as a cause. McAlester (1970) has suggested that the selective nature of many extinctions might be accounted for by differential oxygen tension tolerances among different taxa; those with higher metabolic rates require more oxygen, other things being equal.

A very different factor postulated as a cause of extinction is habitat area (for example, Moore, 1954). The suggestion was based on the observation that widespread regressions over continental platforms often separate different platform biotas, the first having become extinct during the regression. According to Moore, the restriction of shelves would "crowd" species together into small areas and thus extinction would ensue. There is neither ecological support nor substantive evidence for the idea of population crowding in this case. Furthermore, so long as the reduced area contained a representative sample of the habitats present during the transgression, most populations could conceivably persist. One result of smaller area is a reduction in the population size of the average species, which could contribute to extinction if adaptation to the local environment required or favored large population sizes.

The extinction of some lineages that are important to the survival of other species can obviously lead to further extinctions, and such a snowballing effect could disrupt some important community relations (Olson, 1952; Nicol, 1961). Thus, one source of extinction is extinction itself. Other biotic effects that can lead to extinction are competitive exclusion or over-predation. These effects are difficult to demonstrate and would not seem even to have produced the major mass extinctions in the sea, except as proximate causes of extinctions of biotas that had been separate and then permitted to commingle. This is usually a case of reduction in provinciality and is discussed under that heading.

Diversity-dependent factors can cause extinction by lowering the number of species that can be supported at any time. One diversity-dependent factor that may cause extinction by the enlargement of biospace is trophic resource level. Margalef (1968) points out that, frequently, increased food supply within an ecosystem permits certain favored species to expand their population sizes rather than allowing additional species to occur. Experience with polluted ecosystems tends to bear this out, for in pollution by eutrophication the favored species tend to take over the environment and less-favored species become (locally) extinct. From these considerations, it appears that, to the extent that trophic resource levels are implicated in diversity regulation, a rise in level should favor extinctions (Valentine, 1971a). It seems doubtful that this factor alone has ever been responsible for mass extinctions.

Changes in trophic resource regimes have been invoked as a major cause of mass extinctions, however (Valentine and Moores, 1970, 1972). We have already reviewed the probable adaptive strategies required of populations that subsist on fluctuating resources; they are large and inefficient, and communities containing many of them must be of low diversity. If the world's shelves were to change from a relatively high resource stability to a relatively low one, the diversity of shelf biota would fall. Extinctions would be least among generalized, flexible lineages that were preadapted to existence in fluctuating food regimes. The main difference in biotic reaction to trophic resource fluctuations as compared with climatic regime fluctuations is that rediversification is possible and expectable following, or even partly contemporaneous with, the extinctions due to climatic regime changes, for new lineages adapted to the new conditions would soon appear. Rediversification following a change to a fluctuating resource regime is simply not possible, for the extinctions bring diversity into approximate equilibrium with the regime, and restabilization is required to permit another diversity increase.

Continental drift provides one mechanism that might control trophic resource regimes (Valentine and Moores, 1970). If all the continental shelves were to drift into high latitudes, then the resulting increase in the seasonality of solar radiation would cause increasing trophic fluctuations and diversity would drop. If the continents attain a configuration that promotes an increase in continentality, with strong seasonal nutrient fluctuations in shelf waters, a similar result is obtained. Continental emergence, which contributes to continentality by decreasing the sea-surface area and increasing continental area, also tends to favor fluctuating regimes.

A number of investigators have proposed that a reduction in some density-dependent factor would lead to extinctions. This suggestion has been made for trophic resources by Bramlette (1965a, 1965b), McCammon (1969), and Tappan (1968).

Reducing the number of available habitats would certainly reduce diversity, and Newell (1956) and others have suggested that lowered sea levels may have this effect. A rapid sea-level fall to the outer shelf edge might eliminate most rocky-shore and sandy-beach habitats and most firm substrates in shallow water, leading to a decrease in environmental heterogeneity and, consequently, in diversity. Lipps (1970) postulated a decrease in habitats in the pelagic realm owing to decreasing thermal hetrogeneity during times of high-latitude warming as producing extinctions among the plankton.

A reduction in the number of marine climatic zones reduces diversity and causes extinction of the biota adapted to the vanished climate (Valentine 1967, 1968a). This may occur by polar warming or equatorial cooling alone, or in combination to effect an increase in planetary thermal monotony to remove latitudinal provincal barriers. Similarly, when longitudinal provinciality is reduced owing to continental collision or to the appearance of intercontinental or intercoastal dispersal corridors, extinctions ensue owing to competition among analogous species or to predation, and diversity falls (Valentine and Moores, 1970). In general, if a province is eliminated the amount of extinction should approximate the size of its endemic element.

In summary, the most likely causes of mass extinctions appear to be those factors which are natural regulators of diversity under normal circumstances and which have effects that pervade the entire planet. Extraordinary events that cause radical alterations in the habitability of the environment would seem to be less likely causes. The "normal" diversity regulators seem to include trophic resource level and regime and changing temperature level and regime as the more effective factors. Global changes in the environmental states of these factors are easily accounted for by tectonic processes that are proven or can be theorized to be occurring today and to have occurred in the past as part of normal planetary evolution.

Waves of Diversification May Be Treated as Consequences of Evolutionary Invention or as Features of Diversity Regulation

The processes of diversification are the processes of the origin of new taxa. Concepts of the origins of new species present no major theoretical difficulties today, although many unsolved problems of mechanism remain. The origins of taxa in higher categories are regarded by many workers as no different from the origins of species, although a few workers believe that somewhat different internal processes of evolution must be invoked to explain the appearance of higher taxa. At present, we are not so concerned with the internal mechanisms of the origin of taxa as with the external factors that lead to such patterns of diversification as appear in Figs. 9-9 and 9-10. Problems as to why one taxon rather than another diversifies, or why a given burst of diversification occurs at a particular time, seem likely to be strongly related to ecological processes. Here, we shall summarize the ecological events that can provide opportunities for mass diversification.

1. The physical-environmental portions of biospace do not change in size or quality, but include regions that are uninhabited. Evolution produces organisms that are appropriately preadapted to these regions, and these favored lineages radiate into them, producing a diversity rise. A classic example would be the invasion of the terrestrial biosphere, first by plants, then by invertebrates, and finally by tetrapods. In the sea, a good example may be the appearance of infaunal burrowers near the base of the Cambrian, which presumably were invading a suite of habitats that were previously unoccupied by comparable or competitive animals, since burrows are nearly or altogether absent from older rocks.

2. The physical-environmental portions of biospace do not change in size or quality, but competitive superiority is attained by certain lineages that radiate into appropriate parts of inhabited biospace and exclude the occupants. There is little evidence that bears directly upon this point. The successful diversification of any lineage in the face of direct competition has never been satisfactorily demonstrated, even to the degree that is possible for such historical questions. A number of groups display complementary diversity records with one expanding as the other contracts, such as the fishes (expanding) and the eurypterids (contracting); other examples are given by Simpson (1953). Yet even if such records do indicate the replacement of one group by another in the same biospace, the pattern is not necessarily a result of competitive exclusion, but may simply indicate that the expanding group was better able to radiate into biospace vacated by prior extinction of the shrinking group. This is not a result of successful competition in an ecological sense, although it indicates differential adaptation to a certain range of environmental conditions.

3. The physical-environmental portions of biospace do not change in size but in quality, so that extinctions occur which create uninhabited portions of biospace and lineages radiate into these, producing a wave of diversification. Such extinctions could be caused by diversity-independent factors that do not significantly alter environmental heterogenity, such as temperature or salinity shifts.

4. The physical-environmental portions of biospace do not change in size, but a diversity-dependent factor changes in quality to permit a higher level of diversity. The inhabited portions of biospace become progressively more partitioned among species so that the average species ecospace shrinks and permits new species to occupy the vacated ecospace; therefore, diversity rises. This same phenomena could occur among higher taxonomic levels as well. It could be due to an increase in trophic resource stability, permitting more species to coexist in communities, or to an increase in longitudinal provinciality owing to the appearance of dispersal barriers, such as ocean deeps between continental shelves.

5. The physical-environmental portions of biospace increase in size, and lineages radiate into the new environments. Such expansions commonly accompany an increase in spatial heterogeneity, which on the community level provides

more habitats to permit the co-occurrence of more species in the same community, on the provincial level provides more biotopes to permit the occurrence of more communities, and on the biosphere level provides more climatic zones to permit a rise in latitudinal provinciality.

The Major Diversity Patterns of the Phanerozoic Can Be Correlated with Major Patterns of Environmental Change

Having reviewed the more likely causes of the major diversifications and extinctions and of the resulting changes in diversity levels, it is now worthwhile for us to attempt to interpret the quantitative taxonomic patterns recorded during the Phanerozoic. In Chapter 10, we shall consider the qualitative patterns.

In Cambrian time, when fossils first became at all common, each phylum was represented by only a relatively few classes, each class by few orders, and so on, right down to the species level, so far as can be determined from the record. However, there were essentially as many invertebrate phyla as today (and perhaps more). Therefore, although there were relatively few species, the variation in basic ground plan was at least as wide as today. With the strong Cambro-Ordovician wave of diversification, the numbers of classes, orders, families, and no doubt genera and species were greatly increased. Now, the recognition of numerous taxa belonging to higher categories has emphatically *not* resulted from the sort of classification procedures that recognize closely allied lower taxa as separate higher taxa because they eventually evolve into distinctive groups (Fig. 9-11). On the contrary, these early Paleozoic higher taxa are by and large thoroughly distinctive when they appear, judging, of course, from their skeletal morphologies. By and large, the taxa of higher categories in Cambro-Ordovician times represent ground plans or modal morphologies fully as different from each other as do living taxa in similarly high categories.

There are two major alternate interpretations to this pattern of diversification. The first is that we are witnessing the appearance of a variety of distinctive new types of organisms that are invading biospace which was either unoccupied or undersaturated, that is, contained fewer populations than permitted by ecological regulators of diversity because occupants for all the potential habitats and/or trophic roles, had not yet evolved. The second is that we are merely witnessing the development of skeletonization, and that these distinctive lineages had long been evolved. We shall evaluate these alternative possibilities in Chapter 10.

The diversity levels from about Middle Ordovician to Late Devonian time are relatively stable, with little change in the proportions among different taxonomic categories, but there is a major extinction accompanying the late stages of diversification at about Late Ordovician time (Fig. 9-10). The stability of the levels suggests that an equilibrium was approximated between the size of available biospace and the number of species present, and that this relation remained

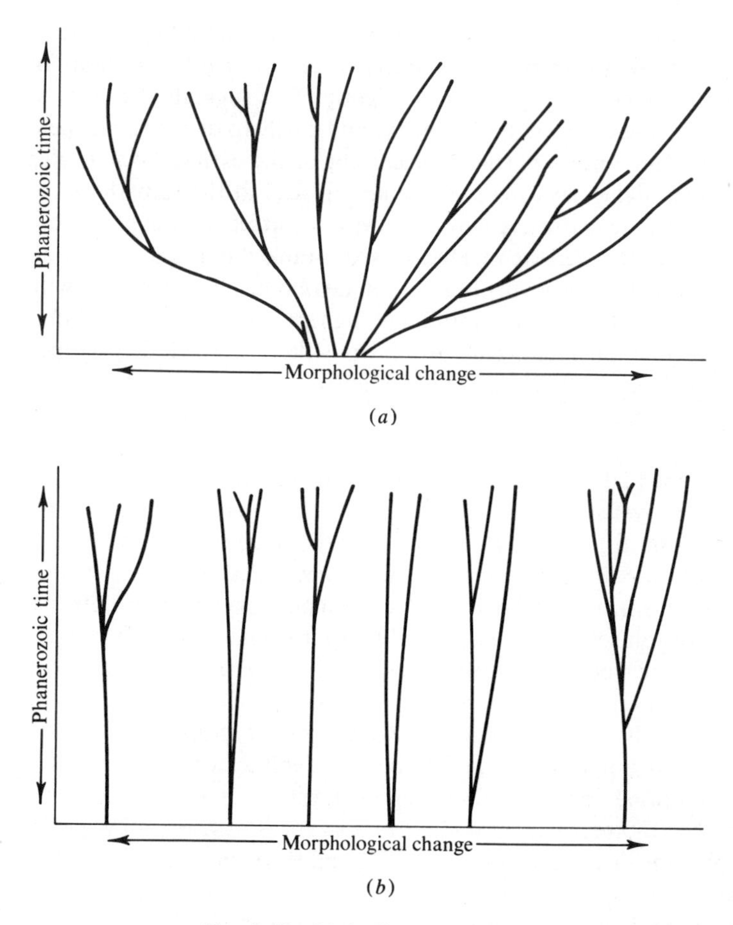

Fig. 9-11 Distinctiveness of major taxa appearing in early Phanerozoic time. (*a*) Closely related groups appear and evolve into distinctive entities, so that early ancestors, although closely allied, are placed in separate higher taxa. This is *not* the common case for early Phanerozoic taxa. (*b*) Distinctive groups appear and continue as distinctive entities. This is more nearly the pattern in the early Phanerozoic.

relatively constant. Of course, there were probably some considerable diversity variations in species, but they cannot yet be traced from the fossil record. The Ordovician extinction does not make much impression on the standing diversity levels, indicating a rapid replacement of the lost lineages. This suggests that the extinction was caused chiefly by diversity-independent factors, if it was accompanied by a reduction in biospace; it must have been of brief duration. The most likely cause of this extinction is a change in climate (Table 9-2), with the diversification due to radiation into biospace that was vacated by extinct lineages.

The next major change recorded in the diversity of the marine biosphere occurred in the Middle and Late Devonian, when a wave of extinction took place, which, although accompanied by some diversification and followed by a wave of diversification in the lower Carboniferous, nevertheless lowered the diversity of classes and orders significantly and families slightly. This suggests that the extinctions were partly due to diversity-dependent and partly to diversity-independent factors, permitting only a partial replacement of the extinct taxa. It also indicates that extinct higher taxa were not being replaced at the same level, so that when a class became extinct, a new class was not necessarily evolved, but that new lineages which invaded the vacated biospace ranked in lower categories. In fact, it is likely that the replacements are often taxonomically piecemeal, with a few species from each of a number of different higher taxa, perhaps even different phyla, radiating into the habitats and trophic positions to which they happen to have access by virtue of their particular adaptive styles. In this event, the replacement of an extinct class might be made even on the generic and specific levels.

Imagine that some higher taxon was eliminated from today's world. Perhaps the echinoderms are systematically hunted down or are ravaged by a terrible echinoderm disease until none is left. Would it be reasonable, then, to expect a new phylum to evolve to fill all the various and diverse roles played by echinoderms? They range from intertidal predatory starfish on rocky shores to detritus-feeding holothurians on soft bottoms. Probably what would eventually happen is that some predatory stock of gastropods or some such group would give rise to species capable of utilizing the trophic resources formerly consumed by starfish, whereas bivalves, crustaceans, or worms might utilize the detritus formerly consumed by holothurians, and so on. The replacement pattern might become very complicated, indeed, with the representatives of a given echinoderm stock being replaced by species from one phylum in one province but by species from another phylum in another province. The category to which the newly evolved "replacement" species would eventually become assigned might vary from group to group. If a large, active, gastropod stock evolved to feed upon rocky-shore starfish prey, and radiated into various provinces around the world, they might form a well-defined group of species that could be classed into one genus, or perhaps several genera classed into one family, if several species groups were involved. The replacement species for holothurians might be drawn from crustaceans in one province and from worms in another. Some of the extinct echinoderms might not be replaced at all, in the sense of having their former energy base utilized by other metazoans.

Diversity levels remained fairly steady throughout the Carboniferous and into the Permian, as both diversification and extinction proceeded at low rates. Then a Middle and Late Permian wave of extinction, which coincided with an all-time low rate of diversification, carried the standing diversity of all categories for which data are available to their lowest levels since the Cambrian, and rediversification did not reach an important level until Middle Triassic. This sequence of diversity events is most easily interpreted as owing to diversity-dependent factors, which first changed in the late Paleozoic so as to supress diversification, and then became much more restrictive to cause a wave of extinction. The processes of diversification were unable to generate replacements until the restriction was removed, which seems to have happened gradually during the Triassic and Jurassic. We have already suggested that the continued Mesozoic-Cenozoic rediversification was primarily owing to increasing provinciality arising from the breakup of continents to produce ocean barriers and therefore endemism, and by the rise in the latitudinal temperature gradient growing from isolation of high-latitude waters from warm low-latitude waters, producing latitudinal provincial chains and, again, increasing endemism.

The Mesozoic-Cenozoic breakup of the continents theoreticalley should have led to a general increase in environmental stability on the world's shelves. After all, the shelves of a supercontinent should, on average, have highly continental climates, whereas continental breakup and separation by sea-floor spreading should raise the average stability of shelf waters by reducing continentality. An increase in stability permits the development of adaptive strategies that lead to more efficient resource utilization by the populations and permit more populations within ecosystems. Therefore, there should be a diversity rise on the community level accompanying the rise on the provincial level, and it should be reflected by changes in the structure and the taxonomic composition of communities.

Other major causes of environmental change include the variations in relative sea level that have led to oscillations in the extent of epicontinental seas. The causes of sea-level variations are themselves certainly varied and interconnected. Some of them rise directly or indirectly from plate-tectonic processes, but it is difficult to assess their relative importance at this time. In general, seas will stand highest on continents when ocean basins have the least volume or when the portion of the crust on which the continent rests lies in a depression; in either case the ocean tends to spill over the continent. Consider a large continental mass that is broken into fragments which rift apart; this implies the appearance of a ridge and of new sea floor between the daughter continents and the presence of a trench system to accommodate the spreading lithospheric plates. Therefore, as the daughter continents first drift apart, they leave relatively shallow ocean floor behind them and migrate down the ridge-flank depressions, which should raise sea level over the spreading continents and, by displacing sea water, effectively reduce the volume of the ocean basin. Sea level therefore rises relative to the continental platforms, and if continental fragmentation is extensive, the consequent sea-level rise could amount to hundreds of meters (see also Russell, 1968; Valentine and Moores, 1970, 1972). There would be a major transgression, enlarging and deepening the shelf seas.

The cycles of sea-level changes accompanying continental fragmentation or assembly would be much more complicated in detail. For example, a rise in continental elevation along the eventual zone of rifting would be an expected prelude to fragmentation, which would cause a regression of any seas overlaying the affected regions. However, the displacement of water in this region would cause a general, although much smaller, rise in sea level (Hallam, 1963; Bott, 1965; Russell, 1968). Furthermore, as a spreading continent approaches a ridge, it would migrate up the ridge-flank rise, forming a local emergence and perhaps a smaller general regression. Increase in spreading rates or other events associated with rising ridges would also affect sea level, even in the absence of continental involvement. Van Andel (1969) has demonstrated a marked correlation between topographic changes on the mid-Atlantic ridge and transgressions and regressions along neighboring Atlantic shores, which seem to be due to the volumetric changes at the ridge. Other causes of sea-level change not associated with global tectonics at all directly (such as glaciation) may cause either local or global sea-level changes, which may mask or reinforce tectonic causes or be entirely separate therefrom.

The climatic results of the spreading of epicontinental seas over the platforms would be to replace an exposed land surface having a relatively low specific heat with water having a higher specific heat, reducing seasonal temperature fluctuations and enhancing general environmental stability. Inasmuch as trophic resource stability would be expected to increase on the average, there would be an increase in the average species diversity permitted within communities. Additionally, there might be an increase in the number of communities owing to increased environmental heterogeneity in the more extensive seas.

If this general model of events is correct, then times of continental fragmentation should correlate with times of rising species diversity within communities, rising community diversity within provinces, and rising provinciality within the biosphere. Times of transgressing seas should also correlate with rising species diversity within communities and rising community numbers within provinces, and might lead either to additional provinciality if partially disjunct or climatically novel arms of the sea were formed, or to lower provinciality if former land barriers were broached by the rising waters. During times of continental assembly or of regression the effects would tend to be in the opposite direction. When continental fragmentation and high sea stands coincide, as at times during the late Cambrian and Ordovician, or during the late Cretaceous, diversity should rise to high levels, whereas when continental assembly and low sea levels coincide, diversity should fall to very low levels indeed.

These relations suggest that the extinctions of the Permo-Triassic were associated with the assembling of continents, formerly dispersed during the Paleozoic, into the supercontinent called Pangaea. A considerable body of geophysical, structural, and petrologic evidence supports the idea of Paleozoic dispersal of some northern continents. The platforms of some continents display different patterns of polar wandering for the Paleozoic, suggesting that those continents had different relative motions and were separated then. Lines of continental collisions may be indicated by mountain systems that lie (or have formerly lain) within continental platforms, such as the Himalayas, which appear to represent the locus of suturing between India and southern Asia. These continents were formerly separated by an ocean basin, which has been collapsed between then by subduction (Fig. 9-12). Other mountain systems that may represent the sites of

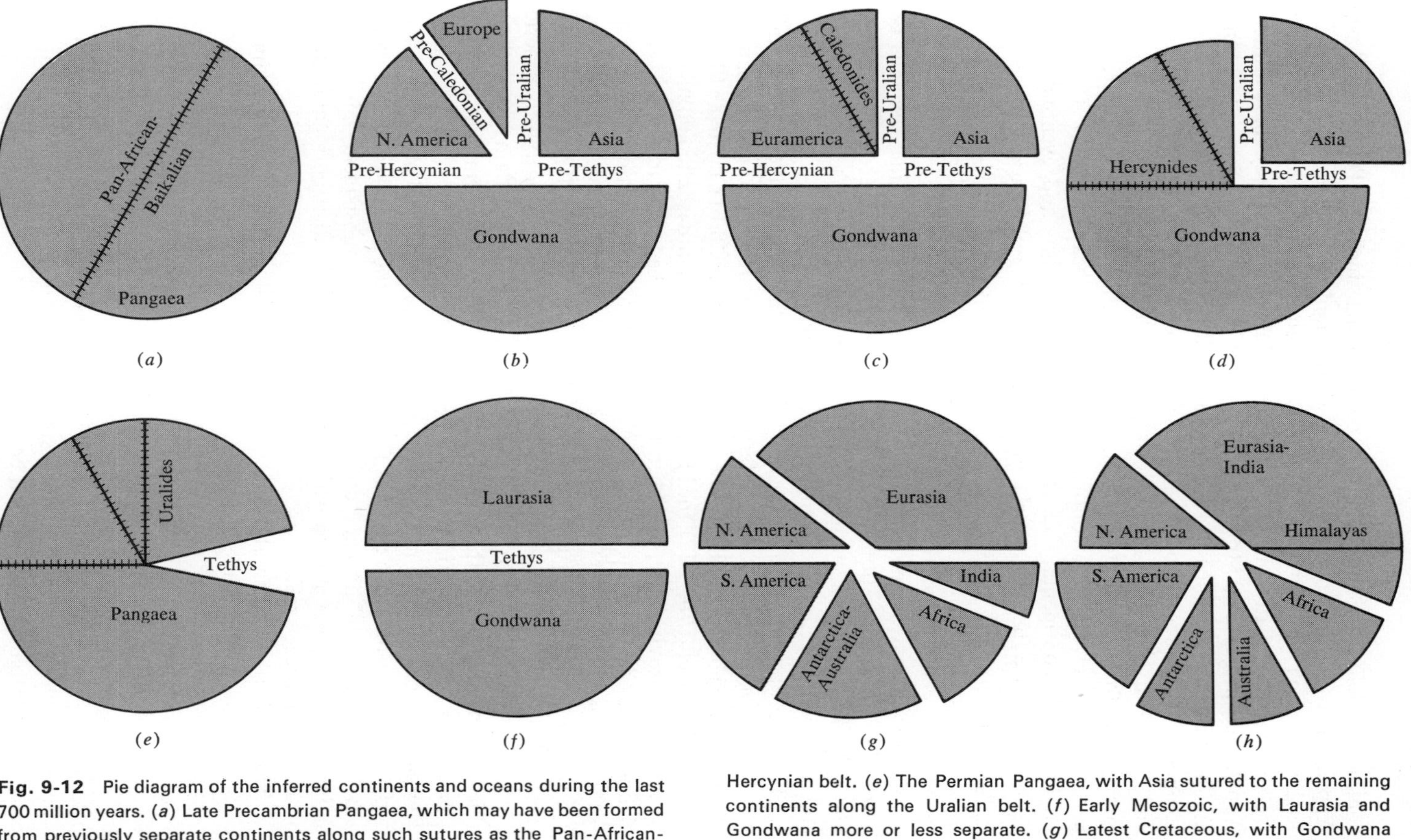

Fig. 9-12 Pie diagram of the inferred continents and oceans during the last 700 million years. (*a*) Late Precambrian Pangaea, which may have been formed from previously separate continents along such sutures as the Pan-African-Baikalian belt. (*b*) Cambrian, showing four continents separated by ocean floor. (*c*) Devonian, showing three continents following the collapse of the pre-Caldedonian Ocean and collision of ancient Europe and North America. (*d*) Late Carboniferous, with Euramerica sutured to Gondwana along the Hercynian belt. (*e*) The Permian Pangaea, with Asia sutured to the remaining continents along the Uralian belt. (*f*) Early Mesozoic, with Laurasia and Gondwana more or less separate. (*g*) Latest Cretaceous, with Gondwana highly fragmented and Laurasia partially so. (*h*) The present, with India sutured to Eurasia. (Modified after Valentine and Moores, 1972; from *The Journal of Geology,* the University of Chicago Press, copyright © 1972 by the University of Chicago.)

continental collisions and collapsed ocean basins include (1) the northern Appalachian-Caledonian Mountains (now separated by a new ocean, the North Atlantic), which were formed in Siluro-Devonian time and united a Paleozoic North American platform with a Paleozoic European one; (2) the southern Appalachian-Hercynian Mountains, formed in Early Carboniferous time and uniting the North America-Europe continent with Gondwana, a large continent composed of the present continents of South America, Africa, India, Australia, and Antarctica, plus some microcontinents (Fig. 9-12); and (3) the Ural Mountain system, formed in Late Permian time and uniting Asia with the European parts of the growing supercontinent to form a Pangaea (Fig. 9-12). Thus, the gradual rise in continental size, reduction in provinciality, and eventual continental emergence that are associated with the assembly of Pangaea may be reflected in the diversity curves, and the concomitance of these diversity-dependent factors may be responsible for Permo-Triassic extinctions as well as contributing to earlier diversity reductions.

The subsequent diversity rise in Mesozoic and Cenozoic time can thus be ascribed to the amelioration of these factors as well as to increased provinciality, and should have been accompanied in low latitudes by increased numbers of species in communities and communities in provinces.

The disproportionate diversification found in families relative to orders and genera relative to families, and inferred for species relative to genera, can be easily explained by this model. As the continents gradually broke up in the Middle and Late Mesozoic (Fig. 9-12), the lineages were broken into isolated populations by the rising provinciality, and then further partitioned as latitudinal provinciality increased. Thus, the rising provinciality provided isolating mechanisms that multiplied the numbers of species enormously; genera represented by one or a few widespread species in the Middle Mesozoic developed numerous new species, a suite of species within each new province that lay within its generic range. In some cases, taxonomists have formed these descendant species into allied groups and ranked them as genera, so that the entire assemblage becomes a family for which the Middle Mesozoic species becomes a common ancestor. The patterns of generic and familial assignment certainly do not correspond perfectly to biogeographic patterns today, owing to the irregularity in the pathways of evolution of lineages once isolated, and to subsequent migrations, extinctions, and replacements, which have generated the distributional patterns of the contemporary taxa.

The extent of continental fragmentation, dispersal, and reassembly is almost certainly considerably underestimated in Fig. 9-12. Present-day Asia, in particular, appears to be a collage of small continents and microcontinents, at least some of which were once part of the Gondwana landmass. In this respect they would resemble India. Their number, former positions, and time of migration have not yet been worked out in any detail. As our knowledge of continental paleogeographies grows, the complexity of the continental dispersion patterns will be more clearly revealed, and it seems likely that many seemingly anomalous paleobiogeographical patterns will be clarified.

CHAPTER TEN

AN APPROACH TO AN ECOLOGICAL HISTORY OF THE MARINE BIOSPHERE

It is generally acknowledged that all organic beings have been formed on two great laws—Unity of Type, and the Conditions of Existence. By unity of type is meant that fundamental agreement in structure, which we see in organic beings of the same class, and which is quite independent of their habits of life. On my theory, unity of type is explained by unity of descent. The expression of conditions of existence . . . is fully embraced by the principle of natural selection. For natural selection acts by either now adapting the varying parts of each being to its organic and inorganic conditions of life; or by having adapted them during long-past periods of time . . . . Hence, in fact, the law of the Conditions of Existence is the higher law; as it includes, through the inheritance of former adaptations, that of Unity of Type.

—From Charles Darwin, *On the Origin of Species,* etc., John Murray, London, 1859.

Despite the growing number of excellent marine paleoecological studies, there have been few attempts to synthesize the trends in ecological processes across the whole course of the Phanerozoic. Much more work remains, both paleontological and biological, before we have a set of hypotheses that is well enough verified to form a solid theoretical framework in which to place the fossil evidence. Nevertheless, a beginning has been made and the discipline has advanced to the point where some of the highlights in the history of life may be examined from a paleoecological perspective. It turns out that some of the highlights are unique or bizarre enough so that the number of ecological explanations at all likely is strictly limited, but in other cases a wide range of hypotheses may still be entertained.

Quantitative aspects of the fossil record of the taxonomic hierarchy seem to be amenable to general ecological interpretation. Therefore, it seems reasonable that qualitative aspects may also be generally interpretable, and that when joined with quantitative evidence the qualitative considerations may lead to rather precise inferences in some cases. Among the more obvious problems are the nature of the Precambrian-Cambrian boundary, the ecological significance of the ground plans of the major taxa, and the patterns of rise and decline in the importance of different taxa. In keeping with previous chapters, the chief concern here is with the shelf biota, but the major features of other marine ecological realms will be discussed briefly. First, it is appropriate to review some of the ideas regarding the origin, and pre-Phanerozoic history, of life.

Acceptable Conditions for the Origin of Life Do Not Require Extraordinary Circumstances But Merely Primitive Ones

Life must have originated under conditions very different from those at present. The first fossils generally accepted as living organisms date from about 3.2 billion years ago (Schopf, 1970). These consist of microscopic remains resembling bacteria, preserved in chert. Thus, consideration of conditions surrounding the origin of life requires the reconstruction of the earth's environment at a relatively primitive stage in its history as a planet.

The characteristics of the early oceans and atmosphere are of particular significance in considering the origin and early evolution of life, but these topics are the center of lively debate. The composition of the early atmosphere, for example, is not estabilshed because the geochemical evidence is equivocal, and there are two main groups of hypotheses and several minor ones.

The earth's gaseous envelope contains little or no representation of the inert gases, such as neon and argon, which are by no means rare in the universe at large, the former being 10 billion, and the latter 100 million, times more common in the cosmos than on earth (Berkner and Marshall, 1964). Presumably, these gases have escaped from the vicinity of earth, owing either to a weak gravity field during the early accumulation of particles that formed the planet, or to heating of the planet during its early compression. Active gases may have been retained by being locked up in solid compounds. The atmosphere that was present when life originated, then, was probably generated largely by degassing of the earth's interior—in other words, from volcanic emanations (Rubey, 1951; Abelson, 1966). Gases that are emitted from volcanic sources include H_2, H_2O vapor, CO_2, N_2, and perhaps NH_4, in addition to sulfur and halogen gases. Free oxygen would have been practically absent at first, for it could be generated only by photolytic dissociation of H_2O and CO_2 and the production by this process would be consumed by oxidative reactions in the reducing atmosphere. The earliest organisms, then, may have originated in an anoxic environment.

The earliest oceans, condensed from H_2O vapor, would have had few or no dissolved ions derived from terrestrial weathering, and would of course have lacked the organic substances that originate as metabolites. They would have contained such elements as sulphur from volcanic sources and much CO_2, implying a rather acidic pH. Garrels and MacKenzie (1971) suggest that the composition of the older Precambrian ocean can be thought of as resulting from an acid leach of basaltic and andesitic rocks, with CO_2 pressure an order of magnitude greater than today and with slightly different ionic concentrations, for example with more dissolved calcium and ferrous iron. Continuous cycling of continental material through the ocean system modified the water, with weathering products providing buffering systems to control the pH, possibly restricting it to the range of 8 to 9 (Sillén, 1961; Abelson, 1966) despite continuing acidic volcanic emanations. Sedimentary rocks became modern in aspect about 1.5 to 2 billion years ago; late Precambrian evaporite deposits show that the atmosphere contained free oxygen, for sulfates were present (Garrels and MacKenzie, 1971).

Much of the radiation that arrives at the earth today is screened off from the lower atmosphere by oxygen (Fig. 10-1), so that the level of ultraviolet radiation would have been much higher when the atmosphere was nearly anoxic. It has been demonstrated experimentally that a number of organic compounds may be generated by the ultraviolet irradiation of gaseous mixtures resembling any of the more probable early atmospheres. When the organic compounds are introduced into an aqueous medium (such as early ocean water), a considerable

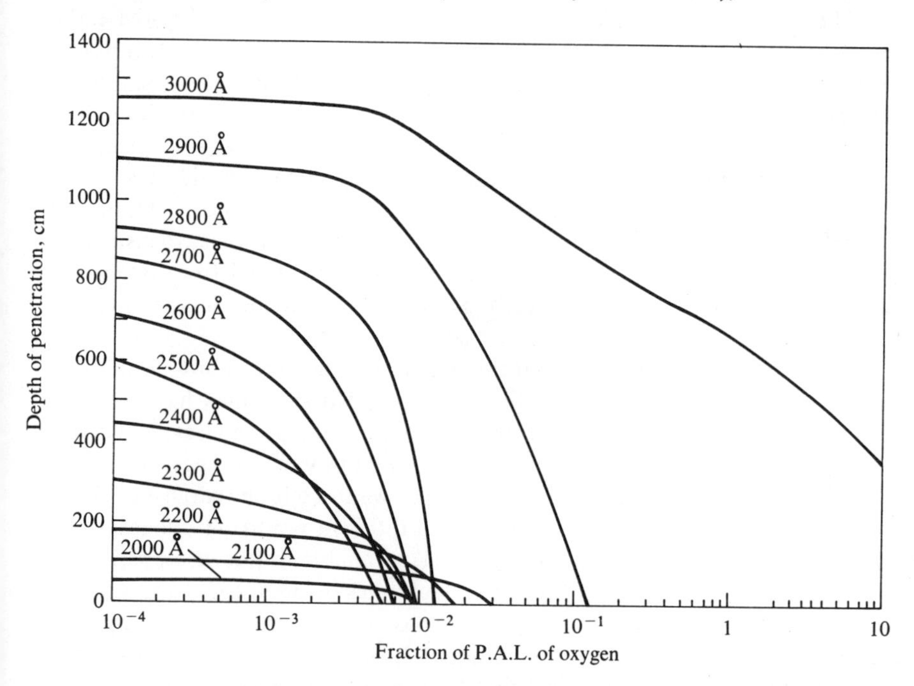

Fig. 10-1 Differences in the depth of penetration of ultraviolet light in water under various conditions of oxygen and ozone concentrations in the atmosphere. (After Berkner and Marshall, 1964.)

range of complex organic substances may be formed, including amino acids, sugars, lipids, and even polypeptides (Ponnamperuma and Peterson, 1965; see Cloud, 1968, for a brief review). The concept of abiotic generation of complex organic compounds under conditions that appear to approximate the primitive earth conditions therefore presents no special problems. Although nucleic acids and enzymes have not yet been synthesized under these conditions, the path of origin of a simple living system that can reproduce itself and its mutants and that can influence other materials can now be imagined (Muller, 1966).

Thus, the early oceans in which life originated may have differed from the oceans of today principally in containing little or no dissolved oxygen, in receiving a high level of radiation at the surface, in containing some ions in significantly different concentrations, in containing a variety of abiotic organic compounds, some of considerable complexity, and in lacking the biochemical circulation and organically produced metabolites that characterize the present marine biosphere.

An Environment Suitable for Metazoa Developed During the Precambrian Partly Owing to Biological Activity

The biochemistry of photosynthesis is sufficiently complex that the earliest organisms very likely employed sources of energy which were much simpler to exploit than sunlight. These early forms were probably heterotrophs that fed upon the abiotically produced organic compounds which must have abounded in the primitive seas (Oparin, 1953). Even after the original pool of organic compounds was depleted, the primitive heterotrophs might have persisted, so long as a sufficient supply of trophic substances was generated abiotically. During this phase, the biochemical complexity of life must have advanced considerably, for eventually photosynthesizing life forms appeared, freed from dependence upon organic food supplies (or perhaps only partly independent, at first) and generating free oxygen.

The sequence of organisms that evolved during the Precambrian is not really known, although the sequence of the properties that they must have possessed can be inferred. The earliest heterotrophs would probably be classed as detritus feeders today; it is possible that predators also appeared, so that some of these early heterotrophs preyed on others. So far as we can tell, the earliest organisms may (or may not) have arisen from several different nonliving sources, so that life may be polyphyletic or monophyletic (Kerkut, 1960). It is also possible that life arose and was extirpated once or several times before the first appearance of the ancestral life forms of living organisms. Chemautotrophs, which derive energy from degrading inorganic compounds, may have evolved next. On the other hand, it is equally possible that true photosynthesizing autotrophs were next, and these may have evolved either once or several times. In any case, these organisms seem to have appeared first in an environment low in free oxygen and with high levels of radiation at the sea surface. It has been suggested that oceanic plankton could not have been present because these high levels of radiation would have been lethal. Instead, primitive photosynthesizers might have been restricted to sites that were protected from much radiation, and may have been only nonmitosing procaryotic types of cells, such as are now represented by blue-green algae. Such types are relatively tolerant of radiation. As oxygen became more abundant in the atmosphere and shielded more effectively against ultraviolet radiation, true oceanic phytoplankton of mitosing eucaryotic types could appear and populate the wide regions of the euphotic zone, thus greatly increasing the rate of O_2 production (Berkner and Marshall, 1964; Cloud, 1968).

It has been argued that the eucaryotic cell arose through the development of symbiotic relationships between different sorts of procaryotic microbes; endosymbionts became organelles, such as mitochondria and chloroplasts (which contain DNA) within eucaryotic cells (see Margulis, 1970). According to this evolutionary model, photosynthesizing eucaryotic protistans were originally heterotrophs, but acquired the ability to photosynthesize when they ingested procaryotic photosynthesizers, probably blue-green algal types. If this view is correct, eucaryotes did not evolve through a step-by-step selection for the development of all their organelles *in situ*, but were amalgamated from several lineages; some of these amalgams gave rise to plants, some to animals. However, this hypothesis is contested (see Raff and Mahler, 1972) and the question is still unresolved.

Protozoa, which are probably polyphyletic, are commonly believed to have evolved from heterotrophic bacterialike organisms (Kerkut, 1960) and to have given rise, in turn, to the Metazoa. The protozoans require oxygen for respiration and thus must have developed after the appearance of free oxygen, although their oxygen demands are modest. Metazoa, on the other hand, require more.

The rate at which oxygen accumulated is a matter of some controversy, one which bears directly on the important problems posed by the appearance of metazoan phyla in the Late Precambrian and Cambrian. So long as atmospheric oxygen concentrations were very low, the evolution of animals would be prevented. Therefore, the appearance of abundant metazoa had been suggested as coinciding approximately with the rise of atmospheric oxygen to a level appropriate to their existence—perhaps to 1 percent of the present level, according to Berkner and Marshall (1964).

Coupled with this suggestion, that early metazoans were permitted to evolve as oxygen levels rose, is the observation that the biosynthesis of cuticle and of skeletons has a relatively high oxygen requirement. It is thus hypothesized that these materials would have had a low priority among early metazoa, since oxygen levels were low at that time, and therefore that the early animals were largely soft bodied and unskeletonized (Towe, 1970). Under this hypothesis, the subsequent rise of oxygen to levels permitting the formation of substances with high oxygen requirements, then, was responsible for the Precambrian-Cambrian boundary. Rhoads and Morse (1971) have formalized these ideas into a model of Late Precambrian-Cambrian oxygen-related evolutionary events by analogy with benthic invertebrate associations living in oxygen-deficient basins today (Fig. 10-2). There are three major conditions in these basins: (1) where oxygen

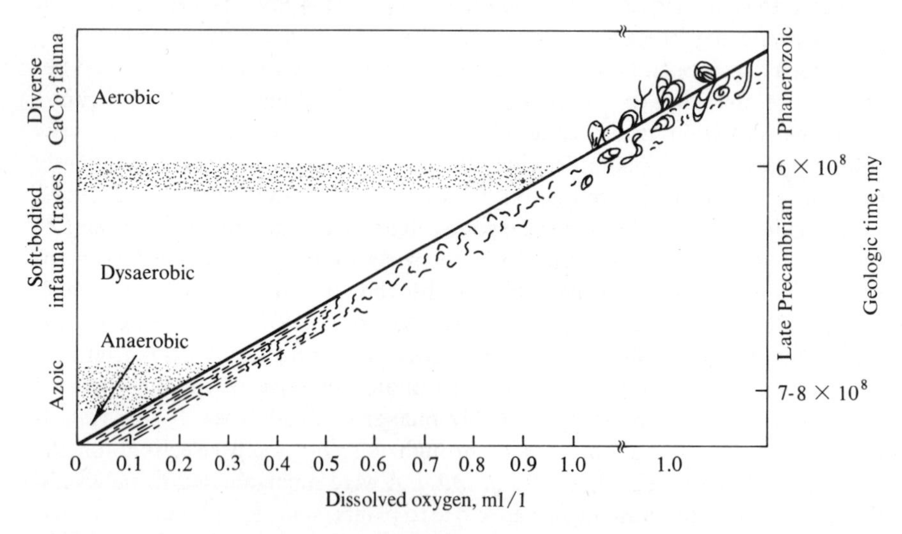

Fig. 10-2 Idealized cross section of a recent marine basin showing the relationship between levels of dissolved oxygen and the benthic fauna, based on the situation in the Black Sea, in the Gulf of California, and in the Santa Barbara and San Pedro Basins off southern California. Inferred oxygen-organism relationships during the Late Precambrian and Cambrian are indicated on the right. (After Rhoads and Morse, 1971.)

levels are below 0.1 milliliter per liter (ml/l) dissolved oxygen, there are no invertebrates; (2) between 0.3 and 1.0 ml/l, there may be a small, soft-bodied infauna; and (3) above 1.0 ml/l, a diverse calcareous fauna may be found. These distinctive associations are suggested to represent conditions (1) before the metazoa arose, (2) during the period of the Ediacaran fauna, and (3) during Early Cambrian times, respectively. If this analogy holds, then the Lower Cambrian atmosphere must have contained about 10 percent of the present level of oxygen, or around 10 times as much oxygen as in Berkner and Marshall's model.

This model is very attractive, although there is as yet no independent evidence that atmospheric oxygen did, in fact, rise to appropriate levels at the times required by the model, and, of course, there is no principle that requires the analogy in the model to be historically valid. Nevertheless, the model does provide an explanation for the relative timing of the early metazoan biotas. The major difficulty, which may be resolved eventually, is that Late Precambrian oxygen levels appear to have been if anything higher than required to limit skeletonization.

Increasingly numerous reports of early plants that have been dated by radiometric methods testify to the widespread occurrence of Precambrian photsynthesizers. Schopf and Barghoorn (1969) have pointed out that by Late Precambrian time, algal stromatolites and microorganisms, almost surely photosythesizers, must have had an oxygen-producing capacity that would lead to relatively high levels of atmospheric oxygen long before the Cambrian. Additional evidence of early photosynthetic activity comes from the presence of organic carbon, presumably derived from photosynthesis, dispersed in most Late Precambrian sediments. Precambrian oxygen accumulation probably lasted at least 1.2 billion years, and may have been as long as 2.5 billion. Therefore, it is possible that the minimum oxygen threshold required for metazoan metabolism was reached much earlier than the first appearance of fossil metazoans.

This possibility is strengthened by isotopic evidence developed by Broecker (1970). Organic carbon contains 2.5 percent less ^{13}C than the carbon in carbonate minerals (Craig, 1953). Today, about 18 percent of sedimentary carbon is in the organic form (Ronov, 1968), presumably originating during photosynthesis. If the oxygen resources in the atmosphere had been enlarged significantly during the Phanerozoic, the reservoir of disseminated organic carbon would have enlarged as well, for each O_2 molecule produced by photosynthesis liberates an atom of organic carbon. Thus, the $^{13}C/^{12}C$ ratio in seawater, and therefore in carbonates, would have been affected, and a Phanerozoic shift in this ratio should be easily detectable. Figure 10-3 depicts the $^{13}C/^{12}C$ ratio measured in marine carbonates, which shows no clear shift during the Phanerozoic. The dashed line indicates the shift expected on the basis of uniform accumulation of organic carbon during this interval. Broecker notes that the weight of this argument depends on the degree to which the assumptions, especially those concerning the fate of the newly liberated oxygen, are valid. If the assumptions are indeed met, then the rise of atmospheric oxygen concentration to near its present level must have occurred before Phanerozoic time, so that the Precambrian-Cambrian boundary was marked by a level far above 1 percent or even 10 percent of the present O_2 concentration. Considering the rapidity with which oxygen can be generated and the length of time that photosynthesis has operated, this appears quite possible, the major unknown quantity being the extent of the oxygen required to oxidize Fe, S, C, and other substances.

Still another indication that free oxgyen may have been present very early and may have been more abundant than suggested by Berkner and Marshall results from the work of Brinkmann (1969), who reconsidered the mechanism of photodissociation of water vapor in the primitive atmosphere. He concludes that an appreciable fraction of the present atmospheric level of oxygen (as much as one-quarter or more) could have been produced even before photosynthesizers had become common. How much of this oxygen would have remained free is uncertain. However, it does appear that the timing of the rise of free oxygen pictured by Berkner and Marshall (Fig. 10-4) may not be very accurate with respect to the evolution of life.

It is not yet certain at what time the metazoa first evolved, although there are no unequivocal metazoan remains before about 700 million years ago. It is pos-

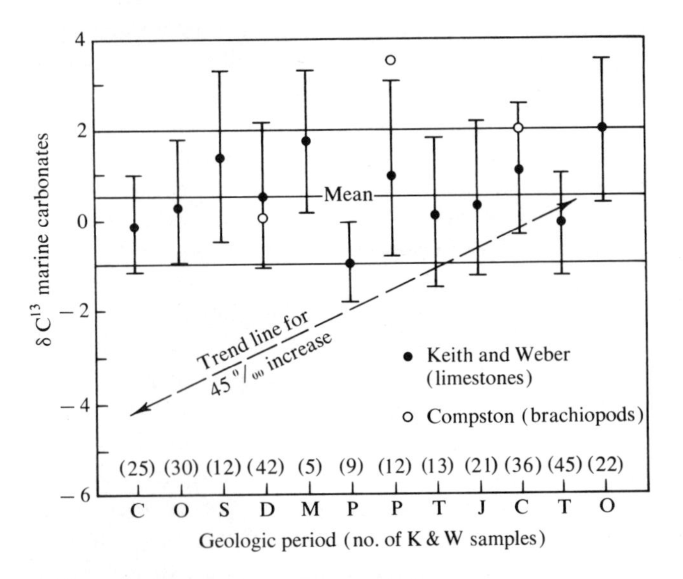

Fig. 10-3 Depicted are $^{13}C/^{12}C$ ratios in marine carbonates during the Phanerozoic. The numbers above the period abbreviations indicate the number of samples used for the point; the error bars associated with each point indicate the standard deviation of individual results from the mean. If the organic reservoir of carbon has accumulated at a uniform rate since Cambrian time, the $^{13}C/^{12}C$ ratio is expected to follow the dashed line; instead, the ratio appears to be relatively constant, suggesting that the organic carbon reservoir and the oxygen reservoir have not increased much during the Phanerozoic. (After Broecker, 1970.)

sible that they were present long before then, but have left virtually no fossils. On the other hand, they may simply not have evolved before that time, because of low oxygen levels or for other reasons. These two possibilities embrace the two leading hypotheses of early metazoan history. The traditional hypothesis, supported by Durham (1971), holds that the metazoa originated at some fairly remote point in time, conceivably many hundreds of millions of years before the Cambrian, and slowly evolved and differentiated into the large number of phyla that we see in Cambrian time. It is pointed out that, on the basis of the evolutionary rates that can be calculated from the Phanerozoic fossil record, the derivation of a new genus from an old one requires a certain time. Inasmuch as the morphological distance between Cambrian phyla is great, the implication is clear that many genera have been evolved during their divergence. Even allowing for some unusually rapid evolutionary mode, Durham estimates an absolute mimimum of 800 million years ago for the origin of the metazoa, and reasonable maximum of 1.8 billion years ago. He tends to favor the earlier age, placing metazoan origins well back in Precambrian time. This pattern is represented diagrammatically in Fig. 10-5.

If this hypothesis is correct, then there must be some special explanation for

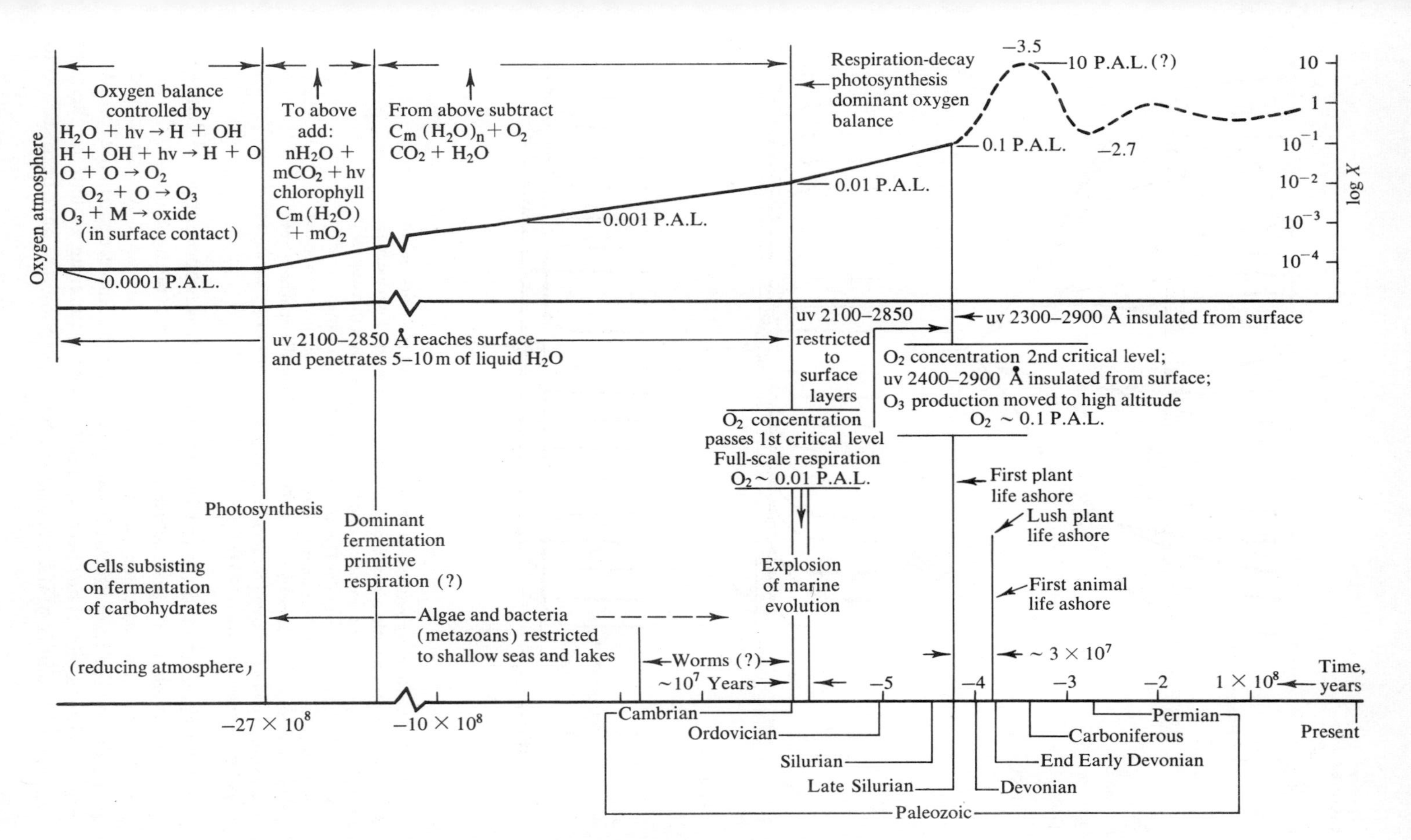

Fig. 10-4 Berkner and Marshall's tentative model of oxygen growth in the atmosphere. (After Berkner and Marshall, 1964.)

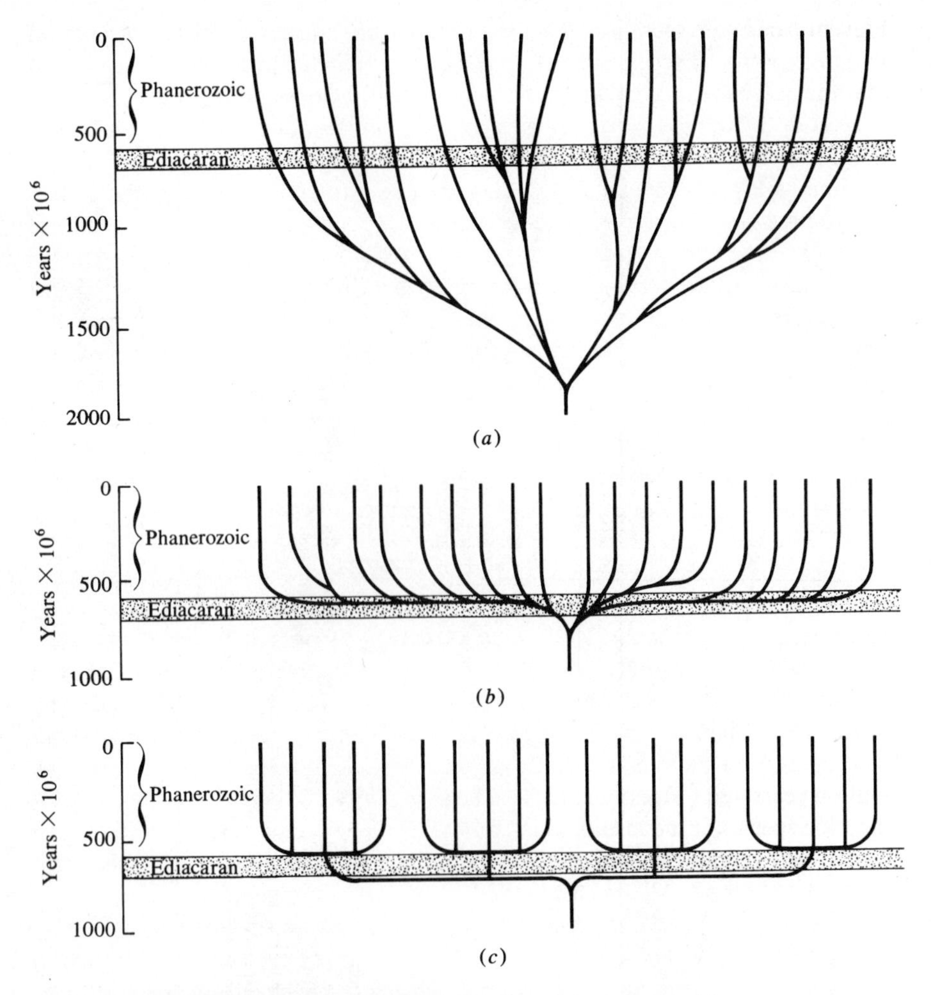

Fig. 10-5 Hypothetical patterns of phyletic evolution of the metazoa. (*a*) A gradually evolving pattern based on successive branching of phylogenetic lineages through time; the common origin of the diverse phyla that appear in the Cambrian is well back in the Precambrian. (*b*) A burst of adaptive radiation in metazoa near the Precambrian-Cambrian boundary, to give rise to the phyla that appear in the Cambrian. (*c*) A hierarchical arrangement of radiations to give rise first to a few Late Precambrian lineages and then to the Cambrian phyla; this pattern is discussed later.

the lack of metazoan fossils in the Precambrian. It was once supposed that there was a great hiatus in the record at this time, representing a universal unconformity (the "Lipalian interval"), but although there was a general regressive phase during the Late Precambrian, there is not a complete gap in the record. A variety of possibilities has been suggested, cheifly centering around the chemical balance and salinity of the oceans, which were supposed to have been such as to prevent carbonate secretion. This is probably not correct; there are carbonates in unfossiliferous Precambrian sequences, and theoretical considerations suggest that the oceans were salty and not especially acidic from an early stage. Another suggestion is that most Precambrian organisms were quite small and that selection for phyletic size increase led eventually to the point where skeletonization became highly advantageous for support and protection (Nicol, 1966). This seems possible, although there is no special supporting evidence, and many of the Late Precambrian trace fossils are not particularly miniature. Still another suggestion is that effective metazoan predators appeared near the beginning of the Cambrian, and skeletons were then evolved as protection (Schuchert and Dunbar, 1933). There is no indication as to why the predators might have appeared then. Nicol (1966) lists some of the other common suggestions for the appearance of skeletonized fossils. As we have seen, the low-oxygen model remains a possible explanation. None of these hypotheses attempts to explain why the Cambrian fossils appeared when they did, rather than earlier or later.

Another hypothesis of the timing of metazoan evolution is due to Cloud (1949, 1968), who suggests that the metazoa have originated only shortly before the beginning of the Cambrian, and that in the appearance of metazoan higher taxa in the fossil record we are witnessing an explosive radiation [Fig. 10-5(*b*)]. Supporting evidence includes the simple paucity of fossils in Precambrian rocks. The only well-authenticated fossil biota that is Precambrian is the Ediacara assemblage, best known from South Australia (Glaessner, 1958; Glaessner and Wade, 1966). Elements of this assemblage have been recorded from several other continents, and radiometric dates suggest that it existed between 700 and 600 million years ago (Glaessner, 1971). The fossils are chiefly coelenterates, including nine species of medusoids classed in seven genera and some pennatulid-like forms, two genera that are probably annelids, and one that is probably a trilobitoid arthropod. All are soft bodied or at least nonmineralized.

Although it is certainly true that soft-bodied organisms do not ordinarily fossilize, there are a number of striking cases where they do, the Ediacaran assemblage itself being a good example. Why, then, are not at least scattered metazoan remains found in earlier Precambrian rocks? Many Precambrian sediments appear to represent the sorts of environments that yield abundant fossils in the Paleozoic, and many careful searches have been made in them, but without success. This suggests that they simply weren't present, and that when they did evolve, late in Precambrian time, their remains do begin to appear, although sparsely. Even if no body fossils were discovered, trails, burrows, and other trace fossils are expectable if a diverse metazoan biota existed. Trace fossils are not uncommon in rocks that yield the Ediacran assemblages, and they become even more common in the Cambrian. In earlier Precambrian rocks, however, they are very rare indeed, so far as is known, and most records are equivocal or can be assigned to nonmetazoan organisms. Glaessner (1969) believes that a structure from the Precambrian Grand Canyon Series, once considered to be a medusoid and originally named *Brooksella canyonensis* by

Bassler, is, in fact, a trace fossil representing a burrowing metazoan that is probably an annelid. The age of the rocks from which it came is not certain, but it may be over 1 billion years. However, there are few other examples of trace fossils of this age, and it is conceivable that this one is incorrectly interpreted. Certainly the possibility that Metazoa do not much antedate the Ediacaran biotas is still strong. The timing for the sudden evolutionary burst that generated the metazoan phyla under this model is usually referred to the rise in oxygen to levels that metazoans require, but, as we have seen, this relation is in some doubt.

Cloud's hypothesis would clearly fit the present fossil evidence best, if the timing of the events in metazoan evolution can be rationalized. This is presumably a paleoecological problem. Before trying to involve ecological considerations in interpreting early metazoan evolution, we must first examine the probable phylogenetic pathways of some of the metazoans and then attempt a functional analysis of these pathways, that is, an analysis of the adaptive pathways that the lineages followed. This requires a closer look at the architecture of a number of metazoan phyla.

The Ground-plan Features of Phyla Were Originally Functional Adaptations

The more common phyla, such as the Mollusca, include members with widely different modes of life. Yet if the phylum Mollusca is monophyletic, which seems quite likely, then this diversity can be traced to a founding lineage that contained an assemblage of features we recognize to be characteristically molluscan. The characteristic features include many polythetic characters that, although not universally present among Mollusca, nevertheless form part of the definitive set of characters for the phylum (such as radulae and ctenidia). Additionally, they include characters that are universal among the Mollusca but that are found in some other phyla as well (such as mesoderm and a coelom). Together, these features can be employed to describe the basic architecture of the molluscan form—the molluscan *ground plan*.

The characters that form the ground plan of a phylum must have arisen through selection for phenotypic fitness, and when they arose were either directly advantageous in the ecological interactions of the population at the time, or were coadapted to such characters so that they constituted phenotypic improvements, perhaps in physiological efficiency or the like, which enhanced the ecological adaptation of the population. In other words, they arose in response to a particular adaptive opportunity, which was associated directly or indirectly with specific environmental relationships. The rise of new ground plans is an important result of anagenesis. Today, most of the features that characterize the ground plans of each phylum are present in the numerous divergent lineages of distinctive adaptive types and diverse ecological functions that compose the phylum;

that is, the features of the ground plan are commonly not associated with one special ecological adaptation. Their original functions have often been modified, and some of the ground-plan characters have become part of the architectural layout that serves as a basis for structural elaboration; the more specialized characters now serve direct ecological functions. Features that were originally adaptive for a special mode of life have become part of a general ground plan. They are adaptive as building blocks that are well integrated with the structure of the organism, but do not necessarily serve in specific ecological interactions.

The original significance of some of the characters of a ground plan is often obscure. It is sometimes inferred from the function of the character in primitive living members of the group or by building a hypothetical conceptual model of an unknown primitive member, by extrapolation of adaptive trends backward within various lineages of the phylum, by analogy with similar features in other phyla, or simply by analogy with mechanical principles. Conclusions as to the original adaptive significance of such characters always retain some elements of conjecture, but, at least in some cases, the evidence is very strong. It is certainly important that we reconstruct the primitive functions of phyla, for it gives us an idea of the quality of the ecospace that was realized in early communities and provinces. Let us examine the major features of the ground plans of two allied but distinctive phyla, Mollusca and Annelida, in the light of hypotheses as to their original functions.

Flatworms Lack Hydrostatic Skeletons and Therefore Creep or "Inch" but Cannot Burrow Well

The Platyhelminths or flatworms may represent the common stock from which both mollusks and annelids have descended. Even if this proves not to be the case, the platyhelminths provide much evidence of the adaptive significance of major features of invertebrate ground plans, and are therefore well worth our attention, before we proceed to the mollusks and annelids themselves.

Platyhelminths are small, generally elongate, bilaterally symmetrical animals that lack a true body cavity (coelom) and therefore are termed *acoelomate*. In this respect, they resemble the Coelenterata, which have no body cavity between their ectodermal and endodermal tissue layers, but have only a fibrous and generally acellular intermediate layer, the mesoglea, and thus are diploblastic. However, the platyhelminths possess a network of cells, a mesenchyme, forming a mesoderm between the ectodermal and endodermal layers, and thus are triploblastic. The organs of platyhelminths are commonly paired about a longitudinal plane of bilateral symmetry and may be serially repeated so that several pairs are present (Fig. 10-6). Such serial organ repetition is called *pseudometamerism*. Serially repeated organs include gonads, protonephridia, gastric pouches or diverticulae, and nerve commissures.

Special respiratory organs are absent in the platyhelminths, oxygen being

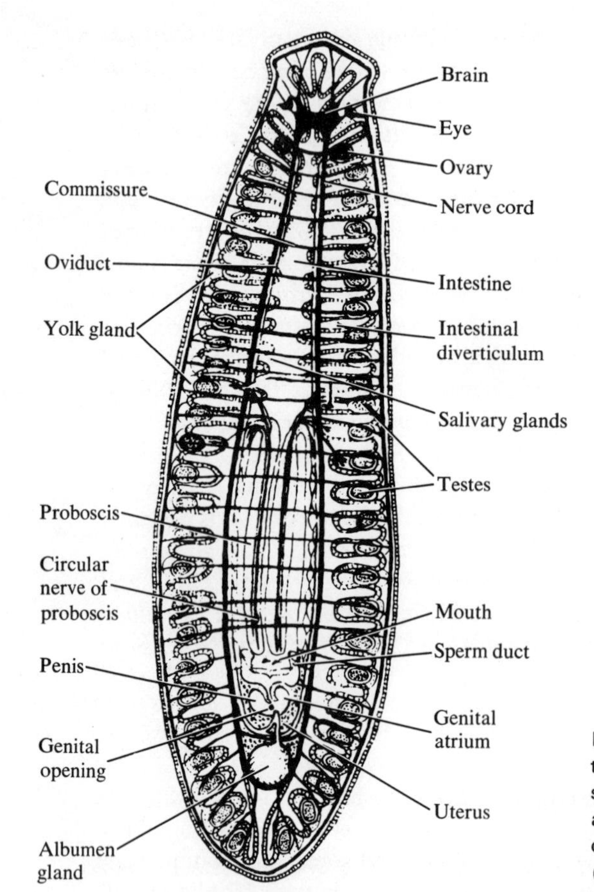

Fig. 10-6 Seriation of a flatworm, the triclad *Procerodes lobata*. Seriation of some characters (commissures, for example) does not match the seriation of others. Excretory system not shown. (From Vagvolgyi, 1967, after Lang.)

supplied through the general body surface. Since there is no circulatory system either, the ratio of their surface area to body volume must be kept very high to provide enough oxygen. This may account for their flattened body form. The development of pseudometamerism in the flatworms is probably also related to the lack of a circulatory system in a long-bodied animal (Clark, 1964), for organs must be multiplied in order to service all parts of the body. For example, some members of the class Turbellaria have an internal digestive space that gives off a whole series of digestive diverticulae to provide nourishment for cells throughout the length of the animal (Fig. 10-6), and, similarly, more than one pair of excretory organs is present to accommodate metabolic wastes. Thus, a serial arrangement of organs is established, and it is natural for the positioning of other organs to harmonize with this plan; gonads, for example, can be related to this arrangement simply by being alternated in position with the diverticulae (Clark, 1964). An integrated psudeometameric ground plan results.

Locomotion in flatworms is by three main methods: by cilia, pedal waves, and "inching." Small organisms swim or creep by means of cilia, but as body size increases, the ability to swim by ciliary action is lost. Owing to their flattened shapes, platyhelminths provide a large basal area, which can support many cilia and thus provide adequate propulsive force for ciliary creeping. At larger body sizes, however, even these copious cilia are insufficient, and muscles in the body walls, far more powerful than cilia, supplement or replace them.

A number of platyhelminths move by means of waves of muscular contractions that pass along their ventral musculature. Parts of the longitudinal muscles of the body wall are contracted at points along the length of the worm, causing a series of three or four slight swellings and alternate hollows to form. The swellings rest upon the substratum, to which they adhere by mucous secretions. The contractions pass backward as waves along the ventral musculature, drawing the body forward over the points of adhesion, which remain at fixed positions on the substrate. New points appear anteriorly as old points are left behind posteriorly. By analogy with gastropods, the waves of contraction are called *pedal waves*.

An additional mode of platyhelminth locomotion also involves body-wall muscles. The body is extended, the forward end attached to the substrate, and then the body is contracted, which drags the posterior parts of the body forward toward the point of attachment. The anterior end is then detached and again extended to a new point of attachment; thus, the worm "inches" along.

Another form of locomotion widespread among the invertebrates is provided by peristaltic waves. These are not known among platyhelminths, but occur in the phylum Nemertea, another acoelomate triploblastic group of wormlike animals that are commonly flattened but are usually far more elongated and somewhat larger than the platyhelminths. Like the platyhelminths, they are pseudometameric, but they do have a simple circulatory system consisting of two or three contractile vessels running longitudinally, and sometimes of transverse vessels as well. The vessels lie entirely beneath the muscle layers of the body wall, so they do not function primarily for respiratory exchange but probably for the transportation of metabolites; they may represent an adaptation to the lengthening of the body (Barrington, 1967). Locomotion of nemertines is sometimes by pedal waves, especially in shorter species (Clark, 1964), whereas some species are provided with suckers and inch along somewhat like platyhelminths. But many long nemertines employ a combination of ciliary activity with muscular contraction in the form of peristaltic waves, a locomotory device widely and possibly primitively employed by the annelids.

The efficacy of peristaltic locomotion depends upon the possession of a hydrostatic skeleton, an internal space filled with fluid or other incompressible material capable of deformation and of the transmission of fluid pressures (Clark, 1964). If several waves of muscular contraction that go right around the body are caused to pass along the body of a worm, the expansion of sections of the body wall between the contractions is assured, because fluid displaced from the actively contracted portions dilates the intermediate sections. Thus, annular waves of alternate bulges and depressions can be created which may be used for creeping, swimming, or burrowing. Although nemerteans do not have a fluid-filled internal body space, their mesenchymal cell network, which serves as mesoderm, is rather incompressible and transmits pressure sufficiently to serve as a hydrostatic skeleton for limited purposes.

Annelids Possess a Coelom, Which Functions as a Hydrostatic Skeleton, Permitting Burrowing

The annelidan ground plan has features in common with those of the platyhelminths and nemertines, but is at once more complex and more regularized (Fig. 10-7). The complexity is partly associated with the presence of a "true" or "secondary" body cavity or *coelom*, a fluid-filled cavity that appears within the mesoderm and is lined with an epithelial cell layer, the *peritoneum*. The coelom often forms a hydrostatic skeleton that is employed to antagonize muscles in locomotory systems. Regularization is provided by the division of most parts of

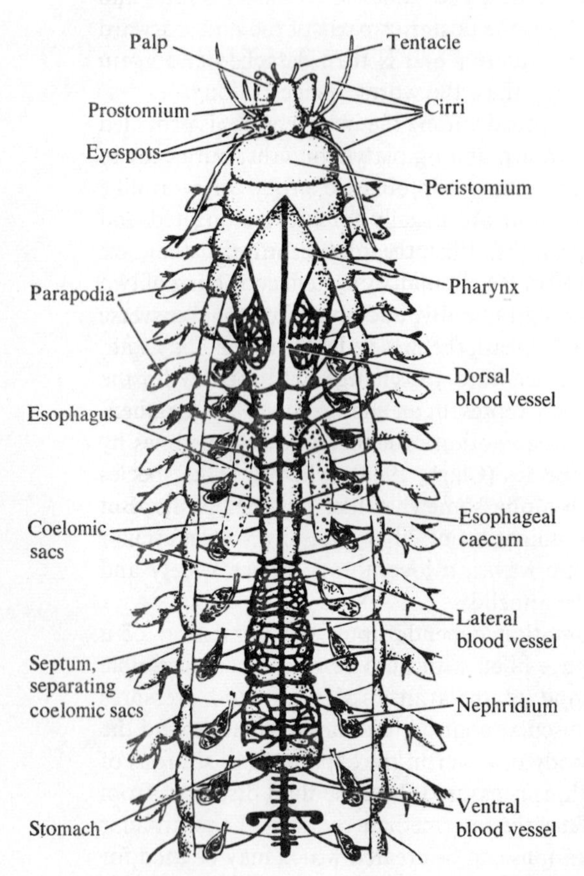

Fig. 10-7 Segmentation of an annelid, the anterior region of the polychaeate *Nereis virens*, which may have about 200 segments in all. Dorsal view. (From Vagvolgyi, 1967, after Brown.)

the body, including the coelom, into a series of longitudinally arranged segments or compartments, each containing a complement of organs. An idealized primitive segment probably contained a pair of coelomic cavities (one on each side); a pair of coelomoducts, which connected each cavity with the exterior; a pair of nephridia, for ionic and osmotic regulation and for excretion; and a pair of nerve ganglia. Segments also frequently bear paired appendages or *parapodia.* The longitudinal musculature of the body wall is also segmented, but not on the same plan as the internal organs; rather, two or three segments are tied together by muscle fibers, thus linking the internal segments into groups (Barrington, 1967). In many annelids, and possibly at one time in all, most of the segments are further separated by transverse septa, which are pierced by openings or *foramina* to allow passage of the nerve cords and blood vessels, and for other purposes. The foramina are closed by sphincter muscles to inhibit fluid leakage between segments. Regularization of organ seriation within segments and association of a coelom and body-wall musculature with the segmentation characterize "true segmentation" or *eumetamerism*, and distinguish it from the pseudometamerism displayed by the acoelomates.

The function of the intersegmental septa appears primarily to permit a degree of mechanical independence of each segment. Selective muscular contractions allow the shapes of the segments to be varied somewhat independently, within limits set by the extensibility of the body wall, the flexibility of the septa, the design of the musculature, and so on. Local pressure gradients are set up within the coelom, for the effects of contraction in one segment are damped in neighboring segments.

In annelids that employ peristaltic waves for locomotion, the localization of pressure changes appears to be adaptive for sustained and efficient burrowing, because high pressures need not be continuously maintained throughout the coelom during peristalsis. Annelids that are relatively sedentary often display a great reduction in septation and in their ability to localize pressure changes; they cannot burrow actively and regularly for long periods (Clark, 1963, 1964). In annelids that creep by the use of parapodia activated entirely by intrinsic muscles, the septa are incomplete, and the hydrostatic skeleton plays no direct role in locomotion. Such creeping is relatively feeble, since the powerful muscles of the body wall do not participate. However, in still other annelids with parapodia, locomotion is accomplished by serpentine motions caused by waves of contractions of longitudinal body muscles alternating on each side, supplemented by correlated parapodial movements. In these forms, the septa tend to be more complete and are advantageous in localizing pressure to help antagonize the longitudinal muscles and also to provide turgor within the parapodia themselves, enabling parapodial flexor muscles to operate efficiently. In sum, it appears that the septa function to localize the hydrostatic effect of the coelom and thereby provide a basis for a variety of locomotory methods. Whenever such localization is not required, the septa tend to be reduced or obliterated, foreshadowing the situation in the Arthropoda, which lack septa.

It seems quite probable that the coelom of the annelids arose as a hydrostatic skeleton, representing a distinct improvement on the mesenchymal tissue skeleton of acoelomates. Peristalsis became more efficient and permitted more extensive burrowing into the substrate, opening the possibility of exploiting food-rich bottom deposits. Burrowing efficiency was further enhanced by the development of segmentation, which may represent a consolidation and regularization of the organ seriation seen in acoelomates. At any rate, the presence of elements of important organ systems in each segment contributed to the independence of the segments, which was secured mechanically by septa. Eumetamerism represented an harmonious adaptive mechanism that was an advance upon the psuedometamerism of the acoelomates. At the same time, it represented a locomotory specialization concerned with obtaining food, and thereby associated with the trophic relations of the protoannelid stock wherein it developed. Problems of the origin and significance of the coelom and of segmentation have been admirably treated by Clark (1964), who should be consulted for further information on these points.

Mollusks Are Probably Coelomate but Employ a Hydraulic Mechanism for Burrowing Involving the Haemocoel

Living mollusks have no definitive traces of eumetamerism, although they are triploblastic with complex organ systems and are coelomate. The living classes are each rather distinctive in plan, and were distinctive when they first appeared as fossils, so to reconstruct a single molluscan ancestor requires the postulation of a hypothetical ancestral form, an *archetype*, which contains the basic molluscan features and which conceivably could have given rise to all the classes by radiation. A popular archetype is depicted in Fig. 10-8. Whether this archetype represents an actual ancestral organism is conjectural, but it does form a useful introduction to the general molluscan ground plan.

The generalized mollusk is composed of two main structural units, one dorsal and one ventral, each of which has two main subunits. The ventral unit consists in an anterior *head* and a *foot*, and is chiefly activated by muscular action; the dorsal unit is composed of a *visceral hump* and an enclosing *mantle-and-shell*, and its activity is chiefly ciliary and secretory. A small unsegmented body cavity surrounds the heart; this pericardial space, together with gonoducts that serve for the discharge of sexual products, is usually assumed to be coelomic. The molluscan nervous system commonly has cerebral ganglia and a nerve loop around the viscera, foot, and mantle that may contain local ganglia connected by commissures. The typical molluscan heart has a ventricle that pumps blood through arteries; the blood returns to paired heart auricles via special body spaces, the haemocoel. Paired gonads and kidneys complete the major visceral organ systems of the archetype.

The mantle overhangs the body wall of the visceral hump and foot, and

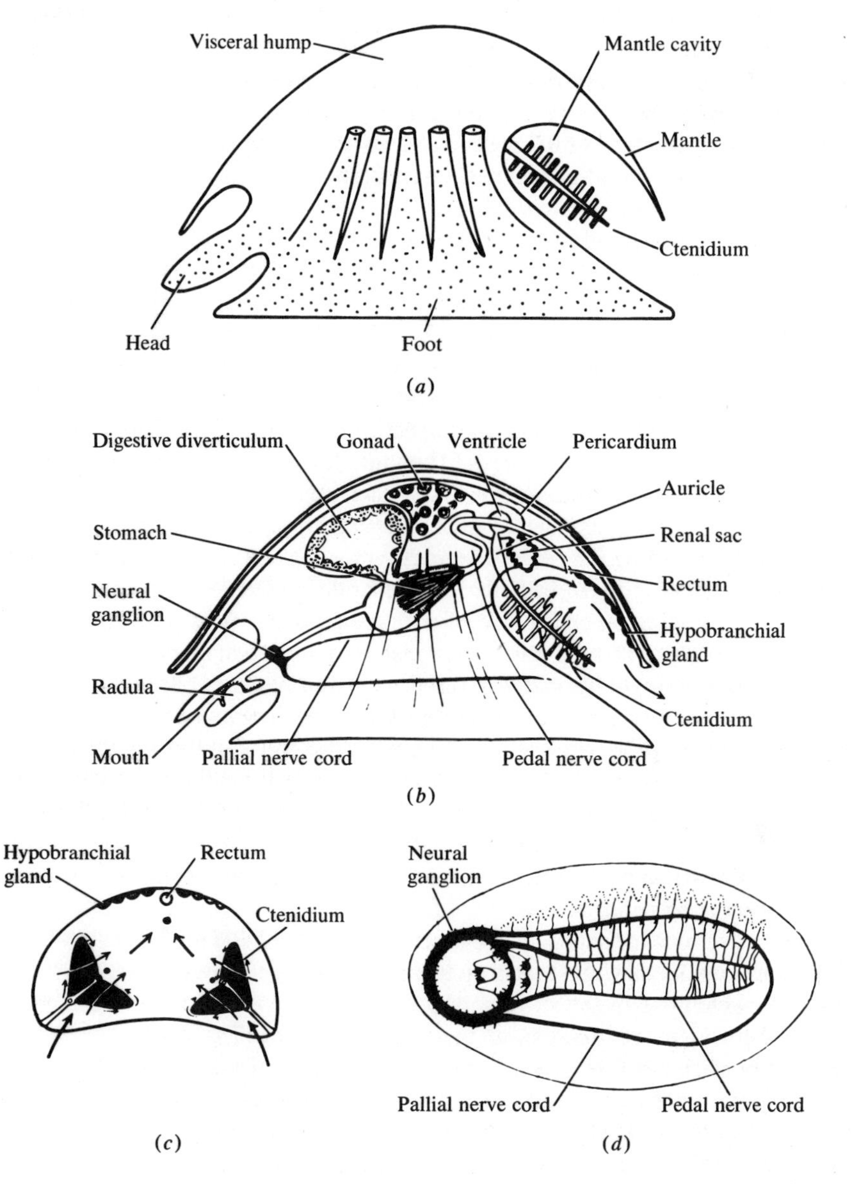

Fig. 10-8 A hypothetical molluscan archetype. (*a*) The major subdivisions of a generalized molluscan body; head and foot, together with their dorsal extensions as shell muscles, are stippled. (*b*) Longitudinal section showing major anatomical features. (*c*) Transverse section of mantle cavity. (*d*) Nervous system of a chiton, presumably a primitive plan. Arrows in mantle cavity indicate current directions. (After Morton and Yonge, 1964.)

the space found within this overhang is called the mantle cavity. In the usual archetype it is posterior and contains paired gills or *ctenidia,* which have a characteristic form in most mollusks [Fig. 10-8(*b*) and (*c*)]. Cilia on the gills and on the surface of the mantle cavity create water currents, which bring oxygen into the mantle cavity and aereate the ctenidial blood vessels. The anus, renal pores, and gonopores also open into the mantle cavity: therefore, the mantle organs must be arranged so that unpolluted water bathes the gills and waste materials are removed by outgoing currents.

Because the living molluscan classes are so divergent and present such distictive modifications of the ground plan, it is useful to review them briefly. The Amphineura or chitons possess nearly all the archetypal features and partly for this reason are considered to be relatively unspecialized and to have a rather primitive organization (Fig. 10-9). The chiton is flattened dorsoventrally and has an anterior head, a broad ventral foot, and a dorsal visceral mass, all overgrown by an enfolding mantle-and-shell. The mantle cavity is posterior but extends forward around both sides of the foot. A series of gills (from 4 to 80) is situated in each lateral mantle groove. The gills are typical molluscan ctenidia. The head contains a mouth, associated with a *radula,* a ribbon of toothlike structures chiefly used for feeding. The digestive tract is more or less median and terminates at an anus situated posteromedially in the mantle cavity. Many internal organs are paired, one lying on either side of the median plane, including the kidneys, which are provided with renal ducts that open into the lateral mantle grooves. The gonad is usually single (probably a fused pair) and median, but communicates with the exterior via a pair of ducts that terminate in gonopores, one in each mantle groove near the renal pores. The shell consists in eight articulated dorsal plates, originating from a single shell gland that becomes partitioned during early ontogeny. Most chitons cling to firm substrates, not unlike limpets, and the segmentation of the shell permits the body to conform to irregular surfaces, and to enroll in order to protect the soft ventral side if detached from the substrate. Locomotion is by pedal waves.

The gastropods, or snails, also possess most of the basic molluscan features, but are unique in that most gastropod larvae undergo a torsion early in their ontogeny which has the effect of rotating their main visceral mass and mantle-and-shell through 180° relative to their head and foot (Fig. 10-9). The mantle cavity is posterior before torsion, but after torsion it lies above the head, and internal organs are twisted around to the new orientation. An additional asymmetry is imparted to the body by the development of the spirally coiled shell, and this leads to the loss of one of the original pair of various organs such as kidneys, gonads, or gills.

Serial repetition of organs is not common in gastropods, although in some patelloids there are series of mantle gills in the pallial grooves, either with or without loss of the ctenidia; these mantle gills are clearly secondary. Gastropod locomotion is chiefly by pedal waves, similar to that which occurs in large platyhelminths, except that at least in some gastropod stocks locomotion

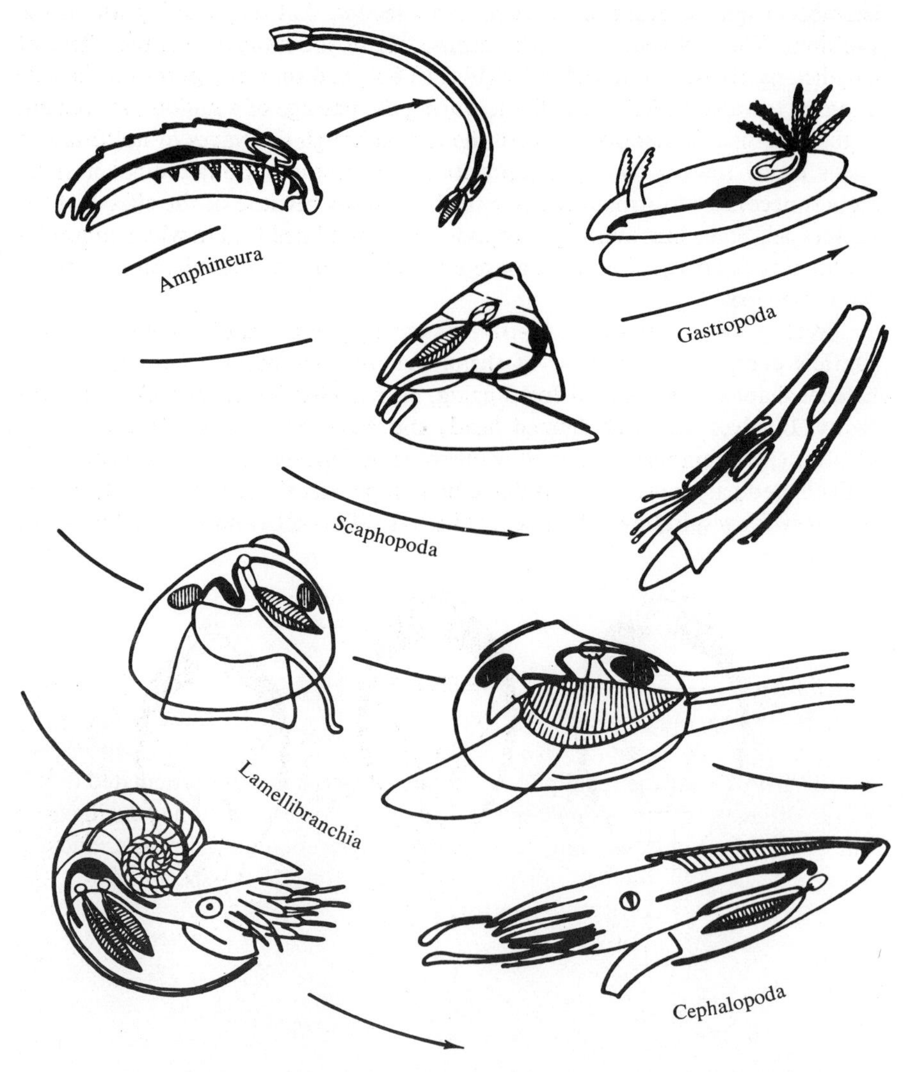

Fig. 10-9 General features of five classes of Mollusca, in cross section. Gills, digestive tract (in black), foot, and shell outline shown. (After Morton, 1958.)

involves a hydrostatic skeleton formed by the haemocoel (blood spaces) instead of the coelom (Trueman, 1969). In the limpet *Patella*, haemocoelic cavities are deformed during the contractions of dorso-ventral muscles that act to create hollows beneath the foot (under *c* in Fig. 10-8*a*), and at the same time a low-pressure region is created beneath the hollow. The alternate swellings and hollows are migrating to the right in Fig. 10-8(*a*); as the muscles relax (under *r* in Fig. 10-8*a*) the sole of the foot is drawn down to the substrate by suction, whereas the haemocoel spaces regain the circular cross section that they display within the swellings. The deformation of the haemocoel in the hollows has the effect of lengthening the sole longitudinally (Fig. 10-8*a*), and thus any point on the sole migrates toward the left in the figure during the passage of a hollow; movement of the organism is therefore toward the left, although the waves of hollows and swells travel toward the right (retrograde locomotion). Longitudinal muscles are not necessary to stretch the sole in the hollows because of the effects of the haemocoel. Some burrowing gastropods have a modified foot in which muscular contraction is antagonized by a more elaborate hydrostatic skeleton formed by the haemocoel.

Bivalvia (Figs. 10-9 and 10-10) are strongly bilaterally symmetrical and laterally compressed; in most of them, the foot has become flattened into a hatchet-shaped organ suited for digging, surrounded by the capacious mantle cavity. Bivalves lack well-defined heads and have no radulae. One group of bivalves (protobranchs) are shallow-burrowing deposit feeders, gathering food with ciliated palp lamellae; ctenidial cilia pump water through the mantle cavity primarily for respiration. Most other bivalves (filibranchs and eulamellibranchs)

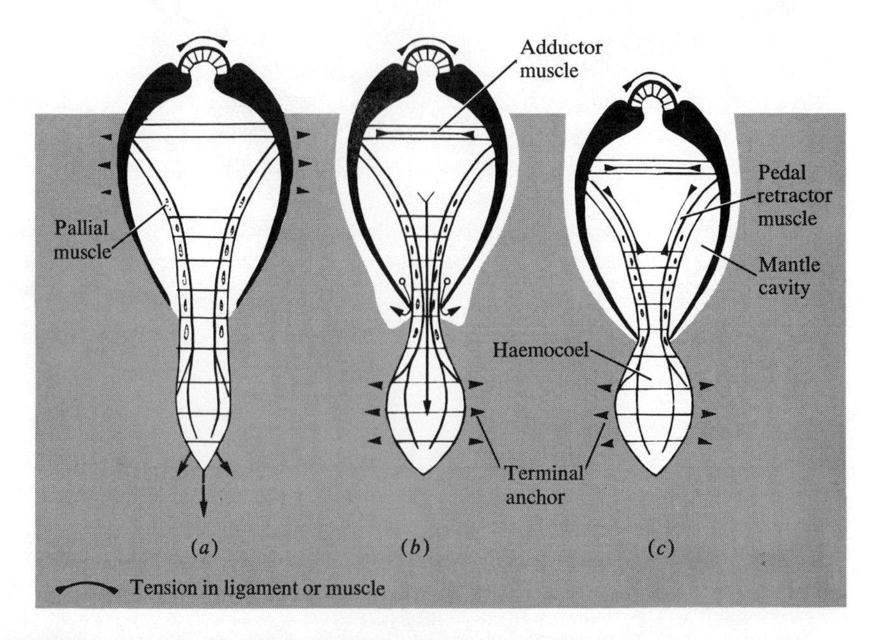

Fig. 10-10 Burrowing mechanism in a generalized bivalve. (*a*) Foot is extended downward into the substrate by internal fluid pressure of the haemocoel. (*b*) Adductor muscles contract, closing valves and distending ventral part of foot owing to compression of haemocoel by water in mantle cavity. At the same time, some water is expelled, liquifying sediment. (*c*) Pedal muscles contract, drawing shell down over foot, which is anchored in sediment by ventral distension. Relaxation of adductor muscles will now permit the valves to open, owing to ligamentary elasticity. (After Trueman, 1968.)

are suspension feeders, however, obtaining food from particles suspended in the ctenidial water currents. Commonly, the currents enter and leave the mantle cavity through tubular extensions of the fused mantle margins called *siphons.* The richly ciliated gills trap and sort particles and transport them to the mouth, if accepted. A specialized group of bivalves (septibranchs) are carnivorous; their gills are modified to form septa across the mantle cavity, and by sudden flexure of the septa they cause a sudden rush of water to enter the cavity, sucking in prey organisms.

Hydraulic systems are widely employed by bivalves, especially for burrowing. Trueman (1966) has shown that in *Mya* (and probably other active burrowers) the valves are employed as a sort of hydraulic pump. To begin a cycle of burrowing activity, the foot is extended, at least partly, by an inflow of blood from the visceral mass. The siphons are then closed by sphincters, and the mantle cavity sealed off as completely as the mantle morphology will permit. Next, powerful adductor muscles close the valves; pressure transmitted by seawater trapped in the mantle cavity compresses the viscera, forcing more blood into the haemocoel of the foot, which is thereby distended and anchored in the substrate. At the same time, some water is ejected from the mantle cavity as the shell closes, flooding into and liquifying the sediment immediately adjacent to the shell. Now the pedal muscles contract; they are inserted into the dorsal part of the valve (Fig. 10-10) and pull the shell down through the liquified sediment and over the distended foot. Finally, the adductors relax and permit the elastic ligament, which has been compressed by adduction, to open the valves (Fig. 10-10). The cycle may now be repeated.

An hydraulic mechanism is also employed to protrude the siphons (Chapman and Newell, 1956). The ends of the tubular siphons are closed by a sphincter muscle, and water in the mantle cavity is forced into the siphonal tube by contraction of the adductor muscles, protruding it. The siphons are retracted by their intrinsic musculature.

The Scaphopoda (Fig. 10-9) are elongated dorsoventrally and enclosed in a highly conical mantle that secretes a tusklike shell. Head and foot are at the larger end of the shell and may be protruded for feeding or locomotion. The scaphopods obtain food directly from the substrate via long organs called captacula. The currents are created either by a pistonlike action of the foot moving within the shell, or by cilia located on the posteroventral portion of the foot. There are no ctenidia, and respiration is performed by general integumental tissues. The heart is also lacking.

Cephalopoda (Fig. 10-9) are so varied and complex that it is difficult to generalize upon them. Some have external shells of conical form that may be straightish to tightly coiled. The problems of utilizing the single external aperture at the large end of the cone for most environmental exchanges, including feeding, respiration, and excretion, have been solved by placing the head dorsally in the aperture and the viscera behind toward the apex, and then bending the gut and associated organs ventrally to run forward to a mantle cavity situated beneath the head. The digestive tract is thus U shaped, but there is no torsion. A rather spacious perivisceral coelomic cavity is often present, apparently an extension of the pericardium. The foot has been greatly reduced and modified to form a funnel with muscular walls to pump water into and out of the mantle cavity, wherein currents are so arranged that the gills are aerated and metabolic wastes removed (Fig. 10-9). The funnel also performs the famous jet propulsion of the cephalopods, and in some lineages the walls of the mantle cavity are themselves muscular, thus enhancing the jet mechanism. One or two pairs of gills lie in the mantle cavity. Armlike tentacles, probably of cephalic origin, extend around the mouth for food gathering and sometimes for locomotion.

There is still another skeletonized molluscan class represented in the living fauna (the Monoplacophora), but before we review its features it is appropriate to re-emphasize the distinctiveness of the molluscan ground plan from that of the annelids. The coelom is inconspicuous in many mollusks, sometimes serving chiefly as a space in which the contractile heart may function, or as ducts. Whenever hydrostatic skeletons have developed, they depend upon the haemocoel rather than the coelom and are commonly supplemented by seawater pressure in the mantle cavity. Most of the hydrostatic skeletal effects are provided by a hydraulic mechanism totally unlike the annelidan system. Serially repeated organs are found, as the shell valves of chitons and the gills of chitons, patelloids, and cephalopods, but seriation is unusual and does not form part of a regular and integrated pattern, such as in the eumetameric annelids. Indeed, seriation is restricted to so few organs in these classes that it cannot be compared to the pseudosegmentation of the acoelomates either.

Mollusca are clearly highly organized invertebrates; the Cephalopoda have been said to "attain a degree of structural complexity and of metabolic efficiency which represent the summit of evolution in the absence of segmentation" (Yonge, 1960). The archetypal mollusk of Fig. 10-8 appears to represent well the basic structure of the classes we have reviewed.

The Monoplacophora, a Class of Mollusks That Appeared Very Early, Exhibits Pseudosegmentation

The Monoplacophora (Fig. 10-11) are clearly a molluscan group. They are known as fossils in strata from the Lower Cambrian to the Devonian, where they are represented by rather simple shells shaped like the bowl of a spoon, or produced into low cones, with a pointed or rounded apex offset toward the anterior. The cone-shaped shells are sometimes strongly curved or even coiled. Monoplacophorans have been known to science since the late 1800's, but their significance has become appreciated only recently.

The spoonlike form of many monoplacophorans is similar to the limpet shape of many gastropods shells. This shape, commonly an adaptation to clinging upon a firm substrate in turbulent waters, has been evolved independently

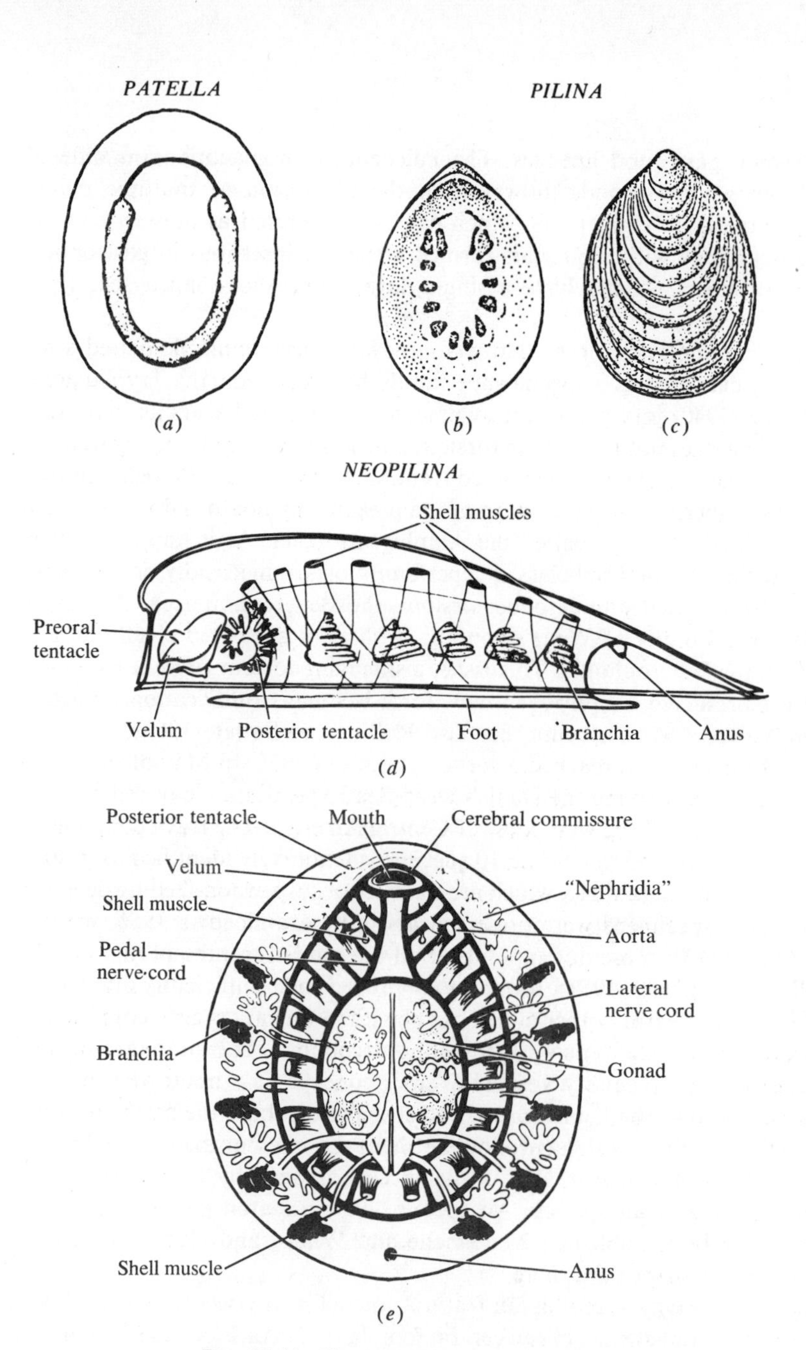

Fig. 10-11 Monoplacophora. (*a*) Muscle scar in a living limpet-shaped gastropod, *Patella*. (*b*) Muscle scars in a limpet-shaped monoplacophoran, the Devonian *Pilina*. (*c*) Dorsal view of *Pilina*. (*d*) Side view of *Neopilina*, a living monoplacophoran, showing some internal organs as if body wall were transparent. Note the serial repetition of shell muscles and branchiae. (*e*) Ventral view of *Neopilina*, showing seriation of the nerve cords, branchiae, gonads, nephridia, and shell muscles. [(*a*) After MacClintock, 1963; (*b*) and (*c*) after Knight and Yochelson, 1960, from *Treatise on Invertebrate Paleontology,* courtesy of The Geological Society of America and The University of Kansas Press; (*d*) and (*e*) after Lemche and Wingstrand, 1959.]

within several gastropod lineages. The Paleozoic monoplacophorans differed from all known gastropods, however, in the possession of multiple paired internal scars, up to eight pairs, which can be interpreted as muscle scars by putative homology with shell scars formed at muscle insertions in gastropods. Gastropod limpets usually have a single, large, horseshoe-shaped scar (Fig. 10-11).

At first, monoplacophorans that were known to have multiple paired scars were usually classified as a special superfamily of gastropods, the Tryblidiacea. Later, Wenz (1940) interpreted the symmetry of the paired scars as indicating that the animal had not undergone torsion, and was therefore to be regarded as dissimilar in basic architecture from conventional gastropods. He believed that this difference merited the separation of the presumably nontorted forms into a special subclass, which he called the Amphigastropoda. As it happens, Wenz included within his new subclass a superfamily of symmetrically coiled shells that may have actually undergone torsion—the Bellerophontacea. A dispute arose over the validity of Wenz's concept, with criticism based chiefly on evidence from the bellerophonts. This issue, a side issue for our present purposes, has not been resolved entirely even now, for it has many ramifications. Yochelson (1967) gives a witty account. See also Rollins and Batten (1968).

The controversy had reached a sort of plateau in 1952. In May of that year, the ship *Galathea*, carrying the Danish Deep-Sea Expedition, recovered an otter trawl from 3570 m off the west coast of Central America. The trawl contained a rich benthic association, including 10 specimens tentatively identified as gastropods of the family Capulidae, which are a family of suspension-feeding deep-sea limpets. These specimens were preserved and duly examined in 1956, when it was realized that they are not capulids at all but modern representatives of the Tryblidiaceae (Lemche, 1957). Indeed, they proved to be sufficiently different in anatomical features from other molluscan classes to warrant their recognition as an entirely separate class, now called Monoplacophora. The announcement of their discovery spectacularly verified Wenz's concept of a nontorted mollusk with a gastropodlike shell, and has revived speculation about the nature of other primitive fossil groups with gastropodlike shells that may or may not belong to the Gastropoda (Yochelson, 1967). The species dredged by the Galathea has been named *Neopilina galathea*. An account of the anatomy of the Galathea specimens has been published by Lemche and Wingstrand (1959) and forms the basis for the description here.

Neopilina has typical molluscan features, including a visceral mass enclosed within a mantle-and-shell, a broad central foot, lateral mantle cavities containing gills, and a radula (Fig. 10-11). In addition, Lemche and Wingstrand have identified numerous features that appear to have homologues within other molluscan classes. Monoplacophorans have strongly bilateral symmetry and, as Wenz inferred from the muscle scars, they have not undergone torsion. The digestive tract of *Neopilina* begins at an anteromedian mouth and the coiled gut ends in a posteromedian anus. Associated with the front and sides of the mouth is a richly ciliated ridge, the *velum*; behind the mouth are paired fan-shaped tufts of tentacles, which are locally ciliated. Presumably, the velum and postoral tentacles are feeding organs, and *Neopilina* is a deposit feeder.

A remarkable feature of *Neopilina* is that many paired organs are serially repeated; there are 10 pairs of nerve connectives along the foot; 8 pairs of large lateral and 8 of large median muscles for the foot, which are inserted together to form 8 pairs of pedal muscle attachments; 6 pairs of nephridia; 5 pairs of gills, and therefore 5 pairs each of the nerves and muscles that serve them; and 2 pairs of gonads. Other organs are also repeated. This serial repetition certainly recalls the pseudosegmentation of the acoelomates.

The coelom of *Neopilina* is more extensive than that of most molluscan classes; in addition to a pericardial cavity and gonoducts, there is a pair of extensive coelomic spaces dorsally beneath the body wall. The coelom is not longitudinally segmented, however, and there is no hint of segmental repetition of the organs in the style of the eusegmented annelids.

The shell of *Neopilina* is naturally especially interesting to paleontologists. Curiously, it does not display muscle scars, probably owing to the thinness of the shell, as Lemche and Wingstrand (1959) have suggested. The shells of many modern deep-sea (and polar) invertebrates are very thin, probably owing to the low water temperatures, which raise the solubility of calcium carbonate. The Lower Paleozoic monoplacophorans that are known as fossils lived in relatively shallow and probably warm water and have thicker shells than *Neopilina*. At any rate, the pattern of muscle insertions on the shell of *Neopilina* corresponds very closely to the scars on fossil monoplacophorans (Fig. 10-11). Shell-inserted muscles on *Neopilina* include, in addition to the eight pairs of large pedal muscles, smaller muscles for the foot, mantle, gills, and radular apparatus. Scars that seem to represent the gill, mantle, and radular muscles can be seen on Paleozoic monoplacophorans.

Mollusks Were Probably Seriated When They Arose

There is embryological evidence that the Mollusca are closely related to the Platyhelminths, Nemertea, and Annelida. All four phyla have similar spiral cleavage patterns on their fertilized eggs, and their developmental similarities continue into larval stages. However, the phylogenetic pathways between them are not definitely known. Because the Platyhelminths are simply organized, and as mollusks display no traces of eumetamerism but may be pseudosegmented, it is hereafter assumed that the mollusks are descended from a primitive acoelomate, which resembled certain turbellaria in flatworms and which was ancestral to or arose from the ancestor of modern Platyhelminths. This is a common but by no means universal view (for a more detailed account, see Vagvolgyi, 1967).

Many of the molluscan features can be associated with an anagenetic trend to larger body size (Fretter and Graham, 1962). Increase in body volume requires additional digestive volume and leads to expansion of the visceral mass with its digestive tract and glands; dorsal expansion of the visceral systems frees the foot for locomotory specialization. At the same time, a circulatory system became necessary to provide transportation of metabolites and respiratory gases, and gills were gradually evolved to enhance blood aeration. The pericardial space must have appeared in conjunction with the localization of contractile segments of the vascular system to form a heart. Some cephalization with development of a radular apparatus to collect and break down food particles completes the assemblage of typical molluscan characters.

One other major step is required to alter the ground plan of the Platyhelminths to that of the archetypal mollusk of Fig. 10-8: pseudometamerism must be obliterated. The presumed original function of the organ seriation, to provide physiological services throughout an elongate organism that lacked circulation, would be bypassed to a considerable extent with the development of a cardiovascular system. It would then be possible for seriation to be reduced whenever it became selectively advantageous to do so.

The phylogenetic pathways of diversification among the molluscan classes are obscure. Often, all the classes are depicted as radiating in different directions from a common source, the archetype of Fig. 10-8. If this is true, then the seriation of the monoplacophorans is secondary, and has developed within the molluscan body plan. A major difficulty with this interpretation is that there appears to be no functional reason for this development. It is more feasible that the monoplacophorans have inherited their seriation from a pseudometameric ancestor. In this event, the other molluscan classes may be descended from a common ancestor in which seriation had become obsolete, such as the archetype of Fig. 10-8, or they may have independently reduced their seriation as parallel evolutionary trends. If seriation were reduced independently, it could account for the retention of some seriation among the chitons and cephalopods. There is some additional although inconclusive evidence on this point from the early Bivalvia.

In the last part of the nineteenth century, at about the time that the Paleozoic monoplacophoran shells were first becoming known to science, an unusual bivalve was described from Middle Ordovician strata of the Bohemian Basin in what is now Czechoslovakia. It was named *Babinka prima* by Barrande, but remained poorly known for over 70 years. *Babinka* is unusual because it has multiple pairs of subequal scars arranged on the shell interior in a shallow arc between the adductor muscle scars (Fig. 10-12). Some living bivalves have multiple pedal muscles, some of which leave multiple scars at their insertions, and a number of early bivalve shells have multiple muscle scars in this area, which have been interpreted as pedal muscle scars (Vokes, 1954). McAlester (1965) restudied specimens of *Babinka*, and found that there are eight scars on each valve which are probably pedal muscle scars, and that, in addition, there are some shell markings otherwise unknown among the Bivalvia. These include a set of smaller scars, numbering about 25 on each valve, arranged beneath the

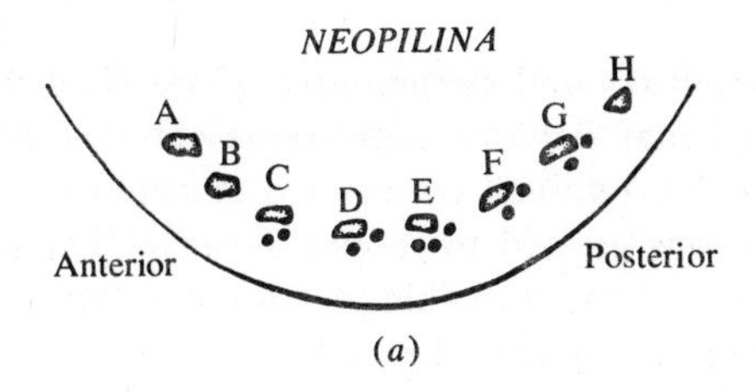

(*a*)

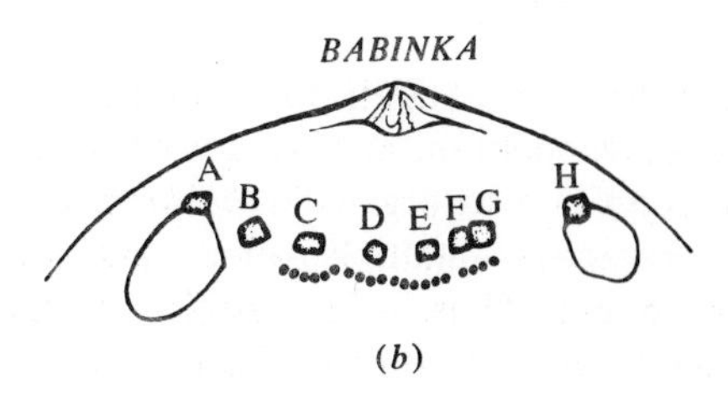

(*b*)

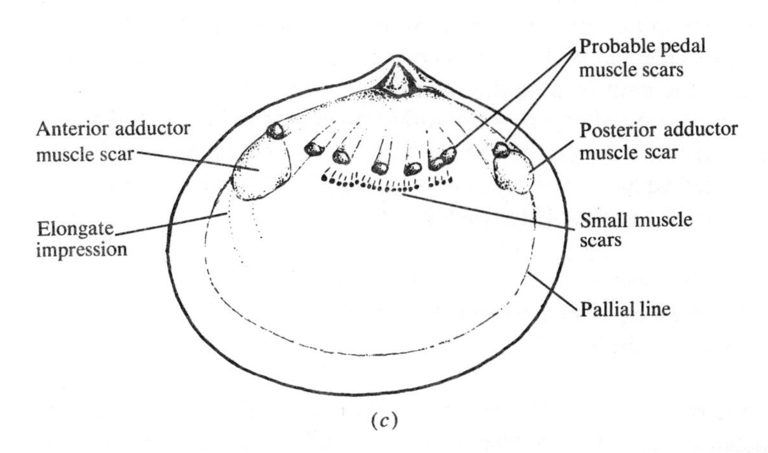

(*c*)

Fig. 10-12 *Babinka*, a probable Silurian lucinoid bivalve. (*a*) Shell muscle and gill muscle insertions on *Neopilina*. Compare with (*b*), muscle scars on *Babinka*. (*c*) *Babinka*, side view, showing typical lucinoid shape and dentition; note, however, the seriated shell muscles and small muscle scars beneath them. (After McAlester, 1965; by permission of the Palaeontological Association.)

third to seventh pedal muscle scars (Fig. 10-12), a position that corresponds well to the gill muscle pattern in *Neopilina*. Possibly the small scars in *Babinka* are gill muscle scars, and if so, there may have been a series of multiple gill pairs in *Babinka*, or there may have been a large gill with multiple supporting muscles (McAlester, 1965).

When *Neopilina* was first described in 1957, and the significance of monoplacophoran scars clarified, it was at first suggested that the monoplacophoran seriation represented eumetamerism, indicating a close structural and phylogenetic affinity between the mollusks and eumetameric phyla (Lemche, 1957). Various authors then suggested that *Babinka* might represent an early or even initial form of the Bivalvia which retained traces of a eumetameric ancestral mollusk. In fact, a new order was erected to receive *Babinka* (Diplacophora; Horný, 1960). However, in view of the probable pseudosegmented ancestry of the mollusks, it now seems more likely that the multiple pedal scars represent either a seriation descended from a pseudosegmented ancestor or simply a secondary seriation of pedal muscles as an adaptation to whatever pedal action was required.

As McAlester has pointed out, *Babinka* certainly possesses the major features of the Bivalvia; it is not a monoplacophoran, and we are not constrained to impute a high degree of serial repetition to organs in *Babinka*. On the other hand, it bears the same sorts of characters—multiple paired scars—that do indicate seriation in the Monoplacophora. Thus, it is possible that the Bivalvia branched off from stocks that led to other molluscan classes before deseriation of a pseudometameric ancestral body plan was completed. The eventual suppression of the remaining seriation may have been associated with increase in size of the bivalve gills and with the posterior concentration of excretory and gonadial pores in the path of outgoing mantle currents.

At any rate, repetition of organs is not unknown among living mollusks, and it seems likely that the archetype of Fig. 10-8 has never existed. Salvini-Plawen (1969) has presented an archetype based on a model of molluscan evolution that begins with a seriated ancestor (Fig. 10-13).

The Function Chiefly Responsible for the Architecture of Ground Plans Is Locomotion

We have briefly outlined major features of the ground plans of four phyla: the pseudometameric acoelomate Platyhelminths and Nemertea; the eumetameric coelomate Annelida; and the coelomate Mollusca, which display a sort of pseudometamerism in one class and occasional seriation in others. In theory, all or nearly all of the characters of these different ground plans should have arisen as anagenetic trends, and should have been associated with the acquisition of ecological adaptations. However, it is not an easy matter to demonstrate the ecological significance of primitive features. A variety of evolutionary models have been proposed for these phyla, and each implies a different ecological pathway; the historical evidence is not yet sufficient to define the evolutionary course. We shall trace pathways that seem plausible on present evidence, but that cannot be confirmed as historical. Nevertheless, they illustrate principles believed to apply to the evolution of ground plans, which are derived from ecological and evolutionary theory.

Probably the acoelomates evolved from a ciliated swimming animal, perhaps from the planula larvae of the coelenterates or from a similar organism (Hand,

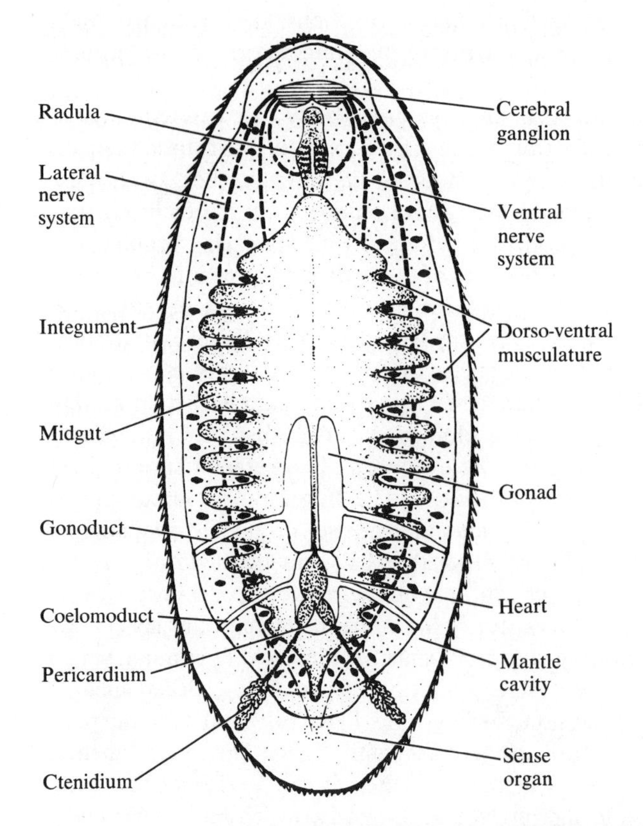

Fig. 10-13 A possible molluscan archetype based on descent from a Platyhelminthlike ancestor, dorsal view. (After Salvini-Plawen, 1969.)

1959; Hyman, 1959). The development of body-wall musculature and of internal mesenchyme as support may be related to selection for the regulation of body shape for swimming; if so, it proved to be preadaptive to an epifaunal crawling habit. It may also have developed as a direct adaptation to epifaunal crawling. At any rate, the primitive acoelomates probably moved over the substrate by ciliary action. Many of the subsequent evolutionary developments in acoelomates seem related to selection for increasing body size, which is so commonly advantageous as leading to robust and well-adapted phenotypes. Larger organisms are also able to exploit resources that are not available to smaller and weaker forms; thus, size increase is commonly preadaptive to new ecological functions.

However, as body size increased, interior tissues became progressively more remote from the environment and from such localized regions of environmental exchange as the mouth. Serial multiplication of visceral organs provided physiological services throughout the interior, and is thus an adaptation in support of the enlarging, elongate, triploblastic body, which is itself adapted to efficient epifaunal crawling and associated ecological functions. The flattened body form of platyhelminths permitted environmental interchange of respiratory gases.

Nemertines continue the trend toward increased body size, and tend also to be rounder, permitting the utilization of the mesoderm as a hydrostatic skeleton for peristaltic locomotion that employs much of the powerful body-wall musculature. The interior thus becomes even more remote from the environment, and a circulatory system is developed to help deal with increasing problems of internal transport. Again, the new features of the body plan appear to be related to crawling modes of life (although other locomotory methods are developed on this plan) and the larger body size, possibly associated with the exploitation of new food resources, such as larger particles or different prey species.

The primitive annelidan ecospace seems to have involved active burrowing and deposit feeding. Evolution of the coelom provided the basis for a more efficient hydrostatic skeletal system, and the evolution of eumetamerism provided the ability for prolonged burrowing activity. The rich food resources buried beneath the surface of the substrate and unavailable or poorly available to acoelomates could thus be exploited. Once the basic annelidan features developed, they were further modified or elaborated to permit the occupation of other regions of biospace. Some lineages developed different locomotory specializations, such as sinusoidal crawling with parapodia or sinusoidal swimming, which must have been coadapted with the exploitation of new foods. Other lineages developed the habit of suspension feeding, and lived in permanent burrows (such as the lugworm *Arenicola*); in these sedentary forms, the intersegmental septa tend to be greatly reduced. Furthermore, there is much evidence that early annelidan stocks gave rise to the phylum Arthropoda, which has undergone a radiation and diversification that far surpasses the annelids.

The molluscan ground plan suggests a creeping habit for the ancestral lineage, much like that of flatworms except with a larger body size as permitted by the coelom and by the development of respiratory and circulatory systems. These developments also allowed the viscera to be placed in a dorsal mass, so that the foot was free for locomotion. Probably the mollusks branched off an early coelomate lineage and returned to the epibenthos (Clark, 1963), where the coelom was not required and was rapidly reduced in some lineages. They radiated into a variety of habitats, which eventually led to the consolidation of the suites of features that now characterize the molluscan classes. The functional paths are still speculative, but it seems likely that if the mollusks were actually derived from deposit-feeding burrowers which evolved into epibenthic forms, the earliest mollusks would be found on soft bottoms, probably as deposit feeders, and they may have been something like the archetype of Fig. 10-13.

From taxa in this habitat, a number of distinctive lineages have arisen that represent invasions of different habitats. Two separate lineages invaded solid substrates and developed into snails and chitons. These two taxa obviously proceeded via quite distinct routes, the chitons perhaps evolving through a wormlike ancestor of aplacophoran type, and the gastropods through an ancestor closer to the monoplacophorans. The shift to hard substrates probably involved a trophic shift as well, to browsing on living plants or larger detritus. Another lineage evolved a pelagic mode of life, perhaps first developing into benthic predators and then into nektobenthic and pelagic forms to become the cephalopods. Some lineages reinvaded the sedimentary substrates, evolving into the scaphopods and bivalves. To judge from their primitive larvae (Salvini-Plawen, 1969), these lineages seem to have descended from rather generalized mollusks that remained in soft-bottom habitats while the chitons, gastropods, and cephalopods were evolving different ecospaces. When the mollusks burrow, they chiefly employ the haemocoel rather than the coelom, inventing a new burrowing technique rather than resurrecting the old.

There is much more evidence and numerous other molluscan types that have not been discussed, but this is hardly the place to become entangled in the details of phylogenetic arguments, as useful as they may be for interpreting molluscan history. For our purposes, it is sufficient to present a model of the origin of a few higher metazoan taxa in order to consider the problem of the relatively abrupt appearances of metazoan phyla during and immediately preceding the Cambrian Period. New ground plans seem to be concerned chiefly with the exploitation of relatively untapped trophic resources, with anatomical changes due to the locomotory requirements associated with this exploitation.

It seems that functional interpretation of ground-plan features can aid in reaching reasonable interpretations of evolutionary pathways. Selective pressures for improved ecological adaptations result in the rise of ground-plan features, and these require a supportive chain of coadaptive characters to produce a well-integrated, efficient organism. These features have arisen in populations as novel adaptations, in the same way that specialized characters appear to arise in advanced lineages. However, some ground-plan features are probably not monophyletic. This possibility seems especially strong for features associated with evolutionary trends that occur frequently as parallel adaptations, such as phyletic size increase. If the lineages were closely allied, and presented similar structural and genetic bases for modification, it would not be surprising if the adaptations supporting the parallel trends were similar. Some ground plans, then, probably represent polyphyletic grades.

Obviously the persistence of a set of structural features causes us to regard them as forming a ground plan. The processes of stasigenesis operate to help a well-integrated set of features endure, and if this set is distinctive with respect to other persisting sets it will be considered to be a higher taxon, perhaps a phylum. Continued anagenetic trends carry the integration and adaptation of the lineage further, chiefly through specializations that involve cladogenetic diversification. Adaptive types that are intermediate between the persistent ground plans must have existed but they served only as anagenetic episodes.

Phyla May Originate Through Adaptive Radiations from Generalized Ancestors

As an introduction to the problems of the Cambrian-Precambrian biotic boundary and the origin of invertebrate phyla, it is appropriate to recount a story of two watchmakers, based on a delightful example by Simon (1962). One of these watchmakers, Tempus by name, assembled his watches from components so constructed that the watch assembly was unstable until it was completed. Once completed, however, it was perfectly stable and ran forever. Now, Tempus was a compulsive telephone-answerer, and he received calls at a high average frequency (his watches were very popular, naturally) but at random intervals. When he answered the telephone, any uncompleted watch assembly would fall apart into its original components, and following the call he would have to begin the assembly over again. Quiet intervals of sufficient length to permit him to complete an assembly were infrequent, and his watch production was consequently low.

The other watchmaker, Chronos, had a different assembly plan. He used stable subassemblies, which could then be fitted together to form intermediate assemblies that were also stable; these, in turn, could be fitted into a final assembly which was stable and ran forever. Even though the subassemblies and intermediate and final assemblies were unstable unless completed, and even though Chronos's telephone rang at the same frequency as Tempus's, the presence of several levels of stability arranged in a hierarchical manner resulted in a far greater production of finished watches by Chronos, for he wasted much less time in reassembling components that had disassembled when he was interrupted by the telephone.

The point of this story for the evolution of distinctive ground plans is that each need not be painstakingly built up from primitive organisms, component by component. On the contrary, once components have been evolved they may be employed as subassemblies in different combinations and arrangements, perhaps together with some new components, to form series of different ground plans. This greatly reduces the time required to generate a new ground plan. It is suggested that most of the components for the phyla that appear more or less fully formed in a relatively short time interval have already been evolved. This implies that the Precambrian-Cambrian boundary does not represent the appearance of all the Metazoa, but merely represents the appearance of opportunities for a radiation of the phyla that we see (and perhaps some that were soft bodied on which we do not have any data) from some basic stocks that were already present. Organisms certainly have a long Precambrian history during which time many of the basic components of higher Metazoa were evolved. These components would include such features as the origin of very large or complex organic molecules, of integrated sequences of biochemical steps that underly the basic metabolic cycles of various types of tissue, and others. With an array of such components available, the potential for the generation of a wide range of morphological types was present once early metazoan worm lineages were freed from the constraints of size and locomotion that were inherent in the acoelomate condition.

The feature that effected this freedom is, of course, the coelom, for it permitted the development of larger body sizes and integrated organ systems that placed coelomates on a new level of capability. Given a coelom and a set of physiological and structural components, their arrangement to form a variety of structural plans is partly just an exercise in "biogeometry," and evolution need not require much time to arrive at quite a large array of body plans from such a beginning. If the evidence suggests that a series of phyla represents not a sequence of organizational levels on an anagenetic trend but rather a variety of different structural plans on the same organizational level, then it is reasonable to believe that they may have radiated from some common generalized ancestral stock in a relatively short time. The reason is that the development of specialized characters limits the extent and direction of subsequent evolution. Thus, although the development of one new ground plan from one specialized old one is plausible if a new habitat is being invaded by the specialist, the radiation of a number of new ground plans usually occurs from a generalized stock.

Thus, the evolution of, say, a fairly specialized burrower into a suspension feeder is easily possible; indeed, it has clearly happened within the phylum Annelida. However, in this case it did not result in the creation of a new ground plan or anything like it. It resulted in modifications in the annelid structure to adapt to the new requirements, but the suspension-feeding annelids are clearly still annelids, presumably because the annelidan ground plan is preadapted to a number of locomotory methods which require structural modifications but not complete restructuring. As mentioned previously, when burrowing, detritus-feeding annelids become suspension feeders, there is a reduction in septa, for the suspension feeders are not continuously active burrowers, although they inhabit a tube or burrow in the sediment. Thus, they must form burrows, but subsequently they simply occupy them and depend upon the movement of overlying water to supply food. Burrowing detritus feeders must constantly be on the move to find new detritus after mining out a given locality; therefore, annelids with these active habits require the advantages of partitioned body segments. However, segmentation involves some disadvantages, such as the trouble of equipping each segment with a set of organs, and the intersegmental partitions seem to be quickly lost and the segmentation reduced whenever annelids develop less active habits. Clark (1964) gives a thorough discussion of segmentation.

If, on the other hand, a burrow-dwelling suspension feeder is evolved not from a fairly specialized and active burrower but directly from a generalized coelomate burrower, a whole new ground plan may be especially tailored for this mode of life. Such a ground plan would probably include only a few coelomic partitions. The oligomerous phyla, which have evidently had two or three separate longitudinal coelomic spaces primitively, may represent just such a ground plan (Clark, 1964). These phyla include the phoronids, brachiopods, and ectoprocts.

As Skeletons Commonly Influence Ground Plans, They May Indicate the Time of Origin of Some Phyla

An important point raised by Cloud (1948) is that a long prior history of many of the Cambrian taxa is not likely if they lacked skeletons. As we have seen previously, skeletons are integral parts of the architectural scheme of many phyla, and there is commonly a skeletal ground plan that is coadapted with the anatomical ground plan; indeed, there is actually one master ground plan that includes "soft" and "hard" parts alike. Therefore, without their skeletons, many higher taxa would not be organized as we find them, but would have to have some other anatomical arrangement. The brachiopods are an excellent example (Cloud, 1948), for their entire feeding and respiratory apparatus and their musculature are all constructed within the framework of a solid exoskeleton. Without the skeleton, they would probably be relatively simple, filtering polypides or worms of some sort, resembling very short phoronids.

As we have seen, some of the molluscan lineages appear to owe many of their anatomical characteristics to coadaptive evolution with skeletons. Gastropods have had their internal organs reduced and consolidated, and their mantle cavity arrangement (a key element in their evolutionary deployment) owes much to shell morphology. Bivalves burrow by employing their shells, which are important functional organs and which at the same time clearly place stringent limits on soft-part anatomy. Cephalopods similarly appear to have developed many of their architectural peculiarities, such as their U shapes, as adaptations to skeletal form.

In other groups, skeletal morphology has had less effect on anatomical organization. The soft parts of scleractinian corals, for example, are essentially identical to the soft-bodied sea anemones. To be sure, many of the habits and habitats taken up by the Scleractinia are possible only because of their skeletons, but this is not reflected by modifications of their soft parts. The only other important skeletonized marine invertebrate phylum with a ground plan that may not be closely affected by the skeleton is the Ectoprocta. Oddly enough, neither the corals nor the ectoprocts appear until at least the late Cambrian. The coelenterates are well known from the Ediacara assemblages, and the ectoprocts could well have evolved during the late Precambrian or early Cambrian, but simply did not become skeletonized until the late Cambrian.

Thus, we must seriously consider that many of the skeletonized invertebrate lineages were not really organized into the ground plans that we now recognize as defining the major phyla until about the time that they appear in the record, and that the development of skeletons contributed heavily to a reorganization of anatomical architectures to define these phyla. Once consolidated, these major invertebrate ground plans have remained in the biosphere down to the present day. For those phyla for which mineralized skeletons do not affect the ground plan, such as the coelenterates, the acoelomates, the coelomate worms, including annelids, and the arthropods (which require a cuticle but which are commonly not heavily mineralized), these arguments have no special force, and other considerations must be brought to bear upon the question of their original development.

The Metazoa May Have Diversified in Stages

Putting all these considerations together, we may arrive at a provisional model of the diversification of selected metazoan phyla and classes, diagrammed in Fig. 10-14(*b*). The phylogenies suggested are those indicated by the work of Clark (1964). For mollusks, they are modified and extended by the discussions of Vagvolgyi (1967) and Salvini-Plawen (1969). The stimulating and controversial views of Jeffries (1967, 1968) have provided a new model of the phylogeny of the chordates and their allies, and it is illustrated here as modified by Eaton (1970). It is likely that the early phylogenies indicated will be modified as new evidence appears, but the evolutionary modes illustrated by Fig. 10-14(*b*) are to a large degree independent of the precise phylogenetic model.

There are five main phases. The first is the origin of the Metazoa, presumably from Protozoa. There is practically no evidence about this phase, and its dates are not known, although it most probably happened after 1.8 billion years ago. In the second phase, the diploblastic phyla evolve and perhaps diversify, and the diploblastic acoelomates arise, presumably by neoteny from a coelomate larva or from a swimming organism evolved from such a larva. The third phase is characterized by the origin of the coelom as a burrowing infauna developed, probably narrowly followed by a radiation of coelomate structure into the basic variations that include metamerous coeloms, which were employed by annelids and reduced in arthropods; pseudometamerous coeloms, which were employed by protomolluscan lineages; oligomerous coeloms, which were eventually employed by Deuterostomes (echinoderms, chordates, and their allies) and by the lophophorate phyla, such as brachiopods; and amerous coeloms, which totally lack partitions and may be represented by the sipunculids today. Because burrows are evidently not common in sediments before about the time of the Ediacaran assemblage, this phase may have begun around 700 million years ago or slightly earlier. Some coelenterates can burrow today; therefore, it is possible that some diploblastic organisms could have been responsible for any burrow-like sedimentary structures of organic origin that are older than this.

Although Fig. 10-14(*b*) makes it appear that several special archetype lineages, each with a distinctive coelomic architecture, are being proposed, it is likely that the stems are actually broken up into numerous lineages. Furthermore,

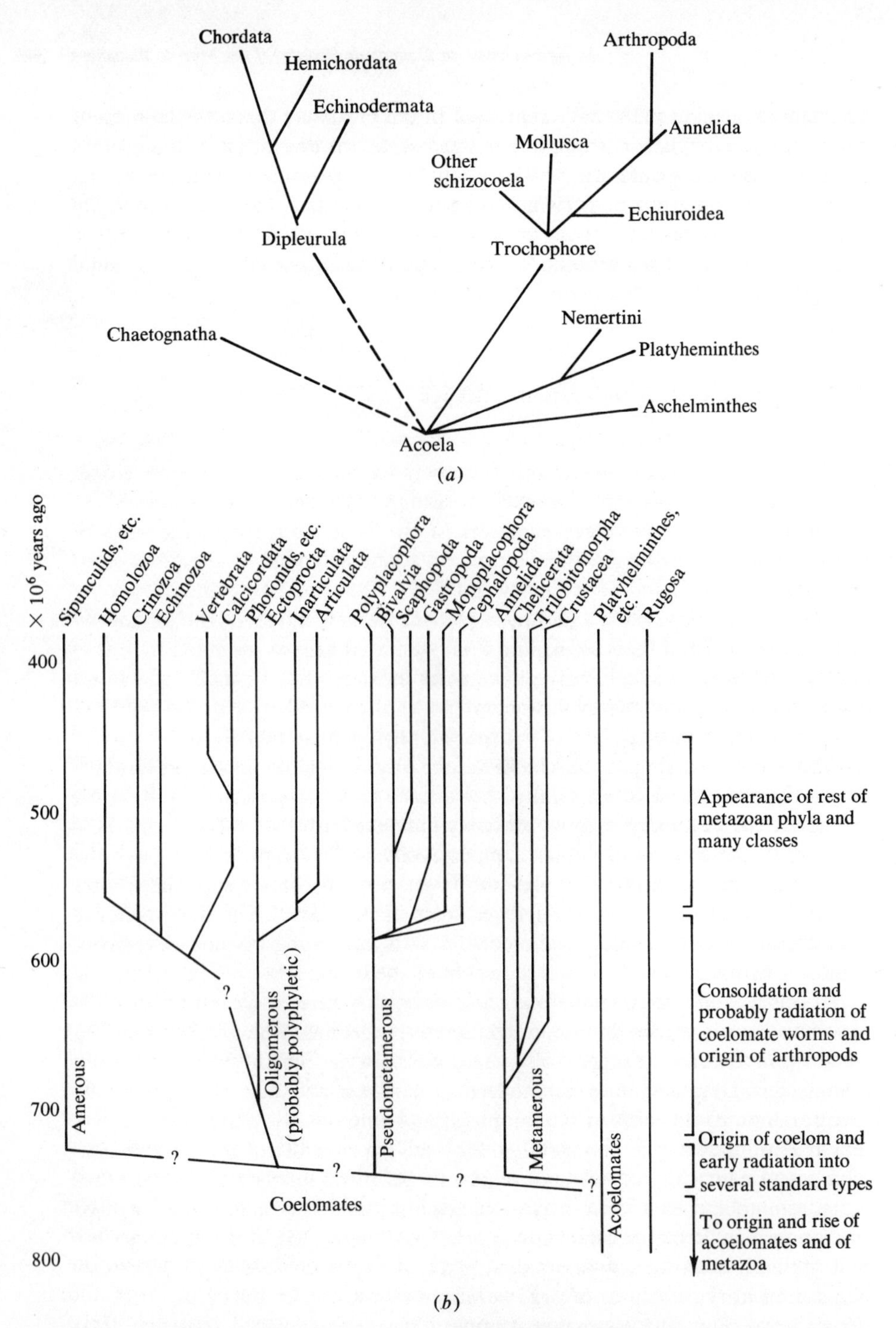

Fig. 10-14 (*a*) A phylogenetic tree of the conventional type, showing the higher Metazoa; the impression is of a striving upward toward the ultimate evolutionary peaks, the chordates (presumably man) and the arthropods (presumably butterflies). (*b*) A phylogenetic model of the radiation of the major shelled invertebrate phyla and some classes. The coelomates radiate from an acoelomate ancestor, and then most coelomate lineages radiate, in turn, into the phyla that we know today, chiefly during the Cambrian. Note that this diagram only extends to about 400 million years ago. [Part (*a*) is after Hyman, 1940.]

the diagram has been drawn as if each of these types were monophyletic. These are merely conventions to simplify construction of the diagram, adopted because the true complexities are unknown in any event and their patterns cannot be drawn. It is quite possible that the coelomates as a whole are polyphyletic, the coelom having arisen more than once from different acoelomate lineages (Clark, 1964), and that several of the archetypes are polyphyletic, this being particularly likely for the oligomerous taxa.

The fourth phase is essentially coextensive with the Ediacaran fauna and represents the continued evolution and consolidation of the coelomate types and probably the radiation of the metamerous annelids and arthropods. The fifth phase includes the development of mineralized skeletons by a variety of lineages, mostly epifaunal, creating the Cambrian fossil record. The phase began about 570 million years ago. Some of these taxa evidently achieved their definitive organizational ground plans as they skeletonized. These would include the Brachiopoda, the Monoplacophora, and the Echinodermata in the Early Cambrian. Other taxa had acquired their ground plans a long time previously but were not mineralized until the Cambrian. These would probably include the Trilobita. Later in the Cambrian, the Cephalopoda and the Gastropoda make their first appearances, and in the Ordovician the Bivalvia appear; all these taxa are probably originating at about the times that they first appear. On the other hand, the Ectoprocts do not appear until the Ordovician, but may well have originated during the Late Precambrian-Early Cambrian radiation, for skeletons are probably not essential to the development of their ground plans. Similarly, the rugose corals do not appear until the Ordovician, but they may represent the beginning of skeletonization of polyps, which stood in the same relation to the Rugosa as modern anemones do to the Scleractinia, and which had flourished in the Precambrian for all we know. By the close of the Ordovician, all the phyla and most of the classes of well-skeletonized invertebrates had appeared. Subsequent collection will surely cause some of the first appearances to be dated earlier than is now known, but this is beside the point.

It is worth re-emphasizing that in postulating the origin of many of the phyla or classes at about the time that their skeletons first appear, we assumed that such rapid radiations are possible because the components of a given ground plan or ground-plan type have already evolved, and that the radiation represents the development of subplans based on variations in the arrangement and differential development of the components, together with the origin of some new characters when appropriate. This type of radiation appears to be extremely common and has long been recognized. In Fig. 10-14(*b*), a temporally localized radiation is inferred for coelomate types to originate sorts of trunk stocks near 700 million years ago, and another wave of radiation occurs at about the Cambrian at the phylum level. This same pattern can be traced in lower and lower taxonomic categories. For example, Fig. 10-15 shows the recorded stratigraphic distribution of subphyla and classes of the Echinoderms. Figure 10-16 shows the recorded stratigraphic distribution of orders and superfamilies of one

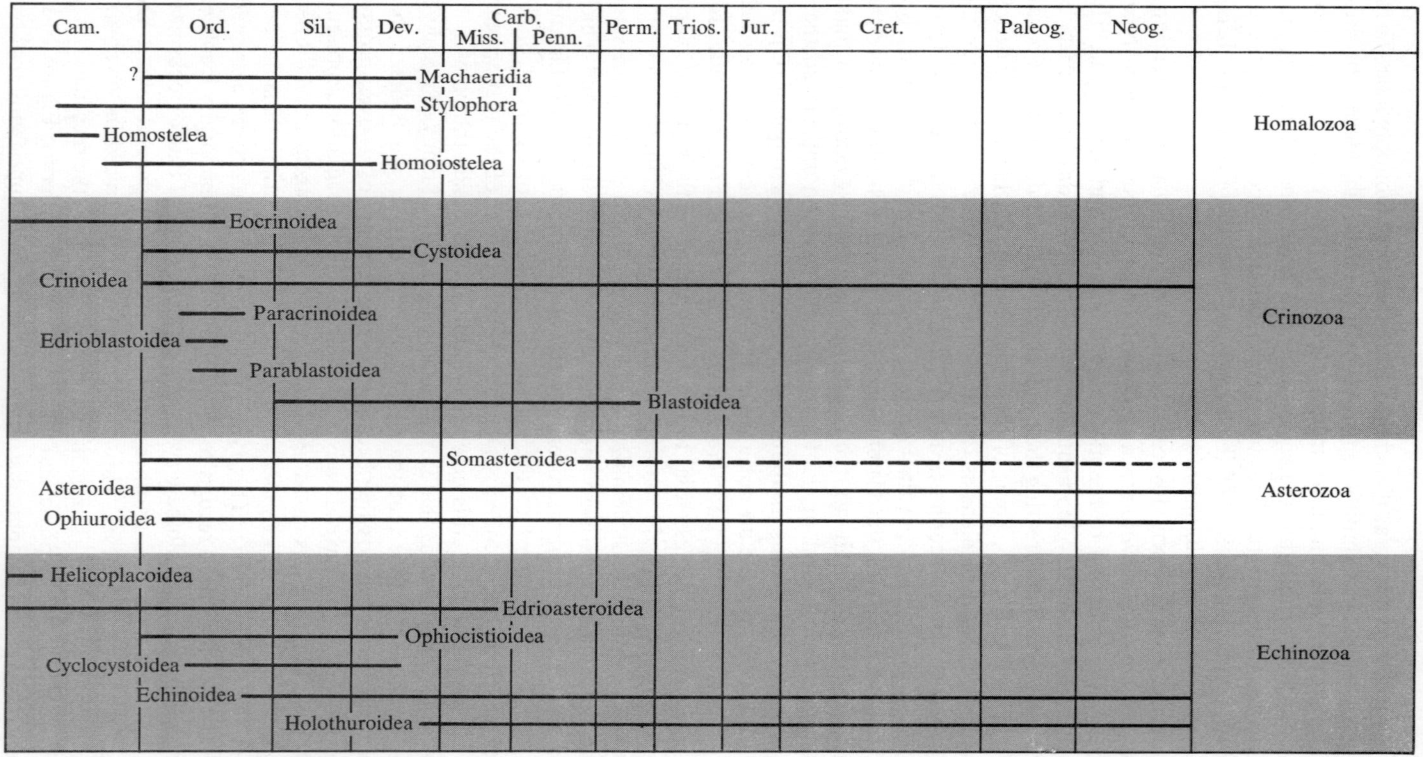

Cam. Ord. Sil. Dev. Carb. Miss. Penn. Perm. Trios. Jur. Cret. Paleog. Neog.

?

Machaeridia

Stylophora

Homostelea

Homoiostelea

Homalozoa

Eocrinoidea

Cystoidea

Crinoidea

Paracrinoidea

Edrioblastoidea

Parablastoidea

Blastoidea

Crinozoa

Somasteroidea

Asteroidea

Ophiuroidea

Asterozoa

Helicoplacoidea

Edrioasteroidea

Ophiocistioidea

Cyclocystoidea

Echinoidea

Holothuroidea

Echinozoa

Fig. 10-15 The appearances and disappearances of subphyla and classes of the phylum Echinodermata. (After Ubaghs, 1967; from *Treatise on Invertebrate Paleontology*, courtesy of The Geological Society of America and The University of Kansas.)

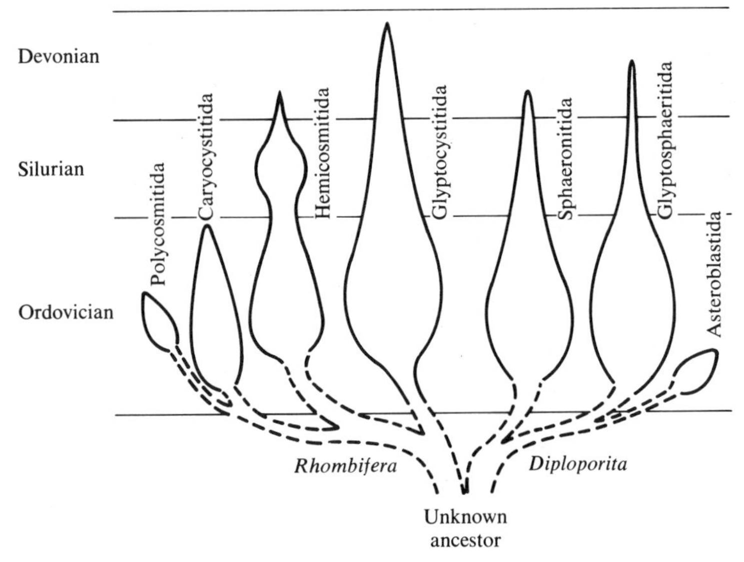

Fig. 10-16 The appearances, times of chief abundance, and disappearances of the echinoderm class Cystoidea. (After Kesling, 1967; from *Treatise on Invertebrate Paleontology,* courtesy of The Geological Society of America and The University of Kansas.)

of the echinoderm classes, the Cystoidea. Figure 10-17 shows the recorded stratigraphic distribution of families of one of the cystoid superfamilies, the Glyptocystitida. The pattern is very similar for taxa in all these categories. It is likely that at each level we are witnessing the sort of radiation that accompanies the invasion of biospace which is fairly free of competitors, so that a number of morphological variations are developed at each level representing rapid adaptive extension of the lineages into a variety of habitats for which they are potentially fit, with some appropriate modifications and elaborations.

The character of the early radiation of coelomate types just before 700 million years ago is, of course, conjectural, for there is no direct evidence that it had just the form indicated in Fig. 10-14(*b*). However, a rapid adaptive radiation of such a new structure, the coelom, is precisely the sort of pattern that is found at lower taxonomic levels when they *do* appear in the fossil record, so that in a sense it is the least conjectural of the possibilities, for it conforms to a well-established evolutionary mode. The timing is based on the evidence from the Ediacaran fauna that metameric ground plans, at least, existed during the time range of this fauna, and from the presumption of descent of the coelomates from triploblastic acoelomates. The presence of significant numbers of trace fossils which demonstrate an infaunal biota during the Ediacaran interval indicates that coelomates were not uncommon.

By Ordovician time, there is clear evidence of reasonably high environmental stability. Reef associations appear (Chazyan Stage) and some benthic communities begin to be fairly diverse. Therefore, it is possible that the Cambro-

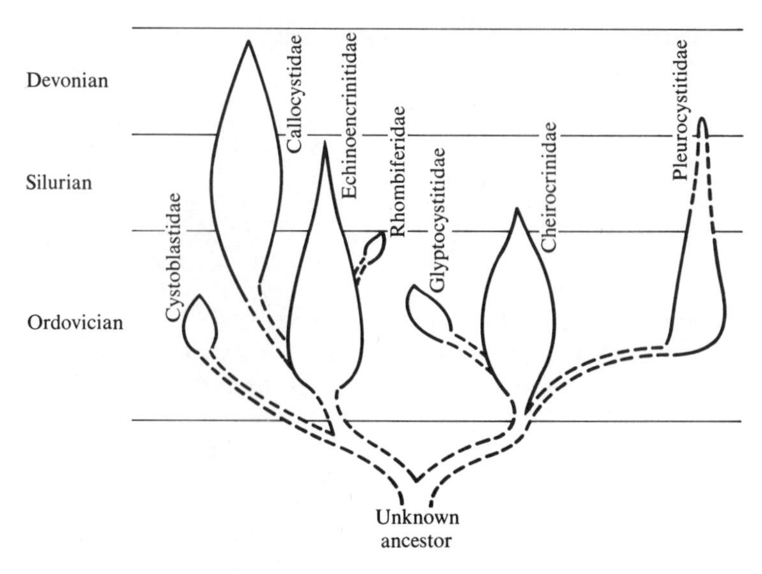

Fig. 10-17 The appearances, times of chief abundance, and disappearances of families of the cystoid superfamily Glyptocystitida. (After Kesling, 1967; from *Treatise on Invertebrate Paleontology*, courtesy of The Geological Society of America and The University of Kansas.)

Ordovician diversification represents not only a radiation by coelomates into relatively unoccupied biospace of the epibenthos but also a time of rising diversity levels due to an amelioration of diversity-dependent factors, probably a trend toward more stable trophic resources. It is reasonable to imagine that this trend began in the Late Precambrian and that it permitted the appearance of a diverse epifauna. A possible bit of independent evidence that continentality was decreasing is the widespread transgressions of the oceans over the continental platforms during Middle, and especially Late, Cambrian time. It does seem likely that environmental stability increased between the rise of the coelomates near 700 million years ago and the appearance of well-skeletonized taxa at about 570 million years ago.

Whatever the reason for the development of a diverse metazoan epibenthos, it seems to have been accompanied by the evolution of a number of new ground plans, which involved the appearance of several skeletal ground plans. There are a number of reasons that mineralized skeletons would develop under these circumstances. First, although peristalsis is an ideal locomotory mode for an infaunal worm that is completely confined by its locomotory substrate, it is inefficient on a surface—only a relatively narrow sector of the circular worm cross section is applied to the substrate (Clark, 1964). With mineralized skeletons, musculature can be organized with leverages and antagonisms appropriate to epibenthic locomotion (L. Fox, personal communication). Second, skeletons can be employed as sorts of suspension-feeding tubes and apertures, replacing the burrows of infaunal suspension feeders, in a manner of speaking. Third, skeletons can be used to support individuals above the substrate, permitting the exploitation of higher water layers; this is useful for solitary organisms and also for those with colonial habits. Of course, mineralized skeletons provide protection. All these points are advantages accruing to epibenthic organisms that skeletonize.

The following summary seems a reasonable ecological model of Late Precambrian and Cambrian events among the higher Metazoa. (1) A metazoan fauna is present, having originated earlier; it includes coelenterates and benthic acoelomates. (2) A rise in trophic resource fluctuations puts a premium on exploiting the more stable resources—chiefly the sedimentary detritus beneath the surface. (3) Infaunal coelomates develop to feed on subsurface detritus, and, while still generalized, undergo an adaptive radiation that is coupled with a significant body-size increase. (4) Increasing environmental stability permits the appearance of a large epibenthos, favoring the development of skeletons for locomotion, feeding, support, and protection. (5) As skeletonization occurs, with coadapted trends in anatomical reorganization, the consolidation of different body plans achieved at this time standardizes a number of ground plans that we now recognize as phyla and classes.

The phylogenetic pattern in Fig. 10-14(*b*) is very unlike a conventional phylogenetic tree, which commonly depicts a trunk stock from which branches split at rather regular intervals. A popular phylogenetic tree is presented in Fig. 10-14(*a*) (after Hyman, 1940). Figures 10-14(*a*) and (*b*) emphasize different lineages and are not perfectly comparable, but nevertheless the difference in their patterns is striking. Actually, the topological similarities between them are great, and the differences they display in the pathways of descent are minor; the important difference is in the patterns. Figure 10-14(*a*) does not have a vertical time scale. Nevertheless, phylogenetic trees drawn in this manner closely resemble the pattern expectable on the basis of a hypothesis of early origin and gradual diversification of the metazoa, as depicted in Fig. 10-5(*a*). It is best to employ a vertical time scale when presenting phylogenetic models, because misapprehension of the limitations of the phylogenetic patterns can easily lead to biased historical concepts.

Paleozoic Species Tend to Decrease in Average Niche Size as Paleozoic Communities Tend to Increase in Diversity, Suggesting Increasing Environmental Stability

Although it seems likely that the development of a skeletonized fauna occurred during a trend toward increasing environmental stability which took place over large regions of the continental shelves and shallow seas, the greatest diversifications evidently did not begin until late in the Cambrian, and Cambrian communities seem rather generalized when compared with those of the Middle and Late Paleozoic. The species seem rather plain, even grubby, and to judge from the taxa that have had their modes of life interpreted with the most confidence, they tended to be detritus feeders, or at least to be relatively generalized trophically. Cambrian assemblages also tend to be similar over broad environmental belts (Chapter 8; Palmer, 1969), suggesting habitat generalization. Cambrian communities do contain some rather strange, exotic-looking forms, but these merely belong to unfamiliar taxa that became extinct in the early Paleozoic, and they may well also be generalized; certainly they are not highly ornamented. It is tempting to ascribe the generalized appearance of Cambrian species and the inferred simple structure of Cambrian communities to the primitiveness of the biota. However, it seems preferable to account for this situation as an adaptation to a rather variable environment; Cambrian biotas are grubby looking because they evolved into a grubby biosphere.

Many of the higher taxa that appear during the Cambrian become extinct during this period or during the Ordovician. This wave of extinction was even higher than is indicated in Fig. 9-8, because some of the early higher taxa that became extinct are poorly known, and their proper categories are in more doubt than usual. For these reasons, they were not even included in the data on which the figure is based. The total number of extinctions is impressive, however. Among the extinct groups is the entire phylum Archaeocyatha; the inarticulate brachiopod orders Obolellida, Paterinida, and Kutorginida; the monoplacophoran order Cambridoida; the peculiar molluscan class Mattheva; some groups of gastropodlike mollusks of the early Cambrian that are not yet assigned to a class of their own; the trilobite orders Redlichiida and Corynexochida and some trilobitoid taxa; and the echinoderm classes Helicoplacoidea and Homostelea. The loss of so many taxa in such relatively high categories is unusual, yet it occurs during a time of general diversification.

Probably it is not as significant an event in the history of the biosphere as it appears at first sight. Although it is not yet proven, the fossil record certainly suggests that higher taxa contained relatively few species during the Cambrian. The extinction of some of those taxa would therefore not involve the elimination of very many populations. As the environment appears to have been changing, and as rapid diversification was occurring within some favored taxa, such as the trilobites and brachiopods, a certain amount of biotic turnover is expectable, and may have been unexceptionable on the species level. The relatively low degree of functional diversity within these higher taxa, which had not yet had the occasion to radiate into a large variety of habitats, would make them especially vulnerable to extinction, for if they became inadaptive or were met by superior competitors within their modal habitats, they might be completely eliminated. A more diverse taxon would have a higher chance of being represented by species in habitats that were relatively unaffected by whatever changes occurred. It is possible that for many of these early taxa the rise of stability and diversity lay at the base of their extinction, for they were probably evolved into a sparsely diversified, unstable environment that tended to contain trophic generalists and scavengers. They may simply have been unable to cope in a biosphere with increasingly diverse and specialized competitors and carnivores.

If this general picture is valid, there should be a significant shift in feeding habits from the Early into the Middle Paleozoic, with suspension feeders and predators of ever-increasing specialization gradually increasing whereas detritus feeders and scavengers became relatively more scarce. Quantitative data are not available to test this hypothesis, but the general characters of the early and mid-Paleozoic biotas appear to be amenable to such an interpretation.

The earliest Paleozoic communities seem to have contained relatively abundant detritus feeders (such as trilobites and, probably, archaeocyathids) with supplementary numbers of epifaunal suspension feeders (such as brachiopods and eocrinoids). By late Cambrian time, and especially in the Ordovician, a new wave of suspension-feeding organisms appeared, raising benthic diversity and altering taxonomic and functional aspects of the benthic communities. Skeletonized carnivores, including cephalopods and corals, also appeared and began to diversify. The algal-coral reef association first appears in mid-Ordovician, indicating an association of benthic algae with browsers or detritus-feeders to provide a source of small animals for the corals. Thus freed from the constraints of phytoplankton resources, these communities were able to employ some of their energy bases in maintaining wave-resistant physical structures. Whether or not zooxanthellae were associated with early reef animals is uncertain. Reef associations arose many times during the Phanerozoic, presumably flourishing in times and at places most conducive both to trophic stability and to the establishment of a high benthic productivity.

Suspension-feeding organisms became noticeably more specialized during the course of the Paleozoic, perhaps prey-specialized and certainly habitat-specialized. For example, Ordovician brachiopods have been judged to range through a wide variety of sedimentary facies (Williams, 1969), implying broad ecospaces; and similar conclusions have been reached for trilobites of this period (Whittington, 1966). Marine arthropods in general became progressively more specialized throughout the Paleozoic (Cisne, 1971). The benthic organisms in Silurian marine communities also appear to have had broad habitat tolerances, insofar as substrate type was concerned (Cocks, 1967).

Cambrian echinoderms tend to live on or near the bottom, whereas by mid-Paleozoic time there are numerous stalked lineages that must have required a dependable food supply well off the bottom. By the Carboniferous, crinoid communities were developed in stories, with species of different column lengths associated so as to exploit different levels in the water column (Lane, 1963). Evidently an appropriate strategy for suspension feeding was to exploit the higher levels of the water column, to get the food first, so to speak, rather than waiting for it to become incorporated into the sedimentary detritus. This implies trophic stability. Bryozoa also display a trend of this sort, beginning in low dense colonies and evolving into rising colonies that include lacy fronds and sheets, such as the well-known *Archimedes*.

Possibly these trends culminated in the reef communities of the Permian, which contained a diverse assemblage of species that appear to be very specialized indeed. The brachiopods, which were very diverse, included such highly specialized forms as the richtohofeniaceans, which were discussed earlier, and also such unusual alliances as the lyttoniaceans and coralloid davidsoniaceans (Rudwick and Cowen, 1968). Reef cephalopods, such as the spiny *Cooperoceras*, appear to have been specialized in habitat, and the large foraminiferan fusilinids also seem to be unusually specialized, although their autecology has not yet been worked out.

Thus, the diversity levels of the Paleozoic correlate in part with levels of specialization of the faunas. The Cambrian faunas were of relatively low diversity and seem to have been generalized, whereas mid-Paleozoic faunas were of somewhat higher diversity and were relatively specialized. In the late Paleozoic, however, diversity appears to drop in intermediate and higher taxonomic categories, whereas specialization appears to have been very great in some communities, notably in reef habitats.

In addition to the Cambrian extinctions, unusually high waves of extinctions occurred during the Late Ordovician and the Late Devonian. These extinctions do not correspond with any obvious major change in the planetary stability patterns. It may be, however, that relatively brief regressions occurred, which increased continentality and reduced shelf area and hence spatial heterogeneity (Newell, 1967), thus reducing diversity. It is also possible that the continents assumed configurations that promoted changes in marine climates along the shelves, causing widespread extinctions and permitting rediversifications of appropriately adapted lineages. In such an event, the most rapidly evolving lineages would be favored, other things being equal.

Thus, these trends and changes in Late Precambrian and Paleozoic faunal quality may be associated with environmental fluctuations and trends. Unfortunately, reconstructions of continental geography are lacking for the Precambrian and are still in preliminary stages for most of the Paleozoic; therefore, the primary determinants of the planetary environment are poorly known. It seems probable that when continental geography becomes better known and climatic inferences can be drawn therefrom, the causes of many of the biotic turnovers will be much clearer. As it is, it is only possible to infer the environmental regimes of the late Precambrian from an evolutionary model of the invertebrates, such as is summarized in Fig. 10-14, and to bring some scattered data to bear upon the Early and Middle Paleozoic regimes. The situation improves in the Late Paleozoic, and thereafter we are on firmer ground. It is thus clear that the following suggestions are speculative, but they are nevertheless worth indulging as an exercise in evolutionary paleoecology.

The evolution of the coelom would be favored by an environment in which detritus feeding was an important adaptation for survival, which implies a highly fluctuating regime. It is therefore possible that this event coincides with the assembly of many of the continents into a supercontinent. Conditions seem to have remained much the same in general until near the Cambrian boundary, when rising stability may have favored diversity increases, and therefore favored the emergence of a larger epifauna and a higher degree of specialization. This implies that the continental configuration reduced the seasonality of shelf waters, either through drift into low latitudes or by continental fragmentation that created more maritime conditions.

Gondwanaland lay on the south pole throughout the Paleozoic, although the polar position shifted from one side of the continent to another. During the Devonian, Gondwanaland was most symmetrical with respect to the pole, and extended northward to within only 20 to 30° of the equator. It is probable that an eastward-trending current encircled the southern hemisphere portions of Gondwanaland throughout the Paleozoic, running southerly along the eastern Gondwanaland coast and northerly along the western coast. The current probably arose in low latitudes as a branch of an equatorial current that crossed the ocean east of Gondwanaland. There, conditioning would presumably create a warm current, which would tend to keep the climates of the Gondwanaland coasts rather mild, like a sort of super Gulf Stream system. There were no major impediments to the transfer of heat from equatorial latitudes to the southern coast of Gondwanaland, so far as is known. Northern high latitudes were evidently essentially free of land, which probably meant that an east-flowing gyre drifted around the pole. This would create a barrier to the northward transfer of heat, and thus it is possible that the northern high latitudes were usually temperate. Below this gyre, however, water temperatures must have been tropical or subtropical. On the other hand, seasonality would presumably have been greater along the shores of Gondwanaland than along the smaller northern continents of Early and Middle Paleozoic times, latitude for latitude.

Glaciation was evidently episodic in Gondwanaland during the Paleozoic, possibly occurring in North Africa during the Ordovician and in Brazil during the Silurian and Devonian, whereas during the interval from Carboniferous to Permian, glaciation seems to have begun in South Africa and shifted as the continents moved over the pole to finish in Antarctica and Australia (Crowell and Frakes, 1970). These episodes would be expected to have affected ocean temperatures; indeed, in the Late Paleozoic there is a well-known southern provincial association, the *Eurydesma* fauna, that is commonly interpreted as representing cool water.

As the Paleozoic closed, the geographic pattern which had characterized most of the era began to change, foreshadowing the appearance of different climatic system. This change is heralded by the northward drift of the portion of Gondwanaland to which the North American-European and Asian continents became sutured, moving those lands into higher and higher latitudes so that a Boreal ocean gradually became differentiated. Although provinciality had been present earlier, perhaps owing in large part to topographic barriers, the appearance of this increasingly enclosed polar ocean created a body of water that was progressively cut off from northward-flowing currents and therefore from a heat supply from lower latitudes. Finally, a sort of Boreal oceanic gulf appeared, which at times was not at all in communication with southerly seas via epicontinental connections. Instead, the only entrances from the south were across high latitudes near and at the pole (see Fig. 9-3), and for a time through the Uralian Ocean, which formed a strait in the late Paleozoic before closing to produce the Uralian Mountains. The highly specialized lineages that were characteristic of the ocean shores and epicontinental seaways bordering the Tethyan region did not penetrate into this gulf, where the fauna contained only cosmopolitan forms that appear to have been somewhat more generalized (Rudwick and Cowen, 1968; Stehli, 1970). The most diverse fauna known from the gulf, and the one which most closely resembled the Tethyan associations, is found on Novaya Zemlya, which lies just at the northern end of what was the Uralian Strait (Stehli, 1970). It is likely that the circulation pattern in the gulf was estuarine, with a net outflow of water from seasonal precipitation and runoff, and that currents along the north European and Asian coasts ran seaward. In this event, a northward drift of relatively warm water through the Uralian Strait is likely, perhaps creating a relatively stable area where it debouched into the gulf.

Late Paleozoic Faunas Were Showy, Whereas Early Triassic Faunas Were Grubby

We have tentatively ascribed the underlying cause of Permo-Triassic extinction to an increase in trophic resource fluctuations arising from greatly enhanced continentality. Of course, even if this is correct, the proximal causes of extinction must have been very different from group to group. The well-documented lowering of sea level that accompanied the close of the Paleozoic certainly had a major effect, for the reef communities were stranded and did not become re-established, presumably owing to unstable environmental conditions. The reefs contained a major share of the specialized, showy lineages of the time.

The lineages that did survive the Permo-Triassic reduction in diversity should have been generalized and opportunistic for the most part according to our model, and well adapted to unpredictable, fluctuating conditions. The trophic and habitat specialists, and any other lineages that depended upon resource stability, would be at a considerable disadvantage and would tend to be eliminated unless favored by some unusual adaptive trait. The communities resulting from this extinction, then, should have low diversity and should contain a high percentage of generalized feeders and ubiquitous and cosmopolitan forms.

It has long been known that the extinctions and survivals of the Permo-Triassic fit this model rather well. The disappearance of lineages with unusual adaptations, as indicated by their elaborate or distinctive skeletons, has been used as evidence for several theories of evolution and extinction, including the idea that these unusual forms indicated racial senescence, and that the lineages died out because of racial old age. A far more respectable variety of this view is that the unusual forms were so highly specialized that they were unable to evolve to meet the requirements of a changing environment, and therefore became extinct. A third view is that the extinctions resulted from a shift from a high- to a low-diversity environmental state, and the specialized forms were differentially eliminated because they were the least fit for life in a low-diversity environment. The second view is roughly equivalent to extinction by diversity-independent factors, and the third to extinction by diversity-dependent factors.

The Early Triassic fossil assemblages are dominated by bivalves of generalized appearance that are rather ubiquitous, occurring in a variety of facies, and also are highly cosmopolitan, being found all over the world. Gastropods and inarticulate brachiopods also survive with relatively little extinction. Articulate brachiopods, which were the dominant forms in so many Paleozoic communities, were heavily reduced in diversity across the Permo-Triassic boundary (Fig. 9-6). The pattern of extinction is puzzling at first, for seemingly unspecialized lineages as well as aberrant lineages disappear. It is easy to account for the extinction of the reef-dwelling forms, but some very common types, such as the productids (Chapter 5), which include reef species but are widely represented in nonreef environments, disappear completely. Richard Cowen has pointed out that the survivors mostly have one thing in common: they are pediculate. Perhaps the productids, many of which are semi-infaunal, were highly specialized by habitat, which proved fatal when generalized habits came to be at a premium. Taxa with pedicles would probably include lineages that were relatively tolerant of different substrate textures, at least to a far greater extent than semi-infaunal forms, which are closely pressed to the substrate.

During the Mesozoic and Cenozoic, erect colonial skeletons are re-evolved by the Cheilostome Ectoprocta, following the extinction of Paleozoic lineages with such structures. The Paleozoic corals were practically extinguished, but another lineage, the Scleractinia, appears for the first time in the Triassic. It eventually develops a wide variety of colonial types, including elongate branching forms such as the staghorn corals that rise well off the bottom, and the extensive sheets, fronds, and reticulations of other tropical forms. In both these cases, the appearance of the new stock is sudden, and obvious immediate ancestors are lacking. In fact, these two phyla probably do not have ground plans that were evolved in coadaptation with skeletons, and it seems likely that the sudden appearance of the new post-Paleozoic lineages is due to the advent of skeletonization in some formerly naked lineages, just as the original appearance of ectoprocts in the early Ordovician was unusually late for higher taxa, and probably corresponded not to the original evolution of these groups but to their first skeletonization. It is possible to imagine that the surviving early Triassic lineages of these groups had opportunistic adaptive strategies, and that in their subsequent differentiation we are seeing the development of specialization. Even among the ammonites, the successful Permo-Triassic passage was made by forms with rather plain shells, conveying the impression of generalists rather than specialists. However, the ammonites do pose a serious problem, for they differentiate rapidly and early in the Triassic, whereas many other groups were still reduced to only a few generalized lineages, yet they were probably predators and perhaps scavengers in part. Certainly their frequent extinctions and radiations suggest that they played a dangerous game of adaptation, and perhaps they were among the first lineages to take advantage of any fraction of the resources that were stabilized to any degree. Even considering the presence of numerous ammonites, the early Triassic faunas have a distinctly grubby appearance, recalling the early Cambrian communities and suggesting that even though their taxonomic compositions were vastly different on all levels, the Cambrian and Triassic faunas had similarities in adaptive strategy and community structure.

Continued northward movement of the North America-Europe-Asian land masses occurred during the Mesozoic, first as the result of a splitting of these northern continents ("Laurasia") from Gondwanaland, and then as a result of the splitting of South America and Africa from the Antarctic continent. Thus, at least two ridge systems promoted northward drifting: one around Antarctica, and one between Laurasia and Gondwanaland. However, Triassic and Lower Jurassic provincialism is very low or nonexistent, possibly because the species living in the more unstable environments, or adapted to exploit the unstable fractions of fairly stable environments, survived the Permian. The specialized lineages that were endemic to Tethyan and other stable ocean margins became extinct, and the resulting biota could live in all latitudes and became essentially world wide. This accords well with the evidence of a Permo-Triassic Pangaea, for tectonic barriers to dispersal do not exist for true Pangaeas because the continents form a contiguous landmass by definition. Tectonic disjunction of shelves may occur only if some fragments are not fitted into the continental assembly. Island arcs or midplate volcanoes may support isolated biotas, of course, and even along the shelves of a Pangaea, habitat disjunction is possible, as when two stretches of rocky coast are separated by a coastal plain of subcontinental extent. In general, however, the shelves are dispersal corridors, and only climatic barriers would ordinarily be effective in producing provinces.

The Permian biota displayed geographic differences that can be interpreted as provincial and based upon climate, with low diversity in the Permian Boreal Gulf. The extinction of the specialized elements in these provinces destroyed the biogeographic differences between them. Thus, although the Boreal Gulf may actually have been more distinctive in climate during the Triassic than during the Permian, the absence of a specialized element prevents the appearance of Triassic provinciality. We have already traced the appearance of another Boreal fauna in the Jurassic (Chapter 8), when endemics appeared in low-latitude biotas. Presumably these were chiefly thermophiles, and their numbers were permitted to increase because continued continental fragmentation resulted in increased stability of the water columns, with a concomitant rise in specialization and in the development of trophic strategies that require a certain resource stability.

Marine Faunas Modernized in the Late Mesozoic

Most of the dominant groups in the present tropics seem to have been differentiated from Tethyan progenitors. However, in high latitudes the Boreal province gradually developed a distinctive fauna that may have contributed some of the more important of the modern taxa. For example, benthic assemblages undergo a reorganization and modernization during the Middle Cretaceous, changing from Mesozoic to Cenozoic aspects, although there are exceptions to this trend. For example, the change is especially notable among the Gastropods. Paleozoic marine gastropods were, for the most part, browsers and probably detritus feeders, which is probably one reason that their passage of the Permo-Triassic boundary was relatively undisturbed. Even Lower Cretaceous gastropod assemblages are composed chiefly of lineages that, judging by their recent descendants and allies, were also herbivores and detritus feeders. Thus, they represented the lowest heterotrophic level, and included both consumers and recuperators. The adaptive types that dominate these assemblages, such as cerithiid gastropods, lend the associations a primitive cast. Although carnivorous gastropods seem to have existed, taxa that were probably carnivorous are generally rare in the record. This is in marked contrast to Late Cretaceous gastropod assemblages, which contain large numbers of species (chiefly Neogastropoda) with modern allies that are first-, second-, and even third-level predators, scavengers, and parasites. Even the Mesogastropods produced a voracious carnivore lineage in the Polinicinae (Sohl, 1969), a shell-boring group. Indeed, the feeding-type ratios among Late Cretaceous gastropods are quite similar to those of today, when similar facies are compared. Furthermore, the association of adaptive types in Late Cretaceous gastropod assemblages has a modern aspect. The gastropod fauna was modernized in a relatively short time during the mid-Cretaceous. The modernization appears to have begun in temperate latitudes and spread into the tropics, for tropical assemblages retained their primitive aspects longest.

The rise of these diverse feeding types on higher trophic levels coincides with a major transgression in the Late Cretaceous, and also probably with the establishment of the Circum-Antarctic Current and consequent cooling of high southern latitudes, and also with the opening of the North Atlantic. Perhaps the growing stability in intermediate and high latitudes permitted diversification of specialized feeders, especially predators, and the gastropods radiated to take advantage of this opportunity. However, many questions concerning these events remain unanswered, as is true of all these sorts of paleobiological happenings. Why did the radiation not begin in the tropics? Was stability there already so high that few new opportunities were created? If so, what organisms were exploiting the trophic resources at these levels? It is likely that such questions can be answered in time.

The end of the Mesozoic was not characterized by a major diversity decline of any duration, although there was a biotic turnover that modernized the assemblages further. The extinction of the ammonites, which serve as the major biostratigraphic tool among the macroinvertebrates, and the essential extinction of the planktonic foraminifera, which serve in a similar capacity among microfossils, have very much impressed paleontologists, who find their most useful guide fossil groups disappearing at about the same time. Important families of sessile benthic taxa also disappear, such as the rudistids (cemented coralloid bivalves) and the inoceramids (large, epifaunal, oysterlike bivalves). The inoceramids are widely used in biostratigraphy; therefore, their disappearance is especially noteworthy. Several families of echinoids also die out—vagrant benthic types. Extinctions occur in other taxa, as well, so that the end of the Cretaceous was considerably higher in extinctions than average.

Hancock (1967) has reviewed the record in detail and shows that the extinctions were usually preceded by a shrinking of the geographic range, and that they occurred in a rather piecemeal fashion over some period of time; thus, they cannot have been caused by any sudden catastrophe. It is difficult to imagine any single environmental change that can account for the extinction of the ecologically and taxonomically diverse groups. Temperature has been suggested frequently. Certainly a general regression occurs near the close of the Cretaceous, which might well account for the extinction of many reef lineages (see Newell, 1971), but why this would eradicate the ammonites, which had weathered so many inclement periods before, is not clear. Qualitative or quantitative changes in trophic resources may be invoked to explain some of the extinctions, but they do not explain them all, and, in fact, require special pleading for many groups that are relatively unaffected. Although the extinctions seem to be real, they are occurring during a time of general diversification and are totally masked in the rising diversity curves; they can hardly be related to the long-term operation of diversity-dependent factors. Perhaps the answer lies in a combination of climatic changes (presumably cooling) with a temporarily lowered capacity for diversity, owing to the Late Cretaceous regression. Certainly, the Mesozoic Cenozoic faunal change is enigmatic.

Late Cenozoic Glaciation Was a Relatively Minor Event in the History of the Marine Biosphere

During the Cenozoic, the major trend in the evolution of ecological structure was the rise in provincialism, described previously. Many tropical communities live in very stable environments and have achieved a degree of diversity and concomitant specialization of their members that probably surpasses that of the reef associations of the Permian. At the same time, the communities in high latitudes face major fluctuations in solar energy and nutrient resources. Species in these communities display a breadth of adaptation that must rival or surpass any required during the entire history of the Metazoa. The glacial stages of the Late Cenozoic, through which the world has just passed or, more likely, is still passing, have greatly altered the temperature distribution in the oceans and also must have affected salinities, therefore changing all the many environmental parameters that partly depend upon these two. These events certainly caused numerous biotic changes. Nevertheless, they are especially noteworthy for what they did *not* cause, and deserve some attention for this reason.

Figure 10-18 depicts the general sequence of oxygen isotopic changes measured in planktonic foraminifera recovered from deep-sea cores. The cores are

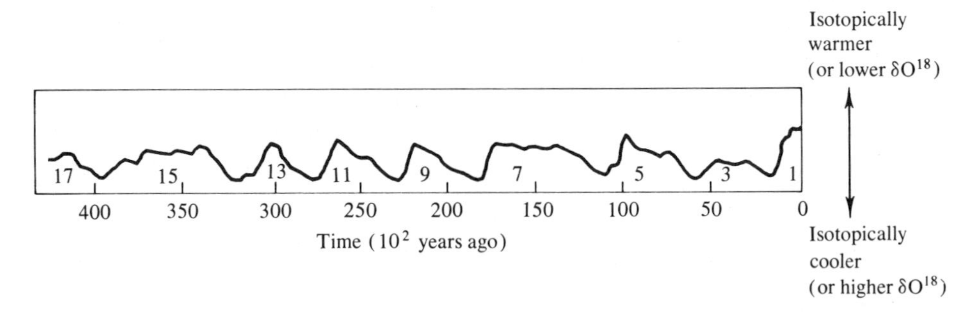

Fig. 10-18 Oxygen isotope paleotemperature curve for low latitudes in the North Atlantic during the last 450,000 years. (After Emiliani, 1966; from *The Journal of Geology*, the University of Chicago Press, copyright © 1966 by the University of Chicago.)

dated radiometrically at certain horizons, and then the ages are extrapolated to other parts of the core, using such criteria as estimated sedimentation rates; they are then correlated from core to core. Because oxygen isotopic ratios in the foraminifera are due not only to temperature but also to variation in oceanic isotope ratios, and because these ratios have obviously been altered to some extent during the great transfers of water from sea to land and back during glacial and interglacial intervals, the absolute temperature significance of the fluctuations is in dispute. Nevertheless, the curves do indicate fairly wide environmental fluctuations that correlate with glaciations and with associated climatic events, occurring rather irregularly during the Late Cenozoic. During warmer intervals, many warm-water organisms extended their ranges to higher latitudes, whereas during cooler intervals, the ranges of these forms were contracted and the ranges of many high-latitude species were extended equatorward. Thus, the climatic fluctuations were characterized by large numbers of range changes, bringing into association species that had previously been isolated from each other, in various patterns.

Epicontinental marine fossils of the Pleisotocene are known from throughout the world. They are chiefly found on marine terraces that roughly parallel the present shorelines. Inasmuch as the cooler intervals generally correspond with times of glaciation, they generally correspond also with low sea levels, because much ocean water is locked up in glacial ice. On the other hand, warmer intervals correspond with high sea stands because the glaciers melt, raising the sea level. Therefore, most of the faunas exposed on marine terraces above sea level seem to have lived at temperatures either much like those of today or warmer than present sea temperatures in their regions. However, uplift has locally exposed marine sediments that were deposited at sea levels below the present sea stand, and these contain fossil associations with species that live in relatively cool water today. By piecing together the faunal sequence in a region, it is possible to erect models to explain the climatic sequence there.

The Mediterranean Sea is a classic region for Pleistocene marine fossils. Some of the assemblages include warm-water elements that are found living today off tropical west Africa, and these occur chiefly in shallow-water associations. Other assemblages include species that live today only well to the north, in the North Sea and Arctic Ocean, and these appear to be chiefly offshore associations that have been elevated tectonically. In each case, there also occur numbers of temperate types of species that are similar or identical to those living in the Mediterranean at present. Some Pleistocene assemblages are composed entirely of these and include no exotic elements.

Mars (1963) has developed a historical interpretation of the Mediterranean Pleistocene that takes account of all these facts and relates them to a hydrographic-climatic model. It focuses upon conditions at the Strait of Gibraltar, the key point through which marine immigrants enter the Mediterranean and where all marine exchange with the Atlantic Ocean occurs. The growth and maintenance of a continental ice sheet in northern Europe would be accompanied by high precipitation and by a southward depression of climate, which would bring coolness and cloudiness to the Mediterranean region. The result of these changes would be to increase the precipitation relative to evaporation, and if precipitation and runoff exceeded the evaporation, which is likely, then the present flow pattern through the Strait of Gibraltar would have been reversed. Surplus surface water would have flowed out from the Mediterranean into the Atlantic, with a reverse flow at depth over the sill, a condition comparable to the Dardanelles at present. With the cooling of Atlantic water temperatures, cool-water species would extend their ranges south and numbers of them must have found their way into the Mediterranean, perhaps via transport at some depth in the inflowing current, which would furnish a supply of water in which they were viable. During intervals between glaciations, the rise in temperature and decrease in precipitation would return the hydrographic regime to its present conditions, and the northern species, cut off from a supply of cool water and caught in a warming basin of increasing salinity, would become locally extinct. With continued warming, surface temperatures along northwestern Africa would rise to permit the northward migration of some shallow-water species now limited to the south by thermal barriers, and these could easily penetrate the strait on the surface inflow and spread across the Mediterranean littoral and shallow sublittoral.

All over the world, similar faunal shifts have occurred, locally complicated by local hydrographic changes accompanying the climatic changes. Populations of many species have disappeared from regions where for a time they dominated one or more communities. The biogeographic changes imply associational changes as well; immigrating populations must have been confronted with different patterns of competition and predation and perhaps with somewhat different requirements for reproduction. Despite this, very few of them, and very few of the species with which they became associated, have become extinct. The Pleistocene extinction rates among invertebrates do not appear to be in any way exceptional, and these impressive environmental fluctuations have not had unusual effects on the structure or composition of the marine biosphere.

Perhaps the reasons for the lack of significant effect of Pleistocene glaciations revolve around the fact that the environmental effects were only variations on an established theme. There is not a major change in the structure of the environment but rather a series of oscillations that carry the present structure to greater or lesser extremes. Once the highest latitudes have reached freezing, the water can get no colder and further climatic deterioration will only cause the cool temperature to spread more widely equatorward. This compresses climatic zones and forces them into low latitudes, but there are no truly new, major climatic regimes created; so long as some low-latitude regions retain their tropical climates, no truly major climatic regimes are eliminated. Therefore, sanctuaries are available for displaced species. To be sure, the new regimes are not precisely like the old ones following poleward cooling, but they are similar. Even during the height of glaciations, the tropics were sufficiently stable and warm that, despite local killing of reefs and displacement of other communities owing partly to sea-level fluctuations, the rich reef communities have survived relatively intact.

The major historical changes that have swept the marine biosphere in the past must have been due to structural reorganizations, fundamental changes in environmental quality, and the appearances of entirely new environmental conditions and/or the eliminations of old ones. These observations are grounds for a strong argument that the biotic revolutions in past oceans are related to environmental restructuring and ecological repatterning. The most likely source for such changes lies in the changing geographic patterns of continental dispersion, fragmentation, assembly, submergence and emergence, and oceanic fragmentation and interconnection.

Pelagic and Deep-sea Ecosystems Must Have Had Different Structural Trends from Shelf Ecosystems

We have applied the model of environmental regulation of ecosystem structure and function almost entirely to the biota of the world's shelves, since these are the best-known marine associations today and in the fossil record. Unfortunately, fossil data are quite scarce for both the pelagic and the deep-sea realms for much of geological time, so that there are many questions to be answered about the course of development of their biotic and taxonomic structures even before explanations can be offered. However, if the processes of the model work for the shelf, they should work for these other realms as well. But using the model in a deductive manner for these realms leads to the conclusion that their histories have been rather different from that of the shelf, owing to their different circumstances.

In the pelagic realm today, the main water masses each support a characteristic biotic association, which is patchily but broadly distributed, and which lacks some species that are found in adjacent water masses and contains some that are not. Although each water mass has characteristic physical parameters and a characteristic regime of nutrient supply and solar radiation, owing to water conditioning, mixing, and local climatic parameters, each is also a highly homogeneous habitat with the chief environmental variations occurring with depth. There is a distinct vertical zonation of organisms, but systematic horizontal variations in species associations that are comparable to the communities of the shelf benthos are not known within water masses, although some species do appear at water-mass boundaries. Rather, each water mass comprises a community, at least at any given depth. Because the water masses are of suboceanic dimension, they tend to have different climatic regimes, and, in fact, correspond in dimension and in the nature of biotic differences to the provinces of the shelf realm. In effect, community and provincial levels are equivalent in the pelagic realm; the ecological hierarchy there contains one level less than on the shelf. However, it is possible to regard the water column as containing a vertical array of communities, although, again, the analogy with the shelf is poor. Variation in environmental partitioning in the pelagic realm, then, would occur primarily when water masses are created or disappear, or when vertical environmental gradients are lessened or enhanced, or when trophic resources become more or less stable.

According to the ecological model, pelagic communities in regions of high or fluctuating productivity should be of lower diversity, whereas those in stable, nutrient-poor regions should be of higher diversity. This pattern is, in fact, well known among pelagic organisms (Ryther, 1969). Now, during times when the continents are gathered into relatively compact masses, oceanic regions are especially broad; during the Permo-Triassic, for example, the Tethyan-Pacific tropical belt extended perhaps 17,000 miles longitudinally. Although coastal waters around supercontinents were probably quite unstable and should have supported relatively few, although very large, pelagic populations in the neritic zone, the vast stretches of open ocean lying beyond significant continental influence must have been superoceanic, with highest stability and low nutrient levels. Therefore, the ecosystems there should have been of the highest diversity and degree of specialization for this ecological realm. In general, then, the trends of change in biotic structure in oceanic and shelf communities should have been in opposite directions, insofar as the primary effects of continental assembly are concerned.

When there are supercontinents, there should be a gradient in environmental stability associated with the continentality which is in addition to the latitudinal gradients arising from variations in solar radiation. Stability from a continental effect will tend to vary from low nearshore to highest antipodal to the center of the supercontinent. In theory there should be a zone encircling the supercontinent, landward of which the effects of the continental assembly result in lower diversity, and oceanward of which the effects result in higher diversity, when compared with a condition involving small, fragmented, scattered continents.

There is a second major cause of variation in the effects of continentality, besides the state of assembly of continents, and that is the extent of epicontinental seas. During highly transgressive episodes the shrinkage in land area must result in an increasingly maritime global climate. Extensive regression would expose more land surface, with its lower specific heat, and create greater seasonal fluctuations in temperature, in storminess, and in associated factors. This effect would decrease away from the continents, but regressions would never act to actually increase stability within the oceanic environment. Of the two major determinants of effective continentality, namely the degree of continental assembly and the degree of transgression, the first affects the stabilities of oceanic and coastal waters in opposite directions, while the second affects them in the same direction.

Clearly, oceanic and coastal plankton communities may be very different in diversity even at comparable latitudes, as they often are today, especially where upwelling or other environmental perturbations are common near the coast. Disparities between the fossil records of oceanic and shelf communities, then, might well arise from this situation. If the diversity or composition of the fauna in these separate realms were interpreted as due to climate, they would appear to be contradictory.

Provinciality within the pelagic realm would be expected to be in phase with shelf provinciality. Increasing latitudinal thermal gradients would tend to speed the average current flow and to make the water masses more distinctive. Presumably, higher endemism would result from this increased environmental discontinuity between water masses. Decreasing gradients would lower the strength of discontinuities and permit more cosmopolitan distributions, Another effect of a high latitudinal thermal gradient would be a steeper vertical thermal gradient, which might enhance the vertical zonation and thus permit more species to inhabit a given water column (Lipps, 1970). This would partly compensate for the lower diversities expected owing to an increase in general vertical instability, which would tend to accompany high vertical thermal gradients. Longitudinal pelagic provinciality should also follow the same trends as shelves, for as continents fragment the oceans created between the daughter continents provide the opportunity for the evolution of endemics. Of course, provinciality in the pelagic realm will never be as high as in the continental shelves. Nevertheless, times of high provinciality should be times of highest overall diversity in the pelagic environment, but not times of highest diversity within individual pelagic provinces.

The deep-sea environment is so poorly known that the pattern of distribution and association there is still obscure. There seems to be no special reason to suppose that the deep sea was not invaded by higher metazoa during the late Precambrian or soon thereafter. The high stability and low nutrient supply of this region would favor high diversity from the start. According to Sanders and Hessler (1969), the major biotic changes are vertical, and perhaps a series of vertical associations on the community level will eventually be recognized. Ridges and other prominences within the deep sea create dispersal barriers; therefore, there is some endemism within deep basins.

The ecological basis of the unusually high diversity of deep-sea associations has not been satisfactorily explained. The deep-sea habitat is relatively homogeneous, and many deep-sea metazoans are generalized feeders that ingest detritus together with small living organisms, a habit termed *cropping* (Dayton and Hessler, 1972). Extensive niche partitioning of food or habitat is therefore not apparent. Dayton and Hessler (1972) suggest that constant cropping limits population sizes, permitting numerous populations to coexist on a given energy base. However, this type of unselective predation is diversity-independent and therefore in theory should not provide an adequate basis for maintaining high diversity levels. Natural selection is expected to circumvent its effects through modifications in the reproductive patterns of some of the populations, permitting them to expand their numbers relative to others. Indeed, it may well be that the mechanism of deep-sea partitioning involves reproductive specializations. The solution to this problem should have an important bearing on ecological theory.

According to the model of resource regulation of diversity, the highest diversity is expectable in the center of the largest (or most stable) basin, and diversity should decline as continents are approached. Thus, the diversity trends in deep-sea ecosystems should generally run counter to diversity trends within shelf communities. The Permo-Triassic deep-sea fauna from the center of the Tethyan-Pacific Ocean may have been of exceptional diversity. Total deep-sea diversity may be highest when there are the most separate ocean basins and thus greatest provinciality, although diversity within any given basin would not be at a maximum at such times.

The Invertebrate Fossil Record May Be Interpreted as an Interplay Between Evolutionary Trends and Environmental Fluctuations

From what we can reconstruct of the record of ancient marine life and ancient marine environments, the larger advances that have marked the development of the present varied array of organisms have resulted from a concomitance of major environmental opportunities with major evolutionary potentials. These circumstances have coincided episodically and have led to a history of life that is divided into distinctive chapters and punctuated by unique events, despite a certain uniformity of process.

The environmental changes having the most profound effect on the evolution of new types of organisms are apparently those that cause changes in the energy budget of the biosphere. One of the fundamental adaptive goals of organisms is surely to secure energy adequate to support their various functions, each of which consumes energy. Consider feeding. An active predator must search out and capture its prey, often spending energy prodigiously when operating at high activity levels, and all of this must be repaid from the prey. Even sedentary organisms require a considerable energy flow, for suspension feeders must pump volumes of water over their food-gathering devices, and detritus feeders must be on the move to new sources of detritus. In addition to repaying the energy requirements of feeding itself, the food must sustain all the other metabolic functions, such as growth and reproduction. The energy intake must be apportioned by selection among the various activities in accordance with their relative contributions to the survival of the organism. It has been postulated that organs or habits that cease to be adaptive are discarded by selection as quickly as possible, in order that the energy they require may be put to positive use to improve adaptation elsewhere.

At any rate, when the trophic resource regime is altered, organisms are required to alter their manner of resource exploitation. Selection may accomplish this in many ways. The lineages may utilize different resources than formerly, so that the system of energy apportionment remains about the same although food-gathering habits and structures might require modification. If it is useful to develop new types of locomotory activities in order to utilize a new food supply then selection might well lead down a pathway to major modification of body architecture. Another way to change resource exploitation is to utilize identical or similar food, but to alter the internal energy apportionment so as to adapt to a new regime of food intake. For example, if food becomes more seasonal, some populations may adapt by altering their seasonal activities and perhaps by consolidating certain high-energy requirements into short periods appropriate to their new energy regimes, such as limiting reproduction to periods following high trophic resource supplies. Usually these sorts of changes would create waves of secondary and tertiary effects, as whole suites of coadapted functions are engineered through selection to achieve a new balance of efficiency and harmony.

Considering the fundamental importance of energy regime to selection, a simple (and perhaps simplistic) explanation for the first-order trends in the history of the marine biosphere, then, can be based upon occasional alterations in the planetary environment that have resulted in fundamental alterations in the trophic resource regime in the oceans. The abundance, stability, and the quality of food supplies are postulated to have altered from one environmental configuration to another. There appear to have been four major changes in the configuration of the energy regime of the biosphere which have had significant effects on the fossil record. The first, in the Late Precambrian, is highly speculative but is envisioned as a shift toward trophic instability, which was instrumental in the evolution of a metazoan infauna. The infaunal organisms employed a coelom as a hydrostatic skeleton, permitting entry into the substrate. The second major change occurred near the start of Cambrian time when a shift toward trophic stability permitted the emergence of an epifauna that diversified spectacularly in Cambro-Ordovician time. The third occurred in the Late Permian, when once again a general shift toward trophic instability culminated. At this time, the substrate already contained a large infauna, and a number of lineages existed that were adapted or preadapted to fluctuating conditions. Therefore, the chief result was to eliminate most of the lineages not so preadapted, and to lower the diversity of infaunal organisms as well. The absolute diversity in all or nearly all environments decreased, whereas the frequency distribution of adaptive types was greatly altered, but no radical new types appeared. The fourth major change was a general shift back toward stability, permitting the rediversification of the Mesozoic and Cenozoic.

This stability has persisted until the present at places in the tropics, although in high latitudes rather unstable and unpredictable shelf environments have developed. This rediversification was accompanied by the appearance of a few taxa in higher categories, presumably as replacements for lineages that were eliminated near the close of the Permian. It has resulted in new, high levels of diversity of species, genera, and families, owing partly to the extreme provincialization of the biosphere by the fragmentation of the continents and by the development of north-south continental configurations correlated with cooling poles and increasing latitudinal thermal gradients. The stability of the tropics and the continued development of new and complex taxa may have raised species diversities in the most stable communities to new highs, also.

The coelom, which originated sometime before it appears in the Ediacaran fauna and may have been present for well over 130 million years before the Cambrian, proved preadaptive to the development of a wide array of body plans employed in a wide range of modes of life. For a while it presumably retained its use as a skeletal organ for the burrowing function, although it seems to have radiated into several different subtypes, depending on the degree of compartmentalization required. Compartmentalization was high when peristaltic locomotory activity was prolonged (see Clark, 1964). In some lineages, the original function was altered to a swimming or running one. In other lineages, the locomotory function of the coelom was replaced by other organs, or the lineages became sessile in adult stages, and the coelom became for the most part simply a feature of the body architecture, functional in design as a receptacle for viscera and to provide ducts. In phyla such as the brachiopods and ectoprocts, it remains partly as a feeding organ, forming the hydrostatic skeleton of the lophophore. In the Echinodermata, the water-vascular system is coelomic, and therefore in echinoids the coelom is associated with a locomotory device—the tube feet—but one that operates on a principle entirely different from that in the early coelomate worms. These examples of the variety of coelomic structures could be multiplied indefinitely. The point is that the coelom has very likely originated as a hydrostatic skeleton to aid in peristaltic burrowing, and it is now employed in this way by some taxa. On the other hand, it is found in a wide variety of taxa wherein it has entirely different functions. It has clearly been retained because it does function in these taxa, and it was such a successful evolutionary development because it combined great architectural and functional utility. At any rate, the possession of a coelom has certainly not hindered lineages from taking up modes of life that differ radically from the early coelomates, although the coelom has remained in many cases as a fundamental part of their ground plans.

This pattern is very common in phyletic evolution, that a structure developed for one function is employed in many others for which it proves preadaptive, or is modified so as to serve other functions. Although it is certain that a ground-plan can impose important constraints on the regions of biospace that can be occupied by a taxon, the ability of taxa to modify their ecospaces is astonishing, and representatives of two taxa that were originally evolved as occupants of two distinctive regions of biospace are commonly found living together in a third region. Consider a subtidal rock encrusted with marine life. One of the chief feeding methods represented will be suspension feeding, which may be carried on by numbers of species that represent a wide range of taxa in higher categories: gastropods, bivalves, cirrepedes, sponges, tunicates, ectoprocts, and brachiopods, as an incomplete example. To what degree the available food is partitioned between these forms is not certain; a review suggests that their feeding tends to be unspecialized (Jørgensen, 1966). At any rate, the partitioning of food between two suspension feeders sharing the same rock seems to be about the same whether they belong to the same genus or to distinct phyla. Thus there is a great array of distinctive adaptive types doing essentially the same thing. To be sure, some of them have vastly different feeding structures and employ different principles to obtain their food, but their functional convergence is remarkably complete. Although some of the ground plans that are represented have been originally evolved to function as sessile suspension feeders, others have definitely not. This suggests that if any of these taxa were eliminated, even those such as ectoprocts, which seem to have originally evolved as suspension feeders, the portion of biospace that they occupy could be quickly filled by other lineages, belonging perhaps to such groups as arthropods or mollusks, which were evolved for and owe their basic ground plans to adaptations for very different sets of functions. Thus, as radiation succeeds radiation and as the ratio of species to phyla grows, the ecospaces of many taxa grow, spreading out through biospace like tentacles as they reach down the pathways of environmental opportunity, until the ecospaces of the major taxa, taxa which were originally evolved to occupy distinctive parts of ecospace and reflect this in their ground plans, nevertheless become hopelessly intertwined like a basket of multidimensional octopuses.

The taxonomic results of repeated diversifications and extinctions have varied through time, partly because the events that caused them have had somewhat different qualities, but chiefly because the great biotic revolutions have occurred each time in a different biosphere. The occurrence of identical events may have vastly different consequences if they occur in vastly different situations, and the configuration of the biosphere has certainly changed in fundamental ways over geological time. The very early development of cells, of heterotrophic organic nutrient feeders and photosynthesizers, all certainly occurred in a biosphere that could not have sustained animal life as it exists today. Conditions surrounding the appearance of the heterotrophic Protozoa and the early Metazoa are not known, but they probably approached those of the Phanerozoic. The very early radiations, if such they were, or at least the early major developments, generated groups of organisms that now form kingdoms or subkingdoms. The early Metazoan radiation that we postulate for the coelomates, in whatever form it may have taken, established stocks that are now differentiated into groups of allied phyla, or perhaps in some cases into separate solitary phyla. In the Cambro-Ordovician radiation, we may be witnessing the origin of most if not all of the remainer of the metazoan phyla. As biospace became more widely and densely populated, any new radiation would be most easily developed by modification of pre-existing lineages, of which there was an ever-growing supply in increasing ecological variety. Thus, subsequent major diversifications during the Paleozoic occurred either during or shortly after extinctions (Newell, 1967), and resulted in the appearance of a gradually decreasing number of taxa in higher categories through time.

It therefore seems likely that early diversifications were far more likely than later ones to produce new ground plans, just as early extinctions were more likely than later ones to eliminate taxa in higher categories. Furthermore, the average degree of complexity of organization of organisms tends to increase through time. This is due to the fact that selection bases solutions to adaptive problems on pre-existing organisms, and that the ensuing modifications commonly involve an elaboration of the older stock. Furthermore, during extinctions, the highly organized, complex stocks are perfectly capable of becoming adapted to, say, fluctuating environments and to life in ecosystems of low diversity. Therefore, the extinctions do not reduce the average level of phyletic complexity, but merely pave the way for its increase or actually cause its increase by subjecting organisms to new adaptive challenges. Even though complex stocks may take on simple roles for a time, the information on which their complexity is based is safely coded in their DNA and is conserved; we thus have a sort of law of conservation of organization, as aptly described by Stebbins (1966).

Although the earlier major radiations seem to have created taxa in the higher categories, they, of course, produced numerous taxa in lower categories as well. Probably taxa that are ranked in lower categories were appearing more or less constantly, owing to the isolation of gene pools leading to speciation and the origin of species groups that may form genera or families. The rates of this sort of low-level cladogenesis must have varied greatly in time, and such speciation is certainly proceeding today in isolated or changing environments.

All this probably means that the average ability of organisms to become highly specialized and to inhabit narrower and narrower ecospaces is continually enhanced. If the communities in randomly sampled standardized environments, wherein the diversity-dependent factors are held constant, are compared over eras of time, then there is expected to be a gradual increase in average diversity, which would probably be most evident in the more stable environments.

Survivors of various periods of extinction are likely to be those lineages that became the best adapted to the "worst" of whatever conditions were eliminating species. Therefore, we should expect that in marine shelf habitats those "living fossils" that date from the Paleozoic, before the Permo-Triassic extinctions, would be adapted to fluctuating environments. Paleozoic marine survivors from stable habitats would be expected chiefly in the deep sea, although some may have weathered the Permo-Triassic events on island chains amid the superocean. *Lingula* is a classic shallow-water holdover from the Paleozoic, and it is well adapted to unstable environments. "Living fossils" from shallow water that arose in the Mesozoic or Cenozoic may be adapted to either stable or fluctuating environments. A marine example from the later Mesozoic is the Australian bivalve *Neotrigonia*, which happens to live in both stable and fluctuating environments today.

In summary, the evolution of both marine invertebrates and marine environments has proceeded hand-in-hand, and the history of their interrelation is recorded in the progressive and episodic changes in the ecological and taxonomic hierarchies, as represented by marine fossils. A sort of moving picture of the biological world with its selective processes that favor increasing fitness and that lead to "biological improvement" is projected upon an environmental background that itself fluctuates, with major changes in configuration occurring over long periods or at long intervals, and with what is probably a spectrum of shorter-term changes superimposed thereon. The resulting ecological images expand and contract, but, when measured at some standardized configuration, have a gradually rising average complexity and exhibit a gradually expanding ecospace.

REFERENCES CITED

ABELSON, P. H. 1957. Organic constituents of fossils. *In* R. S. Ladd [ed.], Treatise on marine ecology and paleoecology. Geol. Soc. Am. Mem. 67, 2: 87–92.

———. 1966. Chemical events on the primitive earth. Proc. Nat. Acad. Sci. U.S. 55: 1365–1372.

AGER, D. V. 1963. Principles of paleoecology. McGraw-Hill Book Company, New York. xi + 371 p.

ALLEE, W. C., A. E. EMERSON, O. PARK, T. PARK, and K. P. SCHMIDT. 1949. Principles of animal ecology. W. B. Saunders Co., Philadelphia. xii + 837 p.

ALLEN, J. A. 1958. On the basic form and adaptations to habitat in the Lucinacea (Eulamellibranchia). Phil. Trans. Roy. Soc. London, Ser. B 241: 421–484.

ARKELL, W. J. 1956. Jurassic geology of the world. Oliver and Boyd, Edinburgh. xv + 806 p.

ARNAUD, P. M. 1970. Frequency and ecological significance of necrophagy among the benthic species of antarctic coastal waters. *In* M. W. Holgate [ed.], Antarctic Ecol. 1: 259–266.

Atkins, D. 1960. The ciliary feeding mechanism of the Megathyridae (Brachiopoda), and the growth stages of the lophophore. J. Marine Biol. Assoc. U.K. 39: 459–479.

Axelrod, D. I. 1967. Quaternary extinctions of large mammals. Calif. Univ. Pub. Geol. Sci. 74: 1–42.

Axelrod, D. I., and H. P. Bailey. 1968. Cretaceous dinosaur extinction. Ecology 22: 595–611.

Baggerman, B. 1953. Spatfall and transport of *Cardium edule* L. Arch. Neerl. Zool. 10: 315–342.

Bandy, O. L. 1960. General correlation of foraminiferal structure with environment. Rept. Intern. Geol. Congr., 21st Sess. 22: 7–19.

Barghoorn, E., and J. W. Schopf. 1966. Microorganisms three billion years old from the Precambrian of South Africa. Science 152: 758–763.

Barker, R. M. 1964. Microtextural variation in pelecypod shells. Malacologia 2: 69–86.

Barnes, H., and H. T. Powell. 1950. The development, general morphology and subsequent elimination of barnacle populations, *Balanus crenatus* and *B. balanoides*, after a heavy initial settlement. J. Animal Ecol. 19: 175–179, 3 Pls.

Barrell, Joseph. 1917. Rhythms and the measurements of geologic time. Bull. Geol. Soc. Am. 28: 745–904.

Barrington, E. J. W. 1967. Invertebrate structure and function. Houghton Mifflin Company, Boston. x + 549 p.

Beckner, Morton. 1959. The biological way of thought. Columbia University Press, New York. viii + 200 p.

Beedham, G. E. 1958. Observations on the non-calcareous component of the shell of the Lamellibranchia. Quart. J. Microscop. Sci. 99: 341–357.

Beerbower, J. R. 1968. Search for the past, 2nd ed. Prentice-Hall, Inc., Englewood Cliffs, N. J. xiii + 512 p.

Beerbower, J. R., and Dianne Jordan. 1969. Application of information theory to paleontologic problems: taxonomic diversity. J. Paleontol. 43: 1184–1198.

Benzer, S. 1962. The fine structure of the gene. Sci. Am. 206, 1: 70–84.

Berkner, C. V., and L. C. Marshall. 1964. The history of growth of oxygen in the earth's atmosphere, p. 102–126. *In* C. J. Brancuzio and A. G. W. Cameron [eds.], The origin and evolution of atmospheres and oceans. John Wiley & Sons, Inc., New York.

Beurlen, K. 1965. Der Faunenschnitt an der Perm-Trias Grenze. Zeitsch. deutsch geol. Ges. 108: 88–89.

Birkett, L. 1959. Production in benthic populations. International Council for Exploration of the Sea: Near Northern Seas Committee 42: 1–12.

Bonner, J. T. 1965. Size and cycle, an essay on the structure of biology. Princeton University Press, Princeton, N.J. viii + 219 p.

Bott, M. H. P. 1965. Formation of ocean ridges. Nature 207: 840–843.

Boucot, A. J. 1968. Origins of the Silurian fauna. Geol. Soc. Am., Progr. with Abstracts, 1968 Meetings, p. 33–34.

Bowen, Robert. 1966. Paleotemperature analysis. American Elsevier Publishing Co., Inc., New York. x + 265 p.

Bramlette, M. N. 1965a. Massive extinctions in biota at the end of Mesozoic time. Science 148: 1696–1699.

———. 1965b. Mass extinction of Mesozoic biota. Science 150: 1240.

Bretsky, P. W., Jr. 1969. Evolution of Paleozoic benthic marine invertebrate communities. Palaeogeogr., Paleoclimat., Palaeoecol. 6: 45–59.

———. 1970. Upper Ordovician ecology of the central Appalachians. Peabody Mus. Nat. Hist. Bull. 34: viii + 150 p.

Bretsky, P. W., Jr., and D. M. Lorenz. 1969. Adaptive response to environmental stability: a unifying concept in paleoecology. Proc. N. Am. Paleontol. Conv., Pt. E: 522–550.

———. 1970. An essay on genetic-adaptive strategies and mass extinctions. Bull. Geol. Soc. Am. 81: 2449–2456.

Brinkmann, R. T. 1969. Dissociation of water vapor and evolution of oxygen in the terrestrial atmosphere. J. Geophys. Res. 74: 5355–5368.

Broecker, W. S. 1970. A boundary condition on the evolution of atmospheric oxygen. J. Geophys. Res. 75: 3553–3557.

Brooks, J. L. 1950. Speciation in ancient lakes. Quart. Rev. Biol. 25: 131–176.

Brown, F. A. [ed.]. 1950. Selected invertebrate types. John Wiley and Sons, Inc., New York, xx + 597 p.

Brown, W. L., and E. O. Wilson. 1956. Character displacement. Syst. Zool. 5: 49–64.

Bruns, E. 1958. Ozeanologie. Bd. 1. VEB Deutsch. Verlag der Wissensch. Berlin. 420 p.

Bullard, E. C., J. E. Everett, and A. G. Smith. 1965. The fit of the continents around the Atlantic. Phil. Trans. Roy. Soc. London, Ser. A 258: 41–51.

Bullock, T. H. 1955. Compensation for temperature in the metabolism and activity of poikilotherms. Biol. Rev. Cambridge Phil. Soc. 30: 311–342.

Buzas, M. A., and T. G. Gibson. 1969. Species diversity: benthonic foraminifera in western North Atlantic. Science: 163: 72–75.

Carson, H. L. 1960. Genetic conditions which promote or retard the formation of species. Cold Spring Harbor Symp. Quant. Biol. 24: 87–105.

Carter, R. M. 1967. On Lison's model of bivalve shell form, and its biological interpretation. Proc. Malac. Soc. London (1967) 37: 265–278.

Chapman, G., and G. E. Newell. 1956. The role of body fluid in the movement of soft-bodied invertebrates. II. The extension of the siphons of *Mya arenaria* L. and *Scrobicularia plana* (da Costa). Proc. Roy. Soc. London, Ser. B 145: 564–580.

Chave, K. E. 1954. Aspects of the biogeochemistry of magnesium. 1. Calcareous marine organisms. J. Geol. 62: 266–283.

Cheatham, A. H., and J. E. Hazel. 1969. Binary (presence-absence) similarity coefficients. J. Paleontol. 43: 1130–1136.

Chia, F. A. 1970. Reproduction of arctic marine invertebrates. Marine Pollution Bull. 1: 78–79.

CISNE, J. L. 1971. The evolution of the taxonomic and ecological structure of aquatic free-living arthropods. Geol. Soc. Am., Abstracts with Programs 3: 525.

CLARK, L. R., P. W. GEIER, R. D. HUGHES, and R. F. MORRIS. 1967. The ecology of insect populations in theory and practice. Methuen & Co., Ltd., London. xiii + 232 p.

CLARK, R. B. 1963. The evolution of the coelom and metameric segmentation, p. 91–107. *In* E. C. Dougherty *et al.* [eds.], The lower metazoa, comparative biology and phylogeny. University of California Press, Berkeley and Los Angeles.

———. 1964. Dynamics in metazoan evolution, the origin of the coelom and segments. Clarendon Press, Oxford. x + 313 p.

CLEMENTS, F. E., and V. E. SHELFORD. 1939. Bio-ecology. John Wiley & Sons, Inc., New York. 425 p.

CLOUD, P. E. 1949. Some problems and patterns of evolution exemplified by fossil invertebrates. Evolution 2: 322–350.

———. 1959. Paleoecology—retrospect and prospect. J. Paleontol. 33: 926–962.

———. 1965. Significance of the Gunflint (Precambrian) microflora. Science 148: 27–35.

———. 1968. Pre-metazoan evolution and the origin of the metazoa, p. 1–72. *In* E. T. Drake [ed.], Evolution and environment. Yale University Press, New Haven, Conn.

COCKS, L. R. M. 1967. Depth patterns in Silurian marine communities. Marine Geol. 5: 379–382.

———. 1970. Silurian brachiopods of the superfamily Plectambonitacea. Bull. Brit. Museum, Geol. 19 (4): 141–203.

COLE, L. C. 1949. The measurement of interspecific association. Ecology 30: 411–424.

COLEMAN, P. J. 1957. Permian productacea of western Australia. Australia Bur. Mines, Mineral Resources, Geol. Geophys. Bull. 40, 147 p., 21 Pls.

CONNELL, J. H. 1961. The influence of interspecific competition and other factors on the distribution of the barnacle *Chthamalus stellatus*. Ecology 42: 710–723.

———. 1970. A predator-prey system in the marine intertidal region. I. *Balanus glandula* and several predatory species of *Thais*. Ecol. Monographs 40: 49–78.

CONNELL, J. H., and E. ORIAS, 1964. The ecological regulation of species diversity. Am. Naturalist 98: 399–414.

COOPER, L. H. N. 1952. Processes of enrichment of surface water with nutrients due to strong winds blowing on to a continental slope. J. Marine Biol. Assoc. U. K. 30: 453–464.

COOPER, L. H. N., and D. VAUX. 1949. Cascading over the continental slope of water from the Celtic Sea. J. Mar. Biol. Assoc. U.K. 28: 719–750.

COWEN, RICHARD. 1970. Analogies between the recent bivalve *Tridacna* and the fossil brachiopods Lyttoniacea and Richthofeniacea. Palaeogeogr., Palaeoclimat., Palaeoecol. 8: 329–344.

CRAIG, G. Y., and A. HALLAM. 1963. Size-frequency and growth-ring analyses of *Mytilus edulis* and *Cardium edule*, and their palaeoecological significance. Palaeontology 6: 731–750.

Craig, G. Y. and Gerhard Oertel. 1966. Deterministic models of living and fossil populations of animals. Quart. J. Geol. Soc. London 122: 315–355.

Craig, H. 1953. The geochemistry of stable carbon isotopes. Geochim. Cosmochim. Acta 3: 53–92.

Crimes, T. P. and J. C. Harper [eds.] 1970. International conference on trace fossils, University of Liverpool, 1970, Seel House Press, Liverpool, 547 p.

Croker, R. A. 1967. Niche specificity of *Neohaustorius schmitzi* and *Haustorius* sp. (Crustacea: Amphipoda) in North Carolina. Ecology 48: 971–975.

Crombie, A. C. 1945. On competition between different species of graminivorous insects. Proc. Roy. Soc. London, Ser. B 132: 362–395.

Crowell, J. C., and L. A. Frakes. 1970. Phanerozoic glaciation and the causes of ice ages. Am. J. Sci. 268: 193–224.

Curry, J. D. 1967. The failure of exoskeletons and endoskeletons. J. Morphol. 123: 1–16.

Danielli, J. F. 1954. Selective active transport of ions. Symp. Exp. Biol. 8: 1–14.

Darlington, C. P. 1958. Evolution of genetic systems, 2nd ed. Oliver and Boyd, Edinburgh. xii + 265 p.

Darlington, P. J. 1957. Zoogeography: the geographical distribution of animals. John Wiley & Sons, Inc., New York. 675 p.

Darwin, Charles. 1859. On the origin of species by means of natural selection, or the preservation of favoured races in the struggle for life. John Murray, London. ix + 502 p.

Darwin, Francis. 1887. The life and letters of Charles Darwin, including an autobiographical chapter. John Murray, London. 3 vol.

Dayton, P. K. 1971. Competition, disturbance, and community organization: the provision and subsequent utilization of space in a rocky intertidal community. Ecol. Monographs 41: 351–389.

Dayton, P. K., and R. R. Hessler. 1972. Role of biological disturbance in maintaining diversity in the deep sea. Deep-Sea Res. 19: 199–208.

DeBeer, Gavin. 1958. Embryos and Ancestors, 3rd ed. Clarendon Press, Oxford, xii + 197 p.

Deevey, E. S. 1947. Life tables for natural populations of animals. Quart. Rev. Biol. 22: 283–314.

———. 1950. The probability of death. Sci. Am. 182: 58–60.

Degens, E. T., M. Behrendt, B. Gothardt, and E. Reppmann. 1968. Metabolic fractionation of carbon isotopes in marine plankton. II. Data on samples collected off the coasts of Peru and Ecuador. Deep-Sea Res. 15: 11–20.

Degens, E. T., R. R. L. Guillard, W. M. Sackett, and J. A. Hellebust. 1968. Metabolic fractionation of carbon isotopes in marine plankton. I. Temperature and respiration experiments. Deep-Sea Res. 15: 1–9.

Degens, E. T., D. W. Spencer, and R. H. Parker. 1967. Paleobiochemistry of molluscan shell proteins. Comp. Biochem. Physiol. 20: 553–579.

Dewey, J. F. 1969. Evolution of the Appalachian/Caledonian orogen. Nature 222: 124–129.

DIETZ, R. S., and J. HOLDEN. 1970. Reconstruction of Pangaea; breakup and dispersion of continents, Permian to present. J. Geophys. Res. 75: 4939–4956.

DOBZHANSKY, TH. 1951. Genetics and the origin of species, 3rd ed. Columbia University Press, New York. x + 364 p.

———. 1970. Genetics of the evolutionary process. Columbia University Press, New York, ix + 505 p.

DOBZHANSKY, TH., and C. EPLING. 1944. Contributions to the genetics, taxonomy and ecology of *Drosophila pseudoobscura* and its relatives. Carnegie Inst. Washington Publ. No. 554: 1–46.

DODD, J. R. 1963. Paleoecological implications of shell mineralogy in two pelecypod species. J. Geol. 71: 1–11.

———. 1964. Environmentally controlled variation in the shell structure of a pelecypod species. J. Paleontol. 38: 1065–1071.

———. 1965. Environmental control of strontium and magnesium in *Mytilus*. Geochim. Cosmochim. Acta 29: 385–398.

———. 1967. Magnesium and strontium in calcareous skeletons: a review. J. Paleontol. 41: 1313–1329.

DOTY, M. S. 1957. Rocky intertidal surfaces, p. 535–585. *In* J. W. Hedgpeth [ed.], Treatise on marine ecology and paleoecology, 1, ecology. Geol. Soc. Am. Mem. 67, 1. viii + 1296 p.

DUNBAR, M. J. 1960. The evolution of stability in marine environments; natural selection at the level of the ecosystem. Am. Naturalist 94: 129–136.

DURHAM, J. W. 1967. The incompleteness of our knowledge of the fossil record. J. Paleontol. 41: 559–565.

———. 1971. The fossil record and the origin of the Deuterostomata. N. Am. Paleontol. Conv., Chicago, 1969, Proc., Pt. H, p. 1104–1131.

EASTON, W. H. 1960. Invertebrate paleontology. Harper & Row, Publishers, New York. 701 p.

EATON, T. H. 1970. The stem-tail problem and the ancestry of chordates. J. Paleontol. 44: 969–979.

EISMA, D. 1966. The influence of salinity on mollusk shell mineralogy: a discussion. J. Geol. 74: 89–94.

EKMAN, SVEN. 1953. Zoogeography of the sea. Sidgwick and Jackson, London. xiv + 417 p.

ELDREDGE, NILES. 1971. The allopatric model and phylogeny in Paleozoic invertebrates. Evolution 25: 156–167.

ELTON, C. S. 1927. Animal ecology. Sidgwick and Jackson, London. xx + 207 p.

EMILIANI, C. 1966. Paleotemperature analysis of Caribbean cores P6304–8 and P6304–9 and a generalized temperature curve for the past 425,000 years. J. Geol. 74: 109–124.

EPSTEIN, S., R. BUCHSBAUM, H. A. LOWENSTAM, and H. C. UREY. 1953. Revised carbonate-water isotopic temperature scale. Bull. Geol. Soc. Am. 64: 1315–1326.

EPSTEIN, S., and T. MAYEDA. 1953. Variation of O^{18} content of waters from natural sources. Geochim. Cosmochim. Acta 4: 213–224.

EVANS, F. G. C. 1951. An analysis of the behaviour of *Lepidochitona cinereus* in response to certain physical features of the environment. J. Animal Ecol. 20: 1–10.

FAGER, E. W. 1957. Determination and analysis of recurrent groups. Ecology 30: 586–595.

———. 1963. Communities of organisms. *In* M. N. Hill [ed.], The sea. Interscience Publishers, New York. 2: 415–437.

FAGERSTROM, J. A. 1964. Fossil communities in paleoecology: their recognition and significance. Bull. Geol. Soc. Am. 75: 1197–1216.

FISCHER, A. G. 1960. Latitudinal variations in organic diversity. Evolution 14: 64–81.

———. 1964. Brackish oceans as a cause of the Permo-Triassic faunal crisis. *In* A. E. M. Nairn [ed.], Problems in palaeoclimatology. Interscience Publishers, New York, xiii + 705 p.

FISCHER, D. H. 1970. Historians' fallacies. Harper & Row, Publishers, New York, xii + 338 p.

FISH, C. J. 1954. Preliminary observations on the biology of boreo-arctic and subtropical oceanic zooplankton populations, p. 3–9. Symposium on Marine and Fresh-water Plankton in the Indo-Pacific, Bangkok.

FISHER, R. A., A. S. CORBETT, and C. B. WILLIAMS. 1943. The relation between the number of species and the number of individuals in a random sample of an animal population. J. Animal Ecol. 12: 42–58.

FORBES, E. and R. GODWIN-AUSTEN. 1859. The natural history of the European Seas. John Van Voorst, London, viii + 306 p.

FORD, E. B. 1964. Ecological genetics. John Wiley & Sons, Inc., New York. 335 p.

FOTHERINGHAM, NICK. 1971. Life history patterns of the littoral gastropods *Shaskyus festivus* (Hinds) and *Ocenebra poulsoni* Carpenter (Prosobranchia: Muricidae). Ecology 52: 742–757.

FOX, W. T. 1968. Quantitative paleoecologic analysis of fossil communities in the Richmond Group. J. Geol. 76: 613–640.

FRETTER, VERA, and ALASTAIR GRAHAM. 1962. British prosobranch molluscs. Ray Society, London. xvi + 755 p.

FRYER, G., and T. D. ILES. 1969. Alternate routes to evolutionary success as exhibited by African cichlid fish of the genus *Tilapia* and the species flocks of the Great Lakes. Evolution 23: 359–369.

GARRELS R. M., and F. T. MACKENZIE. 1971. Evolution of sedimentary rocks. W. W. Norton and Company, New York, xvi + 397 p.

GAUSE, G. F. 1935. La théorie mathématique de la lutte pour la vie. Hermann et cie., Paris. 61 p.

GEORGE, T. N. 1948. Evolution in fossil communities. Proc. Roy. Phil. Soc. Glasgow 73: 23–42.

GERE, G. 1957. Productive biologic grouping of organisms and their role in ecological communities. Ann. Univ. Sci. Budapest. Rolando Eötvös Nominatae, Sect. Biol. 1: 61–69.

GHISELIN, M. T. 1969. The triumph of the Darwinian method. University of California Press, Berkeley. 287 p.

GHISELIN, M. T., E. T. DEGENS, D. W. SPENCER, and R. H. PARKER. 1966. A phylogenetic survey of molluscan shell matrix proteins. Breviora, No. 262, 35 p.

GLAESSNER, M. F. 1958. The oldest fossil faunas of South Australia. Geol. Rundschau 47, 2: 522–531.

———. 1969. Trace fossils from the Precambrian and basal Cambrian. Lethaia 2: 369–393.

———. 1971. Geographic distribution and time range of the Ediacara Precambrian fauna. Bull. Geol. Soc. Am. 82: 509–514.

GLAESSNER, M. F., and M. WADE. 1966. The late Precambrian fossils from Ediacara, South Australia. Palaeontology 9: 599–628.

GLYNN, P. W. 1965. Community composition, structure, and interrelationships in the marine intertidal *Endocladia muricata-Balanus glandula* association in Monterey Bay, California. Beaufortia 12: 1–198.

GOLDSCHMIDT, V. M. 1954. Geochemistry. Clarendon Press, Oxford, xi + 730 p.

GOOCH, J. L., and T. J. M. SCHOPF. 1970. Population genetics of marine species of the phylum Ectoprocta. Biol. Bull. 138: 138–156.

GOULD, S. J. 1965. Is uniformitarianism necessary? Am. J. Sci. 263: 223–228.

———. 1966. Allometry and size in ontogeny and phylogeny. Biol. Rev. 41: 587–640.

———. 1971. Geometric similarity in allometric growth: a contribution to the problem of scaling in the evolution of size. Am. Naturalist 105: 113–136.

GRAN, H. H., and T. BRAARUD. 1935. A quantitative study of the phytoplankton in the Bay of Fundy and the Gulf of Maine (including observations on hydrography, chemistry, and turbidity). J. Biol. Board Canada 1: 279–467.

GRANT, R. E. 1963. Unusual attachment of a Permian linoproductid brachiopod. J. Paleontol. 37: 134–140.

———. 1966. Spine arrangement and life habits of the productoid brachiopod *Waagenoconcha*. J. Paleontol. 40: 1063–1069.

———. 1968. Structural adaptation in two Permian brachiopod genera, Salt Range, West Pakistan. J. Paleontol. 42: 1–32.

GRANT, VERN. 1963. The origin of adaptations. Columbia University Press, New York. 606 p.

GREEN, ROGER H. 1968. Mortality and stability in a low diversity sub-tropical intertidal community. Ecology 49: 848–853.

GRÉGOIRE, CH. 1967. Sur la structure des matrices organiques des coquilles de mollusques. Biol. Rev. 42: 653–688.

GREIG-SMITH, P. 1964. Quantitative plant ecology. Butterworth & Co., Ltd., London. xii + 256 p.

GUNTER, GORDON. 1957. Temperature, p. 159–184. *In* J. W. Hedgpeth [ed.], Treatise on marine ecology and paleoecology, 1, ecology. Geol. Soc. Am. Mem. 67, 1. viii + 1296 p.

HALLAM, A. 1963. Major epeirogenic and eustatic changes since the Cretaceous, and their possible relationship to crustal structure. Am. J. Sci. 261: 397–423.

———. 1965. Observations on marine Lower Jurassic stratigraphy of North America, with special reference to United States. Bull. Am. Assoc. Petrol. Geologists 49: 1485–1501.

———. 1967. The interpretation of size-frequency distributions in molluscan death assemblages. Palaeontology 10: 25–42.

———. 1969. Faunal realms and facies in the Jurassic. Palaeontology 12: 1–18.

———. 1971. Mesozoic geology and the opening of the North Atlantic. J. Geol. 79: 129–157.

HALLAM, A., and N. B. PRICE. 1968a. Further notes on the strontium contents of unaltered fossil cephalopod shells. Geol. Mag. 105: 52–55.

———. 1968b. Environmental and biochemical control of strontium in shells of *Cardium edule*. Geochim. Cosmochim. Acta 32: 319–328.

HANCOCK, J. M. 1967. Some Cretaceous-Tertiary marine faunal changes, p. 91–104. *In* W. B. Harland *et al.* [eds.], The fossil record. Geological Society of London, London.

HAND, CADET. 1959. On the origin and phylogeny of the coelenterates. Syst. Zool. 8: 191–202.

HARBAUGH, J. W., and F. DEMIRMEN. 1964. Application of factor analysis to petrologic variations of Americus Limestone (Lower Permian), Kansas and Oklahoma. Kansas Geol. Surv. Spec. Distrib. Pub. 15, 40 p.

HARDY, A. C. 1954. Escape from specialization. *In* J. S. Huxley, A. C. Hardy, and E. B. FORD [eds.], Evolution as a process. Allen and Unwin, Ltd. London, 367 p.

HARE, P. E. 1963. Amino acids in the proteins from aragonite and calcite in the shells of *Mytilus californianus*. Science 139: 216–217.

HARE, P. E., and P. H. ABELSON. 1965. Amino acid composition of some calcified proteins. Carnegie Inst. Year Book 64: 223–232.

HARRISON, C. G. A., and B. M. FUNNELL. 1964. Relationship of paleomagnetic reversals and micropaleontology in two late Cenozoic cores from the Pacific Ocean. Nature 204: 556.

HARVEY, H. W. 1928. Biological chemistry and physics of sea water. Cambridge University Press, Cambridge, 194 p.

———. 1945. Recent advances in the chemistry and biology of sea water. Cambridge University Press, Cambridge, 164 p.

HATFIELD, C. B., and M. J. CAMP. 1970. Mass extinctions correlated with periodic galactic events. Bull. Geol. Soc. Am. 81: 911–914.

HAY, W. W. 1960. The Cretaceous-Tertiary boundary in the Tampico embayment, Mexico. Intern. Geol. Congr. 21st Copenhagen, 1960, Rep. Session, Norden, Pt. 5: 70–77.

HAZEL, J. E. 1970. Binary coefficients and clustering in biostratigraphy. Bull. Geol. Soc. Am. 81: 3237–3252.

HECKER, R. F. 1965. Introduction to paleoecology. American Elsevier Publishing Co., Inc., New York. x + 166 p. (First published as Wedeniye V. Paleoekologiyu, Moscow, 1957.)

HEDGPETH, J. W. 1957. Classification of marine environments, p. 17–28. *In* J. W. Hedgpeth [ed.], Treatise on marine ecology and paleoecology, 1, ecology. Geol. Soc. Am. Mem. 67, 1: viii + 1296 p.

HESSE, R., W. C. ALLEE, and K. P. SCHMIDT. 1951. Ecological animal geography, 2nd ed. John Wiley & Sons, Inc., New York. xiii + 715 p.

HESSLER, R. R., and H. L. SANDERS. 1967. Faunal diversity in the deep-sea. Deep-Sea Res. 14: 65–79.

HOLMES, R. W. 1957. Solar radiation, submarine daylight, and photosynthesis, p. 109–128. *In* J. W. Hedgpeth [ed.], Treatise on marine ecology and paleocology, 1, ecology, Geol. Soc. Am. Mem. 67, 1: viii + 1296 p.

HOLSER, W. T., and I. R. KAPLAN. 1968. Oxygen balance in the Permian atmosphere. Geol. Soc. Am. Spec. Paper 101: 97.

HORNÝ, R. 1960. On the phylogeny of the earliest pelecypods (Mollusca). Vestn. Geol. Úst. Čsl. 35: 479–482.

HORRIDGE, G. A. 1957. The coordination of the protective retraction of coral polyps. Phil. Trans. Roy. Soc. London, Ser. B 240: 495–529.

HOWARTH, R. J., and J. W. MURRAY 1969. The foraminiferida of Christchurch Harbour, England: a reappraisal using multivariate techniques. J. Paleontol. 43: 660–675.

HUBBS, CARL L. 1952. Antitropical distribution of fishes and other organisms. Proc. 7th Pac. Sci. Congr. 3: 324–329.

HUTCHINSON, G. E. 1957a. Concluding remarks. Cold Spring Harbor Symp. Quant. Biol. 22: 415–427.

———. 1957b. A treatise on limnology. I. Geography, physics and chemistry. John Wiley & Sons, Inc., New York. xiv + 1015 p.

———. 1967. A treatise on limnology. II. Introduction to lake biology and the limnoplankton. John Wiley & Sons, Inc., New York. xi + 1115 p.

HUXLEY, J. S. 1932. Problems of relative growth. Methuen & Co., Ltd. London. xix + 276 p.

———. 1939. Clines: an auxiliary method in taxonomy. Bijr. Dierk. 27: 491–520.

———. 1957. The three types of evolutionary process. Nature 180: 454–455.

HYMAN, L. H. 1940. The invertebrates. I. Protozoa through Ctenophora. McGraw-Hill Book Company, New York. xii + 726 p.

———. 1959. The invertebrates. V. Smaller coelomate groups. McGraw-Hill Book Company, New York. viii + 183 p.

IMBRIE, JOHN. 1955. Quantitative lithofacies and biofacies study of Florena Shale (Permian) of Kansas. Bull. Am. Assoc. Petrol. Geologists 39: 649–670.

———. 1959. Classification and evolution of major adaptive invertebrate types. Rept. Intern. Oceanogr. Congr., p. 278. Washington, D. C., Am. Assoc. Advan. Sci.

———. 1964. Factor analytic model in paleoecology, p. 407–422. *In* J. Imbrie and N. Newell, Approaches to paleoecology. John Wiley & Sons, Inc., New York.

IMBRIE, JOHN, and N. NEWELL. 1964. Approaches to paleoecology. John Wiley & Sons, Inc., New York. viii + 432 p.

IMLAY, R. W. 1965. Jurassic marine faunal differentiation in North America. J. Paleontol. 39: 1023–1038.

IRVING, E. 1964. Paleomagnetism and its application to geological and geophysical problems. John Wiley & Sons, Inc., New York. xvi + 399 p.

ISAAKS, B., J. OLIVER, and L. R. SYKES. 1968. Seismology and the new global tectonics. J. Geophys. Res. 73: 5855–5899.

ISTOCK, C. A. 1967. The evolution of complex life cycle phenomena: an ecological perspective. Evolution 21: 592–605.

IVANOVA, E. A., and I. V. KHVOROVA. 1955. Stratigraphy of Middle and Upper Carboniferous in western part of Moscow Basin. Tr. Paleontol. Inst., Akad. Nauk, SSSR 53: 1–282.

JEFFRIES, R. P. S. 1967. Some fossil chordates with echinoderm affinities. Symp. Zool. Soc. London 20: 163–208.

———. 1968. The subphylum Calcichordata (Jeffries, 1967) primitive fossil chordates with echinoderm affinities. Bull. Brit. Museum Geol. 16: 243–339.

JOHNSON, J. G. 1971. A quantitative approach to faunal province analysis. Am. J. Sci. 270: 257–280.

JOHNSON, R. G. 1960. Models and methods for the analysis of the mode of formation of fossil assemblages. Bull. Geol. Soc. Am. 71: 1075–1086.

———. 1962. Interspecific associations in Pennsylvanian fossil assemblages. J. Geol. 70: 32–55.

———. 1965. Pelecypod death assemblages in Tomales Bay, California. J. Paleontol. 39: 80–85.

———. 1970. Variations in diversity within benthic marine communities. Am. Naturalist 104: 285–300.

JOPE, H. M. 1967. The protein of brachiopod shell. Comp. Biochem. Physiol. 20: 593–605.

JØRGENSEN, C. B. 1966. The biology of suspension feeding. Pergamon Press, Oxford. xv + 357 p.

KAESLER, R. L. 1966. Quantitative re-evaluation of ecology and distribution of Recent Foraminifera and Ostracoda of Todos Santos Bay, Baja California, Mexico. Univ. Kansas Paleontol. Contrib., Paper 10, 50 p.

KAESLER, R. L., and W. L. FISHER. 1969. Population dynamics of *Triticites ventricosus* (Fusilinacea), Hughes Creek shale, Kansas. J. Paleontol. 43: 1122–1124.

KEITH, M. L., G. M. ANDERSON, and R. EICHLER. 1964. Carbon and oxygen isotopic composition of mollusk shells from marine and fresh-water environments. Geochim. Cosmochim. Acta 28: 1757–1786.

KEITH, M. L., and R. H. PARKER. 1965. Local variations of ^{13}C and ^{18}O content of mollusk shells and the relatively minor temperature effect in marginal marine environments. Marine Geol. 3: 115–129.

KEITH, M. L., and J. N. WEBER. 1965. Systematic relations between carbon and oxygen isotopes in carbonates deposited by modern corals and algae. Science 150: 498–501.

KERKUT, G. A. 1960. Implications of evolution. Pergamon Press, Oxford, x + 174 p.

KESLING, R. V. 1967. Cystoids. *In* R. C. Moore [ed.], Treatise on invertebrate paleontology, 5, I(1): 85–267.

KING, C. E. 1964. Relative abundance of species and MacArthur's model. Ecology 45: 716–727.

KING, J. L. and T. H. JUKES. 1969. Non-Darwinian evolution. Science 164: 788–798.

KINNE, O. 1963. The effects of temperature and salinity on marine and brackish water animals. I. Temperature. Oceanog. Marine Biol. Ann. Rev. 1: 301–340.

———. 1964. The effects of temperature and salinity on marine and brackish water animals. II. Salinity and temperature salinity combinations. Oceanog. Marine Biol. Ann. Rev. 2: 281–339.

KNIGHT, J. B., and E. L. YOCHELSON. 1960. Monoplacophora. *In* R. C. Moore [ed.], Treatise on invertebrate paleontology, I(1): 77–84.

KNIGHT-JONES, E. W., and J. MOYSE. 1961. Intraspecific competition in sedentary marine animals. Symp. Soc. Exp. Biol. 15: 72–95.

KOHN, A. J. 1959. The ecology of *Conus* in Hawaii. Ecol. Monographs 29: 47–90.

———. 1966. Food specialization in *Conus* in Hawaii and California. Ecology 47: 1041–1043.

———. 1968. Microhabitats, abundance and food of *Conus* on atoll reefs in the Maldive and Chagos Islands. Ecology 49: 1046–1061.

KOHN, A. J., and P. HELFRICH, 1957. Primary organic productivity of a Hawaiian coral reef. Limnol. Oceanog. 2: 241–251.

KOZHOV, M. 1963. Lake Baikal and its life. Monographiae Biologicae, 11. vi + 344 p.

KUHN, T. S. 1962. The structure of scientific revolutions. University of Chicago Press, Chicago, xv + 172 p.

KURTÉN, BJÖRN. 1953. On the variation and population dynamics of fossil and recent mammal populations. Acta Zool. Fennica 76: 1–122.

———. 1964. Population structure in paleoecology, p. 91–106. *In* John Imbrie and Norman Newell [eds.], Approaches to paleoecology. John Wiley & Sons, Inc., New York.

LACK, D. 1954. The natural regulation of animal numbers. Clarendon Press, Oxford, viii + 343 p.

LANE, N. G. 1963. The Berkeley crinoid collection from Crawfordsville, Indiana. J. Paleontol. 37: 1001–1008.

———. 1964. Paleoecology of the Council Grove Group (Lower Permian) in Kansas, based upon microfossil assemblages. Kansas Geol. Surv. Bull 170, Pt. 5, 23 p.

LANG, A. 1881. Der Bau von *Gunda* segmentata und die Verwandtschaft der Plathelminthen mit den Colenteraten und Hirudineen. Mittheil. Zool. Stat. Neapel 3: 187–251.

LAPORTE, L. F. 1968. Ancient environments. Prentice-Hall, Inc., Englewood Cliffs, N. J. x + 116 p.

LAWRENCE, D. R. 1968. Taphonomy and information losses in fossil communities. Bull. Geol. Soc. Am. 79: 1315–1330.

LEMCHE, HENNING. 1957. A new living deep-sea mollusc of the Cambro-Devonian class Monoplacophora. Nature 179: 413–416.

LEMCHE, HENNING, and K. G. WINGSTRAND. 1959. The anatomy of *Neopilina galatheae* Lemche, 1957 (Mollusca Tryblidiacea). Galathea Report, Copenhagen 3: 1–71, Pl. 1–56.

LE PICHON, X. 1968. Sea-floor spreading and continental drift. J. Geophys. Res. 73: 3661–3697.

LERMAN, ABRAHAM. 1965. Strontium and magnesium in water and in *Crassostrea* calcite. Science 150: 745–751.

LEVENE, H. 1953. Genetic equilibrium when more than one ecological niche is available. Am. Naturalist 87: 331–333.

LEVINS, RICHARD. 1968. Evolution in changing environments. Princeton University Press, Princeton N. J. ix + 120 p.

LEVINS, RICHARD, and R. MACARTHUR. 1966. The maintenance of genetic polymorphism in a spatially heterogeneous environment: variations on a theme by Howard Levene. Am. Naturalist 100: 585–589.

LEVINTON, J. S. 1970. The paleoecological significance of opportunistic species. Lethaia 3: 69–78.

LEVINTON, J. S., and R. K. BAMBACH. 1969. Some ecological aspects of bivalve mortality patterns. Am. J. Sci. 268: 97–112.

LEWONTIN, R. C. 1958. Studies on heterozygosity and homeostasis. II. Loss of heterosis in a constant environment. Evolution 12: 494–503.

LIPPS, J. H. 1970. Plankton evolution. Evolution 24: 1–22.

LISON, L. 1949. Recherches sur la forme et la mécanique de développement des coquilles des lamellibranches. Mem. Inst. Roy. Sci. Nat. Belg., Ser. 2, 34: 3–87.

LLOYD, R. M. 1964. Variations in the oxygen and carbon isotope ratios of Florida Bay mollusks and their environmental significance. J. Geol., 72: 84–111.

LOCHMAN-BALK, C., and J. L. WILSON. 1958. Cambrian biostratigraphy in North America. J. Paleontol. 32: 312–350.

LOEBLICH, A. R., JR., and HELEN TAPPAN. 1964. Foraminiferida. *In* R. C. Moore, [ed.], Treatise on invertebrate paleontology, part C, Protista 2, 1: 55–510a.

LOWENSTAM, H. A. 1954. Factors affecting the aragonite-calcite ratios in carbonate secreting marine organisms. J. Geol. 62: 284–322.

———. 1957. Niagaran reefs in the Great Lakes area. *In* H. S. Ladd [ed.], Treatise on marine ecology and paleoecology, 2, paleoecology. Geol. Soc. Am. Mem. 67, 2: 215–248.

———. 1961. Mineralogy, $O^{18/16}$ ratios, and strontuim and magnesium contents of recent and fossil brachiopods and their bearing on the history of the oceans. J. Geol. 69: 241–260.

———. 1963a. Biologic problems relating to the composition and diagenesis of sediments, p. 137–195. *In* T. W. Donnely [ed.], The earth sciences, problems and progress in current research. University of Chicago Press, Chicago. vii + 195 p.

———. 1963b. Sr/Ca ratio of skeletal aragonite from the recent marine biota at Palau and from fossil gastropods, p. 114–132. *In* H. Craig, S. L. Miller, and G. J. Wasserburg [eds.], Isotopic and cosmic chemistry. North-Holland, Amsterdam. xxv + 553 p.

LOWENSTAM, H. A., and SAMUEL EPSTEIN. 1954. Paleotemperatures of the post-Aptian Cretaceous as determined by the oxygen isotope method. J. Geol. 62: 207–248.

LUDWIG, W. 1950. Zur Theorie der Konkurrenz die Annidation (Einnischung) als fünfter Evolutionsfaktor. Neue Ergeb. Probl. Zool., Klatt-Festschr. 1950, 516–537.

MacArthur, R. H. 1955. Fluctuations of animal populations and a measure of community stability. Ecology 36: 533–536.

———. 1957. On the relative abundance of bird species. Proc. Nat. Acad. Sci. U.S. 43: 293–295.

MacArthur, R. H., and E. O. Wilson. 1963. An equilibrium theory of insular zoogeography. Evolution 17: 373–387.

———. 1967. The theory of island biogeography. Princeton University Press, Princeton, N. J. xi + 203 p.

MacClintock, Copeland. 1963. Reclassification of gastropod *Proscutum* Fischer based on muscle scars and shell structure. J. Paleontol. 37: 141–156.

Macdonald, K. B. 1969. Quantitative studies of salt marsh faunas from the North American Pacific Coast. Ecol. Monographs 39: 33–60.

MacFadyen, A. 1963. Animal ecology, aims and methods, 2nd ed. Sir Isaac Pitman, & Sons, Ltd. London. xxiv + 344 p.

Mackie, G. O. 1963. Siphonophores, bud colonies, and superorganisms, p. 329–337. *In* E. C. Dougherty *et al.* [eds.], The lower metazoa, comparative biology and phylogeny. University of California Press, Berekeley and Los Angeles.

Malthus, T. R. [1798] 1803. Essay on population. London, rev. ed. Reprinted as Population: the first essay. University of Michigan Press, Ann Arbor. 139 p.

Manson, Vincent, and John Imbrie. 1964. Fortran program for factor and vector analysis of geologic data using an IBM 7090 or 7094/1401 computer system. Kansas Geol. Surv. Spec. Distrib. Pub. 13, 46 p.

Margalef, Ramón. 1968. Perspectives in ecological theory. University of Chicago Press, Chicago. viii + 111 p.

Margulis, Lynn. 1970. Origin of eukaryotic cells. Yale University Press, New Haven, xxii + 349 p.

Marine Research Committee. 1953. California Cooperative Oceanic Fisheries Investigations Progress Report, 1951–1952. California Dept. Fish and Game.

Mars, Paul. 1963. Les faunes et la stratigraphie du Quaternaire Méditerranéen. Recent Trav. St. Mar. End., Bull. 28: 61–97.

Mayr, Ernst. 1940. Speciation phenomena in birds. Am. Naturalist 74: 249–278.

———. 1942. Systematics and the origin of species. Columbia University Press, New York. xiv + 334 p.

———. 1954. Change of genetic environment and evolution, p. 157–180. *In* J. S. Huxley, A. C. Hardy, and E. B. Ford [eds.], Evolution as a process. George Allen & Unwin, London.

———. 1963. Animal species and evolution. Harvard University Press, Cambridge, Mass. 797 p.

McAlester, A. L. 1965. Systematics, affinities, and life habits of *Babinka*, a transitional Ordovician lucinoid bivalve. Palaeontology 8: 231–246.

———. 1968. The history of life. Prentice-Hall, Inc., Englewood Cliffs, N. J. viii + 151 p.

———. 1970. Animal extinctions, oxygen consumption, and atmospheric history. J. Paleontol. 44: 405–409.

McCAMMON, H. M. 1969. The food of articulate brachiopods. J. Paleontol. 43: 976–985.

McLEESE, D. W. 1956. Effects of temperature, salinity, and oxygen on the survival of the American lobster. J. Fish. Res. Bd. Canada 13: 247–272.

MEDAWAR, P. B. 1945. Size, shape, and age. *In* Clark, W. E. L. and P. B. Medawar [eds.], Essays on growth and form. Clarendon Press, Oxford, viii + 408 p.

MENARD, H. W. 1969. Growth of drifting volcanoes. J. Geophys. Res. 74: 4827–4837.

MENDEL, GREGOR. 1866. Versuche über Pflanzen-Hybriden. Verhandl. Naturforsch. Vereines Brünn 4 (1865), Abh., p. 3–47. (Translation in C. Stern and E. R. Sherwood. 1966. The origin of genetics. W. H. Freeman and Co., Publishers, xvi + 179 p.)

MENZEL, D. W., and J. H. RYTHER. 1961. Nutrients limiting the production of phytoplankton in the Sargasso Sea, with special reference to iron. Deep-Sea Res. 7: 276–281.

MENZIES, R. J., and R. Y. GEORGE. 1967. A re-evaluation of the concept of hadal or ultra-abyssal fauna. Deep-Sea Res. 14: 703–723.

MEYER, DAVID L. 1969. Functional morphology and living habits of shallow water unstalked crinoids of the Caribbean Sea. Geol. Soc. Am. Abstr. Progr. 1969, Pt. 7, 150–151.

MILLER, R. S. 1967. Pattern and process in competition. *In* J. B. Cragg [ed.], Advances in ecological research. 4: 1–74.

MINOT, C. S. 1908. The problem of age, growth, and death. John Murray, Publishers, Ltd., London. xxii + 280 p.

MOBERLY, RALPH. 1968. Composition of magnesium calcite of algae and pelecypods by electron microprobe analysis. Sedimentology 11: 61–82.

MOORE, H. B. 1958. Marine ecology. John Wiley & Sons, Inc., New York. xi + 493 p.

MOORE, R. C. 1954. Evolution of late Paleozoic invertebrates in response to major oscillations of shallow seas. Bull. Museum Comp. Zool. Harvard Coll. 112: 259–286.

———. 1958. Introduction to historical geology, 2nd. ed. McGraw-Hill Book Company, New York, ix + 656 p.

MORGAN, W. J. 1968. Rises, trenches, great faults and crustal blocks. J. Geophys. Res. 73: 1959–1982.

MORTON, J. E. 1958. Molluscs. Hutchinson University Library, London. 232 p.

MORTON, J. E., and C. M. YONGE. 1964. Classification and structure of the Mollusca. *In* K. M. Wilbur and C. M. Yonge [eds.], Physiology of mollusca 1. Academic Press, New York, xiii + 473 p.

MULLER, H. J. 1966. The gene material as the initiator and the organizing basis of life. Am. Naturalist 100: 493–517.

MUNK, W. H. 1950. On the wind-driven ocean circulation. J. Meteorol. 7:79–93.

MURDOCH, W. M. 1969. Switching in general predators: experiments on predator specificity and stability of prey populations. Ecol. Monographs 39: 335–354.

MUSCATINE, L. and E. CERNICHIARI. 1969. Assimilation of photosynthetic products of Zooxanthellae by a reef coral. Biol. Bull. 137: 506–523.

Nansen, F. 1913. The waters of the north-eastern North Atlantic. Intern. Rev. Ges. Hydrobiol. Hydrogr., Suppl. vol. 4, 139 p.

Nayar, K. N. 1955. Studies on the growth of the wedge clam *Donax* (*Latona*) *cuneatus* Linnaeus. Indian J. Fisheries 2: 325–348.

Newell, N. D. 1956. Catastrophism and the fossil record. Evolution 10: 97–101.

———. 1967. Revolutions in the history of life. Geol. Soc. Am. Spec. Papers 89: 63–91.

———. 1971. An outline history of tropical organic reefs. Am. Mus. Novitates, No. 2465, 37 p.

Newell, N. D., J. K. Rigby, A. G. Fischer, A. J. Whiteman, J. E. Hickox, and J. S. Bradley. 1953. The Permian reef complex of the Guadalupe Mountains region, Texas and New Mexico. W. H. Freeman and Co., Publishers, San Francisco. xix + 236 p., 32 Pls.

Nicholson, A. J. 1957. The self-adjustment of populations to change. Cold Spring Harbor Symp. Quant. Biol. 22: 153–173.

Nicol, David. 1961. Biotic associations and extinction. Syst. Zool. 10: 35–41.

———. 1962. The biotic development of some Niagaran reefs—an example of an ecological succession or sere. J. Paleontol. 36: 172–176.

———. 1966. Cope's rule and Precambrian and Cambrian invertebrates. J. Paleontol. 40: 1397–1399.

———. 1967. Some characteristics of cold-water marine pelecypods. J. Paleontol. 41: 1330–1340.

O'Brien, S. J., and R. J. MacIntyre. 1969. An analysis of gene-enzyme variability in natural populations of *Drosophila melanogaster* and *D. simulans*. Am. Naturalist 103: 97–113.

Odum, E. P. 1971. Fundamentals of ecology, 3rd ed. W. B. Saunders Co., Philadelphia. xiv + 524 p.

———. 1969. The strategy of ecosystem development. Science 164: 262–270.

Odum, E. P., and A. A. de la Cruz. 1963. Detritus as a major component of ecosystems. Am. Inst. Biol. Sci. Bull. 13: 39–40.

Odum, H. T. 1957. Biochemical deposition of strontium. Publ. Inst. Marine Sci., Univ. Texas, 4: 38–114.

———. 1960. Ecological potential and analogue circuits for the ecosystem. Am. Scientist 48: 1–8.

Odum, H. T., and E. P. Odum. 1955. Trophic structure and productivity of a windward coral reef community on Eniwetok Atoll. Ecol. Monographs 25: 291–320.

Okazaki, Kayo. 1960. Skeletal formation of sea urchin larvae. II. Organic matrix of the spicule. Embryologia 5: 283–320.

Olson, E. C. 1952. The evolution of a Permian vertebrate chronofauna. Evolution 6: 181–196.

———. 1957. Size-frequency distributions in extinct organism samples. J. Geol. 65, 309–333.

Oparin, A. I. 1953. The origin of life, 2nd ed. Dover Publications, New York. xxv + 270 p.

Opdyke, N. D., B. Glass, J. D. Hays, and J. Foster. 1966. Paleomagnetic study of Antarctic deep-sea cores. Science 154: 349–357.

Owen, G. 1953. The shell in the Lamellibranchia. Quart. J. Microscop. Sci. 94: 57–90.

Paine, R. T. 1963. Ecology of the brachiopod *Glottidia pyramidata*. Ecol. Monographs 33: 187–213.

———. 1966. Food web complexity and species diversity. Am. Naturalist 100: 65–75.

———. 1969. A note on trophic complexity and community stability. Am. Naturalist 103: 91–93.

Palmer, A. R. 1969. Cambrian trilobite distributions in North America and their bearing on Cambrian paleogeography of Newfoundland. *In* Marshall Kay [ed.], North Atlantic—geology and continental drift, Tulsa, Am. Assoc. Petrol. Geol. Mem. 12, 139–148.

Pannella, G., and C. MacClintock. 1968. Biological and environmental rhythms reflected in molluscan shell growth. J. Paleontol. 42, Suppl. to No. 5, 64–80.

Park, R. A. 1968. Paleoecology of *Venericardia sensu lato* (Pelecypoda) in the Atlantic and Gulf coastal province: an application of paleosynecologic methods. J. Paleontol. 42: 955–986.

Parkinson, D. 1960. Differential growth in Carboniferous Brachiopoda. Proc. Geol. Assoc. 71: 402–428.

Pavlov, A. P. 1924. About some still little studied factors of extinction. *In* M. B. Pavlov, Causes of animal extinction in past geologic epochs. Moscow, 89–130.

Pearl, R., and J. R. Miner. 1935. Experimental studies on the duration of life. XIV. The comparative mortality of certain lower organisms. Quart. Rev. Biol. 10: 60–79.

Pearl, R., and S. L. Parker. 1921. Experimental studies on the duration of life: introductory discussion of the duration of life in *Drosophila*. Am. Naturalist 55: 481–507.

Pearse, A. S., and Gordon Gunter. 1957. Salinity, p. 129–158. *In* J. W. Hedgpeth [ed.], Treatise on marine ecology and paleoecology, 1, ecology. Geol. Soc. Am. Mem. 67, 1. viii + 1296 p.

Pearse, J. S. 1969. Slow developing demersal embryos and larvae of the Antarctic sea star *Odontaster validus*. Marine Biol. 3: 110–116.

Pianka, E. R. 1966. Latitudinal gradients in species diversity: a review of concepts. Am. Naturalist 100: 33–46.

Pielou, E. C. 1969. An introduction to mathematical ecology. Interscience Publishers, New York. viii + 286 p.

Pilkey, O. H., and J. Hower. 1960. The effect of environment on the concentration of skeletal magnesium and strontium in *Dendraster*. J. Geol. 68: 203–216.

Ponnamperuma, C., and E. Peterson. 1965. Peptide synthesis from amino acids in aqueous solution. Science 147: 1572–1574.

Popper, K. R. 1961. The logic of scientific discovery. Science Editions, Inc., New York. 480 p.

PORA, E. A. 1962. Considérations sur l'importance du facteur osmotique et du facteur rapique dans le développement de la vie dans la Mer Noire. Acta Biotheoret. 15: 161–174.

POWELL, J. R. 1971. Genetic polymorphisms in varied environments. Science 174: 1035–1036.

PRESTON, F. W. 1948. The commonness, and rarity of species. Ecology 29: 254–283.

———. 1962. The canonical distribution of commonness and rarity: Parts I, II. Ecology 43: 185–215, 410–432.

PROSSER, C. L., and F. A. BROWN, JR. 1961. Comparative animal physiology, 2nd ed. W. B. Saunders Co., Philadelphia. 688 p.

PURDY, E. G. 1964. Sediments as substrates, p. 238–271. *In* John Imbrie and Norman Newell [eds.], Approaches to paleoecology. John Wiley & Sons, Inc., New York.

RAFF, R. A., and H. R. MAHLER. 1972. The nonsymbiotic origin of mitochondria. Science 177: 575–582.

RAO, K. P. 1953. Rate of water propulsion in *Mytilus californianus* as a function of latitude. Biol. Bull. 104: 171–181.

RAPER, J. R. 1954. Life cycles, sexuality and sexual mechanisms in the fungi, p. 42–81. *In* W. D. Wenrich, I. F. Lewis, and J. R. Raper [eds.], Sex in microorganisms. Am. Assoc. Advan. Sci., Washington, D.C. 362 p.

RAUP, D. M. 1966. Geometric analysis of shell coiling: general problems. J. Paleontol. 40: 1178–1190.

RAUP, D. M., and A. MICHELSON. 1965. Theoretical morphology of the coiled shell. Science 147: 1294–1295.

RAYMONT, J. E. G. 1963. Plankton and productivity in the oceans. Pergamon Press, Oxford. viii + 660 p.

REDFIELD, A. C., B. H. KETCHUM, and F. A. RICHARDS. 1963. The influence of organisms on the composition of seawater. *In* M. N. Hill [ed.], The sea. John Wiley & Sons, Inc., New York. 2: 26–77.

REMANE, ADOLPH, and CARL SCHLIEPER. 1958. Die biologie des brackwassers. E. Schweizerbartische Verlags., Stuttgart. viii + 348 p.

RHOADS, D. C., and J. W. MORSE. 1971. Evolutionary and ecologic significance of oxygen-deficient marine basins. Lethaia 4: 413–428.

RHOADS, D. C., and G. PANNELLA. 1970. The use of molluscan shell growth patterns in ecology and paleoecology. Lethaia 3: 143–161.

RHOADS, D. C., and D. K. YOUNG. 1970. The influence of deposit-feeding organisms on sediment stability and community trophic structure. J. Marine Res. 28: 150–178.

RHODES, F. H. T. 1967. Permo-Triassic extinction, p. 57–76. *In* W. B. Harland *et al.* [eds.], The fossil record. Geol. Soc. London, London.

RICHARDS, F. A. 1957. Oxygen in the ocean, p. 185–238. *In* J. W. Hedgpeth [ed.], Treatise on marine ecology and paleoecology, 1. Ecology. Geol. Soc. Amer. Mem. 67, 1, vii + 1296 p.

Richards, F. A., and R. F. Vaccaro. 1956. The Cariaco Trench, an anaerobic basin in the Caribbean Sea. Deep-Sea Res. 3: 214–228.

Roche, Jean, Gilbert Ranson, and Marcelle Evsseric-Lafon. 1951. Sur la composition des scléroprotéins des coquilles des mollusques (conchiolines). Compt. Rend. Soc. Biol. 145: 1474–1477.

Rollins, H. B., and R. L. Batten. 1968. A sinus-bearing monoplacophoran and its role in the classification of primitive mollusks. Palaeontology 11: 132–140.

Ronov, A. B. 1968. Probable changes in the composition of sea water during the course of geological time. Sedimentology 10: 25–43.

Rosa, D. 1931. L'ologénèse. Nouvelle théorie de l'évolution et de la distribution géographique. Felix Alcon, Paris. xii + 368 p. (Originally published, 1918.)

Roughgarden, Jonathan. 1971. Density-dependent natural selection. Ecology 52: 453–468.

Rubey, W. W. 1951. Geologic history of sea water: an attempt to state the problem. Bull. Geol. Soc. Am. 62: 1111–1147.

Rucker, J. B., and J. W. Valentine. 1961. Salinity response of trace-element concentration in *Crassostrea virginica*. Nature 190: 1099–1100.

Rudwick, M. J. S. 1961. The feeding mechanism of the Permian brachiopod *Prorichthofenia*. Palaeontology 3: 450–471.

———. 1964. The inference of function from structure in fossils. Brit. J. Philos. Sci. 15: 27–40.

———. 1970. The functional morphology of the Pennsylvanian oldhaminoid brachiopod *Poikilosakos*. *In* J. T. Dutro, Jr. [ed.], Paleozoic perspectives: a paleontological tribute to G. Arthur Cooper. Smithsonian Contrib. Paleobiol. No. 3.

Rudwick, M. J. S., and R. Cowen. 1968. The functional morphology of some aberrant strophomenide brachiopods from the Permian of Sicily. Boll. Soc. Paleontol. Italiana 6: 113–176, Fauv. 32–43.

Russell, K. L. 1968. Oceanic ridges and eustatic changes in sea level. Nature 218: 861–862.

Ryther, J. H. 1969. Photosynthesis and fish production in the sea. Science 166: 72–76.

Sackett, W. M., W. R. Ecklemann, M. C. Bender, and A. W. H. Bé. 1965. Temperature dependence of carbon isotope composition in marine plankton and sediments. Science 148: 235–237.

Salvini-Plawen, L. v. 1969. Solenogastres und Caudofoveata (Mollusca, Aculifera): Organisation and phylogenetische Bedeutung. Malacologia 9: 191–216.

Sanders, H. L. 1958. Benthic studies in Buzzards Bay. I. Animal-sediment relationships. Limnol. Oceanog. 3: 245–258.

———. 1968. Marine benthic diversity: a comparative study. Am. Naturalist 102: 243–282.

———. 1969. Benthic marine diversity and the stability-time hypothesis. *In* Diversity and stability in ecological systems. Brookhaven Symp. Biol. 22: 71–81.

Sanders, H. L., and R. R. Hessler. 1969. Ecology of the deep-sea benthos. Science 163: 1419–1424.

Sargent, M. C., and T. S. Austin. 1949. Organic productivity of an atoll. Trans. Am. Geophys. Union 30: 245–249.

Savage, R. E. 1956. The great spatfall of mussels (*Mytilus edulis* L.) in the River Conway estuary in spring 1940. Fish. Invest., Ser. II, 20: 7.

Scheltema, R. S. 1971. Larval dispersal as a means of genetic exchange between geographically separated populations of shallow-water benthic marine gastropods. Biol. Bull. 140: 284–322.

Schindewolf, O. H. 1954. Uber die Moglichen Ursachen der Grossen Erdgeschlichlichen Faunenschnitte. Neues Jahrb. Geol. Palaeontol. Abhandl. 10: 457–465.

Schmalhausen, I. I. 1949. Factors of evolution; the theory of stabilizing selection. Blakiston, Philadelphia. xiv + 327 p. (First published in Russian, 1946.)

Scholander, P. F., F. Flagg, V. Walters, and L. Irving. 1953. Climatic adaptation in arctic and tropical poikilotherms. Physiol. Zool. 26: 67–92.

Schopf, J. W., and E. S. Barghoorn. 1969. Microorganisms from the Late Precambrian of South Australia. J. Paleontol. 43: 111–118.

Schopf, J. W., E. S. Barghoorn, M. D. Maser, and R. O. Gordon. 1965. Electron microscopy of fossil bacteria two billion years old. Science 149: 1365–1367.

Schopf, T. J. M. 1970. Taxonomic diversity gradients of ectoprocts and bivalves and their geologic implications. Bull. Geol. Soc. Am. 81: 3765–3768.

Schopf, T. J. M., A. Farmanfarmaian, and J. L. Gooch. 1971. Oxygen consumption rates and their paleontologic significance. J. Paleontol. 45: 247–252.

Schuchert, Charles, and C. O. Dunbar. 1933. Historical geology. John Wiley & Sons, Inc., New York. v + 241 p.

Scrutton, C. T. 1964. Periodicity in Devonian coral growth. Palaeontology 7: 552–558.

Segal, E., K. P. Rao, and T. W. James. 1953. Rate of activity as a function of intertidal height within populations of some littoral molluscs. Nature 172: 1108–1109.

Segerstråle, S. G. 1957. Baltic Sea, p. 751–800. *In* J. W. Hedgpeth [ed.], Treatise on marine ecology and paleoecology, 1, ecology. Geol. Soc. Am. Mem. 67, 1: viii + 1296 p.

Seilacher, Adolf. 1964. Biogenic sedimentary structures, p. 296–316. *In* J. Imbrie and N. D. Newell [eds.], Approaches to paleoecology. John Wiley & Sons, Inc., New York. 432 p.

Selander, R. K., S. Y. Yang, R. C. Lewontin, and W. E. Johnson, 1970. Genetic variation in the horseshoe crab (*Limulus polyphemus*), a phylogenetic "relic." Evolution 24: 402–414.

Sellmer, G. P. 1967. Functional morphology and ecological life history of the gem clam, *Gemma gemma* (Eulamellibranchia: Veneridae). Malacologia 5: 137–223.

Shaffer, B. L., and S. C. Wilke. 1965. The ordination of fossil communities; an approach to the study of species interrelationships and communal structure. Papers Mich. Acad. Sci. 1: 199–214.

Shelford, V. E. 1913. Animal communities in temperate America, as illustrated in the Chicago region: a study in animal ecology. University of Chicago Press, Chicago. xiii + 362 p.

SHEPARD, F. P. 1963. Submarine Geology, 2nd ed. Harper & Row, Publishers, New York. xviii + 557 p.

SHINN, E. A. 1968. Burrowing in recent lime sediments of Florida and the Bahamas. J. Paleontol. 42: 879–894.

SILLÉN, L. G. 1961. The physical chemistry of sea water, p. 549–581. *In* M. Sears [ed.], Oceanography, Am. Assoc. Advan. Sci. Publ. 67.

SIMON, H. A. 1962. The architecture of complexity. Proc. Am. Phil. Soc. 106: 467–482.

———. 1970. The sciences of the artificial. Massachusetts Institute of Technology Press, Cambridge, Mass. xii + 123 p.

SIMPSON, G. G. 1940. Mammals and land bridges. J. Wash. Acad. Sci. 30: 137–163.

———. 1944. Tempo and mode in evolution. Columbia University Press, New York. xviii + 237 p.

———. 1952. Probabilities of dispersal in geologic time. Bull. Am. Museum Nat. Hist. 99: 163–176.

———. 1953. The major features of evolution. Columbia University Press, New York. xx + 434 p.

———. 1960. Notes on the measure of faunal resemblance. Am. J. Sci. 258a: 300–311.

———. 1963. Historical science, p. 24–48. *In* C. C. Albritton, Jr. [ed.], The fabric of geology, Addison-Wesley Publishing Co., Inc., Reading, Mass. *x* + 372 p.

———. 1968. Evolutionary effects of cosmic radiation. Science 162: 140–141.

SIMPSON, J. F. 1966. Evolutionary pulsations and geomagnetic polarity. Bull. Geol. Soc. Am. 77: 197–204.

SLOBODKIN, C. B. 1960. Ecological relationships at the population level. Am. Naturalist 94: 213–236.

———. 1962. Growth and regulation of animal populations. Holt, Rinehart & Winston, Inc., New York. 184 p.

SMITH, A. G., and A. HALLAM. 1970. The fit of the southern continents. Nature 225: 139–144.

SMITH, R. I. 1964. On the early development of *Nereis diversicolor* in different salinities. J. Morphol. 114: 437–463.

SNYDER, JEREMY, and P. W. BRETSKY. 1971. Life habits of diminutive bivalve molluscs in the Maquoketa formation (Upper Ordovician). Am. J. Sci. 271: 227–251.

SOHL, N. F. 1961. Archaeogastropods, mesogastropods, and stratigraphy of the Ripley, Owl Creek and Prairie Bluff Formations. U. S. Geol. Surv., Prof. Paper 331A. 151 p.

———. 1969. The fossil record of shell boring by snails. Am. Zool. 9: 725–734.

———. 1971. North American Cretaceous biotic provinces delineated by gastropods. Proc. N. Amer. Paleo. Convention 1969, Pt. L: 1610–1638.

SOKALL, R. R., and P. H. A. SNEATH. 1963. Principles of numerical taxonomy. W. H. Freeman and Co., Publishers, San Francisco. xvi + 359 p.

SØRENSON, T. 1948. A method of establishing groups of equal amplitude in plant sociology based on similarity of species content and its application to analyses of the vegetation of Danish commons. Biol. Skrifter Danske Videnskab. Selskab, Vol. 5, No. 4: 1–34.

Southward, A. J., and D. J. Crisp. 1956. Fluctuations in the distribution and abundance of intertidal barnacles. J. Marine Biol. Assoc. U. K. 35: 211–229.

Staiger, H. 1956. Genetical and morphological variation in *Purpura lapillus* with respect to local and regional differentiation of population groups. Année Biol. 33: 251–258.

Stanley, S. M. 1968. Post-Paleozoic adaptive radiation of infaunal bivalve mollusks—a consequence of mantle fusion and siphon formation. J. Paleontol. 42: 214–229.

———. 1969. Bivalve mollusk burrowing aided by discordant shell ornamentation. Science 166: 634–635.

Stanton, R. J., Jr., and J. R. Dodd. 1970. Paleoecologic techniques—comparison of faunal and geochemical analyses of Pliocene paleoenvironments, Kettleman Hills, California. J. Paleontol. 44: 1042–1121.

Stasek, C. R. 1962. The form, growth and evolution of the Tridacnidae (giant clams). Arch. Zool. Exp. Gen. 101: 1–40.

———. 1963. Geometrical form and gnomonic growth in bivalved Mollusca. J. Morphol. 112: 213–231.

Stauber, L. A. 1950. The problem of physiological species with special reference to oysters and oyster drills. Ecology 31: 109–118.

Stebbins, G. L. 1950. Variation and evolution in plants. Columbia University Press, New York. 643 p.

———. 1960. The comparative evolution of genetic systems, p. 197–226. *In* Sol Tax [ed.], Evolution after Darwin, 1, the evolution of life. University of Chicago Press, Chicago. viii + 629 p.

———. 1966. Processes of organic evolution. Prentice-Hall, Inc., Englewood Cliffs, N. J. xii + 191 p.

———. 1968. Integration of development and evolutionary progress, p. 17–36. *In* R. C. Lewontin [ed.], Population biology and evolution. Syracuse University Press, Syracuse, N. Y.

———. 1971. Processes of organic evolution, 2nd ed. Prentice-Hall, Inc., Englewood Cliffs, N. J. xiii + 193 p.

Stehli, F. G. 1970. A test of the earth's magnetic field during Permian time. J. Geophys. Res. 75: 3325–3342.

Stehli, F. G., R. G. Douglas, and N. D. Newell. 1969. Generation and maintenance of gradients in taxonomic diversity. Science 164: 947–949.

Stehli, F. G., A. L. McAlester, and C. E. Helsley. 1967. Taxonomic diversity of recent bivalves and some implications for geology. Geol. Soc. Amer. Bull. 78: 455–466.

Stevens, Calvin. 1971. Distribution and diversity of Pennsylvanian marine faunas relative to water depth and distance from shore. Lethaia 4: 403–412.

Stevenson, R. E., and D. S. Gorsline. 1956. A shoreward movement of cool subsurface water. Trans. Am. Geophys. Union 37: 553–557.

Stokes, W. L. 1960. Essentials of earth history. Prentice-Hall, Inc., Englewood Cliffs, N.J., x + 502 p.

STOMMEL, HENRY. 1948. The westward intensification of wind-driven ocean currents. Trans. Amer. Geophys. Union 29: 202–206.

STRUGHOLD, H., and O. L. RITTER. 1962. Oxygen production during the evolution of the earth's atmosphere. Aerospace Med. 33: 275–278.

SUTTON, R. G., Z. P. BOWEN, and A. C. MCALESTER. 1970. Marine shelf environments of the upper Devonian Sonyea Group of New York. Bull. Geol. Soc. Am. 81: 2975–2992.

SVERDRUP, H. U., M. W. JOHNSON, and R. H. FLEMING. 1942. The oceans. Prentice-Hall, Inc., Englewood Cliffs, N. J. 1087 p.

TANSLEY, A. G. 1935. The use and abuse of vegetational concepts and terms. Ecology 16: 284–307.

TAPPAN, HELEN. 1968. Primary production, isotopes, extinctions and the atmosphere. Palaeogeogr., Palaeoclimat., Palaeoecol. 4: 187–210.

———. 1971. Microplankton, ecological succession and evolution. Proc. N. Am. Paleontol. Conv., Pt. H, p. 1058–1103.

TASCH, PAUL, and J. R. ZIMMERMAN. 1961. Comparative ecology of living and fossil conchostracans in a seven-county area of Kansas and Oklahoma. Univ. Wichita Bull., Univ. Studies 47: 1–14.

TAYLOR, J. D., W. J. KENNEDY, and A. HALL. 1969. The shell structure and mineralogy of the Bivalvia. Introduction. Nuculacea—Trigonacea. Bull. Brit. Museum Zool., Suppl. 3. 125 p., 29 Pls.

TEICHERT, CURT. 1964. Discussion [of Fischer, 1964]. *In* A. E. M. Nairn [ed.], Problems in palaeoclimatology. Interscience Publishers, New York, xiii + 705 p.

TERRY, K. D., and W. H. TUCKER. 1968. Biologic effects of supernovae. Science 159: 421–423.

THOMPSON, D'ARCY W. 1917. On growth and form. Cambridge University Press, Cambridge. xvi + 794 p. (Second ed.: 1942. Cambridge University Press. 1116 p.; abridged ed.: 1961. J. T. Bonner [ed.]. Cambridge University Press. xiv + 346 p.)

THORSON, G. 1950. Reproductive and larval ecology of marine bottom invertebrates. Biol. Rev. 25: 1–45.

———. 1952. Zur jetzigen Lage der Marinen Bodentier-Ökologie. Verhandl. Deut. Zool. Ges. Wilhelmshaven, 1951, 34: 276–327.

TOURTELOT, H. A., and R. O. RYE. 1969. Distribution of oxygen and carbon isotopes in fossils of late Cretaceous age, western interior region of North America. Bull. Geol. Soc. Am. 80: 1903–1922.

TOWE, K. M. 1970. Oxygen-collagen priority and the early metazoan fossil record. Proc. Nat. Acad. Sci. 65: 781–788.

TRUEMAN, E. R. 1966. Bivalve mollusks: fluid dynamics of burrowing. Science 152: 523–525.

———. 1968. The burrowing activities of bivalves. Symp. Zool. Soc. London 22: 167–186.

———. 1969. The fluid dynamics of molluscan locomotion. Malacologia 9: 243–248.

TUREKIAN, K. K. 1963. The use of trace-element geochemistry in solving geologic problems. Roy. Soc. Canada Spec. Publ. 6, p. 1–24.

TUREKIAN, K. K., and R. L. ARMSTRONG. 1961. The composition of fossil shells from the Fox Hills formation, South Dakota. Bull. Geol. Soc. Am. 72: 1817–1828.

TYLER, S. A., and E. S. BARGHOORN. 1954. Occurrence of structurally preserved plants in pre-Cambrian rocks of the Canadian shield. Science 119: 606–608.

UBAGHS, GEORGES. 1967. General characters of Echinodermata. *In* Moore, R. C., [ed.], Treatise on invertebrate paleontology. S, I(1): 3–60.

UFFEN, R. 1965. The evolution of the interior of the earth and its effects on biological evolution. *In* C. H. Smith and T. Sorgenfrei [eds.], The upper mantle symposium, New Delhi, 1964. Intern. Union Geol. Sci., Copenhagen, p. 14–19.

VAGVOLGYI, J. 1967. On the origin of the molluscs, the coelom, and coelomic segmentation. Syst. Zool. 16: 153–168.

VALENTINE, JAMES W. 1961. Paleoecologic molluscan geography of the Californian Pleistocene. Univ. Calif. Publ. Geol. Sci. 34: 309–442.

———. 1966. Numerical analysis of marine molluscan ranges on the extratropical northeastern Pacific shelf. Limnol. Oceanog. 11: 198–211.

———. 1967. Influence of climatic fluctuations on species diversity within the Tethyan provincial system. *In* C. G. Adams and D. V. Ager [eds.], Aspects of Tethyan biogeography. Syst. Ass. Pub. 7: 153–166.

———. 1968a. Climatic regulation of species diversification and extinction. Bull. Geol. Soc. Am. 79: 273–276.

———. 1968b. The evolution of ecological units above the population level. J. Paleontol. 42: 253–267.

———. 1969. Patterns of taxonomic and ecological structure of the shelf benthos during Phanerozoic time. Palaeontology 12: 684–709.

———. 1970. How many marine invertebrate fossil species? A new approximation. J. Paleontol. 44: 410–415.

———. 1971a. Resource supply and species diversity patterns. Lethaia 4: 51–61.

———. 1971b. Plate tectonics and shallow marine diversity and endemism, an actualistic model. Syst. Zool. 20: 253–264.

VALENTINE, JAMES W., and BOB MALLORY. 1965. Recurrent groups of bonded species in mixed death assemblages. J. Geol. 73: 683–701.

VALENTINE, JAMES W., and E. M. MOORES. 1970. Plate-tectonic regulation of faunal diversity and sea level: a model. Nature 228: 657–659.

———. 1972. Global tectonics and the fossil record. J. Geol. 80: 167–184.

VALENTINE, JAMES W., and R. G. PEDDICORD. 1967. Evaluation of fossil assemblages by cluster analysis. J. Paleontol. 41: 502–507.

VAN ANDEL, TJ. H. 1969. Recent uplift of the Mid-Atlantic Ridge south of the Vema fracture zone. Earth Planet. Sci. Letters 7: 228–230.

VAN'T HOFF, J. H. 1884. Études de dynamique chimique. F. Muller & Co., Amsterdam. iv + 214 p.

VAN VALEN, L. 1964. Relative abundance of species in some fossil mammal faunas. Am. Naturalist 98: 109–116.

VERHULST, P. F. 1838. Notice sur la loi que la population suit dans son accroissement. Corresp. Math. Phys. 10: 113–121.

VERMEIJ, G. J. 1970. Adaptive versatility and skeleton construction. Am. Naturalist 104: 253–260.

———. 1971. Gastropod evolution and morphological diversity in relation to shell geometry. J. Zool. 163: 15–23.

VINE, F. J. 1969. Sea-floor spreading—new evidence. J. Geol. Educa. 17: 6–16.

VINE, F. J., and H. H. HESS. 1970. Sea-floor spreading. *In* A. E. Maxwell *et al.* [eds.], The sea. Interscience Publishers, New York. Vol. IV, Pt. 2: 587–622.

VINOGRADOVA, N. G. 1962. Vertical zonation in the distribution of deep-sea benthic fauna in the ocean. Deep-Sea Res. 8: 245–250.

VOKES, H. E. 1954. Some primitive fossil pelecypods and their possible significance. J. Wash. Acad. Sci. 44: 233–236.

VOLTERRA, V. 1926. Variazioni e fluttuazioni del numero d'individui in specie animali conviverti. Mem. Acad. Lincei 2: 31–113. (Translated in Chapman, 1931, Animal ecology, Appendix, New York.)

VON ARX, W. S. 1962. An introduction to physical oceanography. Addison-Wesley Publishing Co., Inc., Reading, Mass. x + 422 p.

WADDINGTON, C. H. 1957. The strategy of the genes. George Allen & Unwin, London. ix + 262 p.

WADDINGTON, C. J. 1967. Paleomagnetic field reversals and cosmic radiation. Science 158: 913–915.

WALKER, K. R., and L. F. LAPORTE. 1970. Congruent fossil communities from Ordovician and Devonian carbonates of New York. J. Paleontol. 44: 928–944.

WALLACE, BRUCE. 1968. Polymorphism, population size, and genetic load, p. 87–108. *In* R. C. Lewontin [ed.], Population biology and evolution. Syracuse University Press, Syracuse, N. Y.

WALLER, T. R. 1969. The evolution of the *Argopecten gibbus* stock (Mollusca: Bivalvia), with emphasis on the Tertiary and Quaternary species of eastern North America. J. Paleontol. Vol. 43, Suppl. to No. 5: 1–124.

WARME, J. E. 1969. Live and dead molluscs in a coastal lagoon. J. Paleontol. 43: 141–150.

———. 1971. Paleoecological aspects of a modern coastal lagoon. Univ. Calif. Publ. Geol. Sci. 87: 1–112.

WATSON, J. D. 1965. Molecular biology of the gene. W. A. Benjamin, Inc., New York. xxii + 494 p.

WEBER, J. N., and D. M. RAUP. 1966. Fractionation of the stable isotopes of carbon and oxygen in marine calcareous organisms—the Echinoidea: pt. 1, Variation of C^{13} and O^{18} content within individuals. Geochim. Cosmochim. Acta 30: 681–703.

WELLS, J. W. 1954. Recent corals of the Marshall Islands. U.S. Geol. Surv., Prof. Paper 260-I: 385–486.

———. 1963. Coral growth and geochronometry. Nature 197: 948–950.

WENZ, WILHELM. 1940. Ursprung und frühe Stammesgeschichte der Gastropoden. Arch. Molluskenk. 72: 1–10.

WHITE, M. J. D. 1968. Models of speciation. Science 159: 1065–1070.

WHITTAKER, R. H., and G. M. WOODWELL. 1972. Evolution of natural communities, p. 137–156. *In* J. A. Wiens [ed.], Ecosystem structure and function, Oregon State University Press, Corvalis, 176 p.

WHITTINGTON, H. B. 1966. Phylogeny and distribution of Ordovician trilobites. J. Paleontol. 40: 696–737.

WILBUR, K. M. 1964. Shell formation and regeneration, p. 243–282. *In* K. M. Wilbur and C. M. Yonge [eds.], Physiology of Mollusca. Academic Press, New York. Vol. I. xiii + 473 p.

WILBUR, K. M., and GARETH OWEN. 1964. Growth, p. 211–242. *In* K. M. Wilbur and C. M. Yonge [eds.], Physiology of Mollusca, Academic Press, New York. Vol. 1. xiii + 473 p.

WILBUR, K. M., and N. WATABI. 1963. Experimental studies on calcification in molluscs and the alga *Coccolithus huxley*. Ann. N. Y. Acad. Sci. 109: 82–112.

WILLIAMS, ALWYN. 1968a. Evolution of the shell structure of articulate brachiopods. Palaeontol. Assoc., Spec. Paper Palaeontol. No. 2. v + 55 p., 24 Pls.

———. 1968b. Significance of the structure of the brachiopod periostracum. Nature 218: 551–554.

———. 1969. Ordovician faunal provinces with reference to brachiopod distribution, p. 117–154. *In* A. Wood [ed.], The Pre-Cambrian and lower Paleozoic rocks of Wales. University of Wales.

WILLIAMS, ALWYN *et al.* 1965. Brachiopoda. Treatise on invertebrate paleontology, Part H, 2 Vols. xxxii + H1–H522; ii + H523–H927.

WILLIAMS, C. B. 1964. Patterns in the balance of nature. Academic Press, New York. vii + 324 p.

WILLIS, J. C. 1922. Age and area. A study of geographical distribution and origin of species. Cambridge University Press, Cambridge. x + 259 p.

WILSON, J. T. 1963. Evidence from islands on the spreading of ocean floors. Nature 197: 536–538.

———. 1966. Did the Atlantic close and then re-open? Nature 211: 676–681.

WOHLSCHLAG, D. E. 1964. Respiratory metabolism and ecological characteristics of some fishes in McMurdo Sound, Antarctica. Antarctic Res., Am. Geophys. Union, Ser. 1: 33–62.

WOODRING, W. P., RALPH STEWART, and R. W. RICHARDS. 1941. Geology of the Kettleman Hills oil field, California. U.S. Geol. Surv., Prof. Paper 195, p. 1–170.

WRIGHT, SEWALL. 1931. Statistical theory of evolution. Am. Statistical J., March Suppl., p. 201–208.

WUST, G., W. BROGMUS, and E. N. NOODT. 1954. Die zonale Verteilung von Salzgehalt, Niederschlag, Verdungstung, Temperatur und Dichte an der Oberflache der Ozèane. Kieler Meeresforsch. 10: 137–161.

YOCHELSON, E. L. 1967. Quo vadis, *Bellerophon*? p. 141–161. *In* Curt Teichert and E. L. Yochelson [eds.], Essays in paleontology and stratigraphy. University of Kansas Press, Lawrence, Kan. 626 p.

YONGE, C. M. 1936. Mode of life, digestion and symbiosis with Zooxanthellae in the Tridacnidae. Sci. Rept. Great Barrier Reef Exped., 1928–1929, Brit. Museum 1: 283–321.

———. 1960. General characters of Mollusca, p. 3–36. *In* R. C. Moore [ed.], Treatise on invertebrate paleontology, Part I, Mollusca 1. xxiii + 351 p.

YONGE, C. M., M. J. YONGE, and A. G. NICHOLLS. 1932. Studies on the physiology of corals. VI. Sci. Rept. Great Barrier Reef Exped. 1: 213–251.

ZIEGLER, A. M. 1965. Silurian marine communities and their environmental significance. Nature 207: 270–272.

ZIEGLER, A. M., L. R. M. COCKS, and R. K. BAMBACH. 1968. The composition and structure of Lower Silurian marine communities. Lethaia 1: 1–27.

ZIEGLER, B. 1964. Boreale Einflüsse im Oberjura Westeuropas? Geol. Rundsch. 54: 250–261.

INDEX